Long-term Changes in Coastal Benthic Communities

Developments in Hydrobiology 38

Series editor
H.J. Dumont

Long-term Changes in Coastal Benthic Communities

Proceedings of a Symposium, held in Brussels, Belgium, December 9–12, 1985

Edited by
C. Heip, B.F. Keegan and J.R. Lewis

Reprinted from Hydrobiologia, vol. 142

1987 **DR W. JUNK PUBLISHERS**
a member of the KLUWER ACADEMIC PUBLISHERS GROUP
DORDRECHT / BOSTON / LANCASTER

Distributors

for the United States and Canada: Kluwer Academic Publishers, P.O. Box 358, Accord Station, Hingham, MA 02018-0358, USA
for the UK and Ireland: Kluwer Academic Publishers, MTP Press Limited, Falcon House, Queen Square, Lancaster LA1 1RN, UK
for all other countries: Kluwer Academic Publishers Group, Distribution Center, P.O. Box 322, 3300 AH Dordrecht, The Netherlands

Library of Congress Cataloging in Publication Data

ISBN 90-6193-637-3 (this volume)
ISBN 90-6193-751-5 (series)

PRINTED IN THE NETHERLANDS

Preface

This volume contains the papers and abstracts of the posters given at the symposium on Long-term Changes in Coastal Benthic Communities organized by the Commission of the European Communities in Brussels, Belgium from 9 till 12 December 1985. The organization of this symposium came to conclude five years of activities in the COST 647 project on Coastal Benthic Ecology, the rationale of which is explained in the foreword by B. F. Keegan.

The importance of this volume is that for the first time special attention is given to long-term data series of relevant biological variables collected in different marine benthic habitats. Many of the data presented here are the result of years of careful data collection by some of the leading scientists in the field of benthic ecology. Some of the series, such as the *Macoma balthica* data from the Wadden Sea or the macrofauna data from Loch Linnhe, to name just those two, are already classics in the marine biological literature. Other data were collected in the framework of a monitoring programme and are now analyzed for the first time in the different perspective of the COST 647 project. Several papers are from related fields where they represent well known case studies; they were chosen in order to see how problems have been tackled elsewhere.

What is the natural variability in benthic populations and communities and how can it be distinguished from man-induced changes? Especially: what long-term natural fluctuations may occur that may mask the effects of pollution? It is hoped that the papers presented here will add to the continuing debate on the use of benthic populations and communities in marine pollution studies. They will certainly add to our basic knowledge of benthic populations and communities in Europe and represent a focus for future research in the field.

C. Heip, B. F. Keegan & J. R. Lewis, Editors

Contents

The COST 647 Project on Coastal Benthic Ecology – A Perspective

COST (abbreviation for 'Coopération européenne dans la domaine de la Recherche Scientifique et Technique') forms a framework and forum for European research co-operation.

Project 647 dates from 1979 and is concerned with coastal benthic ecology. It had its origin in a series of meetings where scientists from different countries considered the then status of benthic ecology and the problems of accurately evaluating biological changes in the marine environment. These meetings recognised that

a) benthic ecological studies in general were short-term, un-coordinated and often non-comparable in their methods;

b) the few extended studies already carried out had greatly expanded awareness of how local physical and biological factors, and especially key species, influenced spatial and temporal variation;

c) the general inability to set data in a geographical context could often make it unclear whether or not changes ascribed to local conditions (either natural or man-made) were in reality part of a broadscale, natural pattern;

d) the dynamic character of communities rendered useless the old narrow concept of 'baselines' and necessitated its replacement by an awareness of the ranges of temporal variation that are entirely natural in different types of communities;

e) while toxicological studies were revealing the extreme sensitivity of larval stages to pollutants, the wide natural fluctuations in annual recruitment to benthic populations suggested that larvae, or other phases in the overall reproduction/repopulation process were also very sensitive to natural variables. The latter appeared to be climatic/hydrographic, but little was known about their operation.

By way of responses, it was agreed that the old and naive 'baseline' concept should be replaced with a much broader range of knowledge and understanding, covering the following:

(i) The spatial and temporal scales of natural variability over as long as possible, to give reasonable assurance that a wide range of climatic/hydrographic events had been observed.

(ii) Sufficient understanding of community dynamics to allow prediction of the consequences to the whole community of dramatic changes in the abundance of individual species as a result of natural factors or selective pollutants. To this end, intensive study of natural variability or experimental manipulation of community composition would reveal the identity of the key species (if they existed in a particular community) and the extent to which community composition was controlled by biological interactions or resulted from chance physical events.

(iii) Sufficient understanding of the specific causes of biological change to permit prediction about the consequences of natural events with some degree of certainty. Special emphasis was to be placed upon the population dynamics of the key species, and particularly upon the extent to which variations in their reproductive cycles, recruitment and mortality could be related to changes in natural physical conditions.

While elements (ii) and (iii) could be dealt with at a local level and were to varying degrees incorporated in much current work, (i) became the hallmark of a new international project, i.e. COST 647. Simultaneous studies of selected communities were proposed, using the same or intercalibrated methods, in a network of recording stations, across the geographical range of each community.

From the many benthic habitats and communities around the Atlantic coasts of Europe, four were selected for collaborative study. Among the criteria used in their selection were: wide geographical distribution around Europe; wide availability in currently non-polluted waters, or relative freedom from frequent physical destruction and a good data base resulting from current studies in one or more countries. Their selection acknowledges the fundamental division of benthic habitats according to type of substratum (rocky or sedimentary) and that both exist subtidally and intertidally.

These four communities were proposed as separate programmes within the overall project and a co-ordinator was nominated to harmonise methods and data analyses within each programme, and to facilitate interprogramme discussions (Table 1). The methods appropriate to each programme would be markedly different but the biological data sought were similar, i.e. community composition and dynamics; the identification of the most important interactions and of 'key species' (if they exist); the population dynamics and lifespans of 'key' or major species with special emphasis upon natural variation in their reproduction and recruitment.

The rationale behind the Project was incorporated into a 'Memorandum of Understanding' which was signed by Belgium, Denmark, France, the Federal Republic of Germany, Ireland, the Netherlands, Norway, Portugal, Spain, Sweden and the United Kingdom.

For its first five-year phase, the Project had 'Category III' status within the COST framework, i.e. where

TABLE 1

Habitat	Community	Co-ordinator
(i) Subtidal rock	Ascidiacea	T. Lundalv (Sweden)
(ii) Intertidal rock	Patella/Trochidae/ Cirripedia	J. Lewis (U.K.)
(iii) Subtidal sediment	Amphiura/Abra	L. Cabioch (France)
(iv) Intertidal sediment	Macoma/Polychaeta	J. Beukema (Netherlands)

Key species in the four habitat programmes:

i) Clavelina lepadiformis, Styela spp., Ciona intestinalis, Corella parallelogramma, Ascidiella aspera, A. scabra, Ascidia mentula, Ascidia virginea, Phallusia mammilata, Dendrodoa grossularia, Boltenia echinata, Pyura tesselata, Halocynthia spp.

ii) Patella depressa (intermedia), Patella vulgata, Patella aspera, Balanus balanoides, Chthamalus montagui, Chthamalus stellatus, Gibbula umbilicalis, Monodonta lineata.

iii) Acrocnida brachiata, Amphiura chiajei, Amphiura filiformis, Abra abra, Abra nitida, Melinna palmata.

iv) Macoma balthica, Arenicola marina, Cardium (Cerastoderma) edulis, Nephtys hombergii, Lanice conchilega, Littorina littorea, Nereis diversicolor, Mya arenaria, Tellina tenuis, Scrobicularia plana, Mytilus edulis.

TABLE 2: National participation in each programme

	Subtidal rock	Intertidal rock	Subtidal sediment	Intertidal sediment
Norway	○	△	▲	n/a
Sweden	●	n/a	○	n/a
Denmark	n/a	n/a	△	●
Germany	△	n/a	●	●
Netherlands	△	△	▲	○
Belgium	n/a	n/a	○	▲
U.K.	▲	●	○	△
Ireland	●	▲	●	▲
France	○	△	○	○
Spain	▲	▲	●	▲
Portugal	▲	●	▲	▲

Key to symbols

● Pre-existing studies which have become the core of that habitat programme or new studies developed specially for COST purposes. In all cases there is specific financial support which variously permits employment ranging from one part-time technical assistant up to 4 full-time scientists.

○ No specific COST funding. Independent research or studies funded for other purposes, but which have been adapted to varying extents to provide data relevant to COST 647. Most data are obtained on a part-time basis.

△ Expressions of interest.

▲ The habitat is available, but there is no activity under COST 647.

n/a The habitat and community are not readily available.

each country funded its own research activity, with the Commission of the European Communities providing a secretariat, but no financial support. National participation over this period was as outlined in Table 2.

This volume contains the proceedings of a symposium organised to mark the close of the first five-year phase and the upgrading of the project to COST 'Category II' status. Most of the contributions derive directly from COST 647; five are invited papers on research relevant to, or with instructive value for, coastal benthic ecology. A synthesis of findings for all four communities is being prepared for publication and should allow an assessment of progress with respect to the objectives of the Project at large.

Continuity for COST 647 has been assured within the provisions of the Community's new environmental research and development programme. As a 'Category II' project, the Commission can now fund essential co-ordinating activities such as workshops and symposia.

Plans are now in train to expand the Project's geographical limits into the Mediterranean. This expansion will not only apply to some of the communities currently under study but is also likely to include a new research programme on the *Posidonia* community.

Brendan Keegan
Chairman COST 647

List of participants

B. KEEGAN (Chairman)
University College GALWAY
Zoology Department,
GALWAY

BELGIUM

M. BERGMANS
Vrije Universiteit Brussel
Pleinlaan 2
B-1050 BRUSSEL

A. BRAECKMAN
Zoological Institute
Marine Biology Section
State University of Gent
Ledeganckstraat 35
B-9000 GENT

J. CRAEYMEERSCH
Zoological Institute
Marine Biology Section
State University of Gent
Ledeganckstraat 35
B-9000 GENT

J. DEREU
Zoological Institute
Marine Biology Section
State University of Gent
Ledeganckstraat 35
B-9000 GENT

C. HEIP
Zoological Institute
Marine Biology Section
State University of Gent
Ledeganckstraat 35
B-9000 GENT

P. HERMAN
Zoological Institute
Marine Biology Section
State University of Gent
Ledeganckstraat 35
B-9000 GENT

A. HUYGEN
Zoological Institute
Marine Biology Section
State University of Gent
Ledeganckstraat 35
B-9000 GENT

R. HUYS
Zoological Institute
Marine Biology Section
State University of Gent
Ledeganckstraat 35
B-9000 GENT

J. NIHOUL
Ministère de l'Education Nationale
Université de Liège
B5, Sart Tilman
B-4000 LIEGE

Y. SCAILTEUR
Université Libre de Bruxelles
24, rue A. Dewinter
B-1160 BRUXELLES

K. SOETAERT
Zoological Institute
Marine Biology Section
State University of Gent
Ledeganckstraat 35
B-9000 GENT

L. THIELEMANS
Zoological Institute
Marine Biology Section
State University of Gent
Ledeganckstraat 35
B-9000 GENT

P. VANHAECKE
Belgian Science Policy Office
Wetenschapsstraat 8
B-1040 BRUSSELS

A. VANREUSEL
Zoological Institute
Marine Biology Section
State University of Gent
Ledeganckstraat 35
B-9000 GENT

DENMARK

P. BRINCH MADSEN
Marine Pollution Laboratory
Skodsborg Strandvej 85
DK-2942 SKODSBORG

S. FOLVING
Roskilde University Centre
ROSKILDE

K. JENSEN
Marine Pollution Laboratory
Skodsborg Strandvej 85
DK-2942 SKODSBORG

N. KINGO JACOBSEN
Geografisk Institut
Øster Voldgade 10
DK-1350 KØBENHAVN K

M. PEJRUP
Geografisk Institut
Øster Voldgade 10E
DK-1350 KØBENHAVN

FRANCE

G. BACHELET
Institut de Biologie Marine –
Arcachon
CNRS et Université de Bordeaux I
2, rue du Professeur Jolyet
F-33120 ARCACHON

C. F. BOUDOURESQUE
Aix-Marseille 2 University
Laboratory of Benthic Ecology
Faculty of Science of Luminy
F-13288 MARSEILLES CEDEX 9

L. CABIOCH
C.N.R.S.
Station Biologique
F-2911 ROSCOFF

J.-C. DAUVIN
Station Biologique
CNRS et Univ. P VI
F-29211 ROSCOFF

M. DESPREZ
Groupe d'Etude des Milieux
Esuariens et Littoraux
IFREMER, 115
Quai Jeanne d'Arc
F-80230 SAINT-VALERY

J. M. DEWARUMEZ
Station Marine
28, av. Foch BP 41
F-62930 WIMEREUX

C. DEWARUMEZ-HOVA
Station Marine
28, av. Foch BP 41
F-62930 WIMEREUX

J.-P. DUCROTOY
Groupe d'Etude des Milieux
Estuariens et Littoraux
IFREMER, 115
Quai Jeanne d'Arc
F-80230 SAINT-VALERY

B. ELKAÏM
Université P. et M. Curie
Laboratoire d'Hydrobiologie
12, rue Cervier
F-75015 PARIS

S. FRONTIER
Université de Lille 1
USTL-SN3
F-59655 VILLENEUVE D'ASCQ

M. GLEMAREC
Institut d'études Marines
Lab. Océanographie biologique
Avenue le Gorgeu
F-29287 BREST

J. MARCHAND
Université Nantes
Lab. biologie marine
2, rue de la Houssinière
F-44072 NANTES CEDEX

H. MASSE
C.N.R.S.
Station Marine d'Endoume
F-13007 MARSEILLE

C. RETIERE
Museum National Histoire Naturel
Laboratoire Maritime
17, Avenue George V
F-35800 DINARD

B. ROBINAU
Université Nantes
Lab. biologie marine
2, rue de la Houssinière
F-44072 NANTES CEDEX

B. SYLVAND
G.E.M.E.L.
Station Marine
Rue de docteur Charcot
F-14530 LUC SUR MER

FEDERAL REPUBLIC OF GERMANY

S. GERLACH
Institut für Meereskunde
Düsternbrooker Weg 20
D-2300 KIEL

A. KÜNITZER
Institut für Meeresforschung
Am Handelshafen 12
D-2850 BREMERHAVEN

H. MICHAELIS
Forschungsstelle Küste
An der Mühle 5
D-2982 NORDERNEY

H. RUMOHR
Institut für Meereskunde
Düsternbrooker Weg 20
D-2300 KIEL

GREECE

P. PANAGIOTIDIS
Institut des Recherches
Océanographiques
Agios Kosmos Ellinikon
ATHENS

A. NICOLAIDOU
University 1
Zoological Laboratory
Panepistimioupolis
Ilisia
GR-17551 ATHENS

M. THESSALOU-LEGAKIS
University of Athens
Zoological Laboratory
Dept. of Biology
Panepistimioupolis
GR-17551 ATHENS

IRELAND

T. BOWMER
University College Galway
Zoology Department,
GALWAY

J. COSTELLOE
University College Galway
Zoology Department,
GALWAY

M. COSTELLOE
University College Galway
Zoology Department,
GALWAY

P. DINEEN
University College Galway
Zoology Department,
GALWAY

A. GREHAN
University College Galway
Zoology Department,
GALWAY

M. LYES
National Board for Science and
Technology
Shelbourne House
Shelbourne Road
DUBLIN 4

B. O'CONNOR
University College Galway
Zoology Department,
GALWAY

M. WHITE
University College Galway
Zoology Department,
GALWAY

ITALY

G. COLOMBO
Istituto di Zoologia
Università di Ferrara
Via L. Borsari 46
I-44100 FERRARA

G. RELINI
Istituto Anatomia Comparata
Università Genova
Via Balbi 5
I-16126 GENOVA

L. TUNESI
Istituto di Anatomia Comparata
Università di Genova
Via Balbi 5
I-16126 GENOVA

THE NETHERLANDS

J. BEUKEMA
Ned. Inst. Onderzoek Zee
P.O. Box 59
NL-1790 DEN BURG
TEXEL

J. COOSEN
Delta Institute for
Hydrobiological Research
Nierstraat 28
NL-4401 EA YERSEKE

F. CREUTZBERG
NIOZ
P.O. Box 59
NL-1790 AB DEN BURG
TEXEL

R. DEKKER
NIOZ
P.O. Box 59
NL-1790 AB DEN BURG
TEXEL

G. DUINEVELD
NIOZ
P.O. Box 59
NL-1790 AB DEN BURG
TEXEL

K. ESSINK
Rijkswaterstaat
Hereweg 99A
NL-9721 AA GRONINGEN

S. J. DE GROOT
Institute for Fish Research
P.O. Box 68
NL-1970 AB IJMUIDEN

P. A. W. J. DE WILDE
NIOZ
P.O. Box 59
NL-1790 AB DEN BURG
TEXEL

P. SEIP
NIOZ – Texel
p.o. RWS-DGW
Grenadierweg 31
NL-4338 PG MIDDELBURG

A. C. SMAAL
Tidal Water Division
Rijkswaterstaat
Grenadierweg 31
NL-4339 PG MIDDELBURG

NORWAY

J. A. BERGE
University of Oslo
P.O. Box 1064, Blindern
OSLO 3

H. CHRISTIE
Ministry of Environment
P.O. Box 8013 Dep.
N-0030 OSLO 1

J. S. GRAY
University of Oslo
Marine Biology Inst.
P.O. Box 1064
N-0316 BLINDERN
OSLO 3

E. OUG
University of Tromsø
Institute of Biology and Geology
P.O. Box 3095 Guleng
N-9001 TROMSØ

T. E. LEIN
University of Bergen
Institute of Marine Biology
N-5065 BLOMSTERDALEN

H. P. LEINAAS
University of Oslo
Dept. of Marine Zool. and Marine
Chemistry
P.O. Box 1064
Blindern
OSLO 3

PORTUGAL

M. J. GAUDENCIO
Instituto Nacional de
Investigaçao das Pescas
Av. Brasilia
P-1400 LISBOA

M. T. GUERRA
Instituto Nacional de
Investigaçao das Pescas
Av. Brasilia
P-1400 LISBOA

SPAIN

F. AGUIRREZABALAGA
Sociedad Cultural de
Investigacion Submarina
INSUB
Apartado 3031
SAN SEBASTIAN

A. BODE
Dep. Zoologia & Ecologia
Univ. Oviedo
Fac. de Biologia
ASTURIAS

O. JUNKAL PENA
INSUB
Apartado 3.031
SAN SEBASTIAN

E. LOPEZ-JAMAR
Instituto Espanol
de Oceanografia
Ap. 130
15080 LA CORUNA

SWEDEN

C. LARSSON
National Swedish Environment
Protection Board
Swedish Ministry of Foreign Affairs
P.O. Box 16121
S-10 323 STOCKHOLM

T. LUNDÄLV
Kristineberg
Marine Biological Station
Kristineberg, 2130
S-450 34 FISKEBÄCKSKIL

UNITED KINGDOM

R. S. BOWMAN
NERC Rocky Shore Group
University of Newcastle
NEWCASTLE

B. BULLIMORE
Skomer Marine Reserve
Subtidal Monitoring Project
6, Mill. St., Penclawdd,
Gower, W. Glam.

J. M. COLEBROOK
Institute for Marine Environmental
Research
23 Seymour Park
PLYMOUTH PL3 5BQ

I. DIXON
Oil Pollution Research Unit
Opry, Orielton Field Centre,
Pembroke, Dyfed, S. Wales
SA71 5HT

P. FOXTON
Natural Environment Research Council
Polaris House
Swindon, Wilts, SN2 1EU

J. HARTLEY
BP International
Britannic House
Moore Lane
LONDON EC2 Y9W

M. A. KENDALL
NERC Rocky Shore Unit
University of Newcastle
Dove Marine Laboratory
Cullercouts, North Shilds
Tyne and Wear NE26 4DG

J. R. LEWIS
62, Station Road
Scalby
SCARBOROUGH YO13 0QG

D. S. McLUSKY
University 1
Dept. Biol. Science
STIRLING EK9 4LA

J. MOORE
Dove Marine Laboratory
Cullercoats
North Shields
TYNE and WEAR
England, NE30 4PZ

P. NEWMAN
Water Research Centre
Medmenham, Marlow,
Bucks SL7 2HD

T. PEARSON
Scottish Marine Biological Association
Dunstaffnage Marine Research Laboratory
P.O. Box No. 3
OBAN ARGYLL
SCOTLAND

H. L. REES
M.A.F.F.
Fisheries Laboratory
Burnham-on-Crouth
ESSEX

N. SHILLABEER
I.C.I.
Brixham Laboratory
Freshman Quarry
Overgang
Brixham, Devon

R. SOENEM
School of Biology
U.C.N.W.
Bangor, N. WALES

U.S.A.

B. C. COULL
Marine Science Program
University of South Carolina
COLUMBIA, SC 29208

D. RHOADS
Yale University
Dept. Geology
New Haven, CT 06511

Commission of the European Communities

Ph. BOURDEAU
H. OTT
H. BARTH
O. RAVERA
A.I. SORS

Hydrobiologia 142: 1–13 (1986)

Latitudinal trends in reproduction, recruitment and population characteristics of some rocky littoral molluscs and cirripedes

J. R. Lewis
(Formerly University of Leeds, England), Now at 62, Station Road, Scalby, Scarborough, YO13, 0QC, United Kingdom

Keywords: geographical distribution, reproduction, recruitment, limpets, trochids, barnacles

Abstract

This paper reviews progress in the COST 647 rocky littoral programme involving three patellids, two trochids and two cirripedes on European Atlantic coasts. Northern geographical limits are set primarily by repopulation failure, and northern populations are characterised by short, mid-summer breeding periods, high frequency of failed or poor recruitments and low density of large, long-lived individuals. Towards the south molluscan breeding periods lengthen and become later, extending in some cases throughout much of the year but with least or no activity in mid-summer. There are probably fewer recruitment failures and thus higher densities, but individuals are of smaller maximum size and shorter life-span.

The cirripedes show similar latitudinal trends in recruitment timing but *Semibalanus* is restricted to a single annual brood throughout its range.

The two species reaching their southern limits show progressive restriction to the lowest tidal levels.

Recruitment failures do not result from inadequate gonad activity. They arise in cirripedes during the planktonic phase and in the molluscs during settlement and early shore life. The temperature sensitivity of molluscan spat is primarily responsible for the north/south gradient in recruitment times from summer towards winter.

Introduction

A principal objective of COST 647 is distinction between the many biological changes which have local physical and/or biological causes and those which are attributable directly or indirectly to broad-scale agencies such as climatic and/or hydrographic anomalies or trends. To achieve this distinction necessitates appropriately broad-scale, and ideally international, data which in turn inevitably lead towards considerations of a) geographical distribution and its control, and b) the possible extents to which population parameters change along the gradients from one geographical limit to the other. These gradients are seldom smooth, being disrupted by local coastal configuration and hydrography, but essentially they are latitudinal and reflect the broad climatic changes between tropical and polar regions. Hence in the region of COST 647 studies, i.e. west and north-west Europe, we are dealing primarily with north (cooler)/south(warmer) gradients.

Accepting the basic role of recruitment in population dynamics and the probably high sensitivity of repopulation processes to climatic fluctuations, the rocky littoral programme of COST 647 has concentrated upon annual recruitment in a few species. These were selected on account of their known ecological significance and/or because their geographical distribution afforded high potential for the investigation of climatic influences (Table 1). The accessibility of this habitat permits mobile investigators to obtain broad-scale data more easily than in other habitats, but nevertheless poor national participation in the programme has restricted international comparisons to data primarily

Table 1. Species in the COST 647 rocky littoral programme and their geographical limits occurring within the study area.

	Southern limit	Northern limit
Patella vulgata L.	N. Portugal	N. Norway
Patella aspera Röding	-	S.W. Norway
Patella depressa Pennant*	-	Wales
Monodonta lineata (da Costa)*	-	N. Ireland Wales
Gibbula umbilicalis (da Costa)*	-	N. Scotland
Chthamalus montagui Southward*	-	N. Scotland
Semibalanus balanoides (L.)	Biscay N.W. Spain	-

Species marked * have additional northern limits variously along the French and British coasts of the English Channel. The last four species will henceforth be referred to by the generic name only.

from Britain, Portugal and, more recently, Spain.

Some latitudinal trends have been discussed previously (Lewis *et al.*, 1982) and the present purpose is to extend earlier data and review progress so far.

Geographical limits

If geographical limits are to provide reliable insights into the role of latitudinal factors, one must question the extent to which these limits are set by the large-scale intervention of non-latitudinal factors such as unsuitable substrata or salinity, decreased habitat area or larval wastage by offshore currents. Such conditions have long been thought to contribute substantially to the loss of several species in our programme (and in effect to the setting of 'north-eastern limits') as one moves eastwards along the Channel coast of Britain: namely *Gibbula, Monodonta, Chthamalus, Patella depressa* (Crisp & Southward, 1958; Lewis 1964).

The other northern limits (see maps in Lewis, 1964) which occur on the more physically complex western coasts of the British Isles, appear to involve both latitudinal and local factors. Thus in the case of *Gibbula*, dying out on the north coast of Scotland (for details see Lewis *et al.*, 1982), a relative scarcity of suitable sheltered habitats on these mainly exposed and steep coasts might be contributing to a decrease in overall abundance and therefore to fewer planktonic larvae, which in turn would fail to find the suitable shores. By contrast, *Chthamalus*, which also reaches its limits around northern Scotland, thrives on open coasts yet still declines to extreme scarcity and ultimate extinction in upper shore sites which seem physically suitable and lack both competitors and predators.

Further south, *Monodonta* fails to bridge the 15–25 km gaps from Northern Ireland to south-west Scotland, presumably because of the strength of tidal currents running parallel to the coast. The other northern limit of *Monodonta* in Britain (N. Wales) roughly coincides with that of *P. depressa*, and here offshore currents coming after intermittent stretches of unsuitable shoreline possibly form the final and unsurmountable barrier for both species.

But away from the geographical limits such non-latitudinal factors seem to be less potent; areas of unsuitable habitat cause gaps in distribution but do not set limits. Individuals become more abundant southwards and situations are occupied that would remain empty further north. It appears, therefore, that where a species is abundant there are sufficient larvae to bridge unsuitable stretches of coastline, but where densities are declining or larvae are regularly scarce for any other reason, the presence of adverse currents or unsuitable habitats may then present unsurmountable barriers to species that are already under pressure. What then are the pressures which have a latitudinal component, and at what stage do they operate? Are population densities reduced and limits finally set by biological factors that increase northwards but which, being biological, must vary in intensity with time and lead to fluctuating limits? Or are there occasional catastrophic mortalities which cut back the range and are then followed by progressive re-occupation of lost ground? Conversely are there less obvious factors which lead directly to an inadequate repopulation rate?

Competition for space is very evident between the lusitanian *Chthamalus* spp and the boreal *Semibalanus*, and their fluctuations have long been linked with broadscale climatic oscillations which favour them alternately (Southward & Crisp, 1954; Southward, 1967). But while *Chthamalus* is usually the poorer competitor in the north its greater vertical range on the shore gives it a high level refuge from *Semibalanus* (and predators) over west and north Scotland. Within these 'safe' levels *Chthamalus* still declines for other reasons. While the barnacles replace each other to north or south, this is not the case with the molluscs. All five species co-exist in northern Portugal, but from south-west Britain northwards one loses first *P. depressa* and *Monodonta*, then *Gibbula*, and finally *P. aspera* until *P. vulgata* alone occurs in west and north Norway. The mid/upper shore trochid microhabitats remain vacant, *P. vulgata* (and possibly *Littorina* spp?) expand into the *depressa* and *aspera* territory, but there are no obvious replacing competitors.

The other alternatives, catastrophic mortality or repopulation failures, have been discussed previously (Lewis *et al.*, 1982) and so I briefly reiterate here the main considerations leading to the conclusion that for our species northern limits are primarily set by repopulation failure. They are:

a) Although severe mortalities occurred in coastal faunas around the British Isles in the winter of 1962/63 they did not affect *Chthamalus* anywhere, nor *Gibbula* at its northern limits (Crisp, 1984); and although *Monodonta* (and to a lesser extent *P. depressa*) did suffer some cut-backs in range there has been no repetition of such mortalities since that date (i.e. in 22 years).
b) Experimental transferences to North Sea coasts (i.e. outside their geographical range) of *Chthamalus* (Southward & Crisp, 1954) and of *Monodonta* and *Gibbula* (Williamson & Kendall in prep.) did not result in increased mortality, even though in 1978/79 the two topshells experienced the most severe North Sea winter since 1962/63.
c) Mortality of *Chthamalus* around north Scotland averaged only 8% over 4 years.
d) In all species, populations close to northern limits have always contained, and sometimes been dominated by, old animals and have frequently lacked young age-classes for a time.
e) A population of *Monodonta* about 80 km from its northern limit, and abundant in the mid-70's, died out by 1983 as more recent recruitments failed to offset the deaths of the '75 and '76 year-classes.

Accepting therefore that repopulation failures appear most likely to set geographical limits to the north, it is most unlikely that the conditions causing failure will be restricted to those limits. Rather they can be expected to occur over varying extents of the adjacent coastline and therefore to affect repopulation well inside a species' range. Geographical-scale data amply confirm that this is so. Fluctuations inside the range are usually synchronous, but less severe, reflections of events at the northern limit. This is shown most strikingly by *Gibbula* (Kendall & Lewis this volume; Lewis *et al.*, 1972), with synchronous patterns of success or failure covering sites up to 700 km from the northern limit. Although it has been suggested above that the north-east limit of this species on the Channel coast is probably set more by local rather than latitudinal factors, the poorest years here (1978, 1981 inputs) nevertheless coincided with the broad-scale recruitment failures elsewhere.

The failures in *Chthamalus* around north and north-west Scotland have been more limited spatially (< 150 km) but of greater frequency, some areas having significant recruitment since 1970 only in 1975 and 1983.

As has been shown by Bowman & Lewis (this volume) populations of *P. aspera* at latitudes, in Britain, near to the northern limit in south-west Norway lacked the 1968 class over virtually the northern two-thirds of the British Isles; while in *P. vulgata* poor recruitment in northern parts of Britain coincided with apparent failure years 2000 km further north at the geographical limit in northern Norway. While this species, in Britain, has not experienced such frequent or widespread failure as the other species it still shows considerable fluctuations that are synchronous over much of northern Britain.

But it is also evident that the controlling factors do not act in smooth north/south gradients of intensity on all occasions. *P. vulgata* and *P. aspera* both show regional asynchrony, with northern areas occasionally being more successful than some further south (e.g. 1977 and 1978, Bowman & Lewis this volume), and *Semibulanus* has shown much local asynchrony within the same region (Kendall *et al.*, 1982).

Despite such irregularities it remains apparent that general declines in abundance northwards primarily reflect an increasingly inadequate rate of repopulation which finally culminates in the setting of the limit of distribution.

Sources of repopulation fluctuations

The successful establishment of an 'O' class in the adult habitat is the end of a long series of consecutive processes or phases, and while overall success doubtless requires that each should be successful, there is increasing evidence that some are more sensitive than others to environmental factors.

Initial expectation of frequently inadequate gonad development at or near the northern limits has not been realised. Crips *et al.*(1981) similarly anticipated that summer temperatures around northern Scotland would seldom allow breeding in *Chthamalus montagui*, but there has been no year since 1970 in which nauplii have not been produced and released. Indeed, the previously mentioned transplants of *Chthamalus* and *Monodonta* to beyond their geographical limits still resulted in apparently normal gonad activity. Nor have levels of activity correlated with annual recruitment fluctuations in *Semibalanus*, naupliar production slowly increasing during a six year period in which annual metamorph numbers fluctuated strongly (Kendall *et al.*, 1985). In *Patella* spp gametic production and output are difficult to assess because releases often overlap with development. Nevertheless, and despite an initial belief of a positive link between gonad size and recruitment in *P. aspera* (see Bowman & Lewis this volume) we are now satisfied that factors other then total gametic output are of much greater importance in determining ultimate recruitment. Indeed, in some British limpet populations it now appears that the major releases of each year's gametes may contribute very little to the next season's 'O' class, this being derived frequently from the relatively few eggs released early in the season (Bowman, 1985).

The planktonic phase is one in which large and annually varying mortalities of larvae can be envisaged (Thorson, 1950), but about which we have been unable to obtain much data. However, the possibility that whatever larvae exist may be driven towards or away from the shoreline by wind-induced currents or drift is more amenable to assessment, and there are several reports of high settlement densities in cirripedes when onshore winds occur (Barnes, 1956; Hawkins & Hartnoll, 1982; Kendall *et al.*, 1985). The generally higher recruitment rates on coastlines facing prevailing winds compared with coasts of different orientation is further circumstantial support (Kendall *et al.*, 1982). If this is correct for cirripedes it is also possible that larvae of the other species are similarly influenced, even though no evidence yet exists. On the other hand much will surely depend upon the duration of the planktonic phase, and cirripedes (4–6 weeks duration) would be more vulnerable than the molluscs (4–10 days). But since wind direction just before and during the settlement period is not only one of the least predictable of environmental factors but also lacks a latitudinal component, we must look elsewhere for a physical gradient that links with geographical distribution.

Settlement and early shore life expose spat/juveniles to new and potentially severe conditions, and our studies point increasingly to the critical sensitivity of these phases, especially in the molluscs. In a species at its northern limit or within the northern part of its range – as most of our species are in Britain – reproduction takes place in summer or early autumn, and the cause of failure is most likely to be temperatures too low for successful establishment. Where there is most information, as in *Patella* spp (Bowman, 1985; Bowman & Lewis, this volume), it is now clear that the strength of the future 'O' class (first recognisable in the field as 1–2 mm spat) depends not on the total numbers of veligers settling on the shore in late summer or autumn but on the numbers which chance to meet sea/air temperatures within the limited range that is suitable initially for metamorphosis, and thereafter for fast growth and attainment of a size sufficient to withstand the coming winter. *P. vulgata* spat from north-east England fail to metamorphose at pool temperatures of 17 °C and above, and growth is best between 10–14 °C, ceasing completely in the field at 6 °C.

Settling veligers of 0.25 mm length and

metamorphosed spat can be studied only by using settlement panels that can be examined microscopically and returned to the field. Such direct observation and experimentation have not yet been possible with trochid larvae, but there are indirect pointers to a similar sensitivity to cold, for the two smallest 'O' classes in eight years have coincided with below-average temperatures during their initial autumn and winter (1978/79, 1981/82, Kendall & Lewis, this volume).

Whereas the molluscan larvae settle (survive?) best in microhabitats protected from desiccation and temperature extremes, cyprids attach to open rock and would appear to be highly vulnerable to aerial conditions. We have not observed *Chthamalus* settlement as it happens and can only report that in the many failure years around northern Scotland nauplii were released each August but no, or few, juveniles could be found next spring. *Semibalanus*, a boreo-arctic species, settles in the spring and there is much heat death of cyprids and new metamorphs, especially in the upper shore (Foster, 1970; Wethey, 1985), but during four years of observation in north-east England the annual variation in initial mortality showed no correlation with metamorph maxima (Kendall *et al.*, 1985). Nevertheless when *Semibalanus* and *Chthamalus* occur together and compete for space, as in north-west Scotland, such heat deaths of *Semibalanus* spat (and adults) in 1975 and 1976 made it possible for the high-level *Chthamalus* to become established lower down the shore (Bowman, in prep.). The fact that *Semibalanus* spat are immediate and fast-growing occupiers of rock surface also influences 'recruitment' data in a manner that does not appear to arise among the molluscs. If circumstances lead to a highly synchronous settlement a high density results, whereas with an extended settling period the growth of the earliest spat reduces the space available for late-comers. And since spat occupy the same habitat as adults, another contrast with *Patella* and the trochids, a high adult density may oblige spat to settle in misleadingly large numbers in the restricted areas of vacant space – including those cleared experimentally! (Kendall *et al.*, 1985).

Latitudinal trends in the molluscs: settlement times

The detailed settlement/survival studies on *Patella* spat have so far only taken place in Britain and are still far from complete there, but the sensitivity of the settlers has considerable latitudinal implications. At their extreme northern limits seasonally reproducing species can be expected on theoretical grounds (Hutchins, 1947) to breed during the summer (i.e. maximum sea temperatures), and if gametes are being produced each year at these limits (as they are in our case) the repopulation failures must be arising during a later phase in the autumn and winter. Admitting our ignorance of planktonic events and dealing only with the spat and juveniles, widespread failure will arise when settlement is too late and/or autumn temperatures are too low too early for the growth needed to withstand the coming winter. It will therefore be advantageous to mature and release some eggs early in the summer; and this will remain so moving towards the south wherever the late autumn and winter temperatures can fall low enough to have strong adverse effects.

But to continue spawning in the mid-summer further to the south exposes the sensitive young to the opposite danger of high temperatures, avoidance of which can come only from later settlement. Ultimately at the extreme southern limits temperatures could be damagingly high for settlers over much of the year, and successful recruitment would therefore have to be accomplished astride midwinter.

Stages in such a north/south switch are clearly discernible in the settling times of British limpets. In north-east Britain the most successful settlements have generally been relatively early in their respective seasons, i.e. July/August in *P. aspera* which is close to its northern limit, and September in *P. vulgata* which is nearing the middle of its range. The spat of both species can first be found in south-west Britain about two months later than in the north-east, and they overwinter in the south-west at smaller sizes as a result.

Patella spp settlement data from Oviedo, North Spain (M. P. Miyares pers commun.) and Portugal (Guerra & Gaudencio, this volume) cover only 2–3 years as yet, and being derived solely from size-frequency analyses without detail of early growth rates they indicate settlement times much less precisely than when settlement plates can be used. Nevertheless, they suggest successful settlement of *P. vulgata* (at its southern limit in Portugal) taking place some time between late October and January, with greatest success towards the end of this period, i.e. up to about two months later than in south-west England. In the other two *Patella* species, not at their southern limits, the situation is more confused, for there are probably several settlements per season, especially in *P. depressa.* Nevertheless, the most discrete emergence of 2–4 mm *P. aspera* spat from cracks, mussel patches etc began in March 1983 and February 1984, suggesting major settlements between October and January. *P. depressa* less than 5 mm are present throughout the year (slow growth or frequent settlement?) but their numbers are highest from January to March, indicating maximum settlement success in the autumn.

For the trochids we are not yet able to record later settlement moving south in Britain, comparable to that in *Patella.* However, preliminary data from northern Spain (Bode *et al.*, this volume) and Portugal (Gaudencio & Guerra, this volume), and again based solely on size-frequency analyses, point to first settlements of *Gibbula* and *Monodonta* occurring from September-November in Spain, and of *Gibbula* about November in Portugal. These compare with August and early September for *Monodonta* and *Gibbula* respectively in Britain (Wales).

Although it would be helpful to have data from North Africa and the Mediterranean, where *P. depressa, P. aspera* and the two trochids reach their southern limits, it is clear that all five molluscs in our programme are functioning as predicted, the most successful settlements becoming progressively later in the year towards the south, apparently in response to the needs of the spat.

Latitudinal trends in the molluscs: gonad cycles

The changes in settling times are accompanied by changes in gonad cycles, but these do not fully conform to expectation based solely upon the times of successful settlement. The commonest pattern is for gametogenesis and spawning at or near northern limits to be completed in 3–4 months astride the summer (in *P. aspera, Gibbula, Monodonta* in

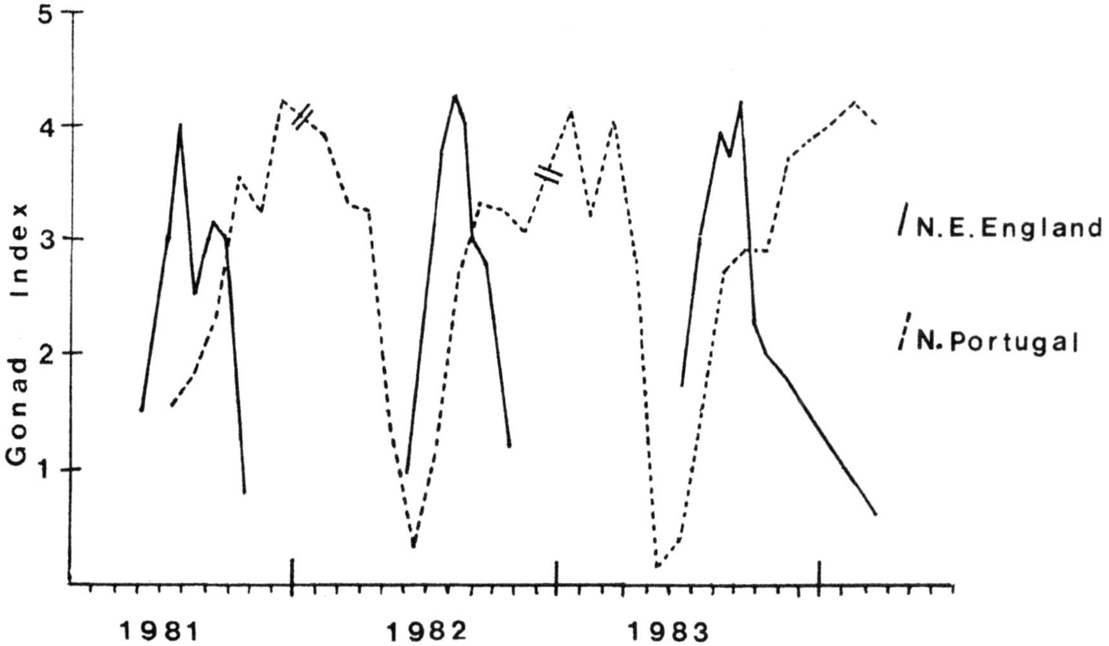

Fig. 1. Patella aspera: changes in time and duration of the gonad cycle from near its northern geographical limits to the centre of its range in Portugal.

Britain; *P. vulgata* in northern Norway). *P. depressa* differs somewhat in that the potential for multiple spawnings and redevelopment is much stronger in this than in the other two limpets, and gonad activity may last from May/June to October/November, with little predictability in the time(s) of spawning.

Moving south there is a common tendency for extension of the breeding season into winter or spring, with or without a change in the onset of gametogenesis (Fig. 1). Marked differences exist between species, however. Thus whereas the breeding season in *P. vulgata* in northern Spain and in Portugal extends from about August to March/April and still leaves a discrete inactive period from April/May to July, *P. aspera* is reproductively active for most of the year and its resting period, not always well defined, is confined to June or July. Erratically fluctuating gonad indices characterise *P. depressa* and suggest frequent spawnings and near-continuous develpment throughout the year with, however, some years showing least or no activity in June/July. This tendency for mid-summer to be the least active period is shown most strongly in both *P. depressa* and *P. aspera* in the southernmost samples available (south Portugal), perhaps foreshadowing a contraction of breeding towards winter only further south. Whether or not this does occur, the almost year-round breeding in Portugal in the middle of the range contrasts strongly with the latitudinal pattern in *P. vulgata*, in which the breeding period just lengthens steadily from north to south.

Monodonta and *Gibbula* similarly show a lengthening of the breeding period, so that although animals in spawning condition first appear at about the same time as in Britain (i.e. July/August) they remain abundant until November/December in northern Spain and October/November in Portugal. Some loss of synchrony is also apparent, especially in *Gibbula*, with ripe individuals occurring both much earlier and later than the main period.

The above brief summaries are based on data from P. M. Miyares (pers. commun.), Bowman (1985), Garwood & Kendall (1985) and from papers in this volume by Bode *et al.*, Bowman & Lewis, Gaudencia & Guerra, Guerra & Gaudencio, Ibañez *et al.*, Kendall & Lewis.

Extension of the period of gonad activity overall or of the period of 'ripeness' does not necessarily mean that spawning is taking place over an equally extended time, either continuously or in short bursts. Spawning which occurs as synchronously as it does in *Patella* spp (Bowman, 1985) or *Gibbula* (Kendall, pers. commun.) requires an environmental trigger, and until it is received ripe *Patella* eggs are stored, other eggs complete their maturation and gonads continue to increase in size. Adaptational logic decrees that spawning should be timed for settlers to meet the optimum conditions, as in the settlement-timing hypothesis of Todd & Doyle (1981); hence it too should be triggered by an environmental stimulus which arrives later towards the south. Spawning stimuli have proved to be very difficult to identify; they are still not known for the trochids and only imperfectly so for *P. aspera* (Bowman & Lewis, this volume), but for *P. vulgata* in Britain there is now overwhelming evidence that it is (or is accompanied by) a fall in sea temperature to below 12 °C. This occurs as a result of the normal autumnal decline or when early gales and swells break the thermoclines and bring colder bottom water to the surface for a time. As sea temperature maxima are higher in the south than the north, falls to below 12 °C (for whatever reason) tend to be progressively later towards the south, and thus account for the broadly north/south pattern of first spawnings around Britain (Bowman, 1985). A further illustration of the same phenomenon which also indicates the geographical scales that may be involved was the major North Atlantic storm of mid-August, 1982, which caused sea temperatures to fall by up to 3 °C as far south as northern Portugal, but only brought them below the critical 12 °C, and caused early spawning, in northern Britain.

Since climatic events do not follow orderly sequences there can be no certainty that the spawning trigger will necessarily be followed at the appropriate time by the optimum or even acceptable conditions for the settlers. Hence the more spawnings per season the better, as this increases the possibility of a good match. To some extent this is a matter of chance – the frequency and severity of storms in any region in any year. But there must also be gametes available for release and here a strong latitudinal difference is probable. In the north, with a short breeding season, one- or at most two-spawnings per season are possible, and these are not always successful; this is especially so for the second one (Bowman, 1985). Where gonads are active longer, in the south or centre of distribution, a succession of storm-induced spawnings and further ripenings could occur (as in *P. depressa* in Spain and Portugal?). Those spawnings which are too close to the summer may fail (in spite of probable selection for larvae and spat with higher temperature tolerances) but others will surely succeed so that the chances of regularly poor recruitment would still appear to be much less than in the north.

When the sensitivity of the settling and juvenile phases first became apparent it seemed likely that recruitment failures could occur throughout a species' range, and not just near the geographical (in this case, northern) limits. This now seems less likely. Nevertheless, fluctuations clearly occur in Portugal but many years' data are required to judge how the frequency and severity of poor years there compare with the north. Meanwhile in *P. vulgata* there have been exceptionally poor years as much as 2000 km from the northern limit, whereas in *Gibbula* the frequency of failure has decreased dramatically within the 6–700 km from northern Scotland to south-west England (Kendall & Lewis, this volume).

Latitudinal trends in cirripede breeding

Both *Semibalanus*, and *Chthamalus* show similarities to the molluscan species in conforming to the expected latitudinal patterns of settlement time. Thus *Semibalanus* settles late February/early March at its southern limit and in mid-summer in the Arctic (Feyling-Hansen, 1953), while *Chthamalus*, breeding in the Mediterranean from February to April (Crisp *et al.*, 1981), settles at its northern limit in Scotland about September. However the similarities now cease, for whereas the sequence in *Patella* spp involves settlement becoming progressively *later* in the autumn from north to south, the cirripede sequence becomes earlier in the spring. There are sound reasons for such differences.

Semibalanus, a boreo-arctic species, is much more constrained by having a single brood per year. Gamete production occurs in summer and autumn, copulation in cued by photoperiod, while embryo development rates during winter and release mechanisms are adapted to releasing nauplii when highly seasonal planktonic food is likely to be most abundant in different latitudes (Crisp, 1959; Barnes & Barnes, 1976). Annual recruitment fluctuations appear to be much less sensitive than in *Patella* to *direct* temperature effects and to arise primarily from mismatches (probably nutritional and/or dispersal) during the planktonic phase (Kendall *et al.*, 1985).

Chthamalus, by contrast, is a warm-temperate species with an opportunistic strategy in which production of successive small broods is both temperature- and food-dependent (Hines, 1978). Northward these conditions become adequate progressively later in the spring and early summer, until at the northern limits around Scotland probably only one brood is possible each year, and this succeeds only in especially warm summers e.g. 1975 and 1983. Since nauplii are released every year the final failure lies in the planktonic or settling phase, but the proximal cause may still be reproductive if gonad and embryo development of a single brood are too slow in cold summers to release larvae sufficiently early for completion of the later phases.

Latitudinal trends in population characteristics

Latitudinal comparisons of life span, size and abundance among near-sedentary animals need caution, not only because many local habitat factors operate with possibly greater effect, but also because of the interactions of these three parameters. Nevertheless it is clear that maximum life-spans decrease from north to south, a feature that has been noted before (Dehnel, 1955, Frank, 1975). This is most strongly shown by our trochids, *Gibbula* ranging from 8–12 years in northern Scotland to about 2 years in Portugal, and *Monodonta* from 10–15 years to 4–5 in Spain (see Lewis *et al.*, 1982 and Gaudencio & Guerra, this volume for further details and references). Current limpet data from Portugal (Guerra & Gaudencio, this volume) possibly do not cover the full range of local habitats, so perhaps longer life-spans occur; but they do suggest that all three species live for only 3–4 years, which is variously about 1–4 years less than in similar habitats in Britain. Barnacle life-spans are so often cut short by predation or competition that comparisons can be especially confusing. We therefore simply place on record that upper shore mortality of *Chthamalus* in northern Scotland (8% in four years), plus other Scottish age/growth/size observations (Lewis, unpublished) suggest that life-spans up to 20 years may not be uncommon in the north. This appears to be the maximum recorded or estimated.

Limpet sizes similarly appear to be smaller in northern Spain and Portugal than in Britain, maximum lengths being up to 5–10 mm less in the fastest growth habitats. *Gibbula* perhaps provides the most interesting data, for while there is an overall decline of maximum base diameter from 20–21 mm at the extreme northern limit to 13–14 mm in Portugal, populations with similar small maxima occur locally throughout Britain and Ireland, apparently as a result of high density and intraspecific competition (Kendall & Lewis, this volume). But it has also been shown in Britain that in broadly similar habitats the density virtually doubles from north to south, and is accompanied by a size reduction from 20–21 mm to 15–17 mm maxima.

We have also seen that settlement failure in *Gibbula* in Britain decreases from north to south, and this could well be the reason for the higher density. Further south still, the potential for more regularly high levels of recruitment increases with the lengthening period of gonad activity; and densities in Portugal are 2–4 times greater than in Britain so that still greater intraspecific competition and smaller maximum sizes could result. While firm evidence of regularly higher recruitment there may still be lacking, we are obliged to accept it does occur; otherwise with a maximum life-span of about 2 years Portuguese populations would be in considerable danger of local extinction.

Thus we have hypotheses that may account for most of the north/south trends. In the north a short breeding period leads to frequent recruitment failures, low density, low intraspecific competition and large maximum sizes. To the south the longer breeding season tends to ensure more regular recruitment, leading to higher densities and small maximum size. But the other major difference, the greater life-span in the north (which almost certainly contributes to the greater sizes there), appears to be less directly related to latitude, for it follows gradients in density which are local as well as latitudinal. The mechanism linking longevity and density remains unresolved.

For the limpets there are less firm data on latitudinal population trends, and indeed the dependance of growth, size and life-span upon local habitat factors (Lewis & Bowman, 1975) suggests that demonstration of latitudinal trends comparable to those in *Gibbula* could be more difficult. But cautious comparisons do point to Portuguese limpets probably being more numerous and smaller than in Britain, and this accords with expectations based on the undoubted latitudinal changes in gonad cycles and settlement times.

Zonation and latitude

While southern limits could be set by excessive heat damaging either settlers or gonad activity per se, the only two southern limits among our species, *P. vulgata* and *Semibalanus*, experience no bar to reproduction in the south (Barnes & Barnes, 1972; Guerra & Gaudencio, this volume). They do show, however, that an additional aspect, the 'zonational environment' in the littoral zone, has also to be considered.

All littoral species are limited upshore by aerial conditions. Sometimes low temperatures are harmful but most are limited by intolerance of heat, desiccation or strong light, and it is commonplace to find upper limits lower down the shore in sunny, quick-drying situations, either locally or geographically. It is therefore no surprise to find that *P. vulgata* at its southern limits in northern Portugal is confined to shaded low-level positions. Even in north-east Britain spat are sensitive to heat and desiccation, and remain in damp micro-habitats for many months in the upper shore (Lewis & Bowman, 1975). In south-west England, Ballantine (1961) reported that 1½–2 years elapsed before small animals moved onto open rock surfaces in the upper levels. There is no specific information about the shore levels within which *P. vulgata* settles and survives in Portugal, but it is logical to assume that the low levels are the best (only?) places for spat to establish in winter, and for all-sized animals to survive in summer.

Semibalanus, in its isolated and fluctuating southernmost populations in south-west France and north-west Spain (Barnes & Barnes, 1966, 1972) and also further north in Brittany, south-west Britain and southern Ireland, is similarly most abundant in shaded or lower shore positions. Although our north-east England studies (Kendall *et al.*, 1985) showed no correlation between heat death of spat and annual repopulation maxima we did record significant heat deaths of adults as well as spat in north-west Scotland in some years. The potential for such deaths must increase towards the south and restrict *Semibalanus* to lower levels. Thus in both species their littoral habitat is being compressed to extinction southwards, and as they vacate the scene their roles, physically and ecologically, are taken over by *P. depressa* and *Chthamalus* spp, processes that perhaps illustrate a common pattern in other boreal species.

For *P. vulgata* the progressive decrease in total population stocks and gametic output southwards must offset to some degree whatever increase in recruitment levels or regularity might result from the longer breeding period, with its potential for more settlements per season (see above); but the resultant between these opposing forces remains unknown. In *Semibalanus*, however, with its single brood, maximum settlement densities in south-west England are well below those further north where the species occupies much of the littoral zone (Kendall *et al.*, 1985).

Littoral gradients may also impinge upon mollusc recruitment more directly. Pools and other damp situations in which *Patella* spat can only survive in mid- and upper-shore levels respond much more quickly to insolation and air temperatures than do the sea or low-level pools. Thus early settlers may fail in the upper shore because pool temperatures become too high by day, whereas late settlers may find low-level pools and sea temperatures too cold and may then succeed, if at all, only in uppershore pools that warm up enough. Thus it is even possible that towards its northern limits *P. aspera*, usually regarded as a lower-shore species extending upshore northwards only as summer temperatures and desiccation decline, is becoming more dependent upon mid-shore pool recruitment than that on open rock at lowest shore levels (Bowman, pers. commun.).

Conclusion

No one species in our programme has been studied in adequate detail throughout its range for a long enough period. Nevertheless the total amount of information about different species in different regions builds up to suggest features that may prove to be broadly similar in all our molluscan species, when future work fills present gaps and puts very local or short-term data into broader perspectives.

Thus northern populations are known to be characterised by short reproductive periods in the summer, frequent recruitment failures and individuals of large size and potentially long life-spans. Towards the south (note than Portugal is *not* the southern limit of four of our species) the partially confirmed expectations are for longer periods of (possibly) less synchronous gonad activity involving more or all of the winter period, a lower frequency of recruitment failures, and probably higher densities of individuals of smaller size and shorter life-span (when allowances are made for effects of very local factors). Despite the different pattern of gonad activity these same population trends apply also to *Chthamalus.*

Closer to the southern limit the most characteristic feature among the molluscs will probably be a marked decline in abundance as the habitable environment retreats downshore.

From the monitoring viewpoint it is important to note that population stability depends to the north on high longevity, which minimises the effects of recruitment failures. On the other hand, the occasional very good settlement may result in a sudden increase in numbers which then persist conspicuously for several years. To the south, continuity depends upon regular recruitment to offset the short life-spans, and densities will primarily reflect the scale of each input. Between these extremes (i.e. over much of a species' range) the various fluctuations in recruitment and mortality rates preclude any characteristic density patterns until a persistent trend develops. But it is important to note that 'northern'-type stability and 'southern'-type annual oscillations can occur in respectively the upper and lower levels of the same shore at the same time in species of wide vertical range, e.g. *P. vulgata* and *P. depressa*, or among topshells in which recruitment is confined to limited 'nursery areas'.

Prediction of recruitment, albeit in terms as broad as 'good', 'moderate' and 'poor' was, and remains, the objective. To an extent this may be done simply at extreme northern limits (e.g. *Chthamalus, Gibbula, Monodonta*) by the now known correlation of warm summers with good recruitment (as in 1975, 1976, 1983) without knowing why. However, for *P. vulgata* and *P. aspera* (even though this is close to its northern limit) the confident prediction of 'failure', which in monitoring terms is the most significant natural event, has necessitated gaining a deep insight into all the phases involved together with accurate environmental data.

Even in Britain (for trochids and *P. depressa*), but more especially in south-west Europe for all species, there is need of precise data on spawning times, storms and sea temperatures; on speed and frequency of gonad re-ripening; on number and exact times of settlements per season and the fate of early or late settlers, together with experimental work on the temperature requirements and tolerances of spat and juveniles. Such data would resolve current uncertainties resulting from the longer breeding seasons which, because they encounter higher temperatures than in the north, suggest either higher values for spawning triggers and settler tolerances, or frequent larval and spat mismatches and wastage, or long retention of ripe gametes before spawning (or a combination of all three).

Finally, how do our current data and hypotheses accord with the generality of data on invertebrate reproduction? The latitudinal reversals of breeding times shown by our *Patella* spp have previously been reported extensively in temperate zone polychaetes by Bhaud (1972). They are also implicit in the different breeding times shown by species of different geographic affinity living in the same location, i.e. summer breeders being at their northern limits and vice-versa (in the northern hemisphere). Such examples are provided by ascidians (Millar, 1958), echinoderms (Falk-Petersen, 1982) and in Australasian mussels, a southern hemisphere obverse (Wilson & Hodgkin, 1967).

The longer breeding period southwards appears to be an accepted principle of biogeography dating back to Orton (1920). When linked with *earlier* spawning to the south (or later northwards depending upon one's own position!) it shows the breeding pattern of a warm-water species in which the temperature and nutritional requirements of the gonad are the main determinants of the timing of the overall repopulating process. Such needs in such species are satisfied sooner and for longer in the south (Seed, 1976).

Clearly, similar needs exist in limpets and trochids, but their food supply is much less seasonally and regionally varying than in the bivalve plankton feeders which figure often in molluscan gonad studies. Hence the longer breeding of *Patella* spp, *Gibbula* and *Monodonta* in south-west Europe probably owes most to the warmer temperatures, even if these operate as much via increased grazing activity as upon metabolic processes. But whatever the environmental or endogenous conditions that initiate gametogenesis and control its duration in our molluscs, the significant point is that the increased duration southwards (or at least to the centres of their ranges) results primarily from extension *later*, i.e. through autumn into winter and the following spring. The accompanying move away from mid-summer breeding in the north finally, as in *P. vulgata*, brings repopulation firmly into the winter period.

A similar pattern of later breeding southwards in the scallop *Argopecten irradians* was attributed by Sastry (1970) to geographical differences in food availability and temperature, and Barker & Blake (1983) elaborated this in terms of energy available for reproduction from the balance of input and reserves relative to latitudinal changes in metabolic rate. Accepting the more critical nutritional requirements of bivalves, and that some workers find that these over-ride temperature or other latitudinal factors (Newell *et al.*, 1982), it is nevertheless particularly interesting in *Argopecten* to observe the latitudinal temperature regimes. Gonad initiation begins in Massachusetts at about 10 °C (April), in North Carolina at 20 °C, and in Florida about 28 °C (July), but spawning temperatures are about 23 °C in the north (July) and were induced at 25 °C (October) in the south. These data indicate that the gonads function adequately over a very wide range of temperatures, whereas the larvae and settlers are meeting very similar conditions throughout the species range.

The existence of latitudinal (by implication temperature-controlled) gradients or reversals in breeding/recruitment times implies a narrow optimum range for some crucial phase, even accepting probable adaptation by local races. Although it is axiomatic that the entire process from gametogenesis initiation to establishment of the 'O' class has evolved to give the best chance of repopulation success under regionally varying conditions, the major research emphasis by far is placed upon studies of gonad development and spawning, with the implication that later phases or events are of little significance. Yet the molluscs in this COST 647 programme point to a different conclusion. Variously within this group we find northern limits set primarily by recruitment not gonad failure, breeding in animals transplanted beyond geographical limits, and recruitment fluctuations and even failures well inside distributional limits that show no correlation with gamete production. Such facts, reinforcing increasing information on the narrow temperature range within which spat are successful, point overwhelmingly to spat being the more sensitive phase and most in need of the latitudinal gradient in settling times. The seasonality of gametogenesis and the shorter duration northwards point clearly to strong temperature sensitivity in gonads, but within the overall reproductive/recruitment sequence the annual timing of gonad cycles appears to have become geared primarily to meeting the needs of spat.

Although the patterns of gonad development and spawning in *Modiolus modiolus* differ greatly from those in *Patella*, Brown (1964) concludes that here too, spawning and larval development operate within a narrower range of temperature regimes than gametogenesis. It seems probable that this will prove to be the case in many other species, provided that studies on 'reproduction' are extended to cover the later and possibly most critical phases within 'repopulation'.

References

Ballantine, W. J., 1961. The population dynamics of *Patella vulgata* and other limpets. Ph.D. thesis, Queen Mary College, University of London. 236 pp.

Barber, B. J. & N. J. Blake, 1983. Growth and reproduction of the bay scallop, *Argopecten irradians* (Lamarck) at its southern distributional limit. J. exp. mar. Biol. Ecol. 66: 247–256.

Barnes, H., 1956. *Balanus balanoides* in the Firth of Clyde; the development and annual variation of the larval population and the causative factors. J. anim. Ecol. 25: 72–84.

Barnes, H. & M. Barnes, 1966. Ecological and zoogeographical observations on some of the common intertidal cirripedes on the coasts of the western European mainland in June-September, 1963. Some Contemporary Studies in Marine Science, pp. 83–105. H. Barnes, Ed. Allen & Unwin Ltd., London.

Barnes, H. & M. Barnes, 1972. Some cirripedes of the French Atlantic Coast. J. exp. mar. Biol. Ecol. 8: 187–194.

Barnes, H. & M. Barnes, 1976. The rate of development of the embryos of *Balanus balanoides* (L.) from a number of European and American populations. J. exp. mar. Biol. Ecol. 24: 251–269.

Bhaud, M., 1972. Quelques donnés sur le determinisme ecologique de la reproduction des annélides polychetes. Mar. Biol. 17: 115–136.

Bode, A., I. Lombas & N. Anadon, 1986. The reproduction and population dynamics of *Monodonta lineata* and *Gibbula umbilicalis* (Mollusca, Gastropoda) on the central coast of Asturias (Spain). Hydrobiologia 142: 31–39.

Bowman, R. S., 1985. The biology of the limpet *Patella vulgata* in the British Isles: Spawning time as a factor determining recruitment success. In Moore, P. G. & R. Seed (ed.), The Ecology of Rocky Coasts. Hodder & Stoughton, London: 178–193.

Bowman, R. S. & J. R. Lewis, 1986. Geographical variation in the breeding cycles and recruitment of *Patella* spp. Hydrobiologia 142: 41–56.

Brown, R. A., 1984. Geographical variations in the reproduction of the horse mussel, *Modiolus modiolus* (Mollusca: Bivalvia). J. mar. biol. Ass. U.K. 64: 751–770.

Crisp. D. J., 1959. Factors influencing the time of breeding of *Balanus balanoides*. Oikos 10: 275–289.

Crisp, D. J. (ed.), 1964. The effects of the severe winter of 1962–63 on marine life in Britain. J. anim. Ecol. 33: 165–210.

Crisp. D. J. & A. J. Southward, 1958. Distribution of intertidal organisms along the coasts of the English Channel. J. mar. biol. Ass. U.K. 37: 157–208.

Crisp, D. J., A. J. Southward & E. C. Southward, 1981. On the distribution of the intertidal barnacles, *Chthamalus stellatus, Chthamalus montagui* and *Euraphia depressa*. J. mar. biol. Ass. U.K. 61: 359–380.

Dehnel, P. A., 1955. Rates of growth of gastropods as a function of latitude. Physical. Zool. 28: 115–144.

Falk-Petersen, I. B., 1982. Breeding season and egg morphology of echinoderms in Balsfjorden, Northern Norway. Sarsia 67: 215–221.

Feyling-Hansen, R. W., 1953. The barnacle *Balanus balanoides* (Linne, 1766) in Spitzbergen. Skr. norsk Polarinst. 98: 1–64.

Foster, B. A., 1971. On the determinants of the upper limit of intertidal distribution of barnacles. J. anim. Ecol. 40: 33–48.

Frank, P. W., 1975. Latitudinal variation in the life history features of the black turban snail *Tegula funebralis* (Prosobranchia: Trochidae). Mar. Biol. 31: 181–192.

Garwood, P. R. & M. A. Kendall, 1985. The reproductive cycles of *Monodonta lineata* and *Gibbula umbilicalis* on the coast of mid-Wales. J. mar. biol. Ass. U.K. 65: 993–1008.

Gaudencio, M. J. & M. T. Guerra, 1987. Observations on *Gibbula umbilicalis* (da Costa, 1778) on the Portuguese coast. Hydrobiologia, this volume.

Guerra, M. T. & M. J. Gaudencio, 1987. Aspects of the ecology of *Patella* spp on the Portuguese coast. Hydrobiologia, this volume.

Hawkins, S. J. & R. G. Hartnoll, 1982. Settlement patterns of *Semibalanus balanoides* (L.) in the Isle of Man (1977–1981). J. exp. mar. biol. Ecol. 62: 271–283.

Hines, A. H., 1978. Reproduction in three species of intertidal barnacles from Central California. Biol. Bull. 154: 262–281.

Hutchins, L. W., 1947. The bases for temperature zonation in geographical distribution. Ecol. Monogr. 17: 325–335.

Ibanez, M., J. Pena & J. Feliu, 1987. Reproduction of *Patella* spp on the Basque coast of Spain. Hydrobiologia, this volume.

Kendall, M. A., R. S. Bowman, P. Williamson & J. R. Lewis, 1982. Settlement patterns, density and stability in the barnacle *Balanus balanoides*. Neth. J. Sea Res. 16: 119–126.

Kendall, M. A., R. S. Bowman, P. Williamson & J. R. Lewis, 1985. Annual variation in the recruitment of *Semibalanus balanoides* on the North Yorkshire coast 1969–1981. J. mar. biol. Ass. U.K. 65: 1009–1030.

Kendall, M. A. & J. R. Lewis, 1987. Temporal and spatial patterns in the recruitment of *Gibbula umbilicalis*. Hydrobiologia, this volume.

Lewis, J. R., 1964. The Ecology of Rocky Shores. English Universities Press, London. 323 pp.

Lewis, J. R. & R. S. Bowman, 1975. Local habitat-induced variations in the population dynamics of *Patella vulgata* L. J. exp. mar. Biol. Ecol. 17: 165–203.

Lewis, J. R., R. S. Bowman, M. A. Kendall & P. Williamson, 1982. Some geographical components in population dynamics: possibilities and realities in some littoral species. Neth. J. Sea Res. 16: 18–28.

Millar, R. H., 1958. The breeding season of some littoral ascidians in Scottish waters. J. mar. biol. Ass. U.K. 37: 649–652.

Newell, R. I. E., T. J. Hilbish, R. K. Koehn & C. J. Newell, 1982. Temporal variation in the reproductive cycle of *Mytilus edulis* L. from localities on the east coast of the United States. Biol. Bull. 162: 299–310.

Orton, J. H., 1920. Sea temperature, breeding and distribution in marine animals. J. mar. biol. Ass. U.K. 12: 339–360.

Sastry, A. N., 1970. Reproductive physiological variation in latitudinally separated populations of the bay scallop, *Argopecten irradians* Lamarck. Biol. Bull. 138: 56–65.

Southward, A. J., 1967. Recent changes in the abundance of intertidal barnacles in S.W. England: a possible effect of climatic deterioration. J. mar. biol. Ass. U.K. 47: 81–95.

Southward, A. J. & D. J. Crisp, 1954. Recent changes in the distribution of the intertidal barnacles *Chthamalus stellatus* Poli and *Balanus balanoides* L. in the Bristish Isles. J. anim. Ecol. 23: 163–177.

Seed, R., 1976. Ecology. In Bayne, B. L. (ed.), Marine Mussels: their ecology and physiology. Cambridge University Press, Cambridge: 13–65.

Thorson, G., 1950. Reproductive and larval ecology of marine bottom invertebrates. Biol. Rev. 25: 1–45.

Todd, C. D. & R. W. Boyle, 1981. Reproductive strategies of marine benthic invertebrates: a settlement-timing hypothesis. Mar. Ecol. Prog. Ser., 4: 75–83.

Wethey, D. S., 1985. Local and regional variation in settlement and survival in the littoral barnacle *Semibalanus balanoides* (L.): patterns and consequences. In Moore, P. G. & R. Seed (eds), The Ecology of Rocky Coasts. Hodder & Stoughton, London: 194–202.

Wilson, B. R. & E. P. Hodgkin, 1967. A comparative account of the reproductive cycles of five species of marine mussels in the vicinity of Freemantle, Western Australia. Aust. J. mar. Freshwat. Res. 18: 175–203.

Hydrobiologia 142: 15–22 (1986)

Temporal and spatial patterns in the recruitment of *Gibbula umbilicalis*

M. A. Kendall[1] & J. R. Lewis[2]
[1]*NERC Rocky Shore Surveillance Group, Dove Marine Laboratory, University of Newcastle on Tyne, Cullercoats, North Shields, NE30, 4PZ, Tyne & Wear, United Kingdom*
[2]*62, Station Road, Scalby, Scarborough, North Yorkshire, United Kingdom*

Keywords: gastropod, prosobranch, *Gibbula umbilicalis*, reproduction, range

Abstract

Size/frequency samples have been taken from populations of the trochid gastropod *Gibbula umbilicalis* at a number of sites around the British Isles each year since 1978. In the North of Scotland, close to the limit of the species distribution, recruitment was generally poor and populations were sparse and dominated by large old individuals. While such poor recruitment may reflect the low density of adults and the shortage of nursery areas, a distinct temporal pattern of recruitment was also evident. At the start of our study, the majority of sites had population structures biased towards young animals implying that conditions for the settlement of larvae or their subsequent survival had been favourable over a wide area. In the years that followed only the enclosed Loch Eriboll regularly received substantial recruitment.
Further to the south, in Wales and S.W. England, recruitment was usually more regular, populations were more dense and individuals smaller.

Introduction

The long term study of trochid gastropods as part of the programme of both the NERC Rock Shore Surveillance Group of the U.K. and COST 647 is based on the proposal by Lewis (1976) that the most cost effective way of gaining an understanding of the processes which regulate coastal ecosystems is to study the reproduction and recruitment of selected key species. Although there are relatively few shores in Britain where trochids are the dominant grazers their importance increases considerably further to the south.

G. umbilicalis reaches its northern geographic limit on the north coast of Scotland (Lewis *et al.*, 1982) although occasional specimens have been collected in the Orkney Islands (Rendall 1956). It is absent from the North Sea although it can be found as far along the French coast of the Channel as Calais (Crisp & Southward 1958) while on the English side, its range extends to the Isle of Wight. The species is present throughout the Atlantic coast of France, Spain (Fischer-Piette, 1955; Fischer-Piette & Gaillard, 1956) and Portugal and probably reaches its southern limit near Cap Blanc on the coast of W. Africa (Fischer-Piette, 1955).

In this paper we consider the temporal changes in the population structure of *G. umbilicalis* towards the north of its range. The data presented come from an extensive long term study aimed at separating geographic patterns in reproduction and recruitment from local scale variation and the identification of those climatic factors which Hutchins (1947) suggested might set the limits of a species distribution. This methodology was used by Kendall *et al.*, (1982) and Bouwman (1985) to detect regional patterns of recruitment in the other rocky shore species.

Methods

This study is largely based on data from five sites in N.W. Scotland, a single site in Wales and two sites in S.W. England which it was aimed to sample 2–3 times annually between 1978 and 1985 although this did not always prove to be possible. This is supplemented by information from a range of secondary sites which were sampled at more irregular intervals. On each visit a sample of about 200 snails was taken (when possible) and returned to the laboratory where the basal shell diameter was measured to 0.1 mm using vernier calipers or the graduated eye-piece of a binocular microscope. Size/frequency histograms were plotted from the data obtained. The single exception was the site at Skerray, at the species virtual northern limit, where searches of one to two hours duration never revealed more than 150 individuals. In view of this extremely low population density all animals were returned after measurement in the field.

Previous experience had shown that the juveniles do not inhabit the full range of adult habitats being most abundant on the undersides of stable stones and cobbles which overly permanently damp clean gravel. Sand or silt appear to be inimical but even in apparently suitable areas the juveniles may be highly aggregated. As our principal aim was the quantification of recruitment, random sampling, which would tend to oversample the adults, was discarded in favour of collections centred on nursery areas. This proceedure tends to over-estimate the relative abundance of spat. As the cryptic behaviour and small size of the newly settled spat prevents accurate assessment of their abundance after their late summer/early autumn settlement all collections were made in the following late spring or early summer.

Most shores occupied by *G. umbilicalis* tend to have a highly heterogeneous structure with many microhabitats within a small area. In such circumstances it is not possible to estimate the true density of the population and thus where comparisons of abundance were made this was on the basis of replicated 5 minute searches by a single operator (M.A.K.) using a standardised technique.

The location of the study sites is shown in Fig. 1. Four major categories can be distinguished: - 1) boulders and stones overlying shallow gravel and bed-rock on sheltered sections of otherwise open coasts (Culkein, Clashnessie, Scourie & Osmington); 2) broken bed-rock with gullies containing boulders, smaller stones and pools (Skerray & Widemouth); 3) Stable clean shingle with small boulders and gravel (Loch Eriboll); 4) Stable shingle on open coasts (Aberaeron).

Results

Preliminary surveys in 1976 and 1977 showed that the majority of populations throughout Britain were dominated by small animals which had settled 1–2 years previously. A similar situation was still evident in N.W. Scotland in 1978 at the start of our detailed studies, at Scourie for example (Fig. 2) the greatest part of the population was less than two years old. The 1978 year class was poorly represented in samples taken during 1979 and in subsequent years we were unable to detect any contribution to the population from spat settling in 1979, 1980, 1981 and 1982. By 1983 the population consisted entirely of large (15–19 mm shell diameter) old animals but in 1984 a small number of the previous years settlers were collected. Prior to the 1984/5 winter it was evident that there had been a successful settlement of the 1984 year class and this was confirmed by the clear bimodality of the 1985 size frequency histogram. A similar pattern of success/failure was recorded at Clashnessie and Culkein, although at the latter site there were always a small number of spat present, even in the worst years.

Although statistical techniques for the comparison of histograms exist (Grant & Tyler, 1983), the methodology necessary to sample *G. umbilicalis* precludes their use to assess the comparative strength of the various year classes. Fortunately, the annual pattern of recruitment success/failure at the majority of sites in N.W. Scotland was sufficiently distinct for such data treatment to be unnecessary. It is also clear that the population structure and recruitment pattern at Skerray and in Loch Eriboll were exeptional. In Loch Eriboll, although the annual intensity of recruitment varied, we were always able to distinguish 'O' class animals. Only 25 km away at Skerray such animals were only identified with confidence in 1985.

Figure 3 shows that populations of *G. umbilicalis* in S.W. England and mid-Wales have more frequent recruitment leading to polymodal size frequency distributions; densities also tend to be greater than in more northerly populations (Fig. 5). Our most extensive studies have been conducted at

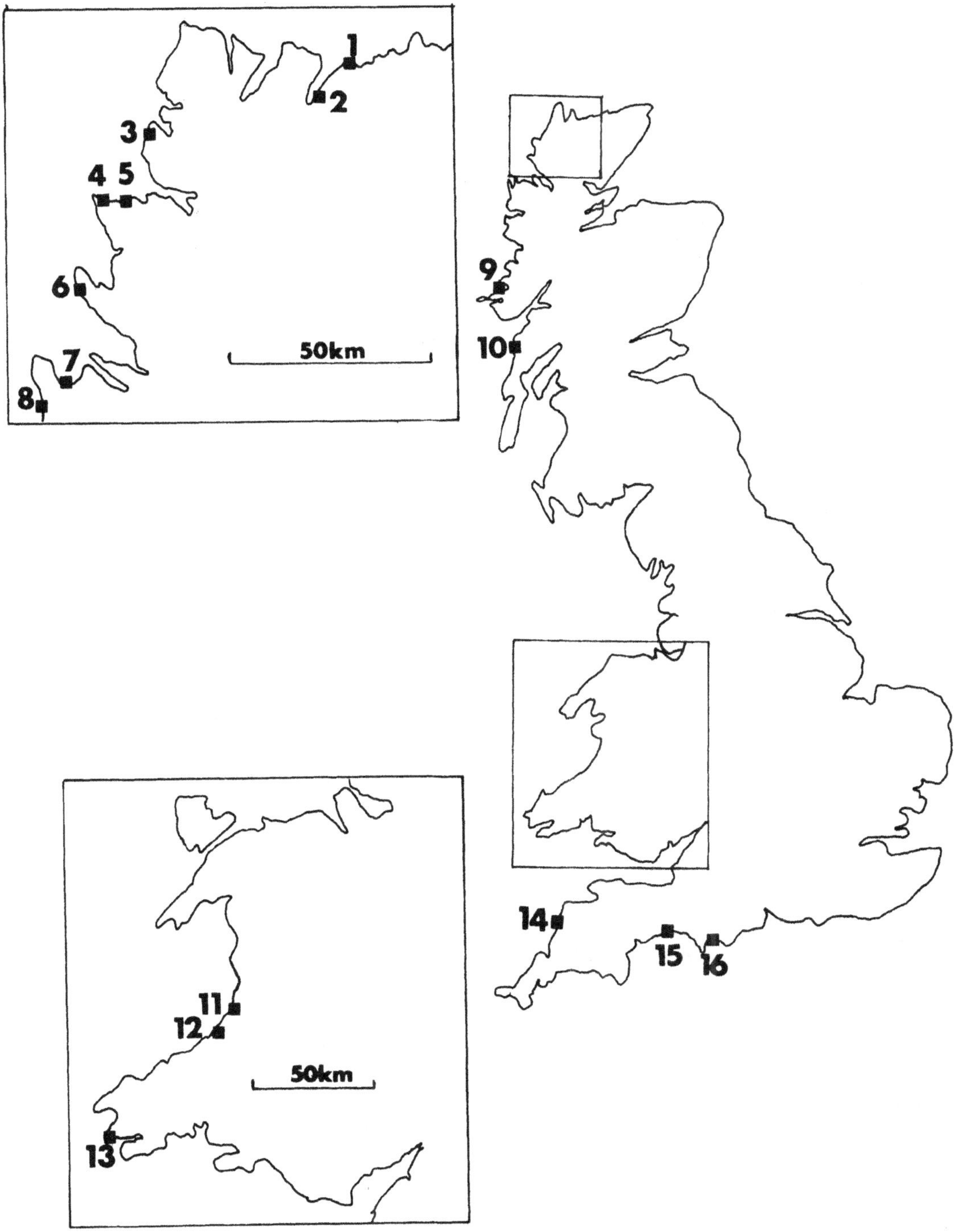

Fig. 1. Map of Britain showing the location of sites referred to in the text. 1)Skerray 2) Loch Eriboll 3)Scourie 4) Culkein 5)Clashnessie 6) Coigach 7) Griunard Bay 8) Loch Ewe 9) Beasdale 10) Loch Melfort 11) Aberystwyth 12) Aberaeron 13) Dale 14) Widemouth 15) Lyme Regis 16) Osmington.

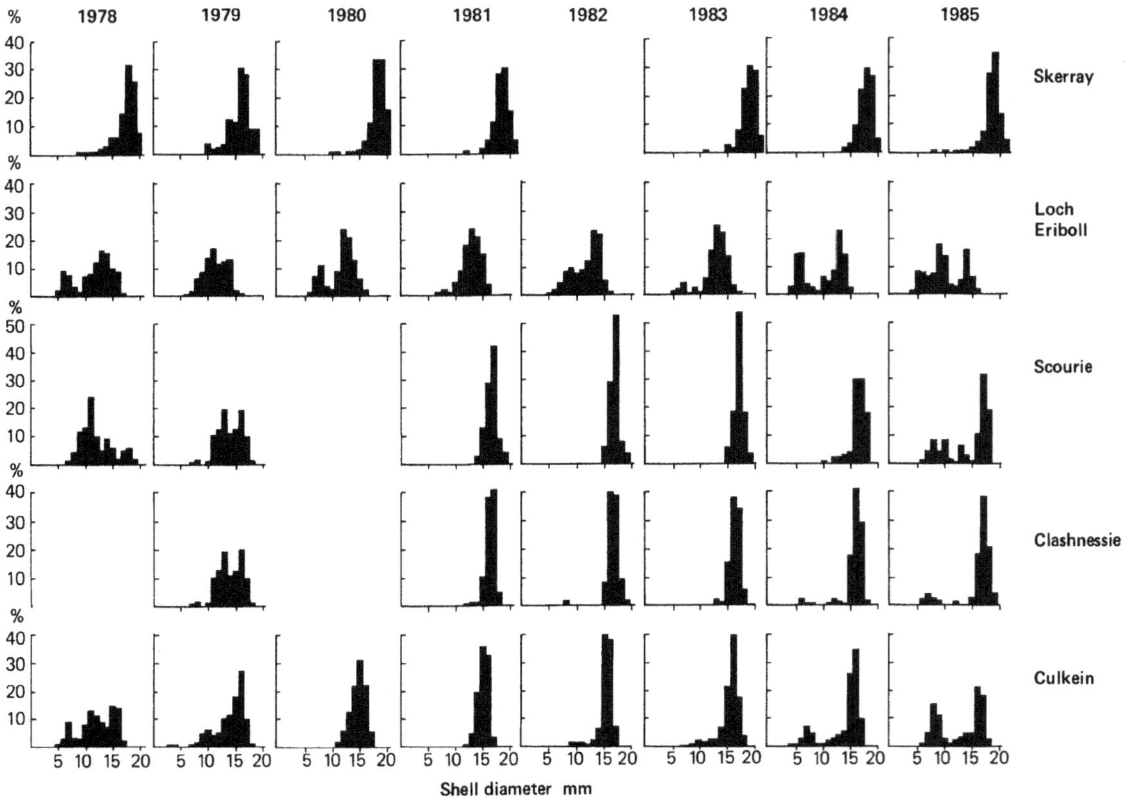

Fig. 2. Size frequency histograms showing the variation in population structure at five sites in N.W. Scotland between 1978 and 1985.

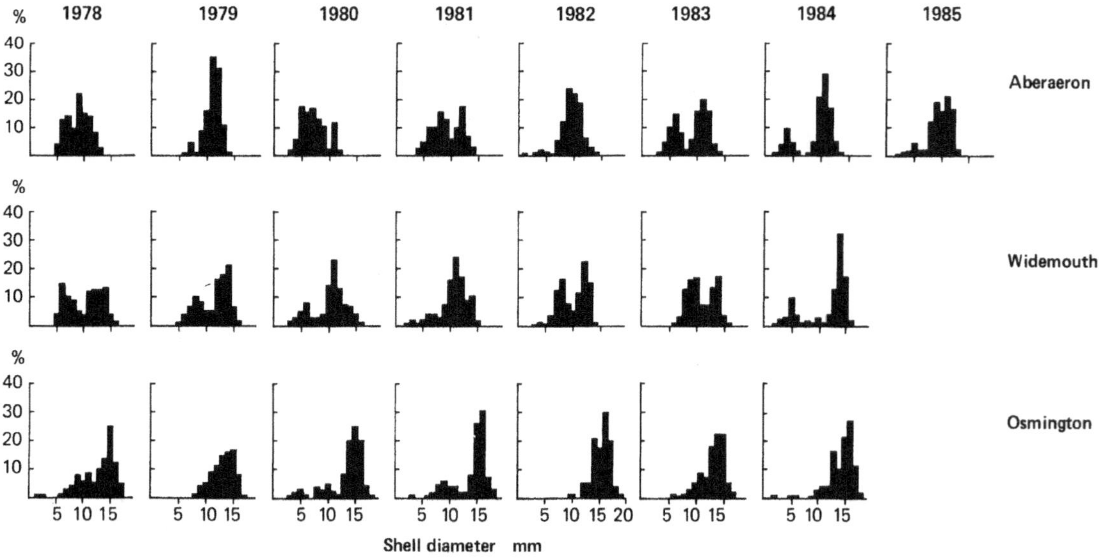

Fig. 3. Size frequency histograms showing the variation in population structure at one site in Wales and two sites in S.W. England between 1978 and 1985.

Aberaeron where, on the basis of timed searches the species is twice as abundant as in Loch Eriboll, the most highly populated of our Scottish sites. At Aberaeron the homogeneous nature of the stable cobble shore facilitates the making of reliable estimates of the density of the population and of recruitment. The full results of these studies will be published elsewhere (Kendall, Williamson & Garwood, in press) but for this paper it is sufficient to record that we have yet to observe a total failure of recruitment. Nevertheless, spat densities recorded in the spring/early summer following settlement have ranged from as few as 0.7 ± 0.21 spat m^{-2} (1978 year class) to 20.7 ± 4.1 spat m^{-2} (1980 year class). Further to the south, Widemouth (Fig. 4) experienced regular recruitment but at Osmington close to the species N.E. limit on the Channel coast 'O' class animals were always rare.

If data from our main study sites are combined with those from the supplementary sites (Fig. 4) we can detect two gradients along which the frequency of poor recruitment increases; from south to north along the west coast of Britain and from west to east along the northern coast of the Channel. At the northern limits, however, the frequency of failure appears to be greater than at the Channel limits. We can also detect widespread consistency in recruitment; thus the pattern of success in 1976/1977 followed by 6 years of failure which was described above for Scourie was common to most of Scotland. It is also clear that (with the exception of Loch Eriboll and Widemouth) settlement/survival was poor throughout Britain in 1981.

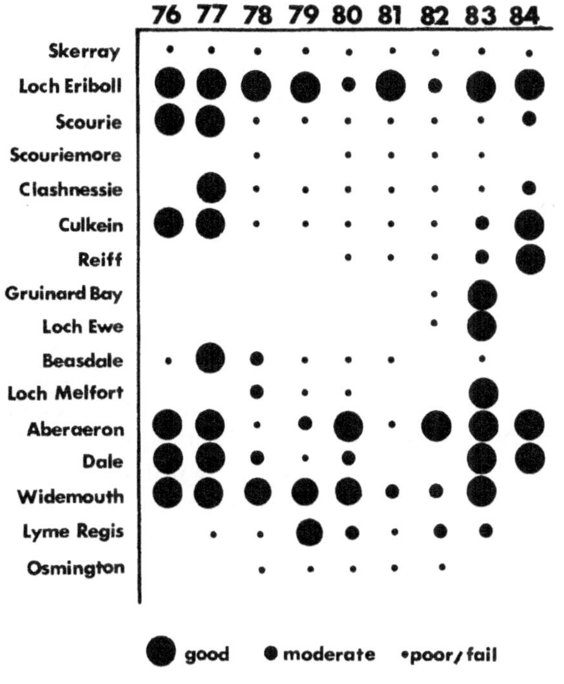

Fig. 4. An assessment, based on samples taken in late spring/early summer, of the strength of *G. umbilicalis* recruitment in Britain between 1976 and 1984. The years referred to are those of larval settlement not the year of sampling.

While it is possible to recognise broad-scale patterns of population structure and recruitment considerable local variation may also occur. A relationship between the extent of exposure to wave action and the abundance of *G. umbilicalis* has long been appreciated (Ballantyne 1961). Those populations which are close to the local limits of any of their physical tolerances have similar characteristics of those close to their geographical limits i.e. they are sparse and dominated by large individuals. The physical nature of the shore may also influence population structure. At Portnancon near the head of Loch Eriboll the population is sparse and dominated by large individuals while that at our regular sampling site 7 km away at the head of the loch is dense and composed of small animals. At the former site the shore is a rock platform with occasional areas of shingle while at the latter site clean shingle predominates. Further to the south a similar relationship exists between the predominantly rocky shore at Aberystwyth and that of stable cobbles at Aberaeron.

Further contrasts between populations in various parts of the United Kingdom are set out in Fig. 5. In general, populations in Wales were more dense than those in the north of Scotland and there was a highly significant negative relationship between abundance and adult size (Spearman rank correlation coefficient $r = -0.96, n = 11, p < 0.001$). Adult size is expressed as the diameter of the shell at the 90th percentile of a cumulative frequency plot, rather than the maximum diameter recorded, in order to minimise any possible distortion which might be caused by sampling within the tail of a size frequency distribution. Objective ageing techniques such as that used for *Monodonta lineata* by Williamson & Kendall (1981) are not suitable for use with *G. umbilicalis* as the shell is frequently worn or covered by encrusting algae. Nevertheless, examination of the degree of erosion of the larger

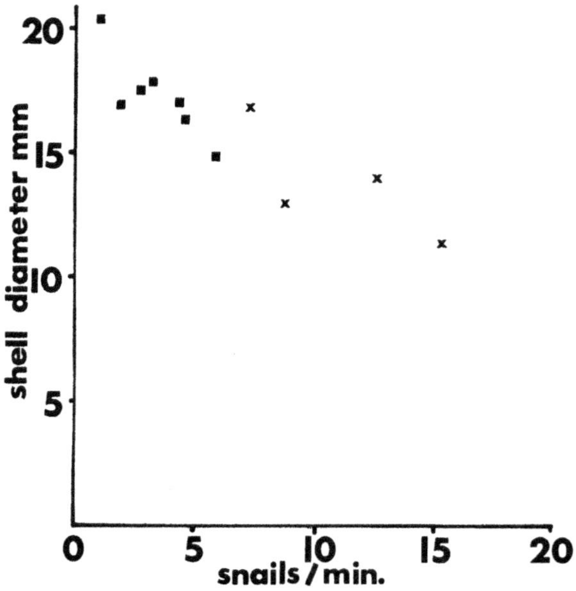

Fig. 5. The relationship between the abundance of *G. umbilicalis* and the shell diameter at the 90th percentile of a cumulative frequency plot. . = sites in Scotland × = sites in Wales.

shells at the more sparsely populated sites and the persistance of the population at Skerray despite minimal recruitment further suggests that individual longevity is associated with low population density.

Discussion

Over the eight years of our study the recruitment of *G. umbilicalis* has failed more frequently in N.W. Scotland than in Wales or S.W. England. This conforms with the predictions of Hutchins (1947) and Lewis *et al.* (1982) that failure would be most frequent close to the limits of a species distribution. These authors also proposed that if the limits were set by climatic/hydrographic factors acting on recruitment then the effect would be detectable well into the species range. The high degree of concordance in the pattern of recruitment and the predominance of large/old animals in N.W. Scotland (Fig. 2 and 4) support this hypothesis, even though we can neither prove that the actual limit of the species distribution is set purely by climatic factor(s) nor identify with certainty the stage of the species life-cycle which is most severely affected.

It is highly probable that such northern limiting factors will be related to an insufficiency of summer warmth which would limit the production of gametes or so delay the process that the juveniles would succumb to low winter temperatures. We are therefore engaged in a sequential examination of the life cycle of *G. umbilicalis* aimed at the identification of the phases in which there is the greatest scope for annual or geographical variation.

Earlier studies (Williams, 1964; Underwood, 1972) suggested that *G. umbilicalis* spawned over a period of three or four months but more recent work (Garwood & Kendall, 1985) has shown that in mid-Wales there is a well synchronised spawning period of about one month with any subsequent gamete release being of minor importance. In 1984 the timing of this major spawning period was the same at two sites in N.W. Scotland and two in Wales (Garwood & Kendall, in prep.). In the absence of complementary settlement studies we cannot be certain that the length of the settlement period was also identical, nor do we have data for other years. In N.W. Spain however, Lombas *et al.* (1984) have shown that the breeding period towards the south of the species range is considerably longer than that in Britain and thus the possibility exists that annual variation in recruitment to Scottish populations may result, at least in part, result from climatically induced variability in the length of the reprodcive season.

We also have evidence from our intensive studies at Aberaeron (Kendall, Williamson & Garwood, in press) that the severity of winter conditions may adversely affect juveniles. The two poorest year classes which recruited during our study (1978 and 1981 settlers) were those which experienced the lowest winter air temperatures, the summed deviations from the long term mean air temperatures for the months of December to March inclusive being −6.4 °C and − 1.7 °C respectively. It is also possible that there is some relationship between the size/age of the juveniles at the onset of winter conditions and their subsequent survival as in *Patella vulgata* late settling individuals usually have a higher mortality rate than those settling earlier in the year (Bowman, 1985). As yet we are unable to separate the effects of processes related to the timing/protraction of spawning from those related to the severity of winter. Thus if low numbers of 6−9 month old spat were to be found at sites over much of N.W. Scotland this could be the result, at one extreme, of poor reproduction/settlement followed by good over-winter survival or, at the other extreme, of a successful reproduction/settlement and heavy winter mortality.

Although climatic factors may ultimately set the limits of a species distribution their effects can be masked by other factors which operate over a much smaller area. Over most of N. Scotland successful recruitment is rare yet in Loch Eriboll, only 25 km from the species effective northern limit at Skerray, there is a thriving population. As yet we are unable to fully explain this anomalous situation although it is known that in some summers water within the loch can be 1–2 °C warmer than the open sea. Such warming might favour the production of gametes and/or the growth of spat. The shelter which permits this local elevation of sea temperature will also minimise the seaward dispersion of the planktonic larvae.

Variation in inshore currents or longshore drift may also lead to substantial year to year differences in local sedimentation. In areas where the breeding period is short, even a temporary covering of sand over nursery grounds can prevent settlement or cause the mortality of a whole year class. Such events resulted in the loss of the 1983 input in parts of Gruinard Bay and of Loch Melfort in N.W. Scotland and of that for 1984 at Llanon, near Aberaeron.

Underwood (1971) suggested that the larvae of *G. umbilicalis* settled passively in areas of low current velocity and this has been supported by our own observations as well as those of Myers & Southgate (1980) who collected the spat by using plastic mesh pan-scrubbers to mimic algal turf. Generally, nursery grounds are found where wave energy is low either due to the sheltered aspect of the shore as a whole or because of the protective effects of gullies or boulder fields in more exposed situations. Thus, while there is an abundance of potential settlement areas on the stable cobble shore at Aberaeron, at nearby Aberystwyth only a comparatively small area of the shore is suitable. Nevertheless, the use of spat collectors has confirmed that the rate of settlement within the gullies at the latter site is no less than that on the open shore at Aberaeron (M. Bedford, pers comm.).

Although the juvenile habitat may be restricted, the tolerances of the adult snails are far broader and hence as animals age they tend to move away from their settlement sites so causing both density and the intensity of intra-specific competition to diminish. On a shore where nursery areas are abundant however, there is little escape from such competitive pressure and thus growth rate and maximum body size will be restricted.

In the British Isles the predictability of recruitment tends to increase towards the south and as population density rises so the maximum size of the individuals declines. Frank (1975) showed that *Tequla* populations towards the south of their range were characterized by a high density of small, short-lived individuals. Further evidence of the importance of density dependant competition in influencing the size/age structure of gastropod populations comes from a study of the neritacean *Nerita atrementosa*. Underwood (1975) showed that at high densities the growth rate of juveniles of this species declined while the mortality rate of adults increased. Similarly, enhanced growth at highly reduced densities has been observed in transplantation experiments on *Monodonta lineata* (Williamson & Kendall, in prep). Such competitive pressures will seldom be of importance towards the northern limit of the species range indeed, the long life span of individuals is often necessary to ensure the continuing existance of these populations. Further to the south, however, the combination of regular recruitment and competition will combine to determine the size/age of many populations.

Conclusion

Our long term studies in N.W. Scotland have strongly suggested the existance of geographically operating factors which might act by influencing the length of the breeding period and/or the overwinter survival of the spat. Close to the species limits of distribution their operation has led to a clear pattern of success and failure of recruitment. Further into the species range the influence of these northern limit setting factors declines and the effects of more local scale physical and biological variables become more clearly evident.

Acknowledgements

We would like to thank the past and present members of the N.E.R.C. Rocky Shore Surveillance

Group who have contributed to discussions or who have collected snails for us. Particular thanks must go to Phillip Williamson and Rosemary Bowman. Work in Wales has benefited substantially from the assistance of Susan Fish, John Fish and Marianne Bedford.

References

Ballantyne, W. J., 1961. A biologically-defined exposure scale for comparative description of rocky shores. Field Studies 1:1–19

Bowman R. S., 1985. The biology of the limpet *Patella vulgata* in the British Isles: spawning time as a factor in determining recruitment success. In Moore, P. G. & R. Seed (eds). The Ecology of Rocky Coasts. Hodder & Stoughton. Sevenoaks: 178–193.

Crisp, D. J. & A. J. Southward, 1958. The distribution of intertidal organisms along the coasts of the English Channel. J. mar. biol. Ass. U.K. 37: 157–203

Fischer-Piette, E., 1955. Repartition, le long des côtes septentrionales de l'Espagne, des principal espèces peuplant les rôchers intercotidaux. Ann. Inst. Ocean. 31: 37–124

Fischer-Piette, E. & J. M. Gaillard, 1956. Sur l'écologie comparée de *Gibbula umbilicalis* da Costa et *Gibbula pennanti* Phil. Journal de Conchyliologie 96: 115–118.

Frank, P. W., 1975. Latitudinal variation in the life history features of the black turban snail *Tequla funebralis* (Prosobranchia: Trochidae). Mar. Biol. 31: 181–192

Garwood, P. R. & M. A. Kendall, 1985. The reproductive cycles of *Monodonta lineata* and *Gibbula umbilicalis* on the coast of mid-Wales. J. mar. biol. Ass. U.K. 65: 993–1008

Grant A. & P. Tyler, 1983. The analysis of data in studies of invertebrate reproduction II. The analysis of oocyte size/frequency data and comparison of different types of data. Int. J. Invert. Rep. 6: 271–283.

Hutchins, L. W., 1947. The bases for temperature zonation in geographical distribution. Ecol. Monogr. 17: 325–335.

Kendall, M A., R. S. Bowman, P. Williamson & J. R. Lewis, 1982. Settlement pattern, density and stability in the barnacle *Balanus balanoides*. Neth. J. Sea Res. 16: 119–126

Kendall, M. A., P. Williamson & P. R. Garwood. Annual variation in recruitment and population structure of *Monodonta lineata* and *Gibbula umbilicalis* populations at Aberaeron, mid-Wales. Est. Coast. Shelf Sci. In press.

Lewis J. R., 1976. Long term ecological surveillance: practical realities in the rocky littoral. Oceanogr. Mar. Biol. Ann. Rev. 14: 371–390

Lewis, J. R., R. S. Bowman, M. A. Kendall & P. Williamson, 1982. Some geographic componants in population dynamics: possibilities and realities in some littoral species. Neth. J. Sea. Res. 16: 18–28

Lombas, I., A. Bode & N. Anadon, 1984. Estudio del ciclo reproductor de *Gibbula umbilicalis* y *Monodonta lineata* en Asturias. Actas do IV Symposio Iberico de Estudios do Benthos Marinho. 2: 103–114

Myers, A. A. & T. Southgate, 1980. Artificial substrates as a means of monitoring rocky shore cryptofauna. J. mar. biol. Ass. U.K. 60: 963–975

Rendall, R., 1956. Mollusca orcadensia. Proc. r. Soc. Edinb. 66B: 131–201

Underwood, A. J., 1971. Behavioural ecology and reproduction of intertidal gastropods. Ph.D. thesis. University of Bristol.

Underwood, A. J., 1972. Observations on the repoductive cycles of *Monodonta lineata*, *Gibbula umbilicalis* and *Gibbula cineraria*. Mar. Biol. 17: 333–340

Underwood, A. J., 1975. Food competition between age-classes in the intertidal neritacean *Nerita atrementosa* Reeve. (Gastropoda; Prosobranchia). J. exp. mar. Biol. Ecol. 23: 145–154

Williams, E. E., 1964. The growth and distribution of *Gibbula umbilicalis* (da Costa) on the rocky shore in Wales. J. Anim. Ecol. 33: 433–442

Williamson, P & M. A. Kendall, 1981. Population age structure and growth of the trochid *Monodonta lineata* determined from shell rings. J. mar. biol. Ass. U.K. 61: 1011–1026

Hydrobiologia 142: 23–30 (1986)

Preliminary observations on *Gibbula umbilicalis* (da Costa, 1778) on the Portuguese coast

Maria José Gaudèncio & Miriam Tuaty Guerra
Instituto Nacional de Investigação das Pescas, Av. de Brasília, 1400 Lisboa, Portugal

Keywords: gastropods, *Gibbula umbilicalis*, breeding cycles, longevity

Summary

Ecological studies on *Gibbula umbilicalis* (da Costa, 1778) were undertaken in Portugal specifically to contribute to the COST 647 programme. This paper describes the reproductive cycle and the annual changes in the population size-structure, and makes comparisons with similar data from the species' northern geographical limits. The major latitudinal changes are a longer spawning period extending later in the autumn, smaller maximum sizes and shorter life spans.

Introduction

Gibbula umbilicalis is an eastern Atlantic species which is abundant on rocky shores in Portugal, where wave energy is low. Although it is sold as food throughout the country under the name 'burrie' its biology has not been previously studied there. Data on its population dynamics and reproductive biology have now been collected as a specific contribution to the COST 647 programme.

Distribution and habitat on the Portuguese coast

Gibbula umbilicalis is found throughout the country at low and mid-levels of the intertidal zone. It favours rocky platforms with a dense algal cover (particularly *Ulva rigida, Corallina mediterranea* and *Enteromorpha* spp) but may also be found in *Lithophyllum incrustans*-lined pools, under stones and on the upper surfaces of boulders. The adults appear to tolerate desiccating conditions, although on bare rock they may group together in cracks or pits or on ledges and fissures on vertical faces.

Juveniles may be found throughout the year, and are confined to damp situations (e.g. among algae or mussels and in small pools and cracks) in the lower shore.

Material and methods

For the determination of size-structure, specimens were collected from sites in the north, centre and south of Portugal (Fig. 1). The initial half-yearly samples, winter 1981/2 to 1982/3, were removed from five one-metre quadrats in the mid-shore and a further five in the low shore. As the population structure at both shore levels was similar, samples after June 1983 were restricted to one square metre at low level.

For the examination of gonads a minimum of 50 specimens were collected monthly between December 1984 and November 1985 from a sheltered shore of boulders with a few small alga-covered rock platforms at S. Pedro do Estoril. All specimens were maintained in sea water for 24 hrs before relaxation in isotonic magnesium chloride and preservation in 4% formalin. The maximum diameter of the shell of each animal was measured to the nearest 1 mm using vernier calipers before extraction of the gonad. In the last two months of the study fresh material alone was examined, experience having shown that both gonad colour and oocyte size and transparency were distinguishable as accurately as in preserved material.

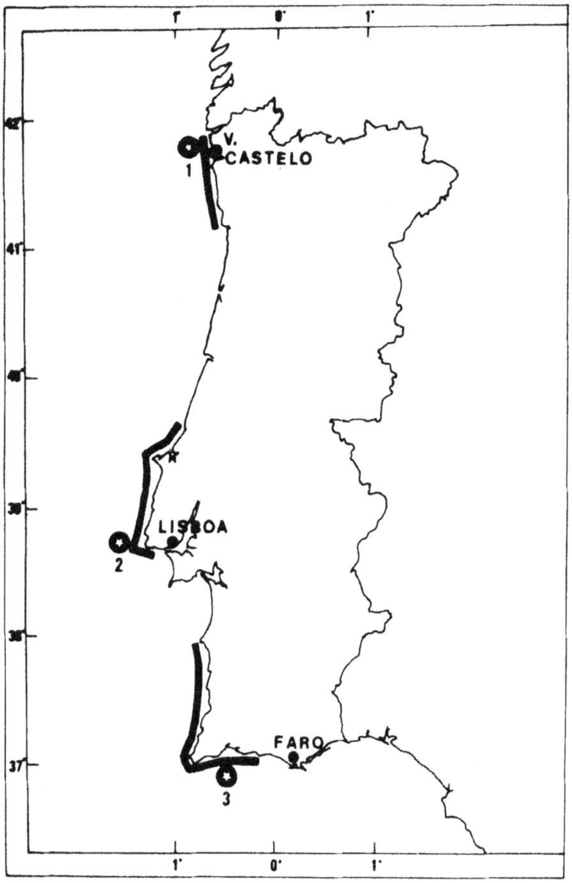

Fig. 1. Distribution of *Gibbula umbilicalis* on the Portuguese coast. ● – Sampling sites: 1 – Moledo do Minho; 2 – Cabo Raso and S. Pedro do Estoril; 3 – Praia da Luz.

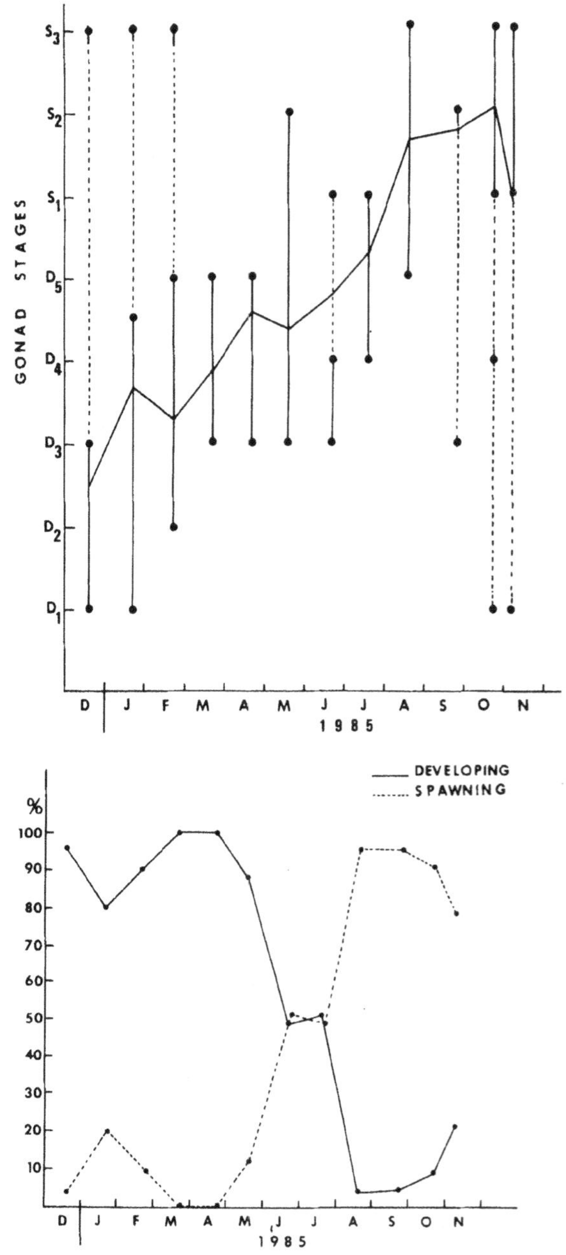

Fig. 2. Gibbula umbilicalis: a – Mean state of gonad development; b – Percentage of developing and spawning gonads.

Reproductive cycle

Fig. 2a shows the mean state of gonad development of *G. umbilicalis* from S. Pedro do Estoril between December 1984 and November 1985.

Between December and May the majority of the gonads were developing, with the whole population in this condition between March and April. After June, the percentage of spawning gonads increased, reaching a maximum between August and October (Fig. 2b). Thus, although spawning animals were found during most months there was a clear pattern of winter/spring development leading to early summer/autumn spawning.

Other authors have reported the infection of gonads of *G. umbilicalis* by trematode cercaria. In mid-Wales, Williams (1964) found that 2.24% of the animals were parasitised while at Plymouth, Underwood (1972) reported a 2% infestation. In N. Spain (Lombas *et al.,* 1984) the frequency was markedly higher at 14.25%. We found only a single infected male (12 mm base diameter) among the 738 specimens examined.

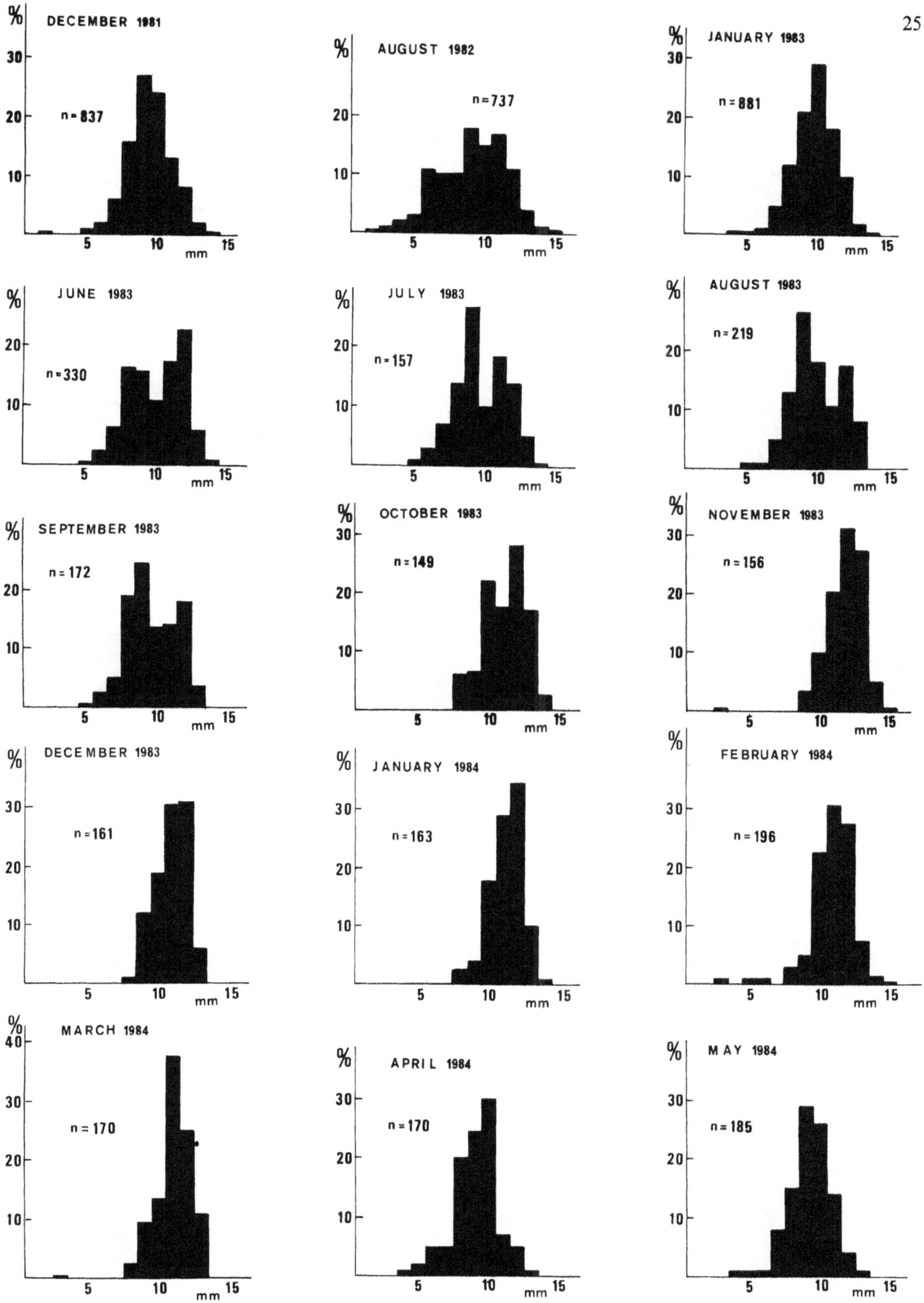

Fig. 3. Size frequencies of *G. umbilicalis* in the north of the country: percentage of individuals in relation to shell diameter.

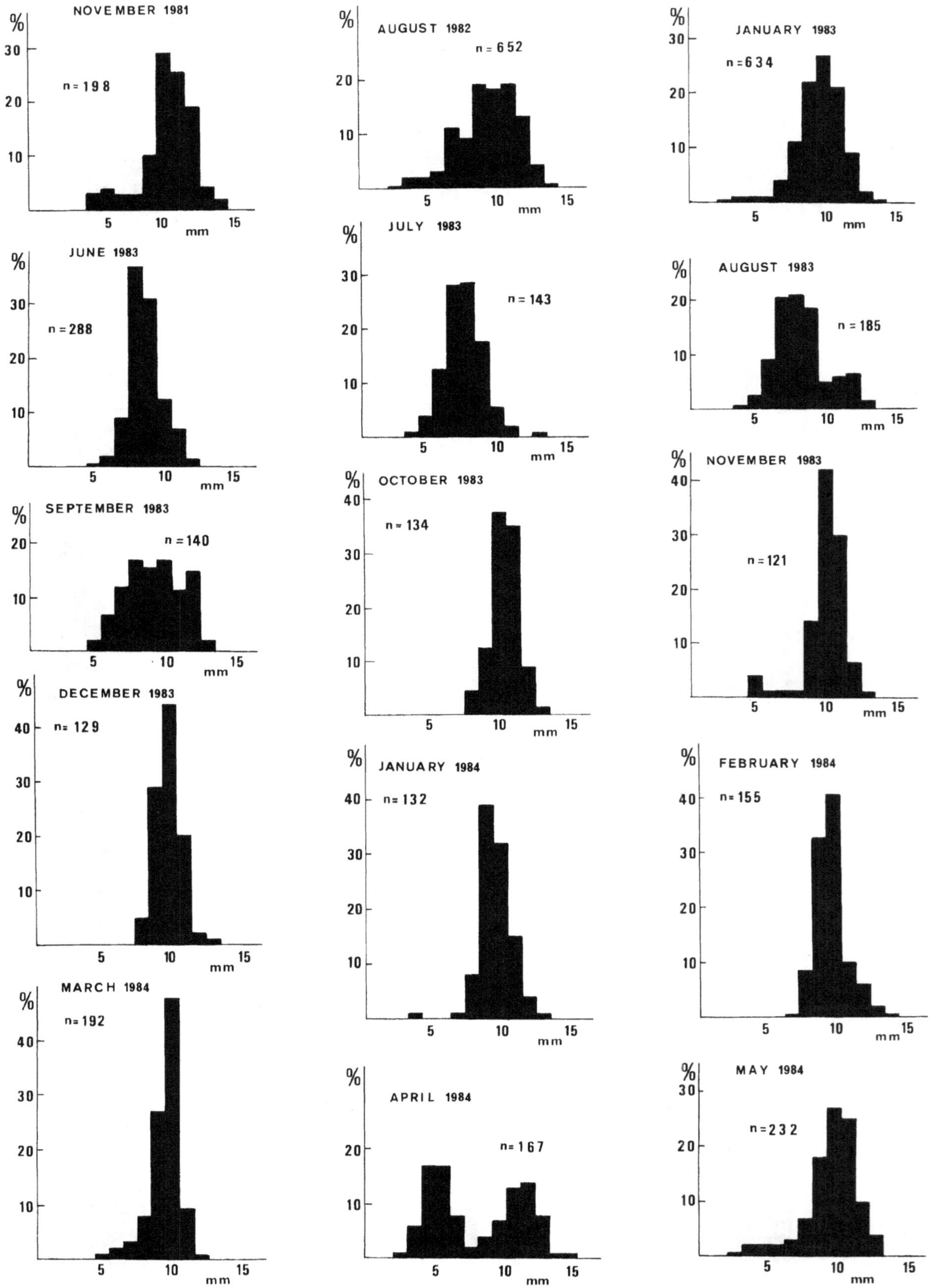

Fig. 4. Size frequencies of *G. umbilicalis* in the centre of the country: percentage of individuals in relation to shell diameter.

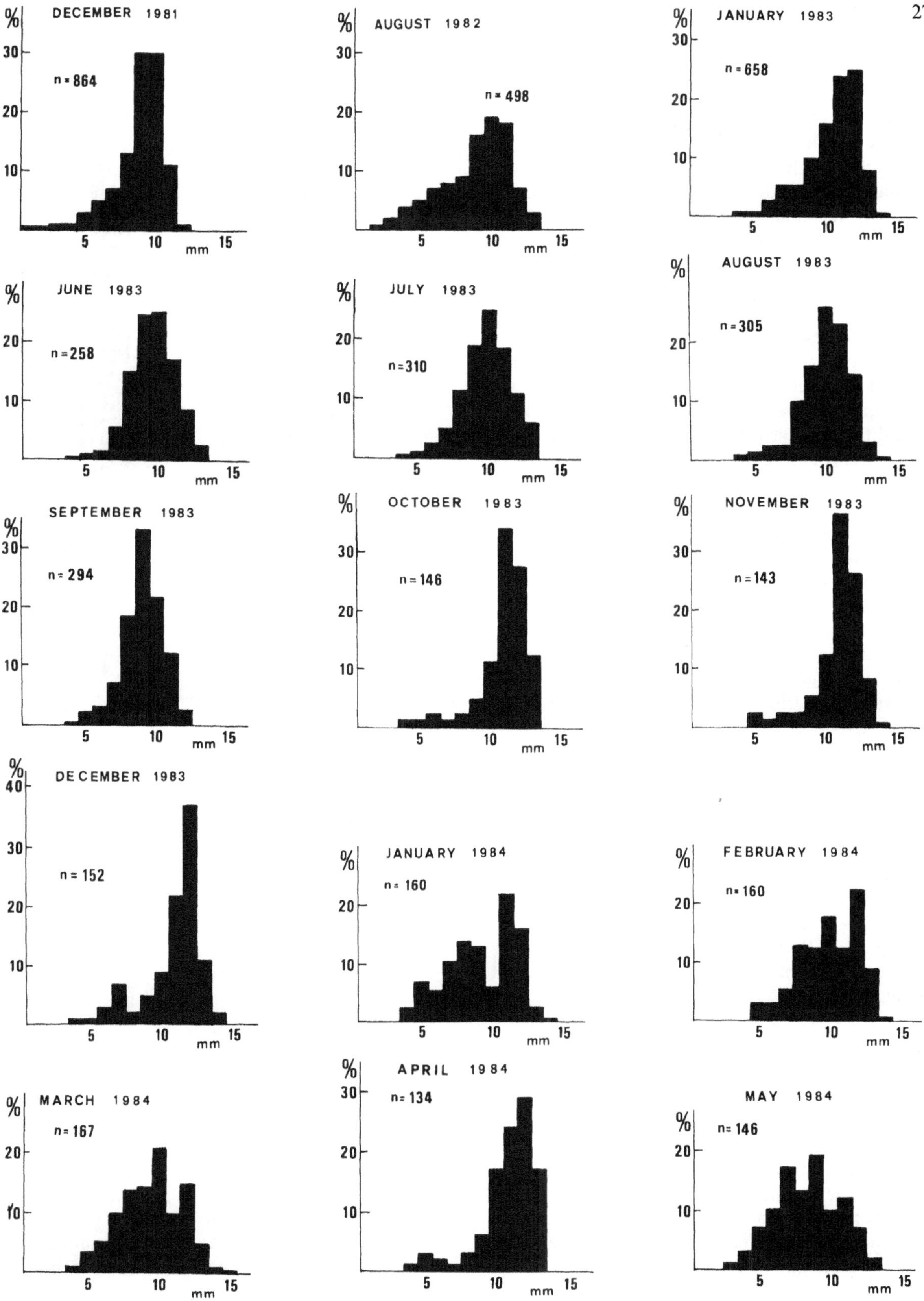

Fig. 5. Size frequencies of *G. umbilicalis* in the south of the country: percentage of individuals in relation to shell diameter.

Size frequencies

Figs 3, 4 and 5 show the size/frequency distribution of samples from the north, centre and south of the country. Clearly, sizes are similar in the three regions, ranging from 3 to 15 mm shell diameter, but with few individuals larger than 12 mm.

Full interpretation of these data in terms of settlement times, growth rate and age classes is not yet possible and may prove difficult because spat live in cracks, mussel patches, among short algae and in other non-desiccating micro-habitats until they are several mm in diameter. The presence of animals 5 mm or less over much of the year probably reflects both spawning over several summer/autumn months and differential growth of spat, the latter giving an impression of more extended settlement than may be the case.

Although several more years' data are needed to provide confirmation, it appears that settlement occurs during autumn/early winter, that 3–5 mm individuals emerge and enter the samples only intermittently during winter but increasingly in spring, occasionally creating very obvious bimodal patterns (April 1984, Figs 4 and 5). The numbers and size-range of juveniles increase during spring and summer, variously resulting in a broader bimodal distribution (e.g. June–August, 1983, Fig. 3) but with usually a tail of slow-growing (or late settling) 4–6 mm individuals. By Oct–Dec the '0' class cohort tends to be absorbed into, or to constitute, the major and often unimodal peak of 10–12 mm animals, while categories less than 5 mm (or even 6 or 7 mm) are lacking or scarce (Oct 1983–Jan '84, Fig. 3; Oct–Dec '83, Fig. 4).

On this tentative basis it seems that maximum life spans in Portugal are about 2–2¼ years, but this should be confirmed by observation of marked individuals.

Sex ratio

Of the 738 specimens subjected to gonad analysis 52.7% were male and 47.3% were female. The modal shell diameter was 12 mm for both sexes; 27.5% of the population (14.9% males and 12.5% females) were within this category (Fig. 6a).

In both sexes gonads began to develop at a shell diameter of 8 mm (i.e. animals within their first year) and became mature by the time a 10–11 mm diameter was reached (Fig. 6b). Among the very few animals of 15 mm only spawning stages were seen, possibly an indication of continuous redevelopment and spawning, but more probably be-

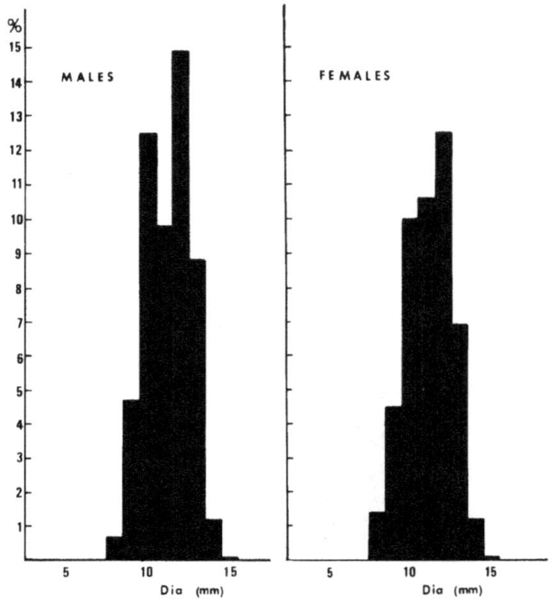

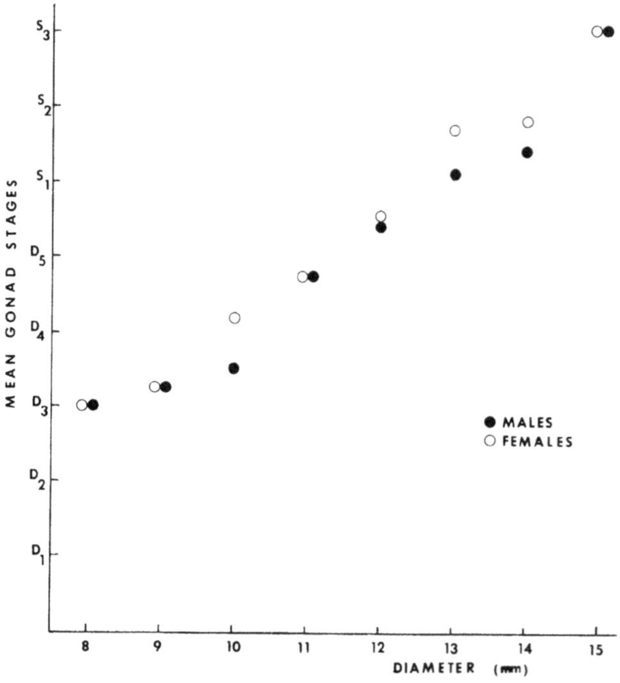

Fig. 6. Gibbula umbilicalis: a – percentage of males and females in relation to shell diameter; b – Gonad development in relation to shell diameter.

cause such sizes are usually attained only by second year animals during their second summer/autumn. Even in animals below 10 mm gonads were in an advanced state of development, and therefore we conclude that they are probably capable of spawning. Similar observations were made in Wales by Williams (1964) who recorded maturation at shell diameters between 7.5 and 8 mm. In Spain gonad development begins at around 9 mm and full maturity is reached at 11 mm (Lombas *et al.,* 1984).

Discussion

Previous studies on the reproductive cycle of *G. umbilicalis* in Britain have produced rather divergent results, probably because different methods were used. Williams (1964) recorded development from November to May/June, spawning until September and a resting period of 1–2 months. Underwood (1972) found a longer resting period (Oct/Nov–Jan/Feb) and spawning from June to September. Recently Garwood & Kendall (1985) have shown that oocyte development may occupy a year overall but the active vitellogenic growth phase occurs only between late April and July or August, and is followed by one or two discrete spawning periods in August and possibly September.

Our Portuguese data differ from the above in the absence of a clearly defined resting period involving the whole population. Although some individuals were in apparent spawning condition in 10 months of the year, and developing stages could be found in all months, there was nevertheless substantial synchrony with most animals developing Dec–June, and most spawning August–Oct/Nov. Similar data have been recorded in Asturias in N. Spain by Lombas *et al.* (1984) although in that study the main development and spawning periods occurred a little earlier than in Portugal.

The presence of 'spawning stages' from late May through to February may prove erroneous if more detailed studies are possible in the future: the first ripe eggs to appear towards the end of the development period are not necessarily about to be released at once, and Garwood & Kendall (loc. cit.) warn against interpreting the persistence of large oocytes throughout the winter as evidence of a very extended season of *effective* spawning. Nevertheless, Fig. 2 shows that at least two major spawnings took place, one in mid-August and the other late October/early November. So by comparison with the data of Garwood & Kendall, Portuguese *Gibbula* have a longer period of effective spawning, extending 1–2 months later into the autumn.

Such differences accord with some of the latitudinal possibilities proposed by Hutchins (1947) and Lewis *et al.* (1982). In Britain (at or near the species' northern limit of distribution) cold restricts spawning to a short period at or just after the period of maximum temperatures. At the extreme southern limits it is suggested that excessive heat will limit some phase of reproduction/settlement to the cooler winter period. Portugal is in the middle of the species' geographical range, and it appears that although spawning is still possible in late summer (i.e. it is still not far enough south to be too hot then) the extension of spawning to October–November is a stage in the transition from northern summer to southern winter. Such a process is seen in *Patella vulgata* in which successful settlement in Portugal, its southern limit, is confined to winter (Bowman & Lewis, 1975; Guerra & Gaudencio, 1986). Whether or not all spawnings of *G. umbilicalis* lead to successful settlement, and especially those in August or before, necessitates more detailed information on settlement times than we now possess.

Both the life span and maximum shell size in Portugal also differ substantially from those recorded in north-west Europe. Size is not simply related to temperature for it depends upon growth rate and life span, both of which may reflect other local or broad-scale influences. Dehnel (1958) and Frank (1975) found that despite slower growth rates higher maximum sizes could be achieved in colder latitudes because of the longer life spans there; and among Portuguese molluscs *Patella vulgata,* here at its southern limit, is both smaller and shorter-lived than in Britain.

G. umbilicalis, at its northern limit, may reach a shell diameter of 21 mm and live for up to 12 years, although at other sites in north-west Scotland most animals have shell diameters between 13 and 14 mm and probably live for between 6 and 8 years (Lewis *et al.,* op. cit.). Both size and life span appear to decrease as one moves to the south; in Wales and N. France the species seldom lives longer than five years (Pelseneer, 1933 in Lewis *et al.,* 1982). In Brittany, Fischer-Piette (1963) found specimens of between 15 and 19 mm shell diameter, although in S.W. France and Spain he failed to record any animals larger than 13 mm. In N. Spain the largest specimens observed by Lombas *et al.* (1984) were of 16 mm shell diameter.

In Portugal, *G. umbilicalis* is closer to its southern limit of distribution than in any of the countries referred to above and the trend towards smaller maximum sizes is carried further, as we seldom found specimens larger than 13 mm diameter. As stated above longevity in Portugal appears to be no more than 2¼ yrs approximately, and whilst the reason for this is unclear there can be no doubt that it limits the size that can be reached. There is also another probable factor: population densities. As Figs 3, 4 and 5 show, densities of $130-200/m^2$ are general and may occasionally approach 300. Comparison with the $50-70/m^2$ maxima in Britain (Kendall, pers. commun.) suggests that a resulting greater competition for food must contribute to the smaller sizes. But these higher densities can be reasonably attributed to the longer spawning period which increases the chances of more successful settlements per season, and the duration of spawning is in turn a consequence of the increasing temperatures from N.W. Europe to Portugal.

But while there appears to be a latitudinal trend, a local Portuguese factor of unknown potency remains: the collection of *Gibbula* for human food. This will surely limit both maximum size and life span.

Acknowledgments

We would like to express our thanks to Dr. J. R. Lewis and to Mr. M. A. Kendall for their help and advice in the accomplishment of this work and for critically reading the manuscript. We also thank our colleague Mr. E. Soares for his help in the achievement of the manuscript.

References

Dehnel, P. A., 1955. Rates of growth of gastropods as a function of latitude. Physiol. Zool. 28: 115–144.

Fischer-Piette, E., 1963. La distribution des principaux organismes intercotidaux nord-iberiques en 1954–1955. Ann. Inst. oceanogr., Monaco 40: 165–312.

Frank, P. W., 1975. Latitudinal variation in the life history features of the black turban snail *Tegula funebralis*. Mar. biol. 31: 181–192.

Garwood, P. R. & M. A. Kendall, 1985. The reproductive cycles of *Monodonta lineata* and *Gibbula umbilicalis* on the coast of mid-Wales. J. mar. biol. Ass. U.K. 65: 993–1008.

Hutchins, L. W., 1947. The bases for temperature zonation in geographical distribution. Ecol. Monogr. 17: 325–335.

Lewis, J. R., R. S. Bowman, M. A. Kendall & P. Williamson, 1982. Some geographical components in population dynamics: possibilities and realities in some littoral species. Neth. J. Sea Res. 16: 18–28.

Lombas, I., A. Bode & N. Anadon, 1984. Estudio del ciclo reproductor de *Gibbula umbilicalis* y *Monodonta lineata* en Asturias (norte de España). Actas do IV Simposio iberico de Estudos do Bentos marinho, Lisboa; 21–25 Maio 1984, 2. Vasco Monteiro Marques, Lisboa: 103–114.

Underwood, A. J., 1972. Observations on the reproductive cycles of *Monodonta lineata, Gibbula umbilicalis* and *G. cineraria*. Mar. Biol. 17: 333–340.

Williams, E. E., 1964. The growth and distribution of *Gibbula umbilicalis* (da Costa) on a rocky shore in Wales. J. anim. Ecol. 33: 433–442.

Hydrobiologia 142: 31–39 (1986)

Preliminary studies on the reproduction and population dynamics of *Monodonta lineata* and *Gibbula umbilicalis* (Mollusca, Gastropoda) on the central coast of Asturias (N. Spain)

A. Bode, I. Lombas & N. Anadon
Departamento de Zoología y Ecología, Facultad de Biología, Universidad de Oviedo, Asturias, Spain

Keywords: reproduction, dynamics, Gastropoda, Trochidae, NE Atlantic

Summary

Populations of *Monodonta lineata* and *Gibbula umbilicalis* were studied at Aramar beach (Luanco, Asturias) from October 1982 to November 1983. Both species show a sexual ratio of 1:1 but differ in their gonadal cycle patterns. *M. lineata* gonadal development stages are found from November to May–June in a significant part of the population. Breeding stages occur from June–July to September, and in some individuals to November, with complete discharge of all spawn at the end of the reproductive season. Some degree of inter-annual variation is detected. Spawning stages of mature *G. umbilicalis* are found through the year but the higher frequencies occur in summer and autumn samples, with development phases predominating in January to May. Animals over 10 mm in shell diameter maintain their gonads close to spawning stages, completely spawned individuals being less than 10 percent. In both species variability in gonad stages for a given month reduces as size increases, indicating a certain synchrony.

The new recruits of *M. lineata* are detected on the bare rock by September to November and are recognizable through their first year. Due to the amplitude of the breeding period and differing growth rates size distribution within an age class is wide. Size class frequencies rapidly decrease after the second year of life. The data for *G. umbilicalis* are difficult to interpret given the scarce number of juveniles and size overlapping.

Introduction

Gibbula umbilicalis (da Costa, 1778) and *Monodonta lineata* (da Costa, 1778) populations occur in a wide belt along the rocky Atlantic intertidal shores. They are very abundant in some localities. The northern limits of their distribution are the Scottish and Welsh coast, and in the south, NW Africa (Fretter & Graham, 1977).

In the British Isles a lot of authors have been studying these populations for a long time (Williams, 1964, 1965; Desai, 1966; Underwood, 1972; Stanbury, 1974; Williamson & Kendall, 1981; Garwood & Kendall, 1985). But references about the species behaviour in central and southern localities of their geographical distribution range are limited.

Thus, some studies were begun in the Cantabrian coast (Anadon *et al.*, 1984) with the object of giving a preliminary picture of their local biology and afterwards making comparison with other localities. It is hoped that useful conclusions about the regional and local factors role in the population composition and function can be derived.

Description of site studied

The Cantabrian coast (N Spain) has a special ecological and biogeographical interest. There is a gradient between the boreo-atlantic conditions in Galicia and more 'mediterranean' ones in the southern part of the Bay of Biscay (see Anadon, 1980; Ibañez *et al.*, 1984). The Asturian coast occupies an intermediate position.

The study was carried out at Aramar beach (Luanco), situated at the eastern margin of Cape Peñas, in the middle of the Asturian coast. The beach is protected from the direct effect of the waves by some rock outcrops, which present a typical zonation pattern of a exposed shore ('sensu' Lewis, 1964). The inner pebble portion is more sheltered and *Fucus spiralis* appears as the dominant vegetation.

Material and methods

Monthly samples, with at least 70 specimens of each species, were collected from October 1982 to November 1983. *M lineata* occurs in a wide zone above E.L.W.S. at the most exposed face. The densest population was of *G. umbilicalis* and was sampled in two different places. From the inner portion of the beach we gathered individuals from pebbles and pools under *Fucus* spp. horizon. While in the exposed zone we collect all the specimens from an area of 3600 cm^2, on bare rock between 1.5 and 2 m above E.L.W.S.

The animals were narcotised in $Mg\ Cl_2$ over a period of 12 h, and fixed in 4% formalin. Afterwards shell width (± 0.1 mm) was measured for individual animals and soft parts of the body were extracted. For the study of the breeding cycle 30 specimens were dissected on each occasion. We used a Gonad Index based on gonad appearance and size under a binocular microscope, similar to that described by Williams (1964, 1965) and Desai (1966). We recognised 4 development stages in *M. lineata* and 5 in *G. umbilicalis,* as well as 3 spawning stages in both species.

Population structure was studied from frequency distribution of size classes (1 mm width) in monthly samples. To separate component populations the data were analysed using probability graph paper (Harding, 1949).

Results

Gonadal studies

Sex ratio agrees significantly with 1:1 for both species (χ^2 test, $p < 0.05$), considering all the sampled individuals.

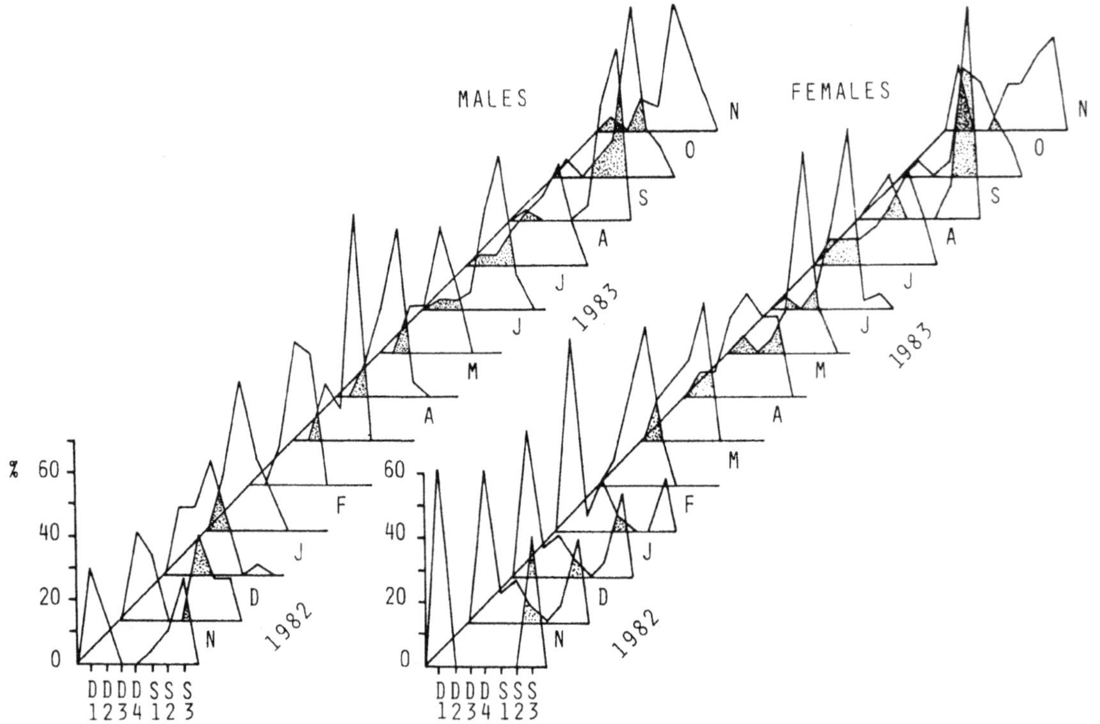

Fig. 1. Distribution of gonadal stages in the M. lineata population through the study period. D1 to D4: development stages; S1 to S3: spawning stages.

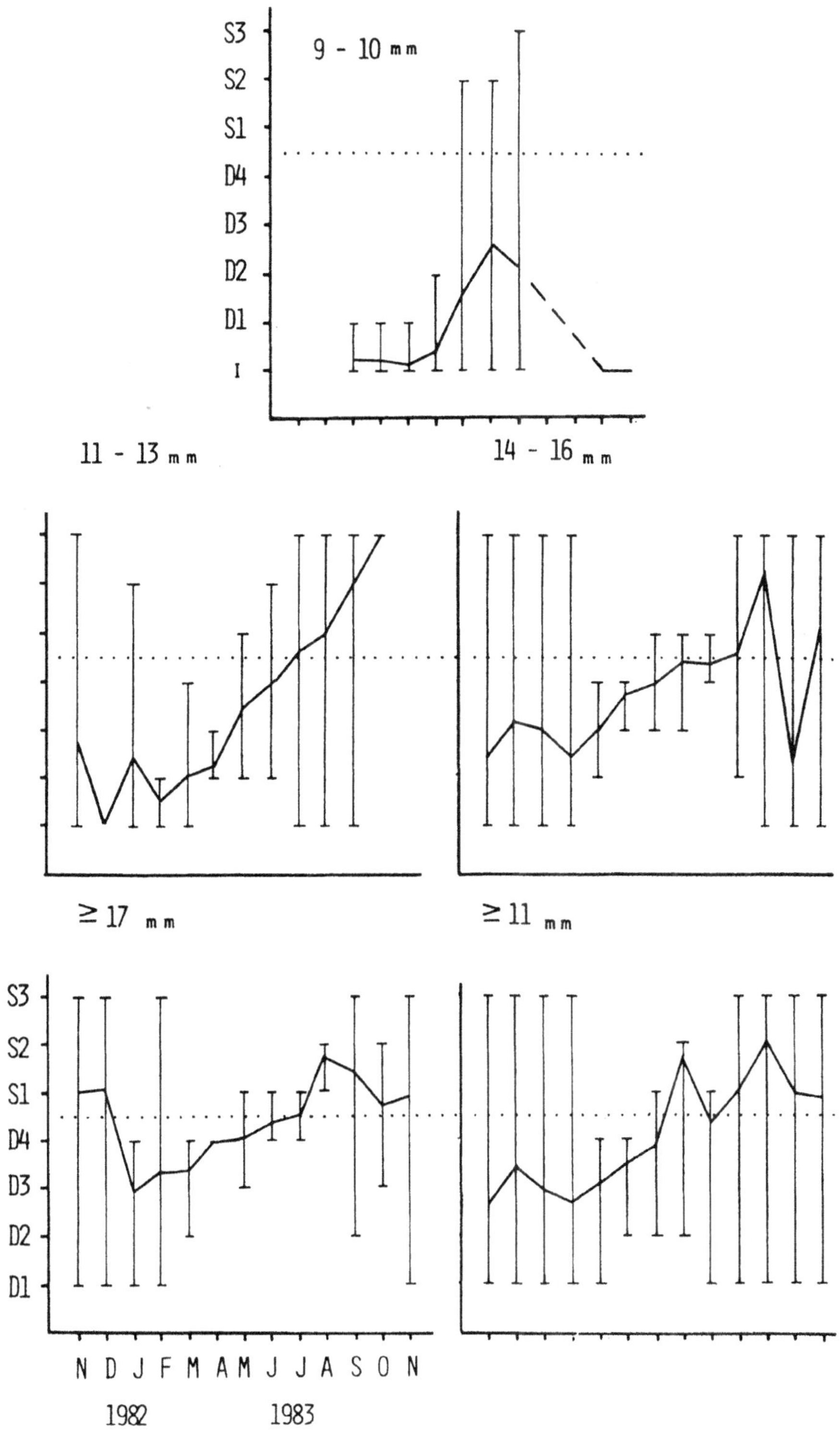

Fig. 2. Mean gonadal stages in some size classes from *M. lineata* samples. Vertical lines indicate the range of variation. Dotted lines show the limit between development (Dl to D4) and spawning (Sl to S3) stages. I: inmature gonads.

M. lineata

This species showed a gradual gonadal development beginning in November 1982. The highest numbers of individuals with ripe gonads were from May and June 1983 samples, both sexes behaving similarly throughout the year (see Fig. 1). Breeding stages reached highest frequencies in summer and autumn with a domination of gonads in spent condition in August and November. A limited recovery of gonads seemed to occur in September (Fig. 1).

Every size class contributed in a different way to the yearly gonadal cycle of the population (see Fig. 2). Individuals with shell diameter lower than 8.5 mm did not show recognisable gonads in any of the monthly samples. Size classes 9 and 10 mm had some specimens with gonads in various stages of development (including spawning), but probably have little effect on the whole population given that their mean gonadal index did not reach spawning values. In larger sizes development of the gonads began in February and proceeded gradually, with less variability in the reproductive stage of each month and more time in spawning stages as size increased.

Considering only sexually mature individuals with significant variations in the gonad index, we can obtain mean values of the index related to spawning stages beginning in May and extending over at least six months (see Fig. 2). Deviations above the mean are attributable to larger animals, spending more time in spawning stages, while lower values correspond to smaller sizes.

Comparison of gonad index values of November 1982 and 1983 samples point out inter-annual variability in development and length of the breeding period (see Figs 1 and 2).

G. umbilicalis

Development stages were most abundant between January and May but limited re-development

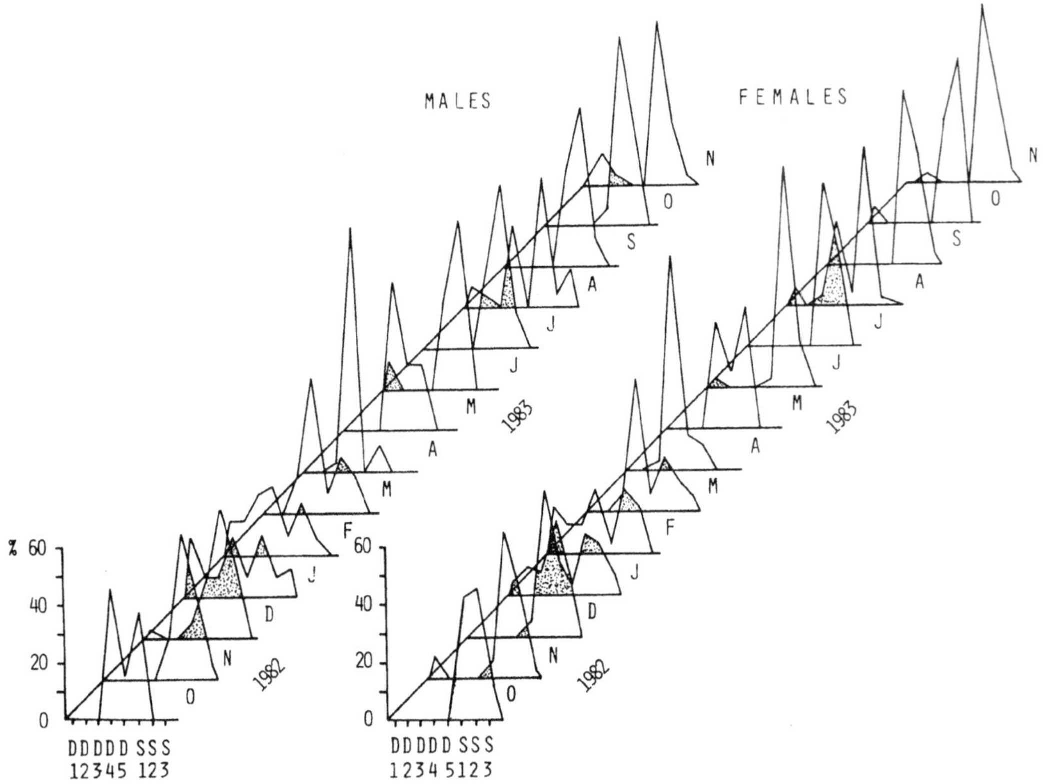

Fig. 3. Distribution of gonadal stages in the *G. umbilicalis* population through the study period. D1 to D5: development stages. S1 to S3: spawning stages.

also occurred in August–November. At the same time the recognised fully mature and spawning stages prevailed in the population for most of the year, being more abundant in June–July and November–December. The samples never had spent gonads with frequencies greater than 10%. Males and females showed similar and synchronous variations (see Fig. 3).

Fig. 4 represents the pattern of change in the gonadal index through the year for various size classes. Individuals of classes 9 and 10 mm did not show mean values of the index related to spawning stages, and gonadal activity was not found in lower sizes, as in *M. lineata.* In the same way there was a reduction in the variability of the index values and the mean increased to spawning stages as size increased.

The whole group of mature individuals maintained their mean gonadal index values close to the limit of the onset of spawning for most of the year (see Fig. 4). Developmental phases occurred mainly between January and May and also in August. Indications of inter-annual variations were not detected.

Population structure and dynamics

M. lineata

The studied population showed three distinctly recognizable components through the year (see Fig. 5). New recruits could be first detected in November 1982 and September 1983. More individuals appeared in the next months and the resulting size distributions have wide ranges dues probably to differences in growth rates and times of emergence from microhabitats. After a slow growing period (December to March) the new cohort was the prevailing mode of the sample size distribution by mid summer. At this time individuals of the preceding recruitments were grouped: *r* class. They had mean sizes larger than 18 mm and were estimated older than 2 years.

As in the gonadal cycle, some displacements of the modes can be noted in samples of a given month in succesive years (see Fig. 5), suggesting variations in the time of maximum recruitment.

G. umbilicalis

The larger classes dominated the size frequency distributions of samples, but individuals lower than 5 mm were found throughout the year (Fig. 6). Separation of components was difficult due to the low representation of the juvenile classes and the high size overlap in the resulting distributions. Nevertheless some age groups could be taken out and studied for some months.

Considering that most spawning occurs between June and November it is reasonable to expect one

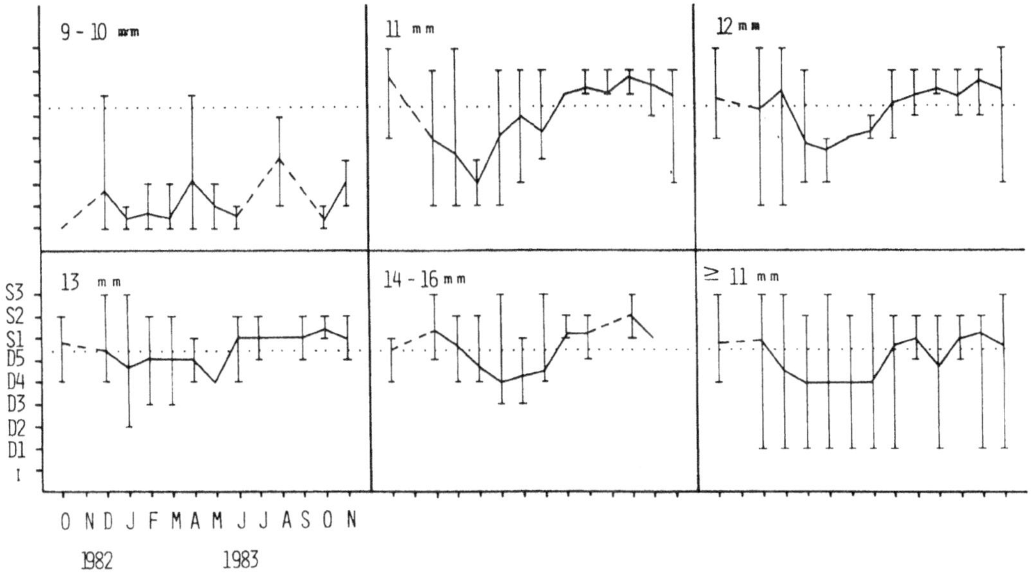

Fig. 4. Mean gonadal stages in some size classes from *G. umbilicalis* samples. Vertical lines indicate the range of variation. Dotted lines show the limit between developmental (Dl to D5) and spawning (Sl to S3) stages. I: inmature gonads.

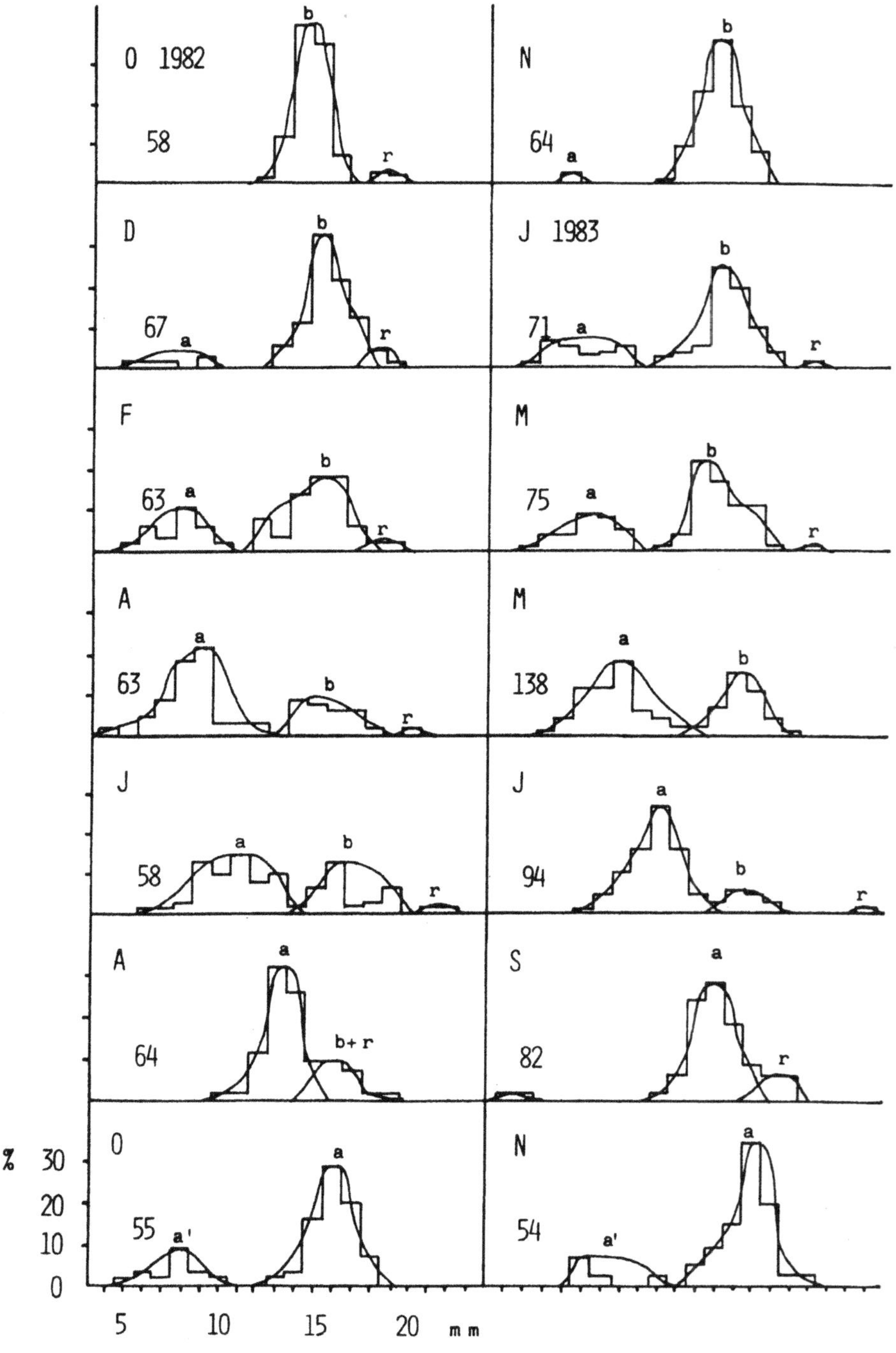

Fig. 5. Size – class frequency distributions from samples of *M. lineata* a', a, b & r: represent different age groups. Number of individuals per sample are also given.

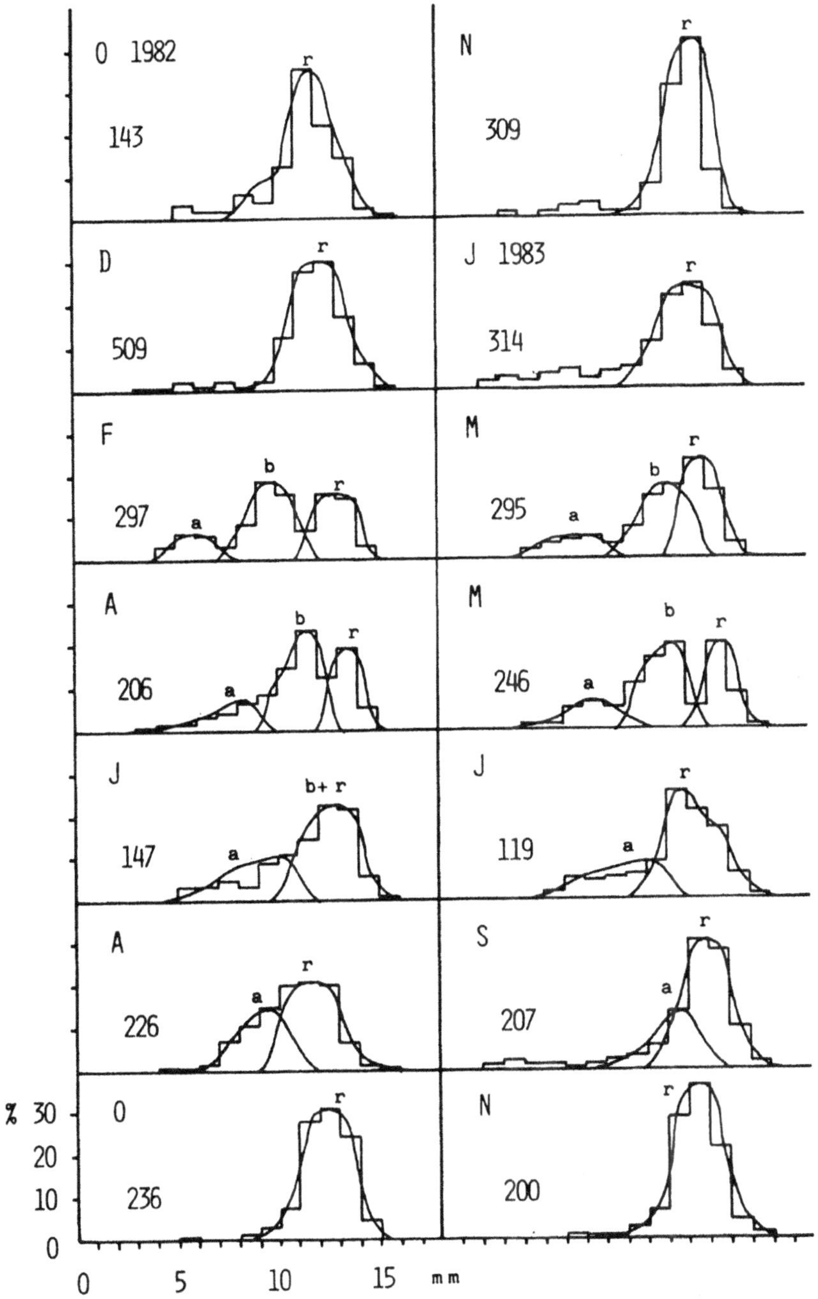

Fig. 6. Size – class frequency distributions from samples of *G. umbilicalis.* Age groups are indicated by: a, b & r. The number of individuals per sample are also given.

main settlement that may extend over several months in the autumn. But it is unclear if the small individuals in the October, November and December 1982 samples belong to the new settled cohort or are the slow growing of the previous year class. Unfortunately the samples of the autumn of 1983 did not solve the question. Indeed the sudden emergence of size group named *b* in Fig. 6 is difficult to interpret.

The growth seems to proceed rapidly in the summer months, and the main size group (*r* class) contains the survivors of the previous year settlement, having mean sizes greater than 10 mm.

Discussion

Our results show that both species have individuals at spawning stages during several months of the year. In comparison, the populations of the British Isles studied with the same methodology as ours, have a shorter breeding season in summer which sometimes lasts until September or October (Williams, 1964, 1965; Desai, 1966; Underwood, 1972). These studies also suggest a parallelism between the sexual cycles of *G. umbilicalis* and *M. lineata,* although we have found important differences.

The observations of Garwood & Kendall (1985) on size and number variations of gametes in these species allowed them to determine more exactly the period of effective spawning within the breeding cycle, as the existence of mature gametes does not imply that spawning will occur. Nevertheless at the moment we can not offer comparable data, we can only suggest that our populations seem to have a large spawning season. This is reflected by the low frequency of the stages with spent gonads (see Figs 1 and 3) and indirectly by the width of size distributions of the new cohorts at the time of their detection on the coast (see Figs 5 and 6). Garwood & Kendall (1985) observed that te synchronisation of spawning individuals of the population is sometimes less marked and a gradual increase of free gametes over a long period of time often occurs.

The observed gonad development did not have a resting phase, as the authors (Williams, 1964, 1965; Desai, 1966; Underwood, 1972; Garwood & Kendall, 1985) indicate for the northern populations, and both species developed ripe gonads in less time. In addition *G. umbilicalis* showed a distribution of gonadal stages that follows the constant formation of mature gametes. Equally we can determine the age of first sexual maturity from the obtained relations between size, gonadal stage and size distributions through the year, assuming that size classes with a starting gonadal activity are about one year old (*G. umbilicalis*) or less (*M. lineata*). Thus our examples seem to mature earlier than those in British populations, which do not reach maturity until 2 years old (Williams, 1964, 1965; Underwood, 1972; Garwood & Kendall, 1985).

A metabolic acceleration with latitudinal decrease (and basically because of temperature increase) is a well known phenomenon in intertidal invertebrates (Kinne, 1970; Newell, 1979). The process of growth and maturation have higher rates in lower latitudes, although there are some factors like competition effects which mask and modify this tendency. In our populations the size of the largest observed individuals are clearly lower than those from higher latitudes (Williams, 1964, 1965; Williamson & Kendall, 1981; Lewis *et al.,* 1982). Although other factors may be involved, it could be a direct consequence of lower longevity.

With a decrease in latitude there is an increase in the length of the molluscs reproductive period (Fretter & Graham, 1964), so that species from temperate climates are not restricted to exact reproductive seasons. Nevertheless, the length of the reproductive season may depend on other conditions, such as the lunar cycles or local dynamics of phytoplankton (Fretter & Graham, 1964; Himmelman, 1975). A consequence of lower latitude (and higher coastal water temperature) could be the extension of the breeding cycle, mainly in our population of *G. umbilicalis*. As the studied zone occupies an intermediate latitude in the distribution range of the species, the populations will not suffer a shortening in their breeding cycle because of high summer temperature, as Lewis *et al.* (1982) suggest.

Often it is difficult to achieve accurate estimations of the number of individuals and date of recruitment at the benthic populations. In our study case detection of the young stages is complicated by their having different habitats to those occupied by adults (pers. obs.), so they are poorly represented in our samples. In spite of this the fact that adult *G. umbilicalis* maintain ripe gonads for most of the year does not imply the absence of spawning as Underwood (1972) suggests. We have no evidence of reproductive failures of this species in the studied locality and there were no extreme climatic fluctuations during the study period (Anadon *et al.,* 1984).

If the studied populations maintain a steady state, we can consider the success of reproduction will be assured because of the high density of mature adults (Anadon *et al.,* 1984). In these circumstances an intensive synchronous spawning will not be essential, nor the release of all of the produced gametes. The less dense population of *M. lineata* (pers. obs.) will need higher gamete emission. The greater number of adults should also allow *G. umbilicalis* to resist higher levels of gonadal parasitism in our locality (Lombas *et al.,* 1984) than British populations (Williams, 1964; Underwood, 1972) without important alterations in the population.

In the climatic conditions of the Asturian coast, warmer than Galicia, the studied trochid populations should occur with a faster turnover rate than ones in localities with colder temperatures, with lower rates of growth and reproduction. So populations should be regulated by competition (mainly intra-specific), rather than the physical environment. Higher metabolic requirements and scarcity of resources (not only the trophic ones) should establish an upper limit to the size of adults in dense populations. The high mobility of *G. umbilicalis* between consecutive low tides (Bode *et al.,* 1984) and through the year (Anadon *et al.,* 1984) may also be a result of this competition.

Acknowledgements

We are grateful for the advise on the early phases of the study by R. Anadon. The whole work group on rocky intertidal of the COST 647 EEC Project provided stimulating discussions and data interchange, particularly J. Lewis and M. Kendall. This work was funded by Fundacion Ramon Areces, Ayuda Puntual 1982.

References

Anadón, N., A. Bode & I. Lombas, 1984. Estudio de la biologia de las poblaciones de *Gibbula umbilicalis* y *Monodonta lineata* (Mollusca, Gastropoda) en la costa de Asturias. Memoria Fundación Ramón Areces (unpubl.) 86 pp.

Anadon, R., 1980. Estructura y dinámica del sistema litoral rocoso de las costas de Asturias. Memoria Fundación J. March (unpubl.), 251 pp.

Bode, A., I. Lombas & N. Anadon, 1984. Distribución espacial y movilidad de *Gibbula umbilicalis* (Mollusca, Gastropoda) en una costa rocosa expuesta. Actas do 4° Simposio Iberico de Estudos do Benthos marinho. Lisboa. 2: 115–124.

Desai, B. N., 1966. The biology of *Monodonta lineata* (Da Costa). Proc. malac. Soc. Lond., 37: 1–17.

Fretter, V. & A. Graham, 1964. Reproduction. In K. M. Wilbur & C. M. Yonge (eds). Physiology of Mollusca. 1: 127–164.

Fretter, V. & A. Graham, 1977. The Prosobranch Molluscs of Britain and Denmark. Part. 2. Trochacea. J. moll. Stud., suppl., 3: 39–100.

Garwood, P. R. & M. A. Kendall, 1985. The reproductive cycles of *Monodonta lineata* and *Gibbula umbilicalis* on the coast of Mid Wales. J. mar. biol. Ass. U.K., 65: 993–1008.

Harding, J. P., 1949. The use of probability paper for graphical analysis of polymodal frequency distributions. J. mar. biol. Ass. U.K., 2: 141–153.

Ibañez, M., A. Romero, J. Feliu, A. Borja, A. Altuna & F. Aguirrezabalaga, 1984. Consideraciones sobre la biogeografía de la costa vasca. Actas do 4° Simposio Ibérico de Estudos do Benthos marinho. Lisboa. 1: 36–45.

Kinne, O., 1970. Temperature. In: O. Kinne (ed.), Marine Ecology. Wiley-Interscience, 681 pp.

Lewis, J. R., 1964. The ecology of rocky shores. Hodder & Stoughton, London 323 pp.

Lewis, J. R., R. S. Bowman, M. A. Kendall & P. Williamson, 1982. Some geographical components in population dynamics: possibilities and realities in some littoral species. Neth. J. Sea Res., 16: 18–28.

Lombas, I., A. Bode & N. Anadon, 1984. Estudio del ciclo reproductor de *Gibbula umbilicalis* y *Monodonta lineata* en Asturias (N de España). Actas do 4° Simposio Iberico de Estudos do Benthos marinho. Lisboa. 2: 103–114.

Newell, R. C., 1970. Biology of intertidal animals. P. Eleck. (Scientific Books) Ltd. London. 555 pp.

Underwood, A. J., 1972. Observations on the reproductive cycles of *Monodonta lineta, Gibbula umbilicalis* and *G. cineraria.* Mar. Biol., 17: 333–340.

Williams, E. E., 1964. The growth and distribution of *Gibbula umbilicalis* (da Costa) on a rocky shore in Wales. J. anim. Ecol., 33: 433–442.

Williams, E. E., 1965. The growth and distribution of *Monodonta lineata* (da Costa) on a rocky shore in Wales. Field Studies, 2: 189–198.

Williamson, P. & M. A. Kendall, 1981. Population age structure and growth of the trochid *Monodonta lineata* determined from shell rings. J. mar. biol. Ass. U.K., 61: 1011–1026.

Hydrobiologia 142: 41–56 (1986)

Geographical variation in the breeding cycles and recruitment of *Patella* spp.

R. S. Bowman & J. R. Lewis[2]
[1] *NERC Rocky-Shore Surveillance Group, Dove Marine Laboratory, (University of Newcastle) Cullercoats, N. Shields, Tyne and Wear. NE30 4PZ. United Kingdom*
[2] *62 Station Rd, Scalby, Scarborough, N. Yorks. YO13 0QG, United Kingdom*

Keywords: gastropods, *Patella* spp., breeding cycles, spawning, recruitment, geographical range

Abstract

The effect of geographical differences in breeding cycles on the recruitment variation of the limpets – *Patella vulgata* L., *P. aspera* Röding and (to a limited extent) *P. depressa* Pennant – has been studied over much of their range in Britain and NW. Europe. In spite of considerable annual and local variation in recruitment success, broader patterns can be distinguished, which can be linked to spawning times and factors affecting the survival of newly-settled spat. The breeding cycles of *P. vulgata* and *P. aspera* differ across their ranges in that, in both species, spawning begins, and gametogenesis ends, earlier in the north and east than in the south and west. The cause of these differences can be correlated with geographical and annual differences in sea temperature over the potential breeding periods, and can be related to the regional incidence of conditions found experimentally to be necessary for successful settlement and survival of spat during a critical stage of their growth. The significance of this 'temperature window' in determining the littoral and geographical distribution of the species is discussed.

Introduction

In the 1970's, a study which had begun in 1965 to explain local cycles of interaction of key rocky-shore species expanded to include the causes of their natural population fluctuations, over as wide a geographical range as was feasible. The rationale behind this study has been extensively discussed (Lewis, 1976, 1984; Lewis *et al.*, 1982), and it now forms the core of the rocky-littoral section of the COST 647 programme. This paper describes some of the latest findings on the relation of differences in limpet breeding cycles to recruitment. Though most data derive from British limpet populations, with less regular data from Ireland and N.W. Europe, the results of the more intensive studies of Portuguese limpet populations, published elsewhere in this volume, have been available since their commencement (Guerra & Gaudêncio, *pers. comm.*).

Of the three species of British patellids, much of the work since 1969 has concerned *Patella vulgata* L., and has been published elsewhere, (Lewis & Bowman, 1975; Bowman & Lewis, 1977; Bowman, 1985). Almost as many data exist for *Patella aspera* Röding, but because of early interpretational problems have not till now been published. Both these species occur throughout the British Isles wherever there is suitable habitat, though *P. aspera* is nearing the northern limit of its range in northern Scotland. The third species, *Patella depressa* Pennant, is restricted to the Channel coast and the south-west of the British Isles (excluding Ireland), reaching its northern limit in North Wales. For logistical reasons, this limpet was not included in the programme until the late 1970s and data are fewer, but are recorded here to allow comparison with the more detailed Portuguese studies (*op. cit.*).

Methods

Recruitment data for *P. vulgata* were obtained by *in situ* observations on marked 1 square metre sites chosen, on the basis of early experience, as being likely to have a high settlement potential and low biological variability. Recruitment levels in *P. aspera* were originally assessed from size-frequency analyses of population clearances from mid- to low-shore pools and 'open' rock. After 1973, low/mid-littoral coralline pools of ca. 0.5–1 sq. m. area and less than 20 cm. depth were found to make suitable sites for *in situ* observations, since not only are they primary settlement areas but they are more stable biologically than adjacent 'open' rock with its fluctuating populations of mussels, barnacles and algae. In SW. England most *vulgata* sites and all *aspera* pools doubled as *depressa* recruitment sites. Almost all sites were located in the mid-littoral of exposed shores, as *vulgata* sites were positioned wherever possible near *aspera* sites, to maximise use of the restricted sampling time.

Figure 1 shows the number and distribution of the observation sites for the three species around

Fig. 1. Map of the British Isles, showing regions studied. Numbers indicate: number of shores studied – total *P. vulgata* observation sites; total *P. aspera* observation sites. Bracketed figure for N. Yorks. indicates total *vulgata* sites 1966–70. Stars show location of sea-temperature recording stations referred to in Fig. 4.

Britain. Each site was visited two to five times each year, mainly in spring and autumn, except in NE. England, where visits were monthly or fortnightly or whenever tides allowed. At each visit the numbers of spat (<1 year-olds), juveniles (1–2 year-olds) and adults (the rest) were counted and their size range noted. The annual recruitment for a site was regarded as the highest number recorded of that spawning-year class. Because of differences in time taken for spat to emerge from settlement crevices, this number was recorded from 2–24 months after settlement. The observation-site data were supplemented by extra counts and occasional clearances for size-frequency measurement, from similar or lower-littoral habitats.

Breeding cycles were studied by regularly collecting 50–100 animals from the same shore level/habitat at each locality, and assessing the breeding state of the population from morphological examination of the gonad, as described by Orton *et al.* (1956). On the basis of size, colour and abundance of ripe (loose) gametes, each gonad was classified as neuter or in one of five stages of development, spawning or re-development, from which a mean gonad index was calculated: G.I.=0 if all animals were neuter, G.I.=5 if all fully developed. Again, frequency of sampling was greater in NE. England, where at the height of the breeding season samples could be taken daily if necessary (for *vulgata* only), to pinpoint spawning time.

Initially, morphological assessment of small early releases was confirmed by histological examination of sectioned gonads; with practice this later proved unnecessary, though since 1980 settlement plates have been used in NE. England to detect spatfalls and thus check for partial spawnings.

Since 1981 an invaluable aid in the understanding of breeding and spawning patterns, and recruitment success, has been the collection of daily sea and pool/shore temperatures by littoral recorders in NE. England, Wales and NW. Scotland.

Results: *Patella vulgata*

Recruitment variation

Table 1 presents the peak annual recruitments for the longest-running Robin Hood's Bay observation sites, which initially showed relative stability of populations other than limpets. However, since

Table 1. P. vulgata: Peak annual recruitment at Robin Hood's Bay observation sites. All sites 1 sq. metre in area.

Year	Site 1	2	3	4	5	6	7	8
1967	13	78	138	(25)				
1968	28	182	278	(90)			1956	
1969	34	145	295	45	58	45	1408	
1970	58	272	336	78	93	93	1908	2516
1971	16	35	44	40	37	37	84	211
1972	15	85	101	16	25	21	53	144
1973	23	179 M+	278	28	38	43	549	M− 372
1974	28	301	172	37	58	37	260	160
1975	10	203	196	24	42	30	228	158
1976	M+ 77	358	515	38	B− 92	56	1728	W+ 1110
1977	86	434 B−	572	62	84	94	509	195
1978	B− 82	528	717	68	84	77	1447	M+ 962
1979	78	281	456	80	97	70	654	603
1980	71	293	369	33	32	29	243	M− 189
1981	B− 27	80	85 B+	17	24	12	319	229
1982	W+ 72	102 B−	161	78	B− 98	W+ 76	496	263
1983	67	268	247	102	153	93	3130	W+ 2088
1984	103	391	332				465	554
1967–75 mean:	25	165	204	42	50	44	805	594
1967–84 mean:	49	234	294	51	68	54	908	650

Key to sites:
1: HL dense barnacles, ca. MHWN.
2: HL barnacles/mussels, ca. MHWN. } exposed north headland of R.H. Bay.
3: ML mussels/barnacles, ca. MLWN.
4–6: ML dense barnacles, ca. MTL.-replicates, moderate exposure, S side bay.
7, 8: ML bare wet rock/mussels, MTL.-replicates, partial shelter, central bay.
M+; −: Mussel % cover doubled; lost. B+; −: barnacle % cover ±25%. W+: appearance or marked increase in weed cover (*Enteromorpha* and *Fucus* spp.).
NB. Year is spawning/settlement year, not that in which peak counts were recorded. Bracketed figures from similar site further S on same reef.

1975, most sites have become increasingly wet, the result of increased mussel cover, loss of old barnacles (and poor recruitment of new), and since 1981 a general increase in algal cover partly because of reduced grazing from declining limpet numbers. Site 1 in particular now bears little resemblance to its original description (see Table 1), being now mostly bare rock with 10% mussel cover, sparse barnacles and thick annual weed cover. In contrast, Sites 7 & 8 proved susceptible to intermittent loss of their seed-mussel cover (and therefore their main limpet spat-settlement habitat) through periodic heavy dogwhelk (*Nucella lapillus*) predation and/or storm scouring. Only Site 3 has not changed appreciably in character, since its wetness derives as much from its lower-shore location as from its mussel etc. cover. Especially on the initially-dry barnacle-dominated sites, a succession of heavy recruitments since 1975 therefore greatly increased the density of adult limpets. For instance, the year-end density (excluding juveniles) on Site 2 was 240 per sq. m. in 1966, 390 in 1974, but rose to >1000 by 1980; poorer recruitment together with predation by oystercatchers (*Haematopus ostralegus*) then reduced the density to 290 in 1984. The increased grazing from the elevated limpet numbers in 1980 was partly responsible for the decline in barnacle cover at these sites, by preventing the establishment of cyprids.

Increased wetness and, particularly, increased small-mussel cover (which provides primary-settlement habitat for limpet spat), have been shown to elevate *vulgata* settlement/recruitment numbers (Lewis & Bowman, 1975), and these changes must have markedly increased the sites' potentials – compare, for instance, the sudden difference in recruitment among the replicate Sites 4–6 in 1976, following increased wetness because of loss of barnacle cover on Sites 5 & 6 but not 4. This, together with differences in timing of site changes, devalues the usefullness of assessing comparative annual settlement/recruitment from deviations from the sites' long-term recruitment means, as was generally possible until 1975. Thus, for the initially-dry Sites 1, 2 and 4–6, though recruitments in 1968–70 were judged 'good' by pre-1976 standards, only those of 1970 exceed the long-term means. 1976–78 were definitely above-average recruitment years but where direct comparison was possible in the absence of major wetness etc. changes, it seems likely that without such changes they might have been more equivalent to 1968–70 than the recruitment figures suggest. This supposition is strengthened by observations on *vulgata* spat numbers in other, unchanged, habitats such as the midshore *aspera* observation pools.

The other difficulty in assessing the relative success of annual settlement is the time elapsing before a site's peak recruitment figure is reached. The sites tabled peak variously after ca. 6 months (Sites 7 & 8), 9–12 months (2 & 3), and 12–18 months (1 & 4–6). The longer the time lag, the more the final recruitment figure can be affected by climatic extremes, or other causes of mortality, and for this reason settlement plates were eventually developed.

This latter problem was greatest at many of the Scottish locations, where peak recruitment was often not recorded until 18–24 months after settlement. All the regional sites were chosen to resemble Site 2 where possible, as (supposedly) combining habitat stability with high settlement/recruitment potential. In practice, they were found to have lower recruitment than that of Site 2, and lower annual variation: compare the mean input per square metre (± S.D.) for the longest-running sites in N. Scotland (62±21); NW Scotland (20±18); W. Scotland (37±23); NE. England (Site 2) (234±137). They also experienced much greater biological variation than the NE sites, because of unstable barnacle and mussel populations, and heavier predation of limpets, barnacles and mussels by oystercatchers, dogwhelks and eiders (*Somateria mollissima*) respectively. The lower limpet densities permitted variable cover of *Fucus vesiculosus* and ephemeral algae and diatoms, and all these factors considerably altered the sites' settlement/recruitment potential. Assessment of relative success was therefore more difficult than in the NE, so, to achieve a better basis for comparison, also taken into account were changes in spat numbers during the build-up to peak recruitment, and the numbers of spat recorded nearer settlement time in less variable habitats where spat emerged earlier, such as mid- and low-shore coralline pools. The raw regional recruitment data are too numerous to be included here (though they are available on request) and would require a lengthy analysis of the sites' vicissitudes. Instead, Fig. 2 presents an interpretation of relative settlement/recruitment success for all regions, based on the above criteria, though the classification has necessarily to be less precise.

The regions in the figure are arranged in order of coldest winter sea temperatures, and the year indicated is the spawning year, not that of peak recruitment (as this might occur up to 18 months after settlement). There is broad similarity in recruitment success between adjacent regions, particularly over much of northern Britain, where the 1976–1978 recruitments were all above-average. In all northern regions the 1975 class was well-represented in spring 1976, but the unexpectedly-low peak numbers recorded (see Table 1 for NE England) almost certainly resulted from high mortality in the abnormally-hot summer of 1976 (Bowman, 1978). Severe cold together with considerable storm damage on sites in northern Britain in the winters of 1978/9 and 1980/1 must also have depressed the eventual totals for these years: in the NE, Sites 7 & 8 suffered severe sand-scouring in storms immediately following the 1980 settlement, which was at least partly responsible for their poor recruitment. In 1976 all regions had good recruitment, and in 1982 there was a gradient of increasing success from NE backing to SW. The limited data we have from outside Britain suggest that the 1980 and 1981 recruitments in NW Brittany were both 'good', as was that of 1982 in Galway, which broadly match the appropriate S. or W. British pattern for those years. The existence of such regional patterns suggests that broadscale climatic, rather than local, factors control recruitment. These might operate at some stage of the breeding cycle, in the planktonic stage, at settlement, or during early shore life.

The breeding cycle

a) NE England At Robin Hood's Bay the breeding cycle has been followed closely since the start of the study, and though the gonads never failed to develop there were considerable annual differences in both peak Gonad Index (ie. maximum mean gonad

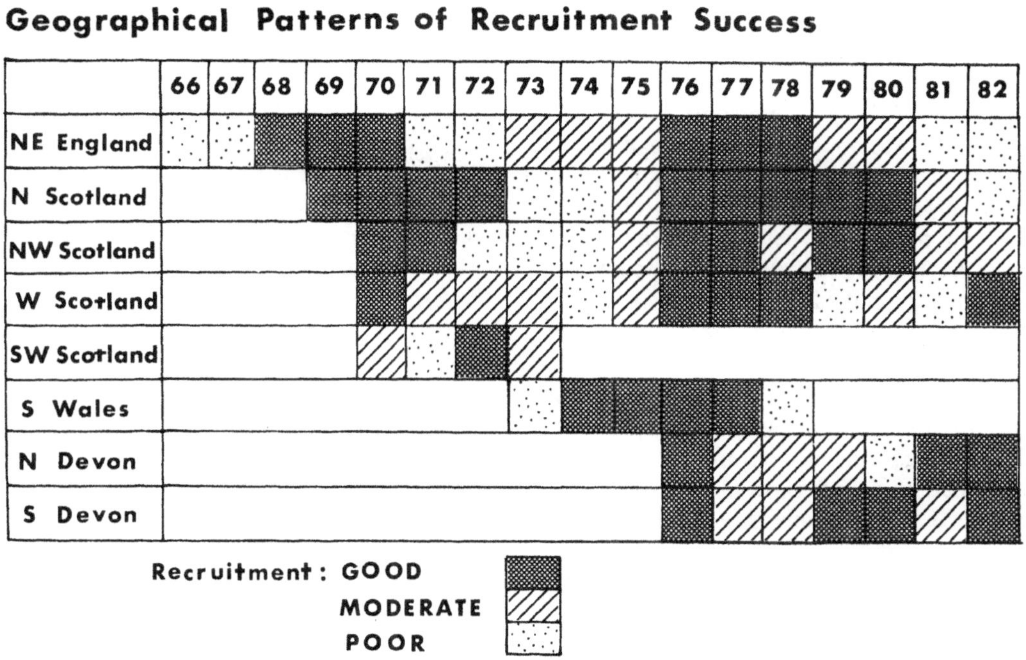

Fig. 2. P. vulgata: Geographical patterns of recruitment success. Assessment of data from size-frequency analyses, settlement pools, and all observation sites (allowing for gross changes in sites' settlement potential caused by biological variation – see text).

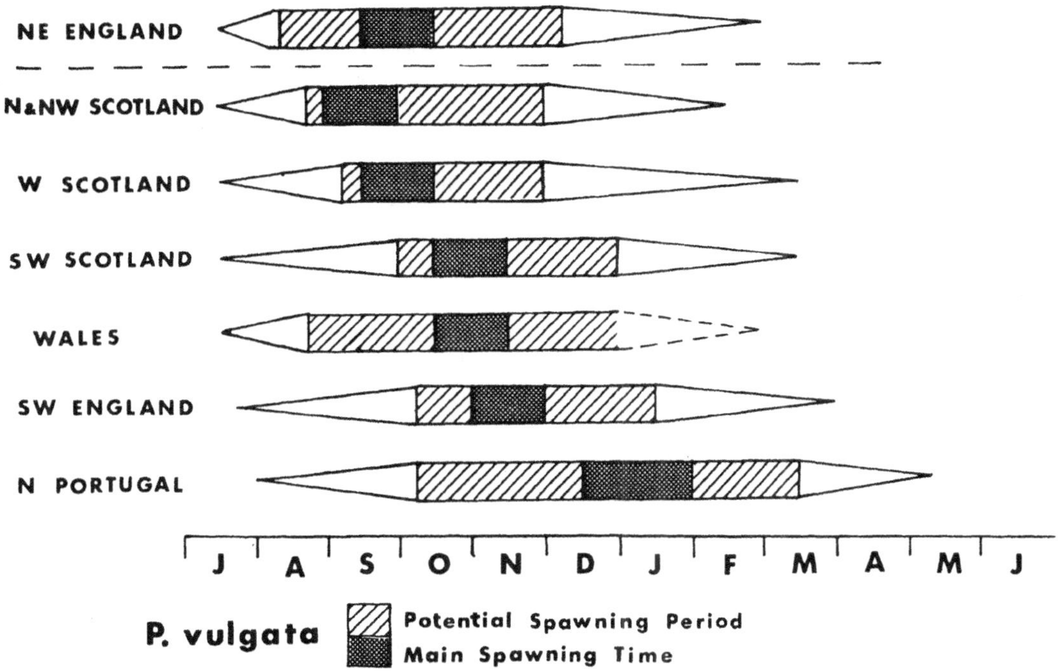

Fig. 3. P. vulgata: Regional differences in breeding periods. Bars indicate duration of breeding period when sex is morphologically distinguishable. Dark shading shows period when most of the gametes are released (main spawning time); light shading shows period when other, usually small, spawnings have occurred: decline of gonads after this is probably through resorption of gametes. Dashed lines indicate only one or two years' data. Period of observations: NE. England, 1968–85; other British regions, 1973–84; N. Portugal, 1974 and 1980–83 (Guerra & Gaudêncio, pers. comm.).

size) and the timing, extent and frequency of spawnings (Bowman & Lewis, 1977; Bowman, 1985). In NE England, the gonads begin to develop in July, as shown in Fig. 3, and some gametes become mature in August. Maximum gonad size is usually reached by October and most gametes are released during September and October. However, spawning can occur well before all gametes are fully-developed (eg. in August), in which case only the ripe gametes are shed. Such releases often do not result in a decrease in gonad size but, with experience and frequent sampling, can be detected by a sudden disappearance from the gonad samples of 'loose' (ie. ripe) gametes. These partial spawnings are followed by continued development of unripe gametes (gonad re-ripening), or occasionally by a new burst of gametogenesis (redevelopment), resulting in an increase in gonad size followed by later spawning, and such gonads can be distinguished by colour or histology from unspawned gonads until ripe gametes are present again. Very high Gonad Indices are therefore attained only when most of the season's gametes ripen before any release occurs, and so tend to be associated with 'late' spawning.

However, the data showed (Bowman & Lewis, 1977) that good recruitment was not attributable to gonad size nor to size of release. Until 1981 there was, however, very good correlation between time of first release and success of recruitment, so that 'early' = good and 'late' = poor. In a Multiple Linear Regression analysis, using 28 variables possibly affecting recruitment, >70% of recruitment variation could be attributed to earliness of release, while much of the remainder related to variation in mussel cover (M. A. Kendall & P. Williamson, *pers. comm.*). This correlation was initially explained by supposing that late-settling spat were likely to be killed by early-winter frosts, but this hypothesis has now been modified after using settlement plates to catch spatfalls and follow their fates directly.

A detailed account of the growth and survival of newly-settled spat is currently in preparation, but their relation to recruitment is summarised in Bowman (1985). It was found that spat are highly susceptible to growth retardation and shell deformities induced by excessively high and low temperatures, as well as to detachment from the substrate during rough seas, during a critical period lasting from settlement until production of the 'true' calcified shell and growth of the mantle tentacles, at a size of ca. 0.7 mm. shell diameter. The temperature range for optimal growth in the laboratory was 10–14 °C, and as growth all but ceases at temperatures below 6 °C, the critical period can vary in length from 6 weeks to 5 months (ie. over the winter). Once past this critical period the spat are much hardier, and a clear relationship has been demonstrated since 1980 between the number of spat passing this stage *before* the onset of winter and the eventual numbers of recruits (Bowman, 1985). Since spat either fail to settle, or die very quickly, when shore temperatures are too high, as in August/September 1981 and 1982, the recruitment at Robin Hood's Bay derives almost entirely from settlements during September/October, and the hypothesis can be reformulated thus: good recruitment follows when spawning results in settlement coinciding with a restricted temperature range, optimal at ca. 10–14 °C, during early shore life.

b) Geographical variation In NE England this narrow range, the temperature 'window', normally occurs during September/October, but there is clearly going to be a north/south gradient in any such climatic factor. Since it must be to the species' advantage to recruit successfully every year, it is perhaps not surprising to find a geographical gradient in time of the main gametic release, from August/September in N. Scotland to November in SW. England. Figure 3 is based mainly on our own data, since in order to prove such differences are geographical it is necessary to compare data for different localities in the same year. Samples received in 1974 from near the European limits of *vulgata*'s range (Tromsø, Norway and Portô, Portugal) suggested that the main release started even earlier (July/August) in Norway, later (after December) in Portugal. The latter observation has been confirmed by Guerra & Gaudêncio (q.v. this volume). Data from S. Britain, Eire and France (Evans, 1953; Orton *et al.,* 1956; Choquet, 1966; Thompson, 1980; Hatch, 1977) and from Orkney (Baxter, 1982) fit into this gradient of spawning times, though hitherto the pattern for SW. England has been regarded as the norm.

Although the earliest spawnings have been recorded in NE. England, this might simply reflect the more frequent sampling there, over a longer period, and the difficulty of widespread synchronous sampling. In the NE, August releases were not recorded until 1981 and 1982 (September spawning being previously earliest), when other localities could not be visited. A very small mid-August release in mid-Wales in 1985 suggests that all British *vulgata* populations might be capable of spawning from August onwards, though where regular runs of gonad data are available over several years it is clear that release normally starts earlier in the north than in the south (Bowman, 1985). Consequently, gonad redevelopment and multiple release were most common in northern Scotland, whereas gonads were regularly found to be more-fully developed in late October in southern Britain, suggesting an absence of early spawning.

The length of the potential breeding period also showed a north/south gradient: except very rarely few ripe gametes remained unspawned by mid-December in the north, whereas in the south major spawnings regularly occurred as late as January, and the Portuguese study (q.v.) extends this into March. After the last release, any remaining gametes were re-absorbed over the resting period, as described by Orton *et al.,* (1956), so that sex became morphologically indiscernible. Whereas 'resting' gonads in other regions resembled those described by Orton in being also histologically unsexable, through complete regression of the germinal epithelium, in NE England this was not the case: the germinal epithelium for the following season's gametes here appeared towards the end of the current breeding season and persisted over the resting period. These over-wintering gonads were therefore of appreciable thickness, in contrast to those of other populations, and their past and/or future sex could usually be identified histologically. In such a condition, sex-change could be seen to have occurred in a high percentage of previously-male gonads. Thus, although there appears to be in any given year no appreciable geographical difference in time of onset of gametic proliferation, in NE England it begins from an already-present germinal epithelium which elsewhere has first to be produced. This might explain the apparently-faster rate of gamete development in NE populations, though it is not easy to compare development rates accurately, because of local and annual differences in energy budgets and partitioning (Workman, 1983; Wright & Hartnoll, 1981).

Spawning & redevelopment

Spawning in *P. vulgata* had previously been linked to rough seas (Orton *et al.,* 1956), but though this was undoubtedly true in many years at Robin Hood's Bay (Bowman & Lewis, 1977) it was not totally convincing, nor did suggested differences in 'ripeness' of gonads/gametes satisfactorally explain the observed spawning differences. There was also no explanation for annual (and regional) differences in the frequency and amount of redevelopment after an initial major release. Yet in some years a very broad pattern of breeding and spawning time existed: thus in 1977 and 1978 spawning began earlier than usual throughout Scotland and NE England, and these breeding periods were unusually prolonged, with considerable redevelopment, especially at Robin Hood's Bay. By contrast, in 1973 and 1974, spawning in all Scottish regions began relatively late and little or no redevelopment followed (Bowman, 1985). The scale of such events suggested that again a climatic factor might be involved.

Comparison of the regions' potential spawning periods (Fig. 3) with the mean sea temperatures in each area (Fig. 4) shows a rough coincidence with the time from the annual peak of monthly mean

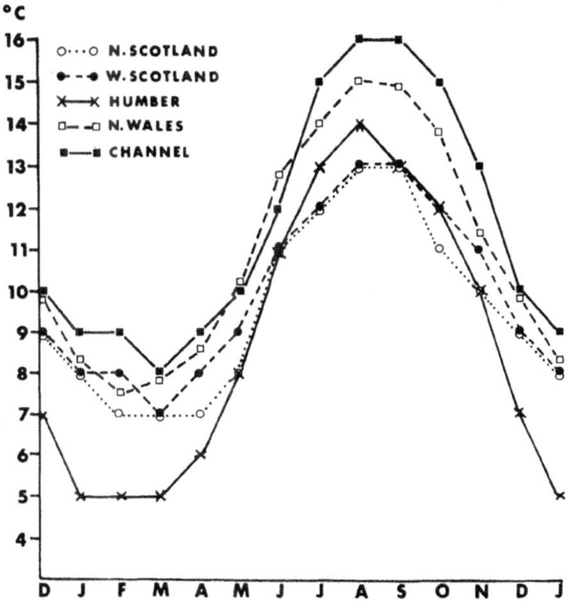

Fig. 4. Inshore sea temperatures around the British Isles. Mean monthly temperatures at the stations shown in Fig. 1, for the period 1966–75 (Meteorological Office data).

temperature to when it drops below about 10 °C. Since on exposed coasts, even in S. England, the peak sea temperature rarely exceeds 16 °C, the spawning period therefore coincides with the proposed temperature 'window' optimal for spat settlement and survival. It also appears that the main spawning time for each region occurs during the period when mean sea temperature drops below 12 °C.

Spawning of molluscs in response to a rise in sea temperature occurs in the limpets *Nacella concinna* Strebel (Picken, 1980; Picken & Allen, 1983) and *Acmaea* spp. (Fritchman, 1962a, 1962b), and several bivalves (Nelson, 1928; Ansell & Lander, 1967; Lammens, 1967). Spawning in response to a drop in temperature past a critical level is rarer, but found in the Bay scallop (Barber & Blake, 1983) and possibly in the South African *Patella oculus* (Branch, 1974). At Robin Hood's Bay the sea temperatures could originally be taken only at low tide in shallow water in the bay, where considerable inshore warming was known to occur. Though storms usually produced a drop in sea temperature, no correlation between this and spawning could be demonstrated. When, however, the nearest available daily inshore sea temperature records (for Humber, about 100 kms south and 0.7 km. offshore) were compared with the Robin Hood's Bay breeding cycles from 1967 to 1980, two connections became obvious (Table 2): a) The season's first spawning was recorded by the first gonad sample taken after the Humber sea temperature first fell below 12 °C. With rare exceptions, spawning had been pinpointed even to the day of occurrence because gonad samples were always taken at the first opportunity after a rough sea, to check for release, and 90% of the spawning drops in temperature accompanied the heavy swell set up by north-westerly gales (either locally or further north). b) After an initial spawning, gonads redeveloped only in years when the sea temperature rose above 12 °C after the spawning drop: later spawning(s) occurred when the temperature next fell to <12 °C. Altogether 20 out of 24 recorded spawnings followed drops to <12 °C, 3 of the remainder being end-of-season spawnings during storms when the sea temperature was already <12 °C.

Table 2. P. vulgata: Relations between Humber sea temperatures (Meteorological Office data), spawning and re-ripening at Robin Hood's Bay, N. Yorks.

	Sea temp. fell <12 °C on:	Spawning first found on:	Sea temp. then rose above 12 °C	Gonads re-ripened
67	18 Oct	10 Nov	No	No
68	12 Oct	14 Oct	Yes	Yes
69	29 Sept	29 Sept	Yes	Yes
70	9 Oct	13 Oct	No	No
71	27 Sept	6 Oct	Yes	Yes
72	21 Oct	23 Oct	No	No
73	26 Sept	27 Sept	Yes	Yes
74	30 Sept	1 Oct	No	No
75	29 Sept	2 Oct	Yes	Yes
76	9 Sept	10 Sept	No	No
77	18 Sept	21 Sept	Yes	Yes
78	19 Sept	21 Sept	Yes	Yes
79	21 Sept	1 Oct	No	No
80	16 Sept	17 Sept	Yes	Yes

Since 1981, temperature recorders installed at MTL on the shore at Robin Hood's Bay have shown 'actual' sea temperature, (ie. that experienced by the mid-littoral for 85% of the high-water period) to be up to 6 °C colder (in summer/autumn) than that taken by hand at the sea's edge. The connection between storms, changes in sea temperature and spawning has therefore been compared directly, as shown in Fig. 5. The figure shows no consistent connection between spawning and temperature drops *per se:* the first spawning occurs only when the sea temperature first drops to <12 °C. This is shown particularly well in Sept. 1984. The August 1982 release was the earliest yet recorded, but small because few gonads had ripe gametes so early in the season. These gonads continued to develop during September, with no further release, despite a few storms, until the next time the temperature fell below 12 °C. It did not rise significantly above 12 °C that season, and redevelopment was minimal, the Gonad Index falling during storms in early October, then at the final release in mid-November. In 1984, after the first (major) release, most of the populations' gonads redeveloped when the sea temperature rose again above 12 °C. The next release followed a subsequent drop to <12 °C, after which the temperature stayed below 12 °C and there was no further redevelopment, only a limited amount of reripening in some males, before the final spawning after a mid-November storm. Similar relationships were found in 1981 and 1983 (Bowman, 1985), so that during 1981–1984 11 out of 14 spawnings were

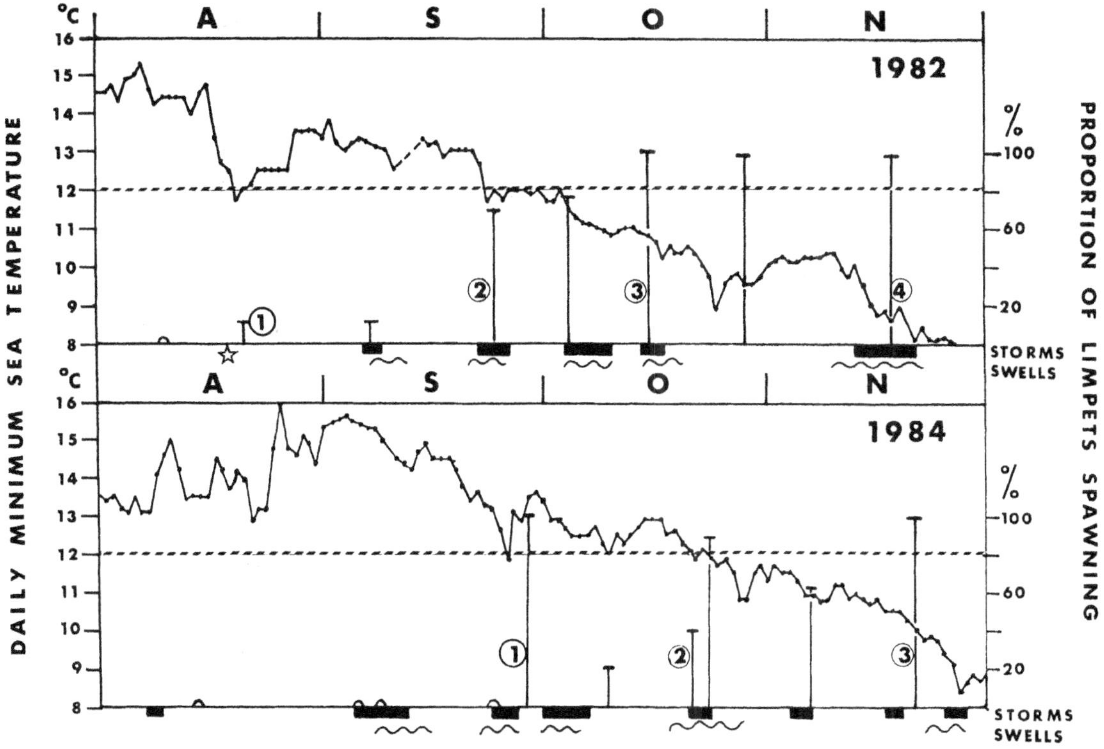

Fig. 5. P. vulgata: Correlation between sea temperature and spawning at Robin Hood's Bay, N. Yorks. Graphs show the daily minimum sea temperature; vertical bars = percentage of spawned limpets in the sample (bumps on the baseline indicate a sample with no spawned limpets) – when this percentage decreases between samples, redevelopment has occurred. Symbols below the baseline indicate occurrence of seiches (star), and duration of storms and heavy swells.

recorded immediately after a drop to <12 °C at Robin Hood's Bay, the remainder being late-season releases when sea temperature was already <12 °C and were recorded after storms.

The connection between sea state and temperature was also clarified. Thus, in August, considerable warming of inshore water could occur during spells of calm sunny (anticyclonic) weather, only to cool as quickly when the calm spells ended, as in August 1984. Similarly, marked drops in sea temperature could occur in late autumn during calm cold anticyclonic spells, often with night frosts, as in late October in 1982 and 1984. These falls in temperature did not result in spawning. Temperature drops causing spawning were almost always associated with heavy swells, set up by NW gales either locally or in more northerly waters (N/NW storms off northern Scotland usually produced at Robin Hood's Bay, three days later, a tidal elevation of 0.2 – 1 m. and a heavy swell even without an accompanying wind, as in mid-August 1982). Such swells presumably brought inshore a body of colder bottom or offshore water, whereas the rough seas produced by local easterly gales presumably resulted in mixing of inshore water which did not always result in a drop in temperature, particularly on the open coast, hence presumably the mismatches with early spawnings.

A comparison of spawning on adjacent exposed and embayed shores suggests that spawning is not induced by a change in some character of the water mass other than temperature. In September 1983 the season's first release occurred simultaneously at exposed shores along 200 km. of the NE coast, when the sea temperature fell to <12 °C for the first time that year. However, in embayed positions, with sea temperatures 4 – 6 °C above 'true' sea temperature, the fall left bay temperatures still >12 °C and spawning did not occur: similar phenomena had previously been recorded in Scotland. A similar event in October 1985 followed a swell which introduced a body of visibly 'different' (less turbid) water into the bays under observation. The marked separation of warm embayed water from cold deep water on either side of Filey Brigg (N. Yorks.) has regularly resulted in asynchronous spawning, the time-lag between first releases on its exposed and sheltered sides has been 48 days (1981), 25 days (1983), ca. 36 days (1984) and 12 days (1985).

Recent data from NW Scotland not only provide further examples of local exposure/shelter spawning differences, and confirm that 12 °C appears to be critical there, but also indicate the widespread effect of a major NW storm. Thus, in September 1983, >80% spawning was found in both NW Scotland and NE England immediately after the same storm, while in early September 1984 a storm causing heavy spawning in N and NW Scotland also caused a temperature drop in NE England (though not to <12 °C on this occasion – Fig. 5). In August 1982, a severe northerly storm in the N. Atlantic not only set up the seiche which produced a large temperature drop and triggered spawning in NE England (Fig. 5, starred) but also caused temperature falls of similar magnitude in northern Scotland and Portugal.

Whether or not 12 °C is critical in the SW remains undecided, because of recorder breakdowns and collection difficulties, but it is suspected that a slightly higher temperature might be involved.

Patella aspera

Recruitment variation and breeding

Unlike *P. vulgata* which occurs throughout the littoral zone on exposed and sheltered shores in the British Isles, *P. aspera* has a much more limited distribution, being restricted to exposed shores and confined to permanently-wet rock and coralline pools in the mid- and low-littoral. In SW. Britain and Ireland the main population and primary settlement areas occur below MLWN, but towards the north of Britain (and therefore in most of our study areas) the greatest numbers and most regular

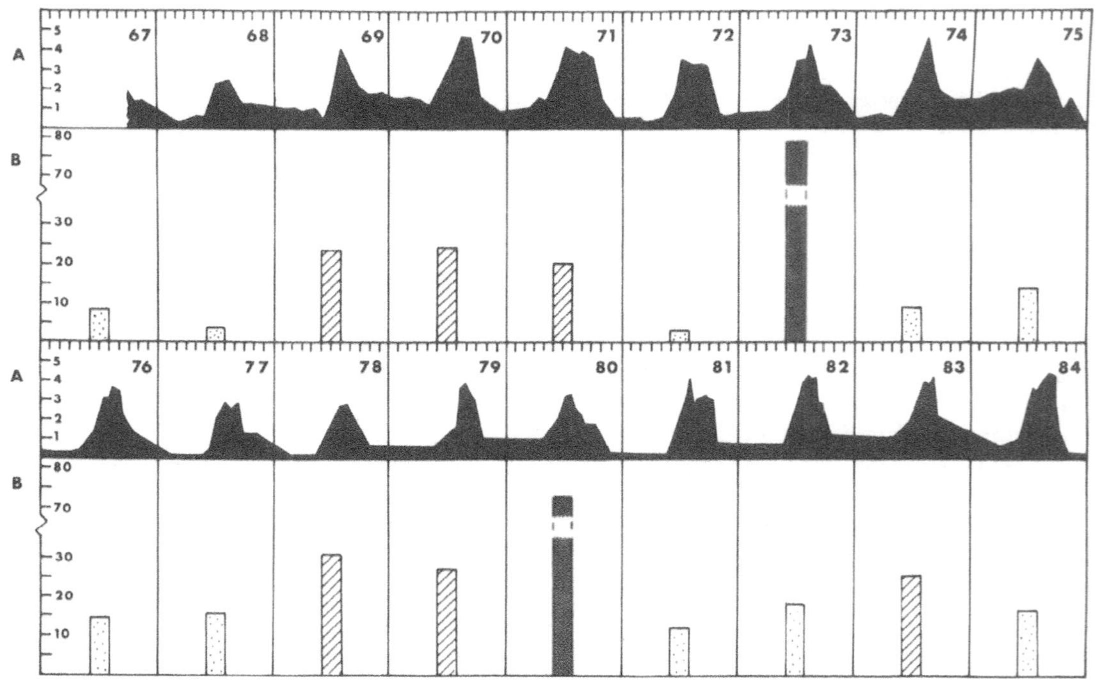

Fig. 6. P. aspera: Annual breeding cycles and recruitment at Robin Hood's Bay, N. Yorks. A: Gonad Index. B: Recruitment – peak number per square metre recorded of each year class. Data from size-frequency analyses, 1967–69, and as the mean of the two main observation pools, 1970–84. Shading indicates percentage deviation of recruitment from the long-term mean: stippled = –25 to –100%; striped = ±20%; hatched = >100%.

recruitment are found in mid-littoral pools. Most recruitment data described here were obtained from such pools, though breeding and recruitment in a variety of habitats was studied wherever possible, and their differences are still being elucidated.

Whereas *vulgata* was regarded as a winter breeder (Orton *et al.*, 1956), *aspera* was reputedly a summer breeder, and, as it approaches its northern geographical limit (S. Norway) in northern Britain, local populations in N. Yorkshire were confidently expected to show frequent recruitment failure, possibly as a result of incomplete gonad development. From 1967 to 1972 this expectation seemed to be confirmed, as there were considerable annual differences in peak Gonad Index – much greater than in *vulgata* – which showed a marked similarity to the size of the subsequent recruitment (Fig. 6).

The size-frequency histograms of the local (NE) population for these years (Fig. 7A) show clearly the prolonged effect of the 'missing' 1968 year-class, exacerbated by a poor 1967 crop, on the population structure. Samples collected from around the British Isles in April/May 1971 (Fig. 7B) showed a similar scarcity of mid-sized animals in most northern and western populations, but those in the south (exemplified by SW Eire in Fig. 7B) had no such missing year-class(es), their size-frequency structure being typical of popula-

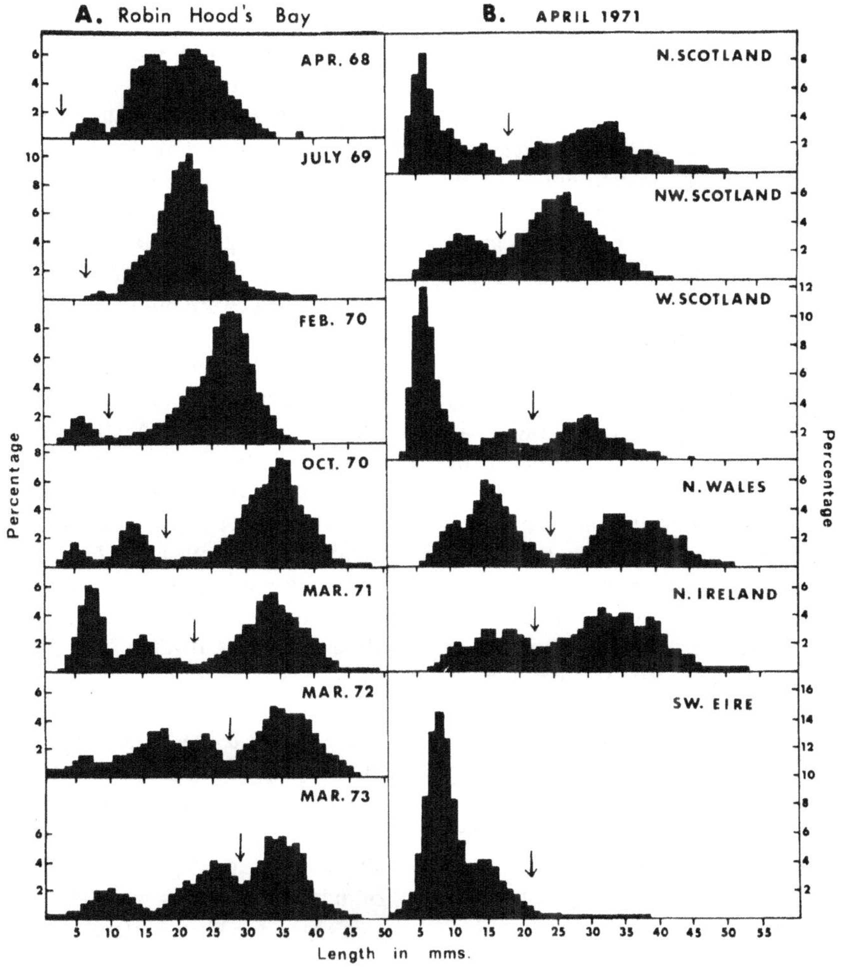

Fig. 7. P. aspera: Recruitment variation as shown by population size-frequency distributions. A: At Robin Hood's Bay, N. Yorks., 1968–73. B: Regional populations in spring 1971. Sample size = ca. 400; arrows indicate position of the 1968 year-class. Samples from N. Ireland and Eire kindly collected by Dr. R. Seed and Dr. G. B. Thompson.

tions with regularly-high recruitment, fast growth and short life (cf. low-littoral *vulgata* populations: Lewis & Bowman, 1975). Such a widespread recruitment failure around the north, and its coincidence with poor gonad development in NE. England in a summer now on record as the coldest in the last 23 years (deviations from monthly mean air temperatures, Whitby, 1968: May −2.7 °C; June −0.8 °C; July −2.5 °C; August −1.3 °C) convincingly substantiated the original prediction.

Though there is probably an element of truth in this explanation of the 1968 recruitment failures, after 1972 the correlation between gonad development and recruitment success broke down, demonstrating (if nothing else!) the need for long-term data in such studies. Fig. 6 shows that there was nothing exceptional about the G.I.s preceding the very good recruitments of 1973 and 1980, or the poor inputs of 1974 and 1981. Though in 1977 (the next-coldest summer after 1968) a low G.I. preceded poor recruitment, the small gonad size here resulted from early spawning of incompletely-ripe gonads, followed by continued development and release of the rest of the gametes, as described earlier for *vulgata.* In fact it became clear that in *aspera* such multiple spawnings are very common, especially in pool-dwelling populations, which tend to be the first to spawn.

When spat first appeared on the shore in early autumn, good recruitment usually followed: in poor recruitment years spat were not found in the observation pools until the following spring, and were often stunted and atypically coloured (Bowman, 1981), continuing to grow unusually slowly and often dying in their second or third year. After the initial regional pattern shown in Fig. 7B, it proved difficult to distinguish any consistent regional similarities. Though the 1973 recruitment was of record proportions in NE. England, N. and NW. Scotland, it failed in W. Scotland; the 1980 recruitment, again very good in NE. England, failed in all other observed regions except S. Devon, Brittany and Portugal. There was a tendency towards similar recruitment throughout the Scottish regions, and (from the limited data available) also throughout S. Britain, Europe and possibly Eire, though the picture was occasionally confused by contrasting recruitment success at different shore levels/habitats. As *aspera* is regarded as a 'southern' species it was expected that hot summers might boost its breeding and recruitment in the north, but in the exceptionally-hot summers of 1975 and 1976 this did not occur: in the NE., although spawning began early (July/August), poor recruitment resulted, which negated a simple 'early = good, late = poor' spawning/recruitment relationship. The Scottish data also seemed anomolous, in that poor recruitment tended to follow hot, sunny summers whereas good recruitment followed dull summers.

Geographical variation of breeding cycles

Compared to *aspera* in NE. England, the size of juveniles in their first spring, and their later emergence onto 'open' rock or in the observation pools, in Scotland and southern Britain (Bowman, 1981) suggested that they settled later and/or grew more slowly over-winter, and this was found to apply to Irish and Breton populations also. Figure 8 shows that *aspera* does breed earlier in NE. England than in the other regions; it also has the shortest potential spawning period. Thus in NE. England >80% of gametes have been discharged by the end of September in most years, whereas in Scotland and SW. England the gonads usually contain abundant gametes until the end of October and November respectively. Since multiple spawnings regularly occur during the breeding season in these other regions, redevelopment must here be able to recur later in the year than is possible in the NE. Nowhere else has a gonad development as poor as that of 1968 in NE. England been recorded: on the contrary, the Gonad Index in other regions, including N. Scotland, regularly peak at a value of >4.0, despite multiple releases. The spawning pattern is therefore not a simple north/south one as in *vulgata,* as the NE. English population is behaving as though it was further 'north' than the Scottish ones.

The other major difference in the regional breeding cycles of *aspera* compared to *vulgata* is that in any given year there is a marked difference in time of onset of gametic proliferation, starting (in Britain) earliest in SW. England and getting progressively later clockwise towards NE. England. Time of onset also varies annually, within each region (see Fig. 6 for NE. England), proliferation seeming to begin later in years with below-average spring sea temperatures. The possibility of a temperature link with gonad development fits the pattern of regional spring sea temperatures (see Fig. 4), which are

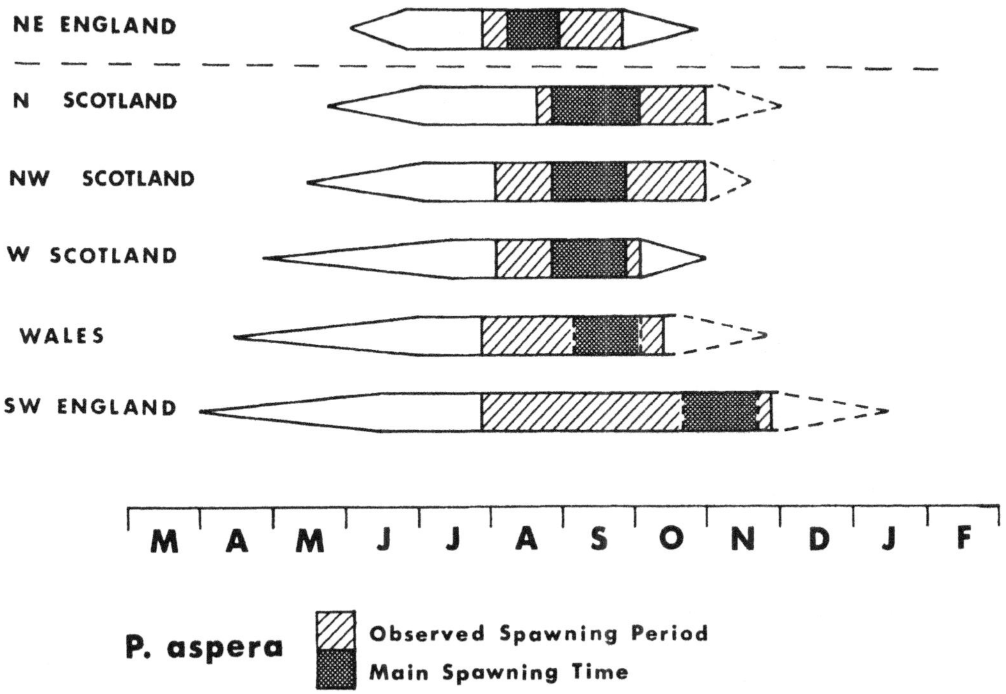

Fig. 8. P. aspera: Regional differences in breeding periods, 1973–85. Key as for Fig. 3.

markedly colder in NE. England than in N. Scotland from autumn to spring. The later maturation of gonads in N. Scotland compared with NE. England could reflect the lower summer temperatures. Despite their later start, gonads apparently mature faster in Scotland and NE. England than in the south, because the germinal epithelium for the next season's gametes is produced towards the end of the current season, as it is in local (NE.) *vulgata*. As in *vulgata* this 'pre-development' enables the gonads to be sexed histologically during the 'resting period', and sex-change can be seen occurring in a high proportion of males – a situation previously only inferred (Thompson, 1979).

Because of the relative inaccessibility of the local (NE) *aspera* populations during neap tides and rough weather, it has proved difficult to pinpoint spawning as accurately as with *vulgata*. Early evidence again suggested storms as a possible trigger (cf. Thompson, 1979), and in recent years the *aspera* have certainly spawned during the same swells as caused first release in *vulgata*. In 1983 first release in *aspera* preceded that of *vulgata*, occurring between 10th and 13th August when a moderate NW wind and heavy swell arose during a very calm period; the sea temperature fell but remained above 12 °C and *vulgata* did not spawn. However, in August 1982, the large temperature drop which triggered *vulgata* spawning was not accompanied by swells or gales (see Fig. 5) and *aspera* did not spawn. Recent evidence from the sea-temperature recorder suggests that spawning requires a protracted (>18 hours?) period of exposure to surf plus continuous (lowered) sea temperature, resulting from swells breaking across the mid-littoral during nominal low-water, but the data are still too few to be conclusive.

Determination of the effect of temperature and settling time on *aspera* spat, using settlement plates, has been hindered by a succession of poor recruitments, so experimental laboratory studies are still lacking. However, studies in progress in NE. England, in littoral pools with known differences in temperature regimes during low water, are indicating that the rules governing *vulgata* spat establishment also apply to *aspera* spat, but that *aspera* spat are less tolerant of temperature variation, and more susceptible to storm damage, than *vulgata* spat. They grow more slowly than *vulgata* spat at any given temperature, growth being optimal between ca. 12–16 °C and effectively ceasing at <10 °C. Low temperatures produce severe shell deformities in *aspera* spat even as large as 2 mm. diameter, so spat in regions with lowest winter sea temperatures must reach a larger pre-winter size to maximise their chances of over-winter survival, sea rather than air temperature being important in this respect as the main *aspera* habitats are rarely uncovered over the winter. Larger pre-winter size is achieved in NE. England when spawning/settlement occurs early in the breeding season when the sea temperature is >12 °C and still rising, and/or in habitats such as pools where temperature, and therefore spat growth, are boosted by insolation, though over-heating can adversely affect growth and survival, as found for *vulgata* spat. Although this evidence is not yet complete, it seems to explain the observed settlement/survival and recruitment variations, and strongly suggests that, once again, geographical differences in breeding times might mirror the time of year when the temperature range optimal for growth is likely to occur.

P. depressa

Recruitment variation

In Britain, good *depressa* populations are confined to SW. England, so this species could be studied only briefly, between 1976 and 1983. Significant recruitments were recorded only on the North Devon shores studied: as shown in Table 3, recruitment varied similarly in the two habitat types monitored, though the range of variation differed. 1979 was clearly an outstandingly-good recruitment year, and (significantly?) also a year when large numbers of very young spat appeared on the shore in late autumn. It is difficult to assess the relative success of the earlier years' recruitment, as primary-settlement observation sites were not established till 1979, when the numbers given were recorded; however, earlier counts and clearances at a variety of habitats and levels suggest that probably only 1976 can be regarded as 'good'. The 1981–1983 recruitments were clearly 'failures'. In June 1982, poor 1980 and worse 1981 *depressa* recruitments were found on various shores in Brittany, which agreed with the Devon pattern. The 1979 crop in Brittany was also sparsely represented; this might not disagree with the Devon pattern as the very good 1979 recruitment which peaked in Devon in March 1980 had virtually disappeared from the sites by October 1980. This disappearance, almost certainly mortality induced by over-heating, highlights both the need for regular, and well-timed, sampling, and the dangers of assessment based on data taken too long after settlement.

Table 3. P. depressa, N. Devon: Mean Peak Annual Recruitment at observation sites. Recruitment expressed as mean numbers per sq. m. for high/mid-littoral pool sites and mid/low littoral rock sites. Brackets indicate counts taken well before or after expected time of peak recruitment.

	Pools	Rock	Rank
1976	(15)	(24)	2
1977	8.0	11.3	4
1978	3.2	4.0	5
1979	170.0	47.0	1
1980	8.6	13.0	3
1981	0	3.0	7
1982	0	3.0	7
1983	(0)	(4)	6
Mean	25.6	15.0	

Breeding cycles

Our gonad data are intermittent, but agree with Orton & Southward (1961) and Hatch (1977) that *depressa* in Britain can breed from May/June to October/November, with multiple spawnings and redevelopment similar to those found in *aspera* and *vulgata* but sometimes so extensive that they resemble separate breeding seasons. Thus in 1978 (and probably similarly in 1980) a large spawning in June/July, dropping the G.I. to 2.9, was followed by marked redevelopment and a later spawning in October/November, whereas in 1979 spawning in early July was apparently not followed by redevelopment. No spat were recorded on the sites from the June 1978 spawning, but that in July 1979 was presumably responsible for the abundant spat found that autumn. In the absence of any data from settlement/survival/growth studies using settlement plates, the fate of spat from these spawnings remains speculative. Better information on spawning times, over a wider geographical range than is available in the British Isles, is needed before any conclusions can be reached for this species, and in this context the Portuguese studies (*op. cit.*) provide a valuable contribution.

Discussion

The breeding cycles of *P. vulgata* and *P. aspera* show shifts in spawning time, relatively earlier towards regions of cold winter-sea temperatures, later towards regions of high summer temperatures. Considering that the temperature tolerance range of the adults of both species is <0 °C to ca. 42 °C (Evans, 1948), the spat of both species apparently have a remarkably low tolerance of temperature variation and a narrow temperature range for optimal growth. Since survival depends on fast initial growth during the vulnerable phase immediately after settlement, it seems unlikely to be a coincidence that regional spawning periods bracket the time of year when the optimal temperature is most likely to be experienced. In the 'north' (not necessarily geographic north), earlier spawning lessens the debilitating effects of severe or prolonged winter cold and rough weather; in the 'south' delayed spawning avoids the equally-disastrous effects of high shore (and sea?) temperatures: particularly for pool-settling spat overheating from strong insolation is the problem, not desiccation. Yet even in 'northern' populations there is evidence that early settlements can fail because of experiencing too-high temperatures, and late settlements in the 'south' could certainly fail in stormy or unusually-cold winters. Because of this, and because of annual climatic differences, an extended breeding period with several releases might increase the chance of some settlement success, particularly towards the limits of a species' range. Since, in the 'north', early settlements have a disproportionately better chance of survival, the breeding cycles here have apparently adapted to produce gametes as fast as possible, by starting next season's gametogenesis before the end of the current season.

Spawning of *P. vulgata* in response to a sea-temperature drop to below the critical 12 °C, whatever the physiological mechanism involved, tends to synchronise local spawning and ensures that the young limpets experience optimal sea temperatures at least while in the plankton. In NE. England, especially during warm summers, such a temperature drop is usually produced only by the strong NW gales which follow in the wake of a deep depression passing over the British Isles. Such storms, which also appear to have been responsible for some *aspera* releases, tend to affect a wide area, and are likely therefore to cause widespread spawning, probably explaining supra-regional patterns of recruitment success, particularly in the north. It is uncertain whether a higher critical temperature is involved in the extreme south of *vulgata*'s continental range, where the winter sea temperature rarely drops below 12 °C. Conversely, it is not clear what triggers spawning in the extreme north if the summer sea temperature does not exceed 12 °C, though very recent data suggest that in such a situation spawning (and rates of gonad development) are unusually asynchronous in different habitats across the shore (for *vulgata* and *aspera*).

The asynchrony of spawning found at adjacent exposed and sheltered shores (in *vulgata*) or at different shore levels (in *aspera*) also increases the spread of the potential settlement period, which may be advantageous in years (or localities) when conditions curtail the period of gonad development in some habitats. Though the evidence indicates that *vulgata* populations tend to spawn first, and early settlements survive better, on exposed shores, it is uncertain whether early or late settlements are more likely to contribute most to sheltered-shore recruitment, where the spat will encounter less wave action but more extreme temperature variation (in summer and winter). Conceivably, the low *vulgata* recruitment/density on sheltered shores, and the absence of *aspera* even if littoral coralline pools are present, indicates that conditions in shelter are much less favourable for settlement or survival of spat of either species.

As the shore is not a uniform environment, the time when conditions favour spat survival must vary with habitat as well as seasonally. For example, a temperature 'window' requiring moderately-warm sea and equable shore temperatures would be more likely to occur in the lower-littoral and in shaded mid-littoral pools in summer; in autumn, if sea temperatures were only marginally warm enough, spat might be able to grow fast enough to survive only in pools whose temperatures were boosted during low-water by solar heating. The littoral distribution of a species over its geographical range would therefore be set by the tolerance range of its spat and the ability of its adults to migrate from the main recruitment areas. This would explain why limpets in the south become progressively restricted to the low-littoral of exposed shores (eg. *vulgata* in Portugal, *aspera* in the Mediterranean), but in the north occupy progressively higher shore levels (eg. *aspera* in NE. England and northern Scotland). Geographical limits of a species' range would then be reached when a temperature 'window' was not available somewhere on the shore with sufficient frequency for recruitment to maintain a population. As limpets seem to have a remarkable ability to adjust their breeding period to cope with the seasonal vagaries of a littoral environment, the onus for setting their geographical limits is thus shifted from breeding failure to spat survival.

Acknowledgement

Many thanks to all the people who have collected limpets for us, in Britain and Europe, and in particular to M. A. Kendall and Dr. P. Williamson for much of the observation site data from southern Britain. Also to the technical staff of the late Wellcome Marine Lab. (Univ. of Leeds), and the Dove Marine Lab. (Univ. of Newcastle), for help with the temperature recorders, and especially to W. T. Ritchie of Achmelvich, Sutherland, for his sea temperature records.

References

Ansell, A. D. & K. F. Lander, 1967. Studies on the hard-shell clam, *Venus mercenaria,* in British waters. III. Further observations on the seasonal biochemical cycle and on spawning. J. appl. Ecol. 4: 425–435.

Barber, B. J. & N. J. Blake, 1983. Growth and reproduction of the Bay scallop, *Argopecten irradians* (Lamark) at its southern distributional limit. J. exp. mar. Biol. Ecol. 66: 247–256.

Baxter, J. M., 1982. Population dynamics of *Patella vulgata* in Orkney. Neth. J. Sea Res. 16: 96–104.

Bowman, R. S., 1978. Dounreay oil spill: major implications of a minor incident. Mar. Poll. Bull. 9: 269–273.

Bowman, R. S., 1981. The morphology of *Patella* spp. juveniles in Britain, and some phylogenetic inferences. J. mar. biol. Ass. U.K. 61: 647–666.

Bowman, R. S., 1985. The biology of the limpet *Patella vulgata* L. in the British Isles: Spawning time as a factor determining recruitment success. In P. G. Moore & R. Seed (eds), The Ecology of Rocky Coasts. Hodder and Stoughton, Sevenoaks (Kent): 178–193.

Bowman, R. S. & J. R. Lewis, 1977. Annual fluctuations in the recruitment of *Patella vulgata* L. J. mar. biol. Ass. U.K. 57: 793–815.

Branch, G. M., 1974. The ecology of *Patella* L. from the Cape Peninsula, S. Africa. 2. Reproductive cycles. Trans. r. Soc. S. Afr. 41: 111–160.

Choquet, M., 1966. Biologie de *Patella vulgata* L. dans le Boulonnais. Cah. Biol. mar. 7: 1–22.

Evans, R. G., 1948. The lethal temperatures of some common British littoral molluscs. J. Anim. Ecol. 17: 165–173.

Evans, R. G., 1953. Studies on the biology of British limpets. The genus *Patella* on the south coast of England. Proc. zool. Soc. Lond. 123: 357–376.

Fritchman, II, H. K., 1962a. A study of the reproductive cycle in the Californian *Acmaeidae* (Gastropoda). Part 2. Veliger, 3: 95–101.

Fritchman, II, H. K., 1962b. A study of the reproductive cycle in the Californian *Acmaeidae* (Gastropoda). Part 4. Veliger, 4: 134–140.

Guerra, M. T. & M. J. Gaudêncio, 1987. Aspects of the ecology of *Patella* spp. on the Portuguese coast. Hydrobiologia 142: 57–69.

Hatch, D. J., 1977. The inter-relationships of three *Patella* spp. on the south coast of Britain. Ph.D. Thesis, University of Southampton, England.

Lammens, J. J., 1967. Growth and reproduction in a tidal flat population of *Macoma balthica* L. Neth. J. Sea Res. 3: 315–382.

Lewis, J. R., 1976. Long-term ecological surveillance: practical realities in the rocky littoral. Oceanogr. & Mar. Biol. Ann. Rev. 14: 371–390.

Lewis, J. R., 1984. Temporal and spacial changes in benthic communities: COST 47 approach. Mar. Pollut. Bull. 15: 397–402.

Lewis, J. R. & R. S. Bowman, 1975. Local habitat-induced variations in the population dynamics of *Patella vulgata* L. J. exp. mar. Biol. Ecol. 17: 165–203.

Lewis, J. R., R. S. Bowman, M. A. Kendall & P. Williamson, 1982. Some geographical components in population dynamics: possibilities and realities in some littoral spp. Neth. J. Sea Res. 16: 18–28.

Nelson, T. C., 1928. Relation of spawning of the oyster to temperature. Ecology, 9: 145–154.

Orton, J. H. & A. J. Southward, 1961. Studies on the biology of limpets. IV. The breeding of *Patella depressa* Pennant on the north Cornish coast. J. mar. biol. Ass. U.K. 41: 653–662.

Orton, J. H., A. J. Southward & J. M. Dodd, 1956. Studies on the biology of limpets. II. The breeding of *Patella vulgata* L. in Britain. J. mar. biol. Ass. U.K. 35: 149–176.

Picken, G. B., 1980. The distribution, growth and reproduction of the Antarctic limpet *Nacella (Patinigera) concinna* (Strebel, 1908). J. exp. mar. Biol. Ecol. 42: 71–85.

Picken, G. B. & D. Allen, 1983. Unique spawning behaviour by the Antarctic limpet *Nacella (= Patinigera) concinna* (Strebel, 1908). J. exp. mar. Biol. Ecol. 71: 283–287.

Thompson, G. B., 1979. Distribution and population dynamics of the limpet *P. aspera* (Lamark) in Bantry Bay. J. exp. mar. Biol. Ecol. 40: 115–135.

Thompson, G. B., 1980. Distribution and population dynamics of the limpet *P. vulgata* L. in Bantry Bay. J. exp. mar. Biol. Ecol. 45: 173–217.

Workman, C., 1983. Comparisons of energy partitioning in contrasting age-structured populations of the limpet *P. vulgata* L. J. exp. mar. Biol. Ecol. 68: 81–103.

Wright, J. R. & R. G. Hartnoll, 1981. An energy budget for a population of the limpet *Patella vulgata.* J. mar. biol. Ass. U.K. 61: 627–646.

Hydrobiologia 142: 57–69 (1986)

Aspects of the ecology of *Patella* spp. on the Portuguese coast

Miriam Tuaty Guerra & Maria Jose Gaudencio
Instituto Nacional de Investigação das Pescas, Av. de Brasilia, 1400 Lisboa, Portugal

Keywords: gastropods, *Patella* spp., ecology, breeding cycles, longevity

Abstract

Three species of the genus *Patella* were studied on the Portuguese coast as a specific contribution to the COST 647 programme. With little prior information existing, this paper presents new basic data on the geographical distribution, relative abundance, habitats, size-structure of populations, and reproductive cycles of *P. depressa* Pennant, 1777, *P. aspera* Roding, 1798, and *P. vulgata* L. 1758. Comparisons with N.W. Europe provide evidence of distinct latitudinal trends in the population dynamics and reproductive biology of these species.

Introduction

Limpets play a fundamental ecological role in controlling the quantity of algae on the shore. As grazers, they destroy large quantities of algae, mainly at the early sporeling stages (Lewis, 1964). They are widely distributed around the coasts of western Europe and the British Isles, variously inhabiting moderately-sheltered to fully-exposed shores, making them ideal candidates for inclusion in the rocky-littoral studies of the COST 647 programme.

In Portugal, limpets are abundant over all the intertidal rocky shore. Together with mussels and trochids, they form part of the staple diet of the poorer populace, especially in the north. Apart from Nobre's work on Molluscs (1931, 1938–40), there were no general studies on limpets in Portugal, and no information on population dynamics. In 1981, the Instituto Nacional de Investigação das Pescas started to collaborate with COST 647 by studying the three limpet species already under investigation by the rocky-shore group in England, i.e. *Patella depressa* Pennant, 1777, *P. aspera* Roding, 1798, and *P. vulgata* L. This paper presents some aspects of the ecology of these species.

Geographical distribution and habitat

The distribution of the three species on the Portuguese coast is shown in Fig. 1.

P. depressa

This species extends from N. Africa to S.W. England and Wales (Fretter & Graham, 1976). In Portugal it replaces *P. vulgata* as the dominant intertidal limpet on exposed and sheltered shores, occurring at all shore levels, on horizontal and vertical surfaces, and in pools lined with *Lithophyllum incrustans* (where it usually co-exists with *P. aspera*).

In the upper shore the shells are usually steeply-conical, and very thick, whereas at lower levels they are polygonal, very flattened, and thin. Shell length also varies according to shore level, thus at high levels, particularly those with dense *Chthamalus* cover, the maximum length is 25 mm, but at low levels it reaches 30–35 mm. On very exposed shores the shells of limpets on verticals are frequently almost completely eroded.

Juveniles are found mainly at low levels and always in damp situations e.g. in small pools or cracks, around the edges of mussel patches, on *Mytilus* or on wet, bare rock.

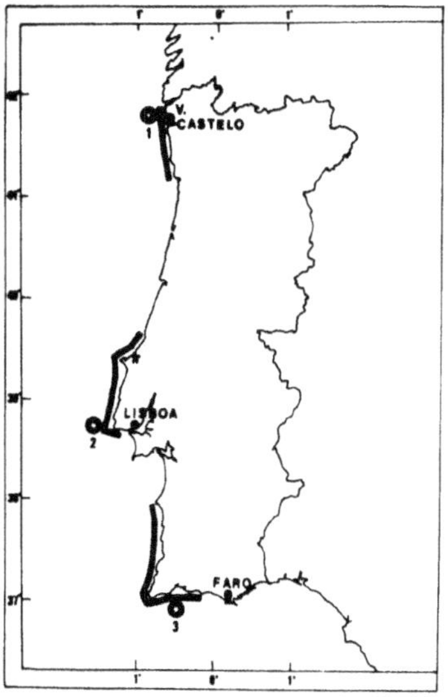

Fig. 1. Distribution of *Patella depressa, P. aspera* and *P. vulgata* on the portuguese coast. • Sampling sites: 1 – Moledo do Minho; 2 – Cabo Raso; 3 – Praia da Luz.

Patella aspera

This is also a southern species, extending from the Mediterranean to S.W. Norway (Lewis *et al.*, 1982). It is also abundant in Portugal, but seems to be less resistant to desiccation than *depressa*, occurring only in the middle and lower shore and almost always co-existing with *depressa*. It is found particularly on rocks exposed to strong wave-action, among mussels and the algae *Corallina, Enteromorpha* and *Lithophyllum* spp., and in the mid-shore it is mainly restricted to coralline pools containing small patches of mussels.

Shells are usually oval in shape and often much larger than in *depressa*. The largest specimens (45–49 mm) were found in the lower shore where their shells were almost always covered by dense algal growth, making them unrecognisable to all but the expert eye.

Juveniles are always found around the edges of mussel patches, or in very small water-filled pits in the *Lithophyllum* covering permanently-wet rock.

Patella vulgata

This is a northern species, extending from N. Norway just into the Mediterranean (Fretter & Graham, 1976). It is thus close to its southern limit on the Portuguese coast, where it is abundant only in the north. Progressing southwards, it becomes increasingly rare until only a few specimens can be found among the other species (mainly *depressa*). It inhabits mainly low shore levels, always in the shade, on damp platforms or in pools with *aspera* and *depressa*. In the mid-shore it is found mainly on vertical walls, in the shade.

The largest specimens are found at low levels, at up to 40 mm in length, the shells being circular to oval and a little flattened, whereas in the mid-shore they are mostly conical.

Juveniles were always found only at low shore levels, in damp sites, e.g. on bare rock having wet pits or cracks where they are able to avoid desiccation during tidal emersion, or among mussels, which also provide shelter from desiccation (Lewis & Bowman, 1975).

Material and methods

Limpets were collected from the north, centre and south of the Portuguese coast (Fig. 1), always from low shore levels, between March 1981 and March 1984. Samples were usually of a minimum of 50 animals per month, per species. After sampling, shell length was measured to the nearest mm, and the gonad state was macroscopically categorised in order to determine the mean gonad condition by the method described by Orton *et al.*, (1956). Areas of 1 m^2 were also established in the same localities, where monthly *in situ* size-frequency analyses were taken for determination of recruitment times and longevity.

Gonad cycles

Patella depressa

Figure 2 shows the gonad cycles of *P. depressa* in the three regions. It is clear that the reproductive pattern differs from north to south. In the north and centre, breeding is very asynchronous, so that during the three years of the study neuter, developing, spawning and spent gonads were found in almost every monthly sample. Such asynchrony suggests that frequent spawnings during the breeding season were followed by bursts of gonad re-ripening or redevelopment, as described for *P. vulgata* in Britain (Bowman & Lewis, 1977; Bowman, 1985).

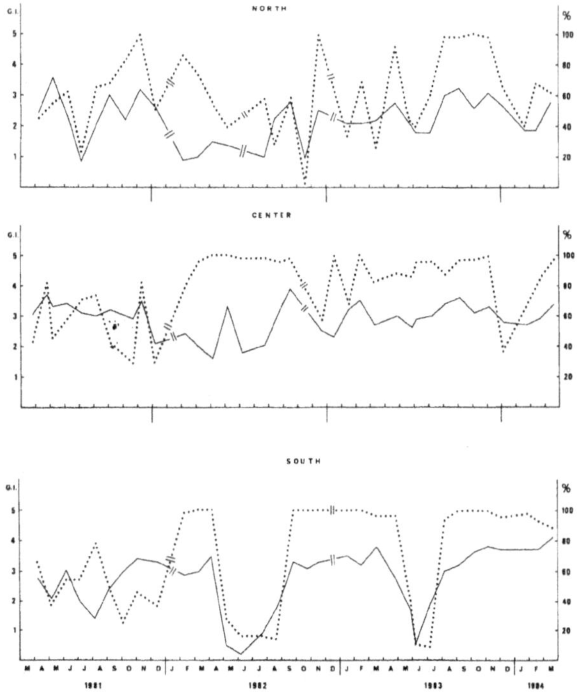

Fig. 2. Gonad index (—) and percentage of gonads in the spawning condition (• • •) of *Patella depressa.*

In the north, the appearance of neuter gonads in midsummer and midwinter suggests that development (and/or redevelopment) briefly ceases during these times in some animals, so that the breeding period appears bimodal.

Rapid spring development is followed by spawning(s) and a brief resting period (June–July) in some animals, followed by a fresh burst of autumn development and spawning, and a somewhat smaller February–March resting period, in fewer animals. This bimodal pattern somewhat resembles that described for *P. depressa* in Britain (Hatch, 1977; Bowman, pers. comm.), where a major spring spawning can, in some years, be followed by considerable redevelopment to a second peak gonad size and spawning, though in Britain the main (very short) resting period occurs over the winter.

In southern Portugal, the summer resting period is much more distinct, and involves most of the population (though this varies annually). Rapid development in late summer is followed by a long over-winter period when the gonads remain fat and ripe, in spite of occasional small releases, presumably because redevelopment is virtually continuous but hard to distinguish in very ripe gonads.

In the central region the breeding situation is confusing, because either the gonads retain large numbers of unspawned gametes over the presumed resting periods, or small releases and redevelopment occur virtually continuously throughout the year.

Patella aspera

Figure 3 shows the gonad cycles of *P. aspera* for the three regions. Development began in late summer, with the main spawning between October and December in the north and centre, December/January in the south. A considerable amount of redevelopment after spawning, particularly in the south, meant that the gonads remained ripe until early spring, when further spawning preceded a brief 'resting' period when the maximum number of neuter gonads were found, although not all animals passed into a resting phase.

Patella vulgata

Data on the reproductive cycle of this species

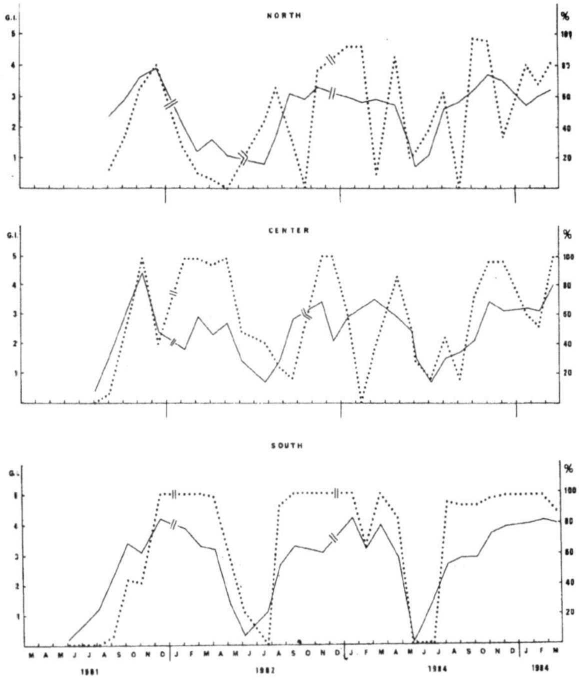

Fig. 3. Gonad index (——) and percentage of spawning gonads (• • •) of *Patella aspera.*

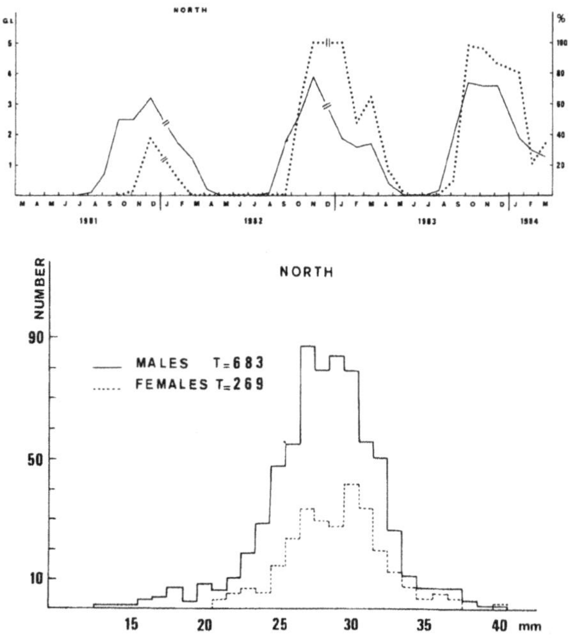

Fig. 4. Patella vulgata: a – Gonad index (——) and percentage of gonads in the spawning condition (• • •). b – Number of males and females in relation to shell length (mm).

come exclusively from the north, since it was impossible elsewhere to find a shore with enough *vulgata* to guarantee a sample of 50 limpets a month. Figure 4a shows the gonad cycles at Moledo do Minho. This species has the shortest reproductive cycle, development starting in July/August. Spawning begins in September/October, though, because of continued gonad ripening or redevelopment, the main spawning does not occur until November/January. By May, all the gonads are in the neuter resting condition and remain so until the onset of the next breeding season.

Size frequencies

P. depressa

Most observations derive from the central region, from a marked 1-metre square in the low-littoral, on 'bare' rock with a great number of crevices and a small patch of mussels. Juveniles (<5 mm) were observed every month (Fig. 5), which can be explained in terms of continuous spawning, but it is difficult to estimate when they settled, as they were not detected until they emerged

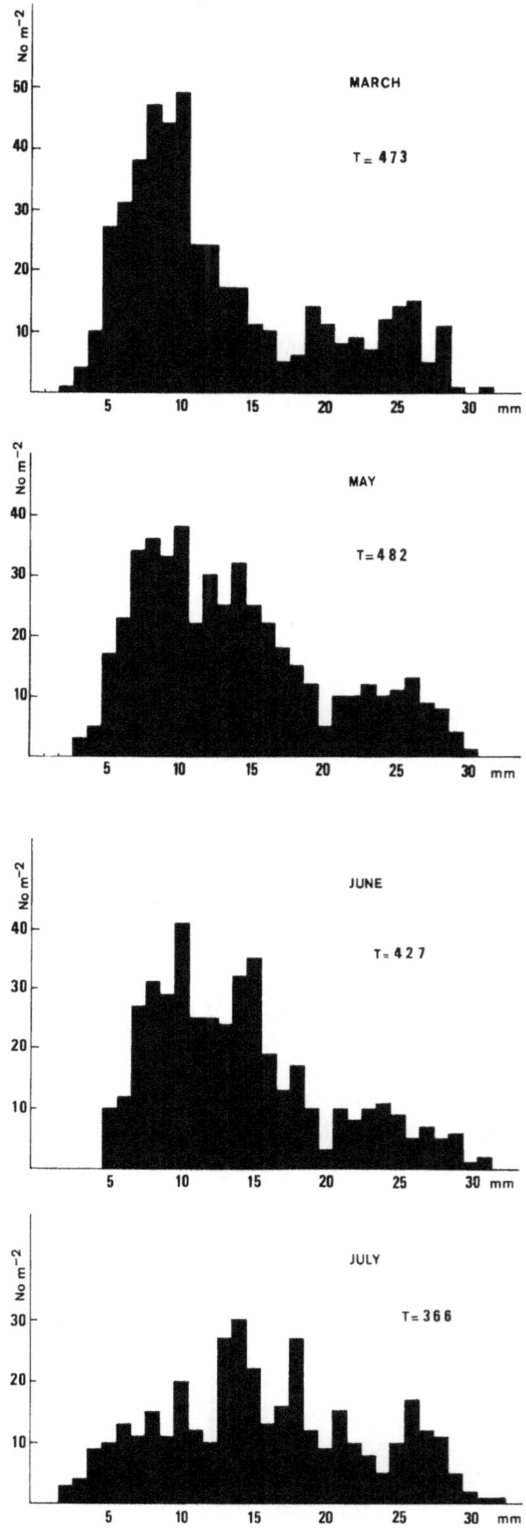

Fig. 5.

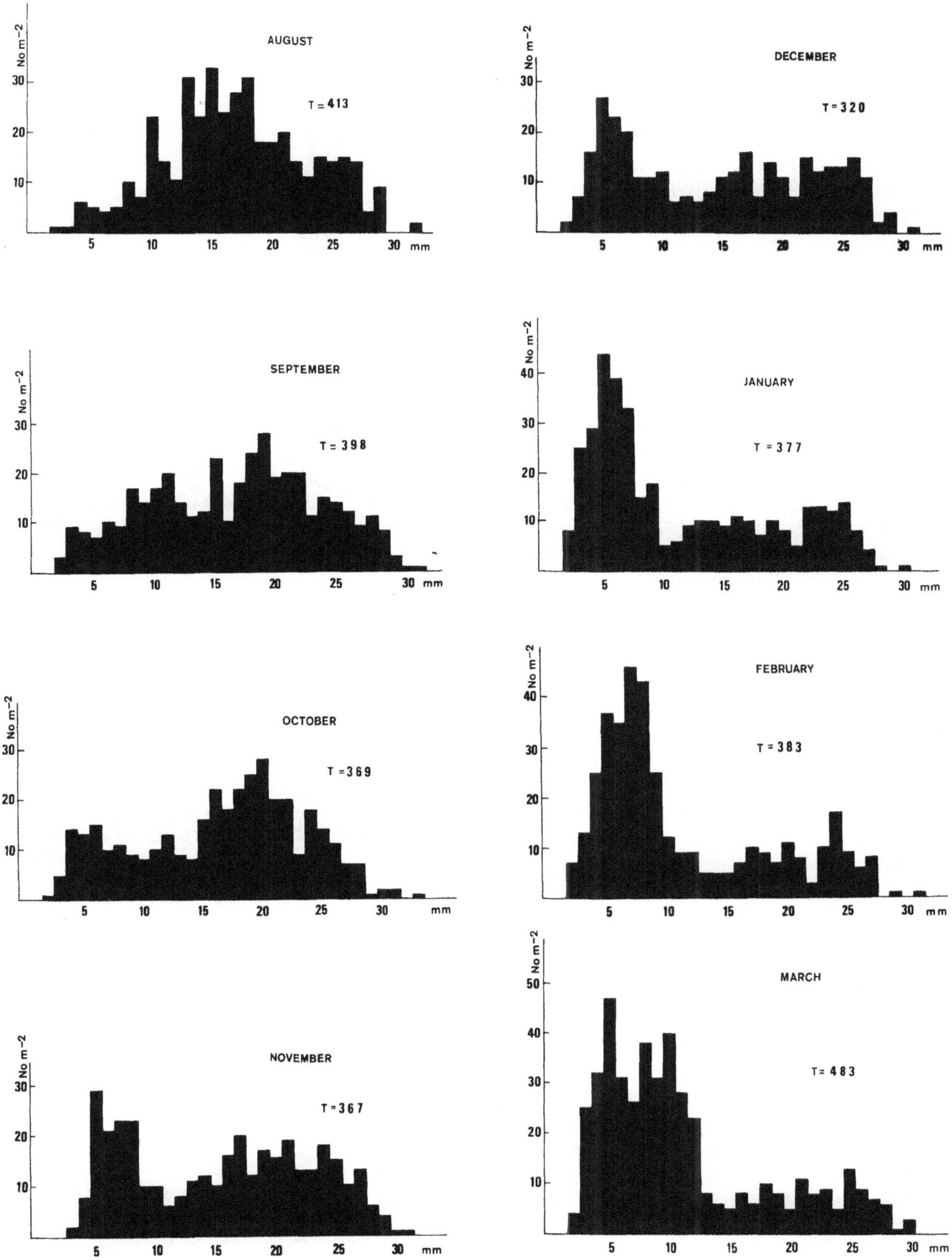

Fig. 5. Size-frequencies of *Patella depressa* in the center of the country, from March 1983 to March 1984 (number of individuals in relation to shell length (mm)).

from their settlement crevices at a size of 2–3 mm, aged probably 2–3 months (cf. problems with detecting *P. vulgata* settlement – Lewis & Bowman, 1975). Peak numbers of juveniles were found from January to March, then a scarcity of very small limpets was followed by further emergence from July to September. From the fast growth rates of limpets in the low-shore it seems likely that the January/March peak resulted from an October/November spawning/settlement, and the summer emergence from possibly a June spawning.

In the south, a few observations were made on a similar low-littoral site, and juvenile numbers peaked in November to January, again possibly from an autumn release.

In the north, the only habitat available was at a high level and *Chthamalus* dominated. Juveniles were scarce, and maximum numbers were found in January and April. On such dry HL sites juveniles often do not emerge until 6 months or more after settlement (Lewis & Bowman, 1975), so it is difficult to assign these *depressa* to one or other of the observed spawning times. Probably both peaks result from the emergence of settlements from various releases during the autumn spawning period.

Settlement of *P. depressa* throughout Portugal is therefore probably most successful during autumn and winter, with early-summer settlement succeeding only at low shore-levels where spat will be largely protected from the adverse effects of high temperatures during low-water.

Analysis of the size-frequency data suggests these limpets have a life-span of no longer than three years.

Patella aspera

Size-frequency data were obtained from the south coast (Fig. 6). The site was a marked 1-metre square in a large mid-littoral pool, lined with *Lithophyllum incrustans* and containing some patches of mussels and *Enteromorpha*. Such a pool must experience high temperatures during low-water in summer, which could adversely affect the growth, survival and rate of emergence of the spat (Bowman, pers. comm., and cf. *P. vulgata*).

Newly-emerged juveniles at 2–3 mm first appear during March/April to June, then, after a lull during the summer, numbers build up (probably through emergence of further limpets from the mussel patches) to a peak in October/November.

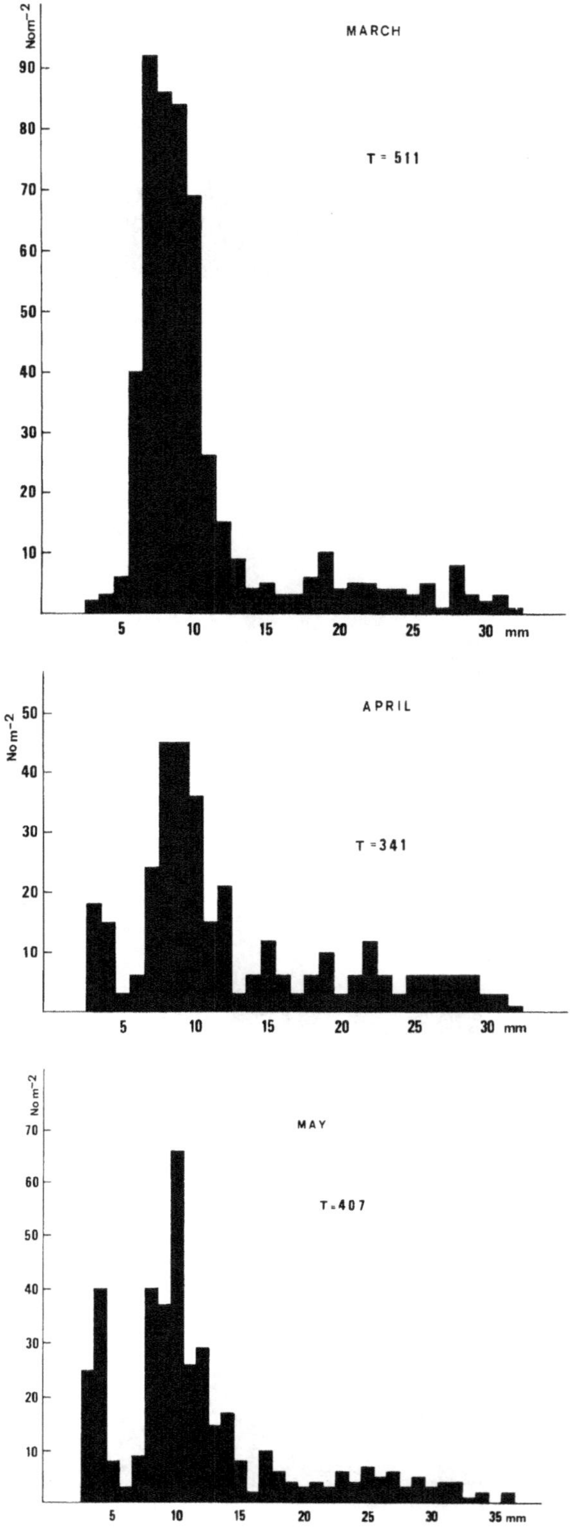

Fig. 6.

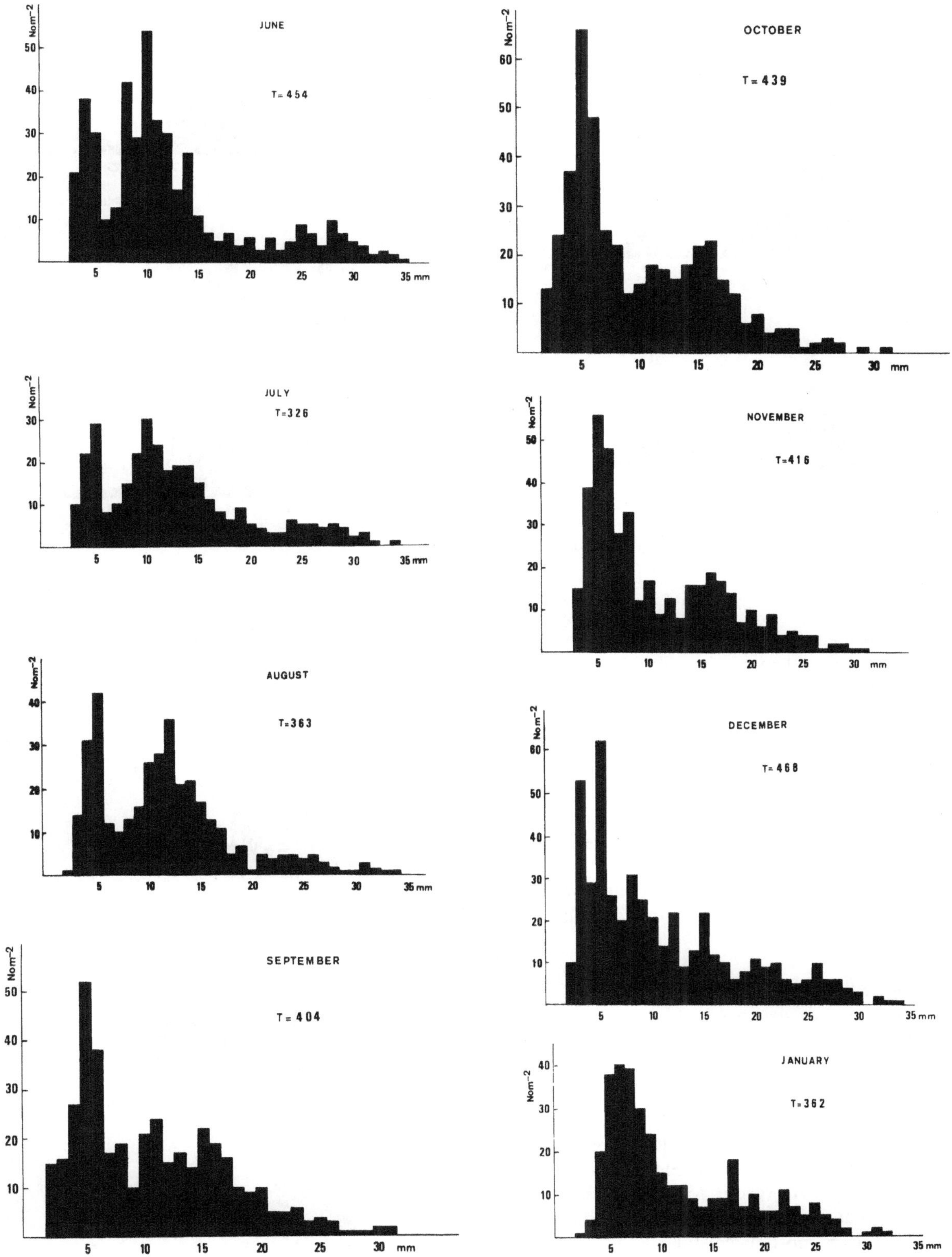
Nom^{-2}
JUNE
T= 454
5 10 15 20 25 30 35 mm
JULY
T=326
AUGUST
T=363
SEPTEMBER
T= 404
OCTOBER
T= 439
NOVEMBER
T=416
DECEMBER
T= 468
JANUARY
T=362

Fig. 6.

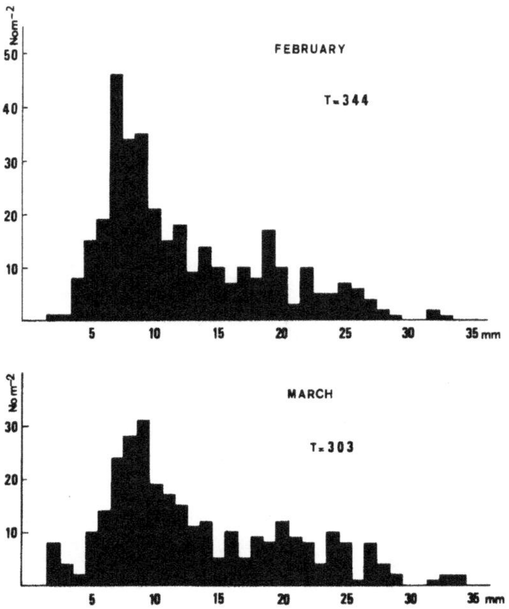

Fig. 6. Size-frequency of *Patella aspera* in the south of the country, from March 1983 to March 1984 (number of individuals in relation to shell length (mm)).

These two emergence periods in 1983 probably resulted from settlements from different spawnings during the breeding season, i.e. the juveniles emerging in April derived from the January/February release (see Fig. 3), and later settlements from end-of-season spawnings, while those appearing in February/March 1984 possibly settled in October 1983. It is clear from the March numbers of 1-year-olds in 1983 and 1984 that the former recruitment was very much larger than the latter.

The size-frequency data suggest that these Portuguese *P. aspera* rarely live for more than four years.

Patella vulgata

Size-frequency data were obtained in the north from a marked 1-metre square in the lower shore. The site was shaded and permanently damp, with a relatively dense cover of mussels, features which have been found to favour settlement and survival of spat (Lewis & Bowman, op. cit.). Juveniles started to appear in late October (Fig. 7), but the large numbers which emerged in March suggest that settlement(s) derived from the later part of the spawning period are most successful.

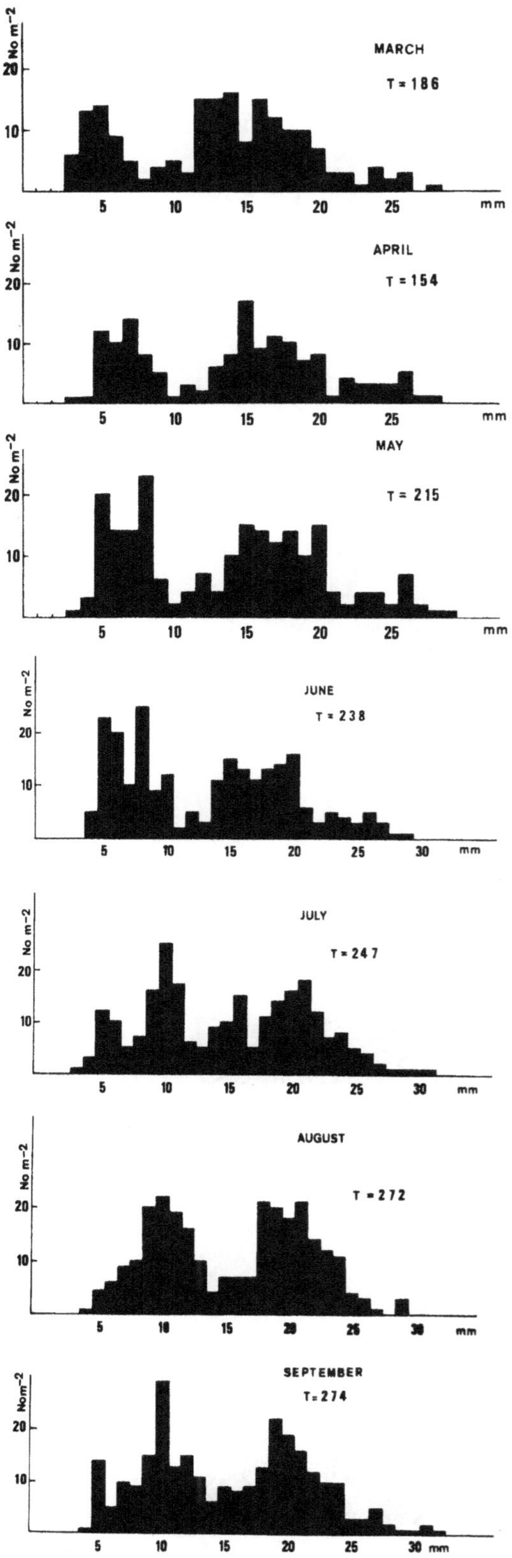

Fig. 7.

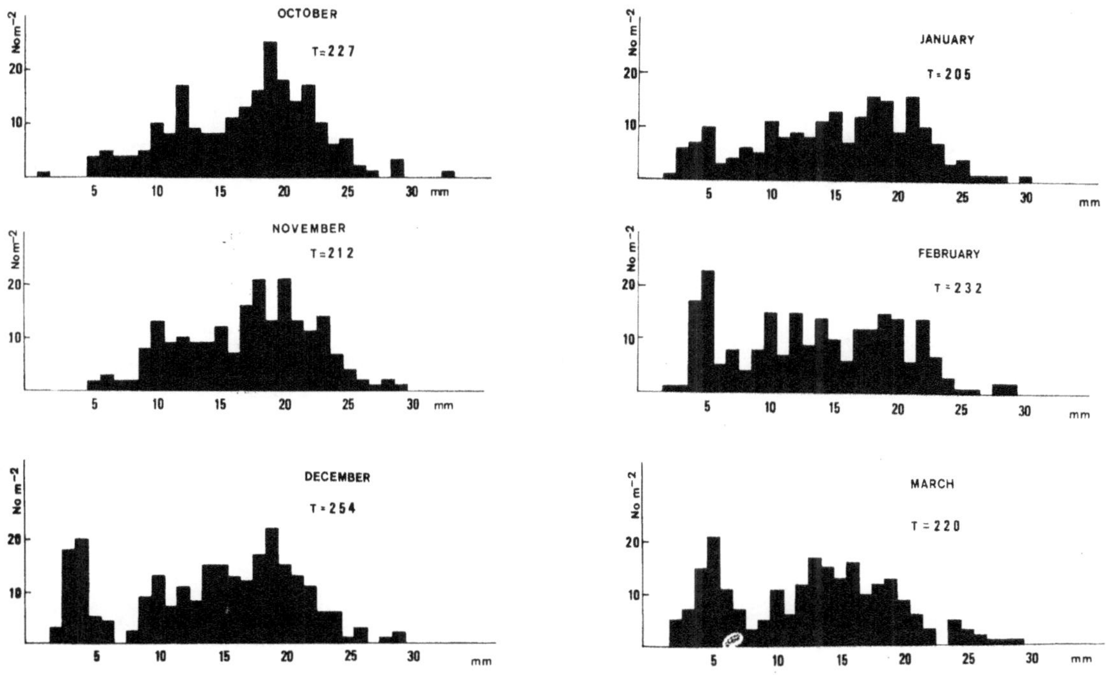

Fig. 7. Size-frequencies of *Patella vulgata* in the north of the country, from March 1983 to March 1984 (number of individuals in relation to shell length (mm)).

Again, the data suggest a life-span of only three or four years. It should be noted that these proposed longevities of the three species, obtained from interpretation of the size-frequency data of the observation-site populations, need to be followed up by *in situ* studies of growth and survival of marked animals, to take into account those animals which migrate off the observation sites into areas of lower settlement/greater food availability, etc.

Sex ratio

All the limpets collected in all regions to determine the breeding state of the populations were also measured, and the size-frequency of these sexed animals suggested the following conclusions on maturation times and the incidence of sex-change in the three species.

P. depressa

Males were more numerous than females, comprising 57% of the total sexable limpets (Fig. 8), and the difference in numbers of males and females at all sizes and in all regions is highly significant ($p < 0.001$). Individuals of both sexes were found over the same size range, i.e. 12/13 mm to 38/40 mm shell length in the three regions, with the bulk of the breeding population at between 24–30 mm. There was no appreciable change in the proportions of the sexes with increasing size (and age?).

These data suggest that most *P. depressa* mature in their second year, and that sex is probably fixed from the outset in at least some individuals, which agrees with the observations of Orton & Southward (1961) on *depressa* in S.W. England. However, in view of the small size (6–10 mm) at which the British *depressa* matured, and their fast growth and extended settlement period in Portugal, it is not impossible that here they might mature before the end of their first year on the shore. To check this it would be necessary to examine animals of all sizes and known age throughout the annual cycle.

P. aspera

In this species, males again outnumbered females significantly, comprising 58% of the sexable population (Fig. 9). However, males were found at sizes from 9–48 mm, females from 13–49 mm, and the proportion of males to females tended to decrease with increasing size/age. This suggests that *aspera* mature in their second year as males, and that sex-change does occur in a proportion of animals in subsequent years. This agrees with the findings of Thompson (1979) for *aspera* in S.W. Ireland, though he found only 7% of animals matured in their second year. He also found that smaller members of an age group matured later in the breeding season than their larger contemporaries, which might also occur in Portugal in view of the extended breeding and settlement periods found. It is, however, possible that some *aspera* might mature as

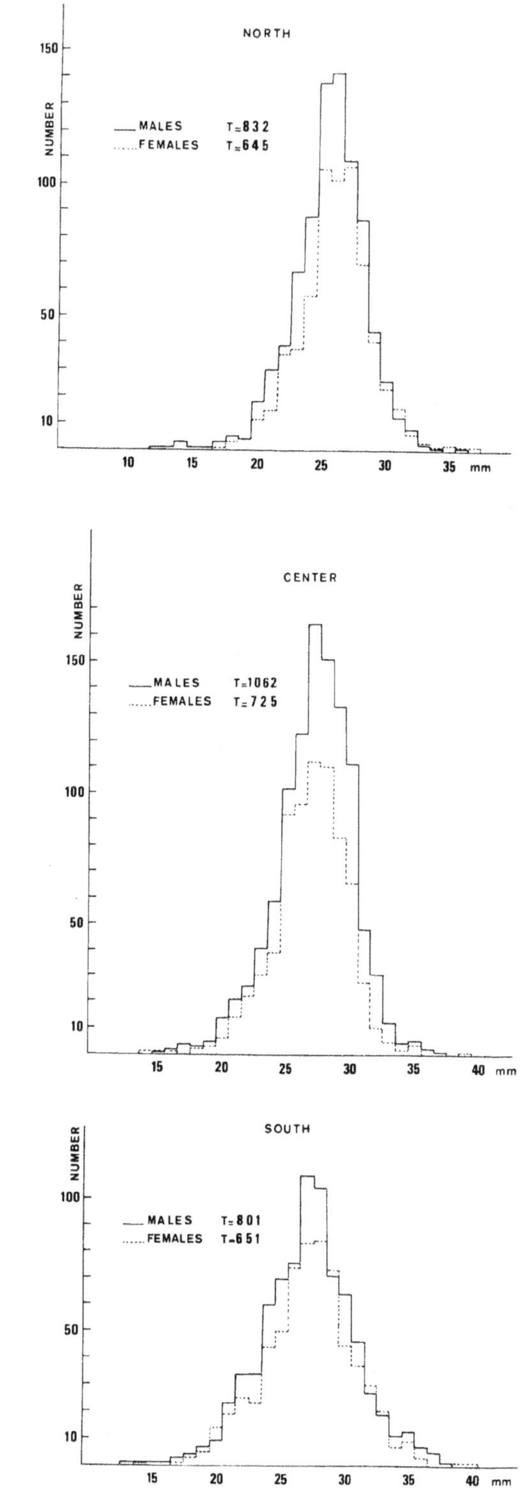

Fig. 8. Patella depressa: number of males and females in relation to shell length (mm).

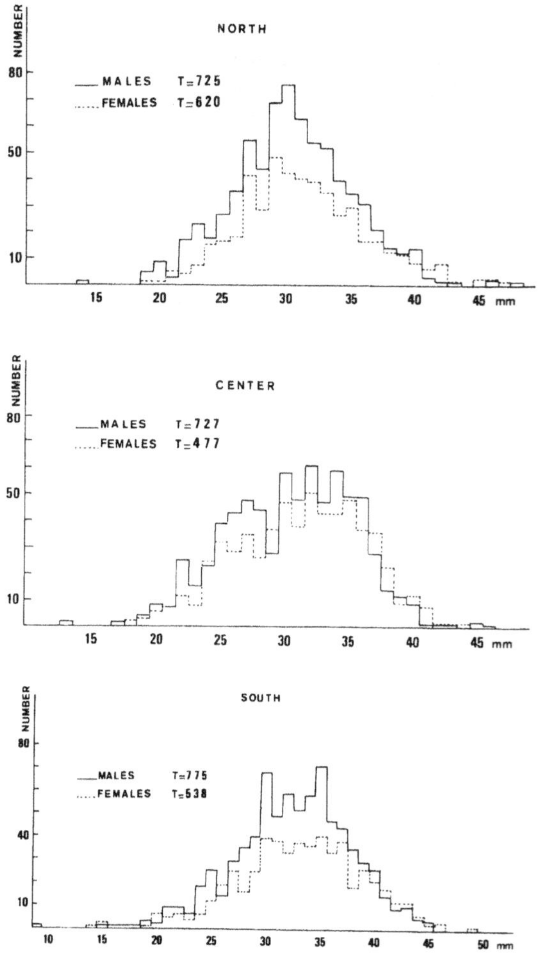

Fig. 9. Patella aspera: number of males and females in relation to shell length (mm).

females in their first breeding period, as the sex/size difference of small animals is not as marked as in Irish and British populations. Examination of a greater number of small individuals of known age is required to check these possibilities.

P. vulgata

In this species, 40% of the animals examined for breeding condition were found to be neuter, because of the long resting period, but of those sexable, 72% were male (Fig. 4b). The size of males ranged from 13–40 mm whereas females were 21–40 mm, and this agrees with the findings of Orton *et al.* (1956) in that *vulgata* maturation begins in the second year of shore life with all animals starting as males and an increasing proportion changing sex in subsequent breeding seasons. The relatively low proportion of females in *vulgata* compared to the proportion of *aspera* females might suggest that relatively fewer *vulgata* change sex, or it might be a result of the shorter lifespan of these *vulgata.* Again, examination of populations of individuals of known age would help to resolve this.

Discussion

The reproductive cycles of the three *Patella* species show some interesting geographical changes along the Portuguese coast, and it is likely that these are linked with the regions' different climatic conditions. *P. depressa* is regarded as a summer breeder in S.W. England, where it is near its northern limit: spawning time there seems to coincide with maximum air temperatures and wave action (Orton & Southward, 1961; Fretter & Graham, 1976), and there is a short winter resting period. In Portugal the reproductive behaviour was different. In the south, nearer the species' southern limit, the decrease in gonad activity coincided with the increase in air and sea temperatures over the summer, and thus seems to reflect the adverse influence of excessive warmth near southern limits postulated by Lewis *et al.* (1982). Late-summer development was rapid, and the number of individuals spawning increased steadily through autumn and winter to early spring, though successful settlements seemed to result only from those spawnings occurring during the falling temperatures of autumn and early winter. In the north, there were several spawning peaks during the year, with minimal gonad activity during the periods of highest summer and lowest winter temperatures. This pattern was similar in the central region in 1981, when more extreme summer and winter temperatures were experienced than during 1982–1984, when in this region spawning and gonad activity was at a continuous high level. It seems unlikely that differences in air temperature are alone responsible for the differences in breeding cycles, particularly in these sampled low-shore populations which are more subject to sea than to air temperatures. There was certainly no connection between spawning times and air temperature, since air temperatures over the spawning periods ranged from 6–27 °C in the north, 9–23 °C in the centre and 7–26 °C in the south. The mean sea temperatures (1956–75) for northern and central Portugal are similar in the summer but higher in the winter compared to those in S.W. England (Bowman, 1985), i.e. 12.2 (Jan.)–16.2 (Oct) °C in the north, 13.0 (Jan.)–17.0 (Sept.) °C in the centre; those in the south are markedly higher: 14.1 (Jan.)–20.4 (Aug.) °C. It is therefore possible that in the central region the equable sea and less extreme air temperatures might explain the continuous gonad development, with spawning occurring on periods of rough seas, whereas the high summer air and sea temperatures in the south, and the low winter air temperatures in the north, might slow the rate of gonad development until temperatures ameliorated.

The spawning of *Patella aspera* occurs while temperatures are falling, in Britain, e.g. during August/September in N.E. England (Bowman, 1981), where it is near its northern limit; August to November in S.W. Ireland, where spawning has been closely linked with gales and wave action (Thompson, 1979). In Portugal the species shows several peaks of spawning and development, between August and early spring, with gonads remaining ripe later in the south (i.e. into March/April) than in the north, but in all regions spawning reaches its maximum during the gales and low temperatures of winter. Since spawning appears to involve falling water temperatures and wave action, the delay in the spawning period on the Portuguese coast in relation to N.W. Europe, might possibly be explained by the progressively higher summer sea temperatures towards the south.

The limpets have thus to wait for sea temperature to decrease sufficiently for storms to stimulate spawning. The frequency of storms on the Atlantic coast of Portugal is higher than that on the south coast, so limpets in this region might therefore take longer to release all their gametes.

Patella vulgata was formerly regarded as a winter breeder, but in northern Britain spawning regularly starts in August/September, earlier still (July/August) at its northern limit in N. Norway (Bowman & Lewis, 1977). Bowman (1985) has shown that in N.E. England and N.W. Scotland spawning starts when the sea temperature first falls below 12 °C, which is linked with the heavy swells produced by north-westerly gales, and also that redevelopment after such spawnings occurs only when the sea temperature rises above 12 °C again. In Portugal the monthly mean sea temperature falls to around 12 °C only in midwinter (Dec.–Feb./Mar.), and does so only rarely in the south. This would explain the extended breeding season in Portugal compared to that in northern Britain. It has not been possible to study daily variations in sea temperature and limpet gonad states as described by Bowman (*op. cit.*), but from the Portuguese temperatures observed it seems possible that here spawning could be in response to a drop to below a slightly higher sea temperature. However, the Atlantic coast of Portugal is notorious for sudden storms, which could bring inshore colder Atlantic water and thus provide the necessary temperature drops and stimulate spawning at any time during autumn and winter. The incidence of these storms might well be responsible for much of the annual and local variation in spawning of the three species of *Patella* in Portugal, but more detailed meteorological and reproductive data are needed to investigate this further.

In all regions, settlements of the three species seemed to be most successful if they occurred in late autumn/early spring, and this is possibly connected with the apparent sensitivity of newly-settled limpets to temperature variations during their early shore-life (Bowman, pers. comm.). Thus the temperatures thought to be necessary for successful establishment of settlements of *P. vulgata* and *P. aspera* would most likely be encountered in Portugal in Nov./Dec., and Mar./Apr., and in the south only then in the lower shore. Data on direct settlement from settlement plates, as described by Bowman (1985), are needed to check on settlement periods, and to determine the relative success of different spawnings/settlements.

It is clear that in Portuguese patellids, growth is fast, and longevity less than is generally found elsewhere in N.W. Europe. Lewis *et al.* (1982) have suggested that species near their northern limits may attain greater sizes and live longer than they do further south, features that appear to be linked with infrequent recruitment and low population densities. Thus *P. depressa* in S.W. England usually reaches 45 mm shell length (Orton & Southward, 1961), *P. aspera* in S.W. Ireland reaches almost 60 mm shell length and has a potential life-span of longer than ten years (Thompson, 1979), and *P. vulgata* in Britain reaches 57 mm, while longevity ranges from 4–5 years in a low-littoral bare-rock habitat to 15–17 years in a high-littoral dense-barnacle habitat (Lewis & Bowman, 1975). Growth rates on British coasts are very variable, particularly in juvenile limpets, apparently reflecting competition for food within a year-class rather than between adults and juveniles, as well as food availability, algal production, shore level, feeding times etc. and also settlement time (Bowman, 1981). In Portugal, shore densities of *P. depressa* and *P. aspera* are high, presumably as a result of more regular recruitments than in Britain, but early growth rates are also higher than those recorded from similar habitats in Britain and Ireland, and maximum size and longevity are less. *P. vulgata,* close to its southern limit in Portugal, occurs at lower densities, with poorer recruitment, yet also shows rapid growth but smaller maximum size and shorter life-span than in equivalent habitats in Britain (although it is possible that the natural life-span/maximum size of all limpet spp. in northern Portugal might be lowered by the continuous heavy human predation of the largest available animals). In general, therefore, it seems likely that, in limpets, growth rates are more a function of temperature and metabolic rate, which are higher in the south than further north and therefore result in shorter life and smaller maximum size. This could have important implications regarding the breeding strategies of the species, and might possibly be responsible for the apparently-earlier maturation of Portuguese limpets compared with those in Britain and Ireland.

Conclusion

Populations of three species of *Patella* in Portugal have shorter life-spans and smaller maximum size than those in N.W. Europe. Their reproductive behaviour seems to be influenced by warm sea and air temperatures, which provide the potential for long breeding periods, of which only a limited period might result in successful recruitment, depending on the habitat and location of the populations.

Acknowledgements

Our thanks are due to Dr. J. R. Lewis for his help and advice in the accomplishment of this work, and to Miss R. S. Bowman who guided our first steps in the study of limpet biology and also kindly criticised the manuscript. We also thank our colleague Mr. E. Soares for his support in the completion of the manuscript.

References

Bowman, R. S., 1981. The morphology of *Patella* spp. juveniles in Britain, and some phylogenetic inferences. J. mar. biol. Ass. U.K. 61: 647–666.

Bowman, R. S., 1985. The biology of the limpet *Patella vulgata* L. in the British Isles: spawning time as a factor determining recruitment success. In P. G. Moore & R. Seed (eds.), The Ecology of Rocky Coasts. Hodder & Stoughton, Sevenoaks (Kent, U.K.): 178–193.

Bowman, R. S. & J. R. Lewis, 1977. Annual fluctuations in the recruitment of *Patella vulgata* L. J. mar. biol. Ass. U.K. 57: 793–815.

Fretter, V. & A. Graham, 1976. The prosobranch molluscs of Britain and Denmark. I. Pleurotomariacea, Fissurellacea, and Patellacea. J. moll. Stud. Suppl. 1, 37 pp.

Hatch, D. J., 1977. The inter-relationships of three *Patella* spp. on the south coast of Britain. Ph.D. thesis, University of Southampton, U.K.

Lewis, J. R., 1964. The Ecology of Rocky Shores. The English University Press Ltd, London. 323 pp.

Lewis, J. R. & R. S. Bowman, 1975. Local habitat-induced variations in the population dynamics of *Patella vulgata* L. J. exp. mar. Biol. Ecol. 17: 165–203.

Lewis, J. R., R. S. Bowman, M. A. Kendall & P. Williamson, 1982. Some geographical components in population dynamics: possibilities and realities in some littoral species. Neth. J. Sea Res. 16: 18–28.

Nobre, A., 1931. Moluscos marinhos de Portugal, 1. Imprensa portuguesa, Porto. 466 pp.

Nobre, A., 1938–40. Fauna malacologica de Portugal. 1. Moluscos marinhos e das aguas salobras. Porto. 806 pp.

Orton, J. H. & A. J. Southward, 1961. Studies on the biology of limpets. IV. The breeding of *Patella depressa* Pennant on the north Cornish coast. J. mar. Ass. U.K. 41: 653–662.

Orton, J. H., A. J. Southward & J. M. Dodd, 1956. Studies on the biology of limpets. II. The breeding of *Patella vulgata* L. in Britain. J. mar. Ass. U.K., 35: 149–176.

Thompson, G. B., 1979. Distribution and population dynamics of the limpet *Patella aspera* (Lamarck) in Bantry Bay. J. exp. mar. Biol. Ecol. 40: 115–135.

Hydrobiologia 142: 71–80 (1986)

Comparative trends and ecological patterns of rocky subtidal communities in the Swedish and Norwegian Skagerrak area

Tomas Lundälv* & Hartvig Christie**
**Kristineberg Marine Biological Station, S-450 34 Fiskebäckskil, Sweden*
***Ministry of Environment, P.B. 8013, N-0030 Oslo 1, Norway*

Keywords: rocky subtidal community, long-term, fluctuations, recruitment, predation, common trends

Abstract

Inside the COST 647 Rocky Subtidal Programme, as well as before this came into operation, a co-operation has been developed between Swedish and Norwegian groups working with rocky subtidal ecology in the Skagerrak area. Along a coastline of more than 300 km, and additional large fjord systems, natural fixed-site rocky subtidal communities have been stereophotographically recorded more or less seasonally over a considerable number of years (15 in Sweden and 7 in Norway). Additional experimental studies have been undertaken *in situ.*

Inter-site comparisons revealed many examples of co-variation in long-term population fluctuations. A common 7-year cyclic pattern was clearly illustrated for the ascidian *Ciona intestinalis* and the echinoid *Echinus esculentus.* Similar long-term trends were found in populations of *Asterias rubens, Ascidia mentula, Sabella penicillus, Crania anomala* and *Protantea simplex* over the time periods available for comparison. Locally deviating populations were also found, however.

Major structuring factors, as well as key species, appeared to be largely the same over the geographic range studied. By comparing all of our results (mostly based on already published data), recruitment and predation were found to be the most important locally organizing factors. Recruitment strategies and patterns influenced both community structure and long-term trends. Predation by echinoderms and gastropods modified the structure of established communities. The common key predator in the area, *Asterias rubens,* largely influenced community structure by selective predation on the dominant competitors *Ciona intestinalis* and *Mytilus edulis* which, in the absence of predation, appeared to be highly capable of substrate monopolization in various parts of the studied communities, depending on depth and exposure.

The results indicate the existence of large-scale structuring factors, probably linked to climatic events. This is considered as an important pre-requisite for the future recognition and understanding of local deviations, e.g. as results of pollution. This study also pinpoints some species and processes that should be focussed upon in monitoring programs.

Introduction

The rocky subtidal is one of the least studied marine environments with respect to quantitative ecological aspects. This is no doubt largely due to inaccessibility in combination with lack of suitable sampling techniques in the past (Lundälv, 1971). Beginning in 1969, a technique for non-destructive quantitative studies on rocky subtidal communities, based on stereophotographic recordings of fixed sites and photogrammetrical analysis, was developed in Sweden (Lundälv, 1971, 1974, 1985a; Torlegård & Lundälv, 1974). The utilization of this technique permits similar studies subtidally that have since long been carried out in the rocky intertidal and that have contributed substantially to ecological theory (e.g. Connell, 1972; Dayton, 1971; Paine, 1966; Lewis, 1964, 1976, 1977).

For long-term recording of biological change, rocky-bottom epibioses offer a number of interesting possibilities compared to most other biological systems in the sea (Lundälv, 1971, 1974, 1985a; Christie, 1985). Most important of these is perhaps the scope for non-destructive sampling of permanent test areas dominated by sessile organisms, eliminating many of the uncertainties associated with most random sampling programmes. While the best practical opportunities are offered by intertidal communities, due to accessibility and ease of study, their usefulness in long-term monitoring programmes may be questioned on the basis of their position in an extreme environment and exposure to a highly variable physical environment in many parts of the world. In the geographic area of the present study (Skagerrak) random and short-term climatic extremes are likely to have less influence subtidally, thereby facilitating the detection and interpretation of long-term trends. Still, however, certain important questions have to be answered before the usefulness of these communities for monitoring purposes can be considered fully established. One of the questions relates to the representativeness of fixed sites as mirrors of large-scale and long-term events. In a specific study on this problem Lundälv (1985a) concluded that 1.5 m^2 sites over a geographical range of approximately 60 km were indeed closely reflecting large-scale events in a small number of considered species populations. There was little evidence that small-scale random processes were significantly influencing the long-term patterns observed.

Another pre-requisite for the successful interpretation of results yielded by monitoring programmes is a reasonable understanding of natural processes causing change in the studied communities (e.g. Lewis, 1976, 1984). As pointed out by Lewis (1984), some factors of possible significance, such as long-term climatic influences, are difficult to isolate and evaluate within a small-scale national programme. Access to intercalibrated data over a wider geographical range could greatly improve the scope for accurate interpretation. This line of thought forms the basis for a European research co-operation in benthic ecology termed COST 647. This co-operative programme includes a subprojekt directed towards rocky subtidal communities. The longest record of collaboration within this programme as well as before it came into operation is between Norway and Sweden.

In this paper we present some of the data currently available for comparison and covering a geographical range of approximately 300 km between the south of Norway and the Swedish west coast. The basic questions asked are whether or not there were common trends in population dynamics and community regulation over the temporal and geographical scales studied. Further we try to (1) summarize what is currently known about the most important common trends and regulating mechanisms in rocky subtidal communities in the Skagerrak area, (2) identify possible common key-species, which by their presence or absence are capable of determining much of the dynamics in the communities where they exist, and (3) evaluate the implications for the design and utilization of future monitoring programmes.

Material and methods

The basic technique used for data aquisition was photogrammetric analysis of stereophotographs from fixed sites on subtidal rock walls.

Fixed sites were marked by underwater rock-drilling utilizing a small pneumatic drilling machine. Each site normally consisted of a horizontal strip of six consecutive quadrats, each measuring 0.5×0.5 m (making a total of 1.5 m^2). The test areas were defined by two plastic dowels, separated by a horizontal distance of approximately 3 m, from which a rod could be suspended. The rod in turn provided support for a photography frame (0.5×0.5 m) with attached stereo-cameras that could be moved along the rod to six well-defined positions. In Sweden, two synchronized 70 mm Hasselblad SWC cameras equipped with corrective front ports and motor drives were used most of the time. In Norway, two synchronized Nikonos III cameras equipped with 15 mm lenses were used.

The stereophotographs were analysed in a Wild Stereocomparator MSTK (Sweden) or in stereoscopes constructed from two aligned stereomicroscopes (Green, 1980) (Norway and Sweden). Determination of faunal densities were made by careful systematic examination (by aid of grid lines) of the colour transparancies at magnification rates between 12 and 25×.

For estimates of cover (%), point sampling of slides was used. For further studies of processes, *in situ* settlement- and predator/prey experiments were performed in addition to the stereophotographic recordings. More detailed accounts of the methods and instrumentation used are given in Christie (1980, 1983b), Christie *et al.* (1985), Green (1980), Lundälv (1971, 1974, 1985a, b) and Torlegård & Lundälv (1974).

Analysis of correlation between density variations in Norwegian and Swedish populations was performed by means of Spearmans rank correlation coefficient (cf. Siegel, 1956). The analysis was based on pairs of interpolated data in order to eliminate effects of varying sampling times. Four interpolated values per year (weeks 13, 26, 39 and 52) were used. Their derivation was based on the assumption of linear change between actual sampling occasions. Adaption of cyclic variations in *Ciona intestinalis* (L.) and *Echinus esculentus* (L.) to harmonic series was performed according to models described in Bliss (1970).

Characterization of sites studied

The map (Fig. 1) shows the sites studied in the Swedish (S2, G4, G6) and Norwegian (H1, D1, T1, F4, Arö) part of this co-operation. The sites cover in distance a coastline of almost 300 km with additional large fjord systems and archipelagos.

The Swedish sites were recorded each year since 1969, the Norwegian recordings started on one station (F4) in 1976 and between 1978 and 1980 on the others. All sites are on near vertical rocky surfaces, and each station was sampled at different depths between 10 and 25 m. All sites were situated within roughly the same hydrographical regime (c.f. Svansson, 1975). While the deeper Skagerrak waters are of oceanic character, the upper layer is influenced by the brackish Baltic current slowly increasing in salinity while it moves north along the Swedish west coast and further along the Norwegian Skagerrak coast. The archipelagos and fjords provide local gradients of exposure, hydrography and pollution.

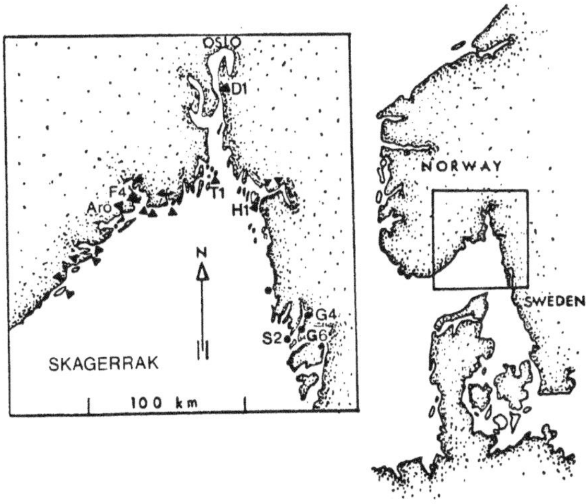

Fig. 1. Map showing localities covered by subtidal rocky-bottom monitoring in the Skagerrak area. The Swedish sites are marked by dots and the Norwegian ones by arrows. The sites referred to here are labelled by their station names.

The rocky subtidal habitats in the area are characterized by a stable surface of free rocky space (or microfoulers) and encrusting algae. Upon this 'surface' species settle and occur with differing patterns of fluctuations (Christie, 1980, 1983b; Lundälv, 1985a; Svane, 1983; Svane & Lundälv, 1981, 1982a, b). Christie (1980) recorded more than 60 species throughout one year in one community. Only a few were abundant or of presumed importance as structuring elements while most belonged to the category 'secondary species' (Lewis, 1976; Christie, 1980). The sites included in this co-operation will totally host much more than 60 species, but of interest for our comparisons there will only be the few species that are: abundant and characteristic, dominating, or of importance in community organization ('key species'). As recommended in the COST 647 core programme (Lundälv, 1985b, COST 647 Activity report), this paper largely deals with ascidians and echinoderm predators.

Results and discussion

The parallel sampling of rocky subtidal communities in Norway and Sweden has resulted in comparable data on population fluctuations. Some examples are given below.

Cyclic fluctuations

Figure 2 shows the fluctuations of *Ciona intestinalis* at four Swedish sites throughout 14 years compared to the one Norwegian site (F4) recorded over the last 6 of these years. The observed data are plotted in Fig. 2A which shows both long-term and

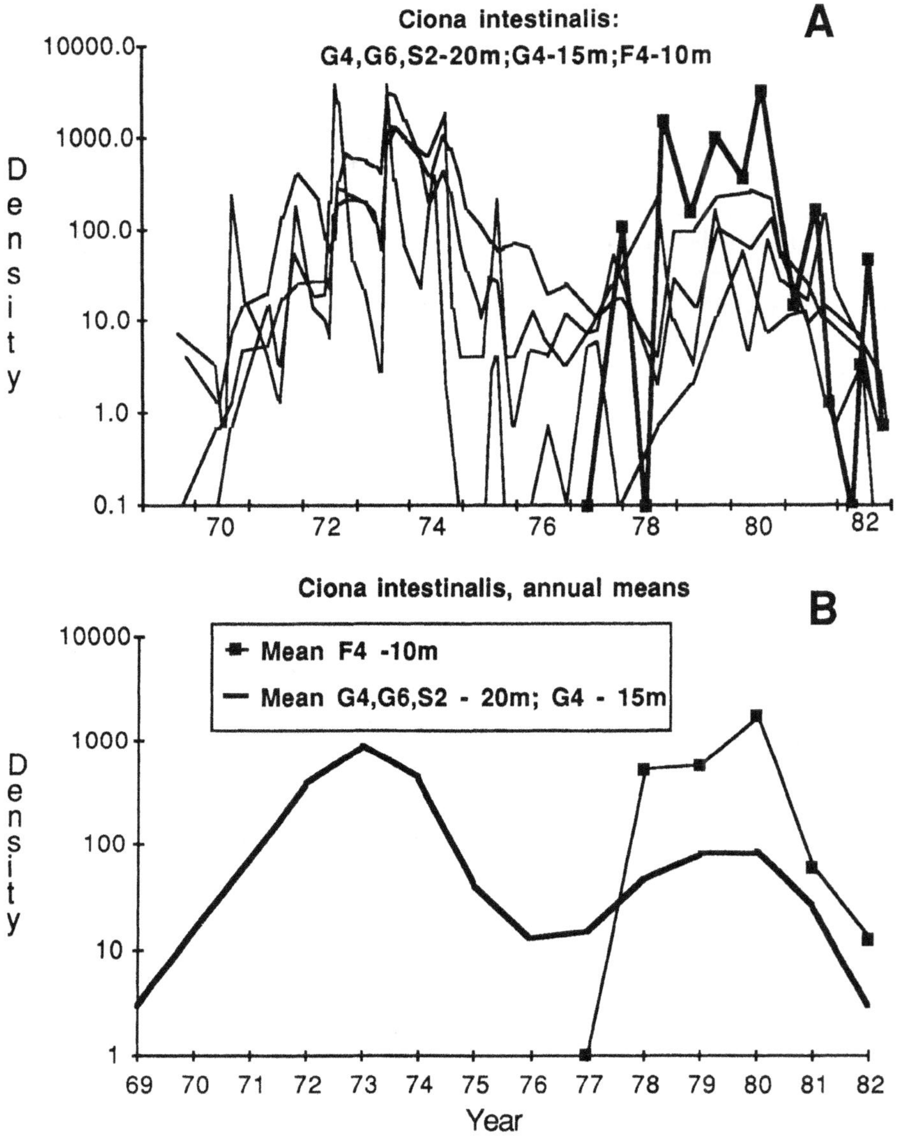

Fig. 2. Fluctuations of *Ciona intestinalis* (ind. m^{-2}). A: Observations plotted from four Swedish and one Norwegian (the bold line) sites. B: Annual means of the observed data.

seasonal fluctuations, while Fig. 2B illustrates the annual means. The observations presented were fluctuating in a cyclic pattern best fitted to a period of 7 years (based on calculations of periodic regression, Bliss, 1970). The amplitudes of the fluctuations are somewhat different between the different sites and between the two different 7-yr periods, but all data (as far as the recordings permit) fits in with a 7 yr cyclic pattern with maximum densities in 1973 and 1980, and minimum values in 1969 and 1976.

A similar (and almost parallel) long-term (7 yr) cyclic pattern was also found for *Echinus esculentus* populations from sites in both countries (Fig. 3). As far as these data permit calculations of harmonic series, they fit in with a common 7 yr cyclic pattern where calculated maximum densities are found to occur between the 1973 and 1974 ob-

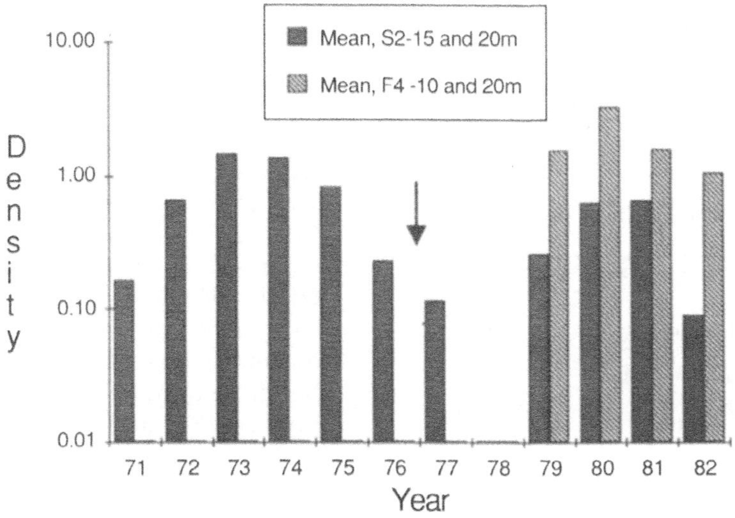

Fig. 3. Annual means (ind. m^{-2}) of *Echinus esculentus* from two depths on the Swedish station S2 and the Norwegian station F4. The arrow indicates the first year of Norwegian recordings.

servations and between 1980 and 1981, and with a calculated minimum value in 1977.

These two species showed comparable cyclic fluctuations over both shorter and longer geographic ranges, and also between different depths and between coastal and fjord sites. Hydrographic cycles of the same (7 yr) and the half (3.5 yr) period (most typical for temperature) have been shown from different stations in the Skagerrak area (Gray & Christie, 1983). Cyclic fluctuations in populations trends are rewiwed by Gray & Christie (1983) and among a variety of cycles, there seems to be several species in the North Atlantic that have been observed to fluctuate in a cyclic pattern with a 7 yr period. It is likely that these cycles were triggered by hydrographical events (see also below). The same cyclic fluctuation of *Echinus* following ½ – 1 yr after the *Ciona* fluctuation could also occur due to the fenomenon that prey organisms (here *Ciona*) attract predator organisms (*Echinus*), as found for the *Mytilus–Asterias* interrelationship from the same area (Christie, 1983a; Lundälv *et al.*, this volume).

The cyclic fluctuations found here are not predictable according to the conclusion of Gray & Christie (1983). We will then not predict the future fluctuations of the two species mentioned, but state that our observed data on the two species do fit to a cyclic fluctuation with a 7 yr period and that the same cyclic pattern was observed all over the geographic range studied.

Long-term trends

In addition to the 7 yr cycle described above, other cycles of longer duration have been suspected to influence population fluctuations along with shorter ones (e.g. Gray & Christie, 1983). Our data series were generally of too short duration to give any indication of cyclic phenomena with a period longer than c. 10 yrs. It is thus, at present, not possible to judge whether trends in our material lasting over the entire observed period are part of long-term cycles or represent true uni-directional trends in response to some 'permanently' (e.g. caused by anthropogenic activity) changing environmental factors.

Most of the characteristic species (dominant in number or individual size) compared between our rocky Skagerrak sites seemed to fluctuate along roughly common trends. Comparisons of Norwegian and Swedish data for the 5 species *Asterias rubens* (L.), *Ascidia mentula* (O.F. Müller), *Sabella penicillus* (L.), *Crania anomala* (O.F. Müller), and *Protanthea simplex* (Carlgren) are shown in Fig. 4. For the first two of these species (Fig. 4A, B) a

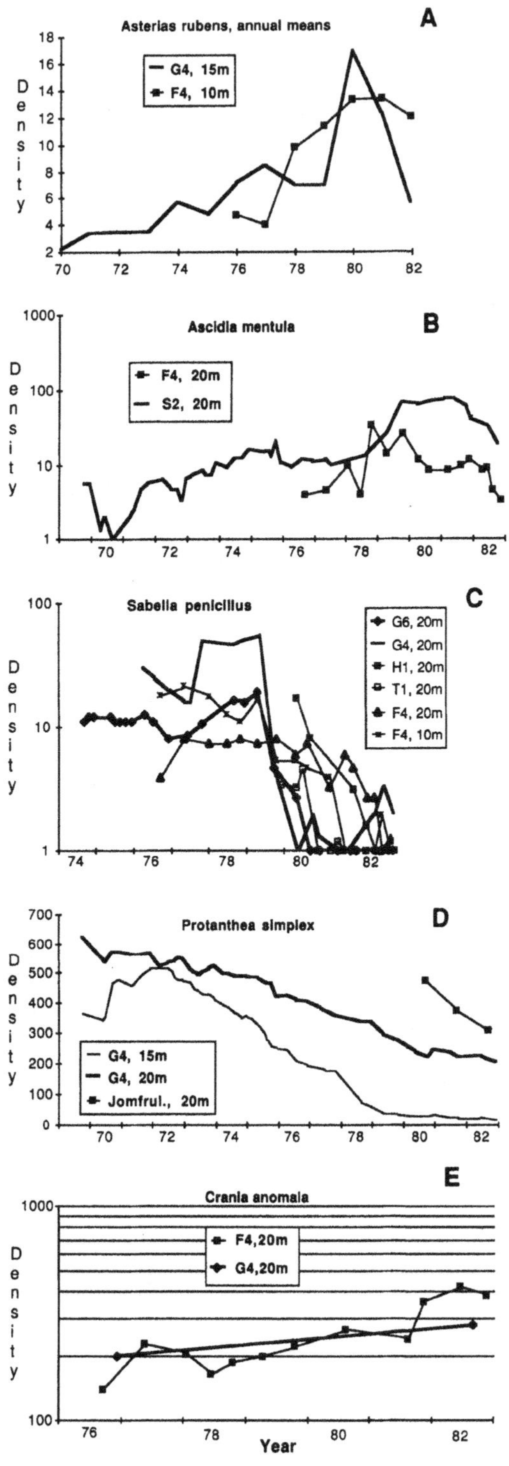

Fig. 4. Comparisons of long-term trends (ind. m^{-2}) in 5 species from different Swedish (G4, G6, S2) and Norwegian (F4, H1, T1, Jomfrul. = Arö) sites.

Spearmans' rank correlation analysis yielded significant correlation between Norwegian and Swedish data ($p < 0.02$), although the fit was not perfect. In the case of the sedentary polychaete *Sabella penicillus* (Fig. 4C), all sites compared (4 Norwegian and 2 Swedish) exhibited a marked drop in densities down to virtual extinction over the period 1979–82. The decline came somewhat later in Norwegian sites as compared to the two Swedish sites (G4 and G6, 20 m).

The anthozoan *Protanthea simplex* was one species exhibiting a very marked long-term trend of decline in the Swedish sites (Fig. 4D). This trend also applied to three additional sites not illustrated here (c.f. Lundälv, 1985a). Norwegian data pertaining to this species only covered the period 1980–82, but still fitted a possible trend of decline.

The brachiopod *Crania anomala* was one species that exhibited a relatively steady trend of increase in the available Norwegian data over the period 1976–82 (Fig. 4E). A Swedish analysis of a *Crania*-population at the beginning and end of this period may also fit with a similar trend of increase.

When considered together, the observations reported above strongly indicate that most major trends in population fluctuations did occur in a roughly synchronized pattern over the entire geographic range studied. The same conclusion was reached on a smaller geographic scale (c. 62 km), but based on considerably longer time-series of data, after a comparison of population fluctuations in five different species between varying numbers (2–7) of the Swedish sites (Lundälv, 1985a).

Comparisons of long-term population fluctuations over geographic ranges have rarely been made in benthic communities. A few examples of species where common trends seem to occur over a geographic range have been reviewed by Gray & Christie (1983). Short-term population fluctuations have been recorded much more often. The importance of access to data over a wide geographic range, before interpreting such data in relation to local pollutants, has been demonstrated by Bowman (1978).

In addition to the populations treated above, and showing varying degrees of co-variation, it should be noted that there were also a few populations compared that did not adapt well to a common pattern. One example was the *Ciona*-population at F4, 20 m (not illustrated), which did not fall into the cyclic pattern exhibited by all other *Ciona*-

populations illustrated here. The same was true for a Swedish *Ciona*-population at station G6, 25 m, and it may well be that these deviations were associated with the vertical separation of different hydrographic regimes, which may also be different between Norwegian and Swedish sites. We have not had access to data, however, permitting a detailed hydrographic characterization of all studied sites.

Physical factors

Since our studied sites were situated in an area largely influenced by the same broadscale climatic and hydrographical conditions, it is reasonable to relate the population fluctuations to the physical factors. Figure 5 shows temperature observations from one Swedish and one Norwegian site flanking the whole area of studies. The temperature data from both countries shows a similar pattern, e.g. a period of high temperatures during the mid of the 1970's, and several smaller comparable peaks and depressions. Gray & Christie (1983) have listed various broad-scale shorter- and longer-term hydrographic fluctuations, also based on the Norwegian salinity and temperature data, and concluded that they generally are reflected in population fluctuations. Svane (1984) has found a relation between temperature and recruitment in one ascidian species. This gives an indication of how temperature fluctuation may cause a population fluctuation,

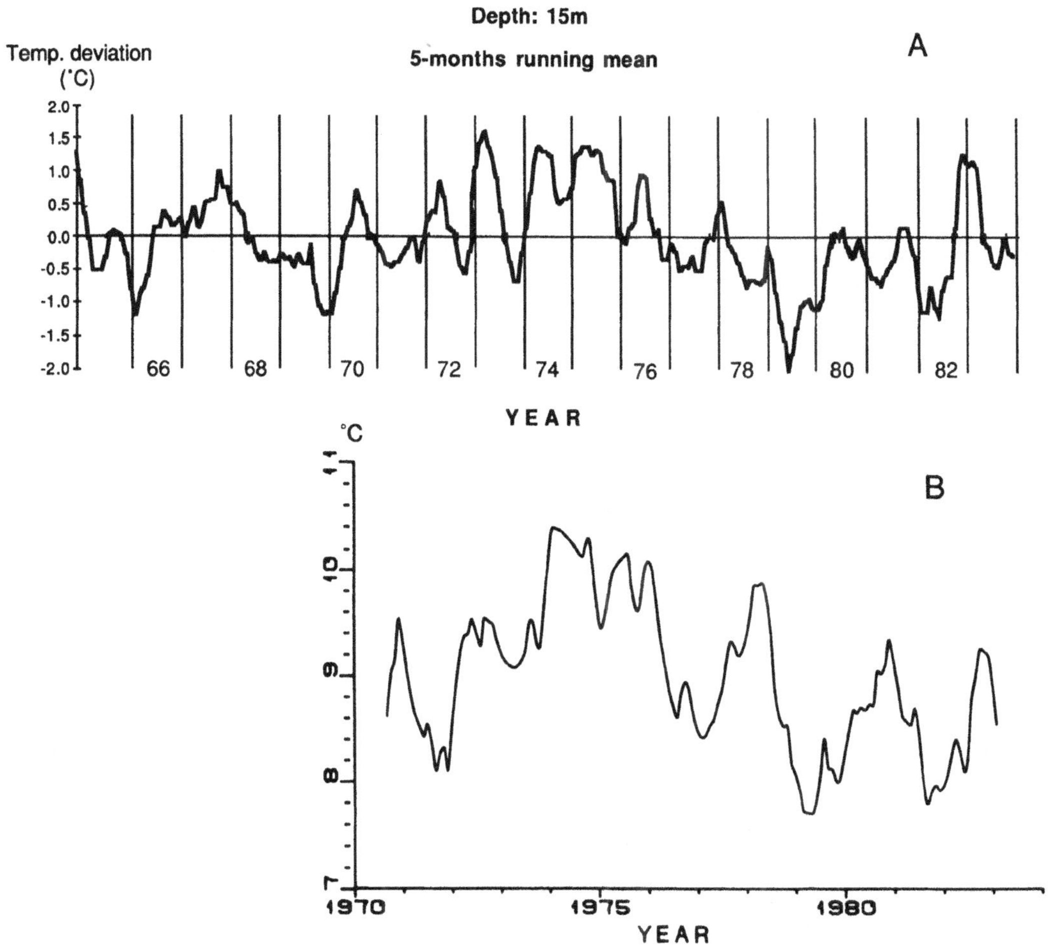

Fig. 5. Mean values of long-term temperature (°C) registrations from the Swedish (upper curve) and Norwegian (lower curve) area. Swedish data from 15 m depth at Bornö Hydrographical Station, Gullmarfjord (Courtesy Fishery Board of Sweden). Norwegian data from 10 m at Torungen station (provided by the Norwegian Oceanographic Data Centre).

but still our knowledge is sparce concerning these interrelationships.

It is likely that the Swedish sites were influenced by less saline waters than the Norwegian ones. This could be one reason why some representative (dense) populations of the subtidal species recorded at the 15–20 m level in Sweden were best reflected at the 10 m level in Norway. Physical factors also affect community structure. By classification Christie (1985) showed a similarity in community structure between all coastal sites on 10 and 20 m depth. These were clustered apart from the fjord sites. Swedish studies also indicate a similar pattern in population dynamical features (Lundälv, 1985a; Svane, 1984).

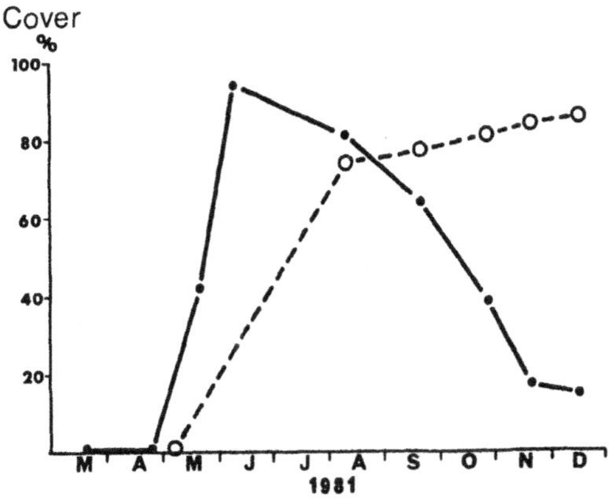

Fig. 6. Total percent cover of organisms on settlement panels. Solid line: control panels. Dotted line: predator exclusion panels (From Christie, 1983b).

Community organizing mechanisms

In addition to the long-term fluctuations presented here, our data show considerable short-term (mostly seasonal) fluctuations (e.g. Fig. 2A). Christie (1983b) found that most species (on station D1) showed regular (predictable) seasonal cycles organized by a seasonal recruitment followed by predation or grazing. Judging from Swedish and Norwegian data, recruitment and predation were the most important structuring factors in the rocky subtidal communities in the area (Christie, 1980, 1983a, b; Svane, 1983, 1984; Svane & Lundälv, 1981, 1982a).

By regular community monitoring and by the use of immersed artificial settling panels, settlement was found to take place all through the year, but the recruitment of significance for community structure took place only during a few months in the summer-autumn season (Christie, 1980, 1983b; Svane & Lundälv, 1981). The structuring impact of a heavy seasonal settlement is obvious and might be illustrated by the increasing parts of the peaks in Fig. 2A. Settlement has been considered as an important factor in controlling rocky community development and fluctuations (Lewis, 1964; Sutherland & Karlson, 1977). The rapid decrease of most of the peaks in Fig. 2A was found to be a result of predation. How the predator *Asterias rubens* controls populations of dominant species has been documented in both the Norwegian and Swedish studies. Figure 6 illustrates the results of one of our experiments (Christie, 1983b). The curves show occurrence of organisms on panels exposed to normal densities of benthic predators compared to panels where benthic predators were excluded. After an intense period of recruitment, *Asterias rubens* entered the predator-exposed panels and preyed down the sessile organisms from an occurrence of almost 100% cover to approximately 20%. On the predator exclusion panels organism occurrence increased, and at the end of the experiment the panels were totally dominated by *Mytilus edulis* (L.). A similar experiment was described by Lundälv (1985b, COST 647 Activity report). Panels protected from *Asterias* predation were totally dominated by *Ciona,* while on panels exposed to normal *Asterias*-predation *Ciona* never reached very high numbers and more species co-existed.

Data from Sweden also illustrate *Asterias* predation controlling *Mytilus* dominance. Lundälv *et al.* (this volume) recorded a sudden increase in *Mytilus* populations on the smallest depths (5 and 10 m) in one of their localities leading to total *Mytilus*-dominance by the end of the 1970's. The increase in *Mytilus* was followed by an increase in *Asterias* that preyed on the *Mytilus*-population and eliminated most of it during the early 1980's. The identical phenomenon was observed in shallow waters along the Norwegian Skagerrak coast, but was not quantified further than recording an identical trend of *Asterias* as shown in Fig. 4A. The same trend of *Asterias* has also been shown from the Norwegian fjord site D1 during the same years (Christie, 1983b).

These results illustrate that *Asterias* all over the

area is the key predator preventing monopolization of the substrate by the two competitively dominant species *Mytilus* and *Ciona*. The fact that dense *Mytilus*-populations attract *Asterias* was shown experimentally by Christie (1983a), and aggregations of *Asterias* are known to cause heavy predator impact on mussel beds (e.g. Sloan & Aldridge, 1981). Data on the importance of *Asterias*-predation on *Ciona* has also been provided by Gulliksen & Skjaeveland (1973). The phenomenon of echinoderm key predators structuring rocky communities by providing space for diverse recruitment when preying down dominant competitors has been widely documented (e.g. Paine, 1966; Karlson, 1978; Peterson, 1979).

Of lesser significance than *Asterias*, but also of common importance in structuring the rocky subtidal communities of the Skagerrak are nudibranchs (*Coryphella* sp) and sea urchins (*Echinus esculentus, Psamechinus miliaris Gmelin) (Christie, 1980, 1983b; Hernroth & Gröndahl, 1985; Lundälv, 1985a).*

Conclusions

– The species of importance in dominance, abundance or functional roles were mainly the same all over the studied Swedish and Norwegian Skagerrak rocky subtidal communities.

– A similar (and parallel) 7 year cyclic pattern was illustrated for the species *Ciona intestinalis* and *Echinus esculentus*.

– Significantly correlated or otherwise similar long-term curved or linear trends in population fluctuations were observed in a number of abundant or otherwise important species. In cases where a species could be recorded representatively, the common trends were often found in all sites over the geographic gradient, fjord gradients and depth gradients.

– Community structure all over the studied area seemed to be organized by the same regulating mechanisms. Both seasonal and long-term fluctuations of different populations were structured by common patterns in settlement/recruitment and, in some cases, a following predation by the same species of predators.

– The potentially dominant faunal competitors of the area, *Ciona intestinalis* and *Mytilus edulis*, were effectively controlled by the common key predator, *Asterias rubens*.

– Common patterns in hydrographic conditions and fluctuations, as well as large-scale climatic events must not be overlooked as important initiators of the trends.

– The results provided by the collaboration inside COST 647 has strongly contributed to increase the ecological understanding of communities along our coast and also inside the fjords.

– The existence of large-scale structuring factors and trends is an important pre-requisite for a recognition and understanding of local deviations e.g. as a result of pollution. The empirical experiences from this study pinpoint species and processes that should be carefully looked for in a monitoring context. Then a large-scale study (or collaboration) should be a great advantage for objective explanations of reasons behind strange fluctuations or patterns than might be observed in our coastal waters, influenced by increasing eutrophication, or in the polluted fjords of the Skagerrak area.

Acknowledgements

We give special thanks to Norman Green (Norway) and to A. Holme, C. Larsson, K. Lindahl and I. Svane (Sweden) for years of cooperation in underwater sampling and analysis of biological data. Valuable suggestions were provided by Professor John S. Gray and Professor J.-O. Strömberg. We also thank the Norwegian Oceanographic Data Centre (NOD) and the Swedish Fishery Board for access to hydrographic data from the Skagerrak. Financial support for the Swedish studies was provided through many research contracts with the National Swedish Environment Protection Board and the Swedish Natural Science Research Council, and for the Norwegian studies by The Norwegian Marine Pollution Research and Monitoring Programme (FOH), which is gratefully acknowledged.

References

Bliss, C. I., 1970. Statistics in biology Vol. 2, McGraw-Hill, New York, 639 pp.

Bowman, R. S., 1978. Dounreay oil spill: major implications of a minor incident. Mar. Pollut. Bull. 9: 269–273.

Christie, H., 1980. Methods for ecological monitoring: Biological interactions in a rocky subtidal community. Helgoländer Meeresunters. 33: 473–483.

Christie, H., 1983a. Use of video in remote studies of rocky subtidal community interactions. Sarsia, 68: 191–194.

Christie, H., 1983b. Natural fluctuations in a rocky subtidal community in the Oslofjord (Norway). Oceanol. Acta, Proceedings 17th EMBS, 69–73.

Christie, H., 1985. Ecological monitoring strategy with special reference to a rocky subtidal programme. Mar. Pollut. Bull. 16: 232–235.

Christie, H., R. A. Evans & O. K. Sandnes, 1985. Field methods for *in situ* subtidal hard bottom studies. In George & Lythgoe (eds), Underwater Photography and television for scientists, Oxford University Press, pp. 37–47.

Connell, J. H., 1972. Community interactions on marine rocky intertidal shores. Ann. Rev. Ecol. Syst. 3: 169–192.

Dayton, P. K., 1971. Competition, disturbance, and community organization: The provision and subsequent utilization of space in a rocky intertidal community. Ecol. Monogr. 41: 351–389.

Gray, J. S. & H. Christie, 1983. Predicting long-term changes in marine benthic communities. Mar. Ecol. Prog. Ser. 13: 87–94.

Green, N. W., 1980. Underwater stereophotography applied in ecological monitoring. Report 1. Methods and preliminary evaluation. Norwegian Institute for Water Research, Oslo, Report OF-80613: 1–99.

Gulliksen, B. & S. H. Skjaeveland, 1973. The sea-star *Asterias rubens* (L.) as predator on the ascidian *Ciona intestinalis* (L.) in Borgenfjorden, North-Tröndelag, Norway. Sarsia 52: 15–20.

Hernroth, H. & F. Gröndahl, 1985. On the biology of *Aurelia aurita* (L.). 3. Predation by *Coryphella verrucosa* (Gastropoda, Opistobranchia), a major factor regulating the development of *Aurelia* populations in the Gullmar Fjord, western Sweden. Ophelia 24: 37–45.

Karlson, R., 1978. Predation and space utilization patterns in a marine epifaunal community. J. exp. mar. Biol. Ecol. 31: 225–239.

Lewis, J. R., 1964. The ecology of rocky shores. Engl. Univ. Press, London, 323 pp.

Lewis, J. R., 1976. Long-term ecological surveillance: Practical realities in the rocky littoral. Oceanogr. Mar. Biol. Ann. Rev. 14: 371–390.

Lewis, J. R., 1977. The role of physical and biological factors in the distribution and stability of rocky shore communities. In: Keegan, O'Ceidigh & Boaden (eds), Biology of Benthic Organisms, Pergamon Press, Oxford: 417–424.

Lewis, J. R., 1984. Temporal and spatial changes in benthic communities: COST 47 approach. Mar. Pollut. Bull. 15: 397–402.

Lundälv, T., 1971. Quantitative studies on rocky-bottom biocoenoses by underwater photogrammetry. A methodological study. Thalassia Jugosl. 7: 201–208.

Lundälv, T., 1974. Undervattensfotogrammetri – nytt hjälpmedel for biodynamiska studier i marin miljo. Svensk Naturvetenskap, Årsbok 1974, 13–20 (Sw., eng.).

Lundälv, T., 1985a. Detection of long-term trends in rocky sublittoral communities: Representativeness of fixed sites. In Moore & Seed (eds), The ecology of rocky coasts. Hodder & Stoughton Educational, U.K., 329–345.

Lundälv, T., 1985b. Rocky Subtidal Programme. In: COST 47, Coastal benthic ecology. Report on the period 1977–1984, Annex 2: 1–27. Commission of the European Communities, Brussels, 89 pp.

Lundälv, T., C. S. Larsson & L. Axelsson, 1987. Long-term trends in algal-dominated rocky subtidal communities on the Swedish west coast – a transitional system? Hydrobiologia: this volume.

Paine, R. T., 1966. Food web complexity and species diversity. Am. Nat. 100: 65–75.

Peterson, C. H., 1979. The importance of predation and competition in organizing the intertidal epifaunal communities of Barnegat Inlet, New Jersey. Oecologia 39: 1–24.

Siegel, S., 1956. Nonparametric statistics for the behavioural sciences. Kosaido Printings, Tokyo, 312 pp.

Sloan, N. A. & T. H. Aldridge, 1981. Observations on an aggregation of the starfish *Asterias rubens* (L.) in Morecomb Bay, Lancashire, England. J. Nat. Hist. 15: 407–418.

Sutherland, J. P. & R. H. Karlson, 1977. Development and stability of the fouling community at Beaufort, North Carolina. Ecol. Monogr. 47: 425–446.

Svane, I., 1983. Ascidian reproductive patterns related to long-term population dynamics. Sarsia 68: 249–255.

Svane, I., 1984. Observations on the long-term population dynamics of the perennial ascidian *Ascidia mentula* (O.F. Müller) on the Swedish west coast. Biol. Bull. 167: 630–646.

Svane, I. & T. Lundälv, 1981. Reproductive patterns and population dynamics of *Ascidia mentula* (O.F. Müller) on the Swedish west coast. J. exp. mar. Biol. Ecol. 50: 163–182.

Svane, I. & T. Lundälv, 1982a. Population dynamics and reproductive patterns of *Boltenia echinata* (Ascidiacea) on the Swedish west coast. Neth. J. Sea Res. 16: 105–118.

Svane, I. & T. Lundälv, 1982b. Persistence stability in ascidian populations: Long-term population dynamics and reproductive pattern of *Pyura tessellata* (Forbes) in Gullmarfjorden on the Swedish west coast. Sarsia, 67: 249–257.

Svansson, A., 1975. Physical and chemical oceanography of the Skagerrak and the Kattegatt. I. Open sea conditions. Report 1, Fish. Bd. Sweden, Inst. Mar. Res.: 1–88.

Torlegård, K. & T. Lundälv, 1974. Underwater analytical system. Photogramm. Engn. 40: 287–293.

Hydrobiologia 142: 81–95 (1986)

Long-term trends in algal-dominated rocky subtidal communities on the Swedish west coast – a transitional system?

Tomas Lundälv, Christer S. Larsson & Lennart Axelsson
Kristineberg Marine Biological Station, S-450 34 Fiskebäckskil, Sweden

Keywords: rocky subtidal communities, algae, eutrophication, long-term change, predation, Skagerrak

Abstract

Rocky subtidal communities on the Swedish west coast were monitored over approximately 16 years (1970–1985) by means of stereophotographic recordings, at intervals, of fixed sites.

Dominating elements of community structure were studied at the 5- and 10 m depth levels of an outer archipelago locality. Excluding seasonal variations, the basic community structure, largely dominated by the perennial phaeophycean *Halidrys siliquosa* (L.), remained relatively stable over the first 7–8 years of study, but with an increasing trend in cover of filamentous epiphytic red algae. Later, the structure of the studied communities changed radically. The changes initially involved intense settling of the blue mussel *Mytilus edulis* L. and reduced cover of red algae. The mussels settled largely on *Halidrys* plants which eventually broke down almost completely. Later, there was intensive predation on *Mytilus* by the starfish *Asterias rubens* Linné. After the elimination of *Mytilus,* the studied test areas were dominated by filamentous ephemeral red algae for the rest of the study period, but with some regeneration of *Halidrys* at the 5 m level.

The observed changes were thought to be indicative of a major disturbance in the coastal ecosystems off the Swedish west coast during the studied period. This notion is discussed in relation to studies of other components of the ecosystem during the same time period and in relation to possible causes. Among the latter may be mentioned a suspected eutrophication process, long-term temperature anomalies and resulting effects on biological interrelationships such as competition, predation and chemical defense.

The observed trends included a shift in dominance from slow-growing perennial algae to fast-growing ephemeral algae. If this shift was a general phenomenon, for which there is some circumstantial evidence, it could have important repercussions on the cycling of nutrients and organic matter in the coastal ecosystem. Possible effects are discussed on hypothetical terms in relation to observed anomalies in phytoplankton blooms and benthic oxygenation.

Introduction

The increasing awareness that man's activities may be causing large-scale and long-term changes in the marine environment has led to an increasing interest in the study of long-term biological changes in the sea over the last decades (e.g. Hartnoll & Hawkins, 1980; Lewis, 1976, 1984). Still, however, the number of long-term (>5 years) studies are few (e.g. Lewis, 1976, 1980) and have largely been directed towards a limited number of biological systems, such as fish populations (e.g. Cushing, 1975), sediment communities (e.g. Beukema *et al.,* 1983; Buchanan *et al.,* 1978; Josefson, 1985) and plankton communities (e.g. Colebrook, 1982). This selection is probably related to the availability of standard techniques for quantitative assessment as well as to economic importance (fish).

The subject of this study, rocky subtidal communities, belongs to the ecologically least studied biological subsystems within the continental shelf realm. In contrast, intertidal epibenthic communities have been the subject of many important contributions in marine ecology (e.g. Bowman & Lewis, 1977; Connell, 1972; Dayton, 1971; Lewis, 1964, 1976; Paine, 1966). They have also been suggested as suitable for long-term monitoring purposes (e.g. Lewis, 1976). Positive features include accessibility, possibilities of field experimentation (e.g. Connell, 1974) and the use of non-destructive sampling techniques (e.g. Hawkins & Hartnoll, 1983). They also have the advantage of including both primary producers and a number of higher trophic levels. It may still be argued, however, that rocky intertidal communities are unsuitable for such studies in many parts of the world, since they may be heavily affected by various short-term natural influences that could easily conceal man-made chronic effects (c.f. Lewis, 1980). In the area of the present study (Swedish west coast), examples of such influences are occassional ice-scoring during the winter, heavy rains or storms and long periods of low water during hot or cold spells associated with high air pressure. Hawkins & Hartnoll (1983) also argued that intertidal test areas of manageable size are often not representative for general phenomena, although this conclusion may be challenged on the basis of their relatively short period of observation (Lundälv, 1985).

The effects of natural short-term perturbations could be expected to decrease with increasing depth, thus rendering subtidal habitats more suitable for long-term monitoring purposes. Utilizing a technique based on stereophotography of fixed sites and photogrammetric analysis of biological processes (Lundälv, 1971, 1974, 1985; Torlegård & Lundälv, 1974) the possibilities for studies in the rocky intertidal can largely be transferred to the subtidal. Experience to date also indicates that fixed sites of the type employed in this study reflect large-scale processes reliably (Lundälv, 1985; Lundälv & Christie, this volume).

At the Swedish west coast, monitoring of rocky subtidal communities by underwater photogrammetry was started in 1969. Some of the more dramatic examples of structural changes in the studied sites over the period 1969–85 relates to communities dominated by algae in relatively shallow water (5–10 m). In this paper we describe these changes in one locality thought to be representative for the open coast system. The changes are discussed in relation to possible causes and to possible interactions with other parts of the coastal ecosystem showing signs of abnormal conditions simultaneously. Similar changes in communities dominated by macroalgae have earlier been described from the Oslo Fjord (Rueness, 1973) and the Baltic (e.g. Kangas *et al.*, 1982) and have variously been ascribed to eutrophication (op. cit.), hydrographical changes (Hällfors *et al.*, 1984) or simply 'pollution' (Rueness, 1973).

Materials and methods

Sampling design

The basic sampling principle, employed in this study, was stereophotographic recording of fixed sites, followed by photogrammetric analysis (for detailed accounts see Lundälv, 1971, 1974, 1985; Torlegård & Lundälv, 1974). Test areas were marked by underwater pneumatic rockdrilling and plugging with plastic dowels into near-vertical subtidal rockfaces. One site was defined by two dowels separated by a horizontal distance of c. 3 m. The dowels formed a support for a long rod with 6 fixed positions for a photography frame, measuring 0.5×0.5 m ($0.25\ m^2$). The photography frame, in turn, held a support for underwater stereocameras and strobes, the design of which have varied over the years covered by this study. Originally, a single Hasselblad SWC camera was moved between two well-defined positions on the support in order to obtain a stereoscopic pair of photographs. From 1977, a motorized and synchronized pair of Hasselblad cameras were employed and from 1985 a synchronized pair of Nikonos IV cameras fitted with 15 mm in-water lenses were employed. By moving the frame with attached stereocameras along the supportive rod, a total of six well-defined test quadrats, covering an area of $1.5\ m^2$, were recorded on each site and sampling occasion.

Sites were marked every fifth meter between 5 m and 25 m depth (to an accuracy of 0.5 m) at the locality considered here. The sites were revisited and recorded photographically at intervals over the time period late 1969–1985. The frequency of recordings varied and was generally higher during the period 1970–75 (4–7 times per year) than during the period 1976–85 (2–3 times per year). This account covers the depths 5–15 m.

Analyzing techniques

The stereophotographs were analyzed in a Wild Microstereocomparator, MSTK, or in stereoscopes constructed from two aligned stereomicroscopes and equipped with a co-ordinate table for synchronized movement and analysis of the two stereophotographs from each test quadrat. A variety of analyzing techniques can be used (Lundälv, 1971, 1985; Torlegård & Lundälv, 1974). This study was primarily based on estimates of cover (%) obtained by a point-sampling technique. Plots of 100 systematically dispersed points on a transparent film were imposed over each 0.25 m^2 quadrat and the photographs were analyzed at magnification rates of 12–25×. Organisms penetrated by a vertical axis through the points were registered, and in this way 600 points from each site (1.5 m^2) were analyzed on each sampling occasion. Cover was estimated as the percentage of these points that penetrated a certain species. Cover was estimated at three different levels: (1) The first level was primary substrate occupance and related to portions of the organisms attached directly to primary (dead) substrates. (2) The second level was secondary substrate occupance and related to portions of the organisms attached to other living organisms. (3) The third level was denoted as canopy and related to portions of the organisms not in contact with the substrate that were penetrated by a point. In the results presented here, these three levels were added. Total estimates of cover in excess of 100% were therefore normal.

Most workers employing point-sampling techniques have used plots of random points (e.g. Sutherland, 1974). By comparison of random and systematic plots, we found that 100 systematic points gave approximately equivalent results to 300 random points in 10 different tests, the reason being that randomization introduces variation. Care has to be taken, however, that investigated organisms are not distributed in a pattern similar to the systematic plot (cf. also Green, 1980). By systematic pointsampling of test-figures with known cover, we empirically estimated the error in our sampling technique to be less than 3%. This relates to relative figures obtained by one person for well-defined objects. It must be anticipated that differences between persons could be greater due to differential basic judgements.

Characterization of the studied sites

The locality (Station S1) mainly considered in this paper is situated in an exposed group of islands (Väderöarna) outside the northern part of the Swedish west coast (Fig. 1). There were no significant domestic or industrial effluents within a range of more than 40 km and the locality should be representative for the general conditions in the outer archipelago. The sites were positioned on a steep rock wall facing SSE on the island St. Sundskär. Some shelter from the open sea and predominant SW- and W-winds was provided by islands to the west and southwest. The locality could still be characterized as highly exposed, especially to winds and swell from the south and southwest.

The area is characterized by vertical stratification of hydrographical properties. The mean depth of the main halocline is between 10- and 15 m but the depth of the halocline is highly variable. The water above the halocline is influenced by run-off from the Baltic through the Baltic current (variable salinity $< 30^0/_{00}$) following the Swedish west coast to the north. The water below the halocline ('Skagerrak water') originates from the Skagerrak and ultimately the North Sea (salinity $31-35^0/_{00}$). At the locality considered here, the surface water can also at times be directly influenced from the Skagerrak and the North Sea through the Jutland current, originating in the North Sea. Sea surface temperatures normally range between −1 °C – +21 °C. The range of variation is continually dampened with increasing depth.

Results

5 m-depth level

In Fig. 2, changes in cover (%) over a 15-year period are shown for four dominating components of the community at St. S1, 5 m. Over a long period of time (1970–77) this site was largely dominated by the perennial phaeophycean *Halidrys siliquosa* (L.), with a cover varying between 50–80% (Fig. 2A). Short-term variations over this period can be explained partly by small seasonal variations, but also by the fact that individual plants close to the test-area borders can be variously positioned inside or outside the test areas due to ran-

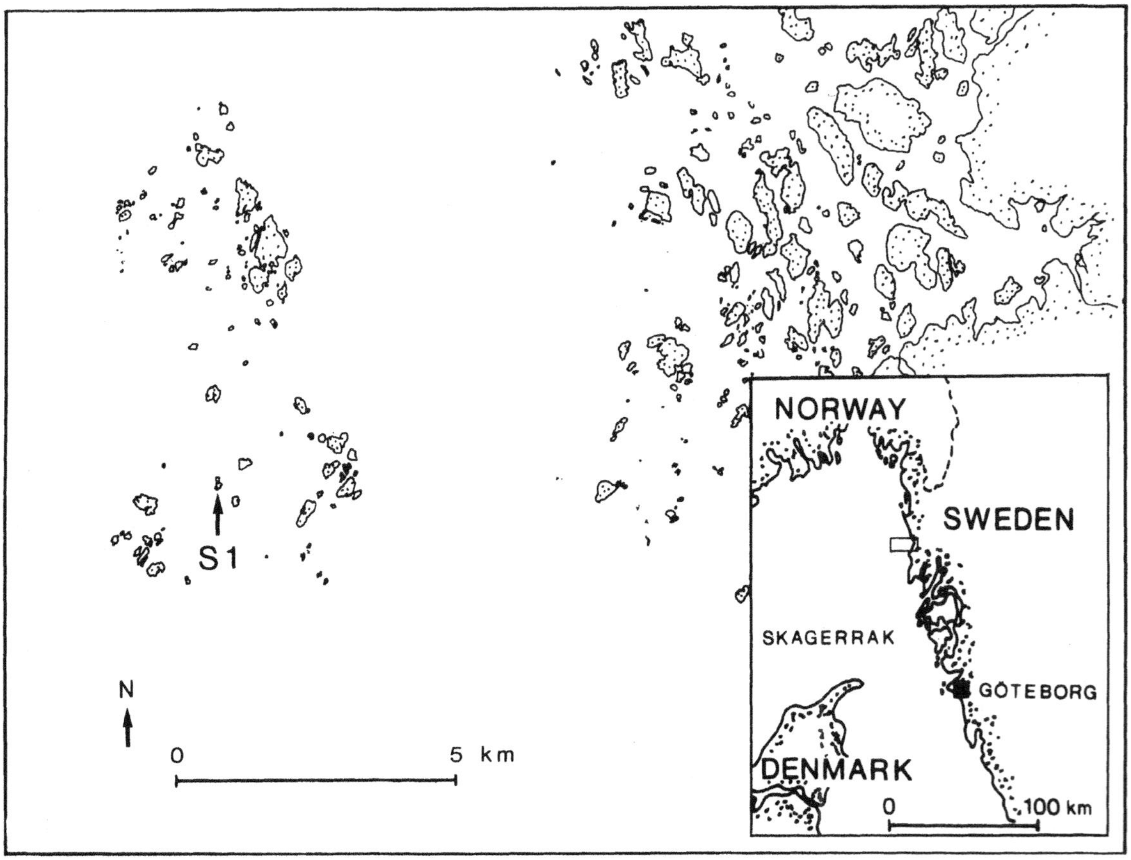

Fig. 1. Map showing the position of the studied locality (Station S1) on the island St. Sundskär in the island group Väderöarna. Inset: Map showing the position of the study area (rectangle) on the northern part of the Swedish west coast.

dom swaying caused by water turbulence. Over the period 1970–75 there was a weak trend of increase in *Halidrys.* Over the period 1977–81, there was a general trend of decreasing cover in this algae, with marked drops in 1978 and 1979–80. In 1981, the cover of *Halidrys* reached a minimum of 11% and then showed a steady trend of increase towards July 1985, when 45% cover was measured. The increase in cover was largely brought about by growth and branching of remaining *Halidrys*-individuals in bad shape at the minimum in 1981. Some recolonization of new individuals was also noted, however.

Coinciding with periods of rapid decline in *Halidrys,* high cover of the blue mussel *Mytilus edulis* Linné was registered. At 5 m depth, the first massive settling was registered in December 1977 (70% cover – Fig. 2A). After a decline in *Mytilus*-cover during 1978 and early 1979, a new massive settling was registered in October 1979 (97% cover). The initial intense *Mytilus*-settling was observed prior to any marked declines in *Halidrys*-cover.

The bulk of settled *Mytilus*-individuals were confined to the rock itself and to crust-forming or low-growing algae (especially *Lithothamnion sp.* and *Corallina officinalis* L.), thus forming a continuous layer over large areas of the substrate. The patches often completely surrounded the attachment plates of *Halidrys.* While the *Halidrys* plants themselves were seldom heavily infested by *Mytilus,* it was frequently observed that small portions of the plants supported dense aggregations of *Mytilus.* These portions usually appeared unhealthy and inactive by discoloration in comparison with normal parts of the plants.

Changes in cover at the 5 m-level of a third species, the predatory starfish *Asterias rubens* Linné,

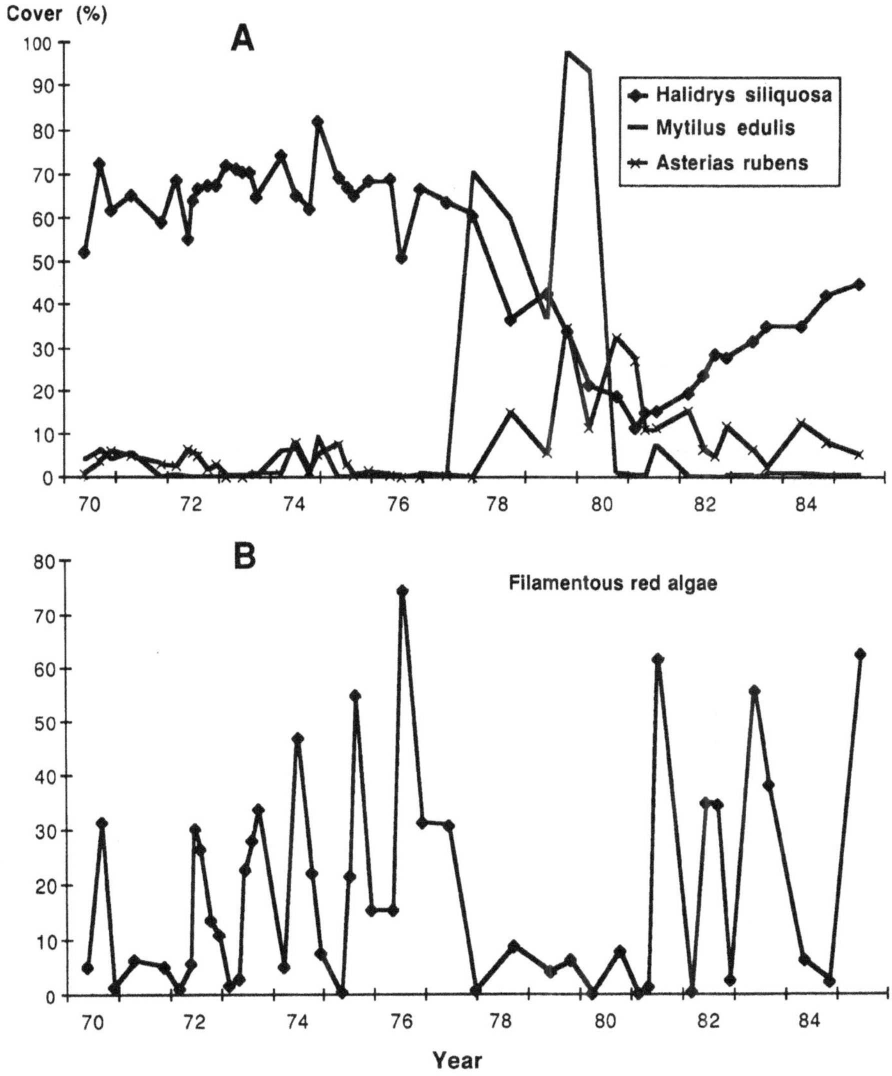

Fig. 2. Variations in cover (%) of dominating organisms at the 5 m depth level of Station S1 over the period 1970–1985.

is also documented in Fig. 2A. The cover of *Asterias* remained below 10% during the time-period prior to intense *Mytilus*-settling. A marked elevation of *Asterias*-cover (15%) was first observed in September 1978. Due to long intervals between photoregistrations at this time, it was not possible to follow the timeschedule of *Asterias* build-up in detail. Available data suggest, however, that there was a time-lag of at least 6 months between the occurrence of dense *Mytilus*-settling and marked elevations of *Asterias* quantities. Excluding one observation in June 1979, *Asterias*-cover remained high (>10%) during the remaining period of high *Mytilus*-cover as well as during a period of more than one year thereafter. The maximal observed *Asterias*-cover of 35% in October 1979 corresponded to a density of at least 235 ind. m^{-2}. Minor peaks in *Asterias*-cover were also observed during the period 1970–75. These peaks seemed to be associated with minor peaks in *Mytilus*-cover, but with a tendency of some retardation. There was, thus, in general, a good correlation between occurrences of *Mytilus* and *Asterias*.

Variations in cover of other quantitatively important floristic components on the 5 m-site are illustrated in Fig. 2B. These data were based on summations of several different species of annual filamentous red algae. They do not include crust-forming, clacareous or other coarse forms. The reasons behind the summation of different species were (1) the difficulty of identifying filamentous red algae to the species level from photographs and (2) our belief that the general type of algae was the most important characteristic in relation to the problems discussed in this paper. It may be noted, however, that *Ceramium spp.* highly dominated the group of unidentified algae. Since the occurrence of these algae was highly seasonal, with marked maxima in late summer, it should be noted that sampling times are critical for their quantification. In years with low frequency of sampling, it must therefore be assumed that peak levels of cover were often overlooked. Certain reliable comparisons could still be made, and especially over the period 1972–76 there was a good seasonal cover of recordings.

The variations in cover of filamentous red algae (Fig. 2B) reveal a few significant patterns. Disregarding seasonal variations, there was a general trend of increasing cover over the period 1971–76, resulting in more than a doubling of the quantities. The red algae occurred both on the primary substrate and, increasingly, as epiphytes on *Halidrys.* During the second half of 1977, and coinciding with massive settling of *Mytilus,* the filamentous red algae were practically eliminated on the 5 m test area and then remained on a low level during the period of high *Mytilus*-cover. In the summer of 1981, the cover of red algae increased again and has remained at a high level up to the present time (1985). It should be noted that the low levels of cover recorded in 1984 could be explained by sampling times outside peak occurrences during the summer.

10 m-depth level

At the 10 m depth level (Fig. 3), the basic variation patterns were similar to those at 5 m but with some noteworthy discrepancies. Initially, *Halidrys* showed a similar, relatively stable, occurrence varying between 45–80% cover and showing a trend of increase over the period 1971–77, interrupted by a short set-back in late 1974 (Fig. 3A). While dramatic reductions in *Halidrys*-cover commenced in 1978 at the 5 m-level, the main decrease was retarded by at least one year at 10 m. Over the period 1979–81 *Halidrys*-cover went down from almost 60% to a few %. Contrary to the situation at 5 m, there was no significant regeneration of *Halidrys* over the period 1981–85.

The occurrence of *Mytilus* at the 10 m-level differed from the 5 m-level mainly by the fact that there was no sign of the intense settling noted in 1977 at 5 m (Fig. 3A). Intense settling, however, was recorded in 1979 (>90% cover) and was again coinciding with a sharp decline in *Halidrys*-cover at this depth. Settling of *Mytilus* on *Halidrys* was notably more intense at the 10 m-level than at 5 m.

The intense settling of *Mytilus* at 10 m was rapidly followed by a marked build-up in *Asterias*-cover. The maximum cover recorded was 36% in October 1980, corresponding to a density of at least 327 ind. m^{-2}. At this time, most *Mytilus* individuals had already been eliminated by *Asterias* predation. High, but declining, figures of *Asterias*-cover were recorded over a year after the virtual elimination of *Mytilus.*

Filamentous red algae at the 10 m-level (Fig. 3B) initially showed a similar trend of increase over the period 1971–76 that was recorded at 5 m. Unlike the 5 m-level, a sharp decline was not recorded until 1979–80, again coinciding with high *Mytilus*-cover. Relatively high cover was reached again over the period 1982–85.

The substrate choice in the group of filamentous red algae is a matter of possible significance. Generally, these algae appeared on a wide variety of substrates such as bare rock, various crust-forming and calcareous algae and epifauna present (e.g. *Balanus spp.* and bryozoans). The dominating phaeophycean, *Halidrys siliquosa,* remained relatively free of epiphytes during the first years studied. However, during the years of maximal red algal cover, especially in 1975–76, massive epiphytic growth of red algae was also noted on this species, both at the 5- and 10 m levels.

15 m-depth level

Variations in cover of filamentous red algae were also recorded at the 15 m depth level (Fig. 4). At this depth, the red algae were quantitatively less important and also largely confined to a shorter period in the summer. There was no clear trend in the

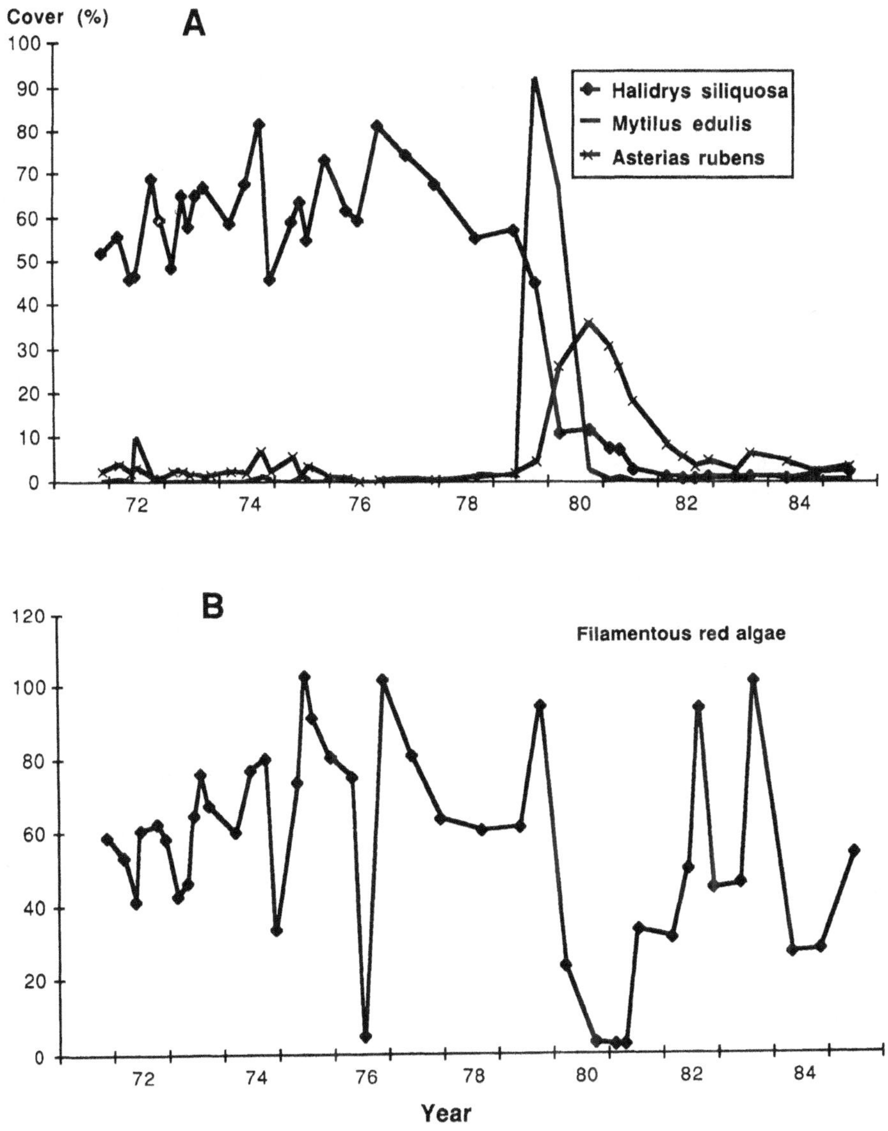

Fig. 3. Variations in cover (%) of dominating organisms at the 10 m depth level of Station S1 over the period 1971–1985.

development of red algal cover during the first 5 years studied, but a maximum was recorded in 1976, largely corresponding to a maximum at the 5- and 10 m levels. Reduced frequency of photoregistrations during the following years limited the possibilities of detailed recording of quantitative change. It still seems clear, however, that red algal cover remained at a very low level over the period 1977–81 and then exhibited a marked trend of increase over the period 1982–85.

Discussion

The results given above were indicative of major structural changes, involving dominants in the studied communities, especially at the 5- and 10 m depth levels, over the studied period. When considering possible causes and repercussions of these changes, it should be noted that anomalies were also observed in other parts of the coastal ecosystem of western Sweden during roughly the same

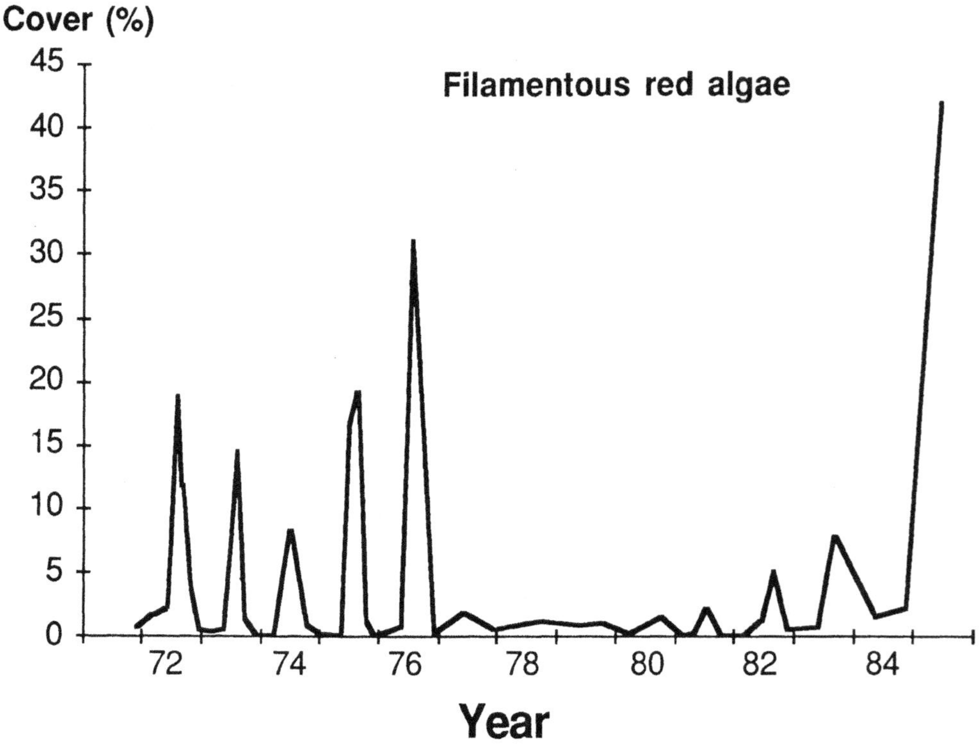

Fig. 4. Variations in cover (%) of filamentous red algae at the 15 m depth level of Station S1 over the period 1971–1985.

period of time. Thus, marked phytoplankton blooms, that were thought to be ab-normal, were observed in the Gullmar Fjord in the years of 1978, 1979 and 1981 (Lindahl & Hernroth, 1983). A similar bloom was observed further south in the Laholm Bight in 1980 (Edler *et al.,* 1982). In connection with these blooms there were reports of varying degrees of adverse effects to marine life. The described effects ranged from severe mortality in benthic organisms due to anoxia in stagnant water bodies (e.g. Josefson, 1981; Rosenberg & Edler, 1981) to mortality in wild and caged fish as well as littoral invertebrates coinciding with discoloration of the water ('red tides') (e.g. Lundälv, 1982). Long-term changes in benthic sediment communities, which were believed to be associated with organic enrichment, have been described by Josefson (1981). A number of reports on phenomena indicative of a general eutrophication process along the Swedish coasts, including the Skagerrak coast, have also appeared (e.g. Rosenberg, 1985).

In the following we discuss, largely on a speculative basis due to lack of pertinent basic knowledge, possible causes for the changes described in this paper as well as possible effects of these changes on other parts of the ecosystem.

Possible causes for the described changes

1. Eutrophication and related effects

There is hardly any direct evidence of eutrophication along the northern Swedish west coast. In partly enclosed water bodies, such as the Gullmar Fjord and the water system inside the islands of Orust and Tjörn, which have been more intensely studied, several pieces of indirect evidence have been reported, however. These include increasing occurrence of oxygen depletion in semi-stagnant water bodies (e.g. Josefson, 1981; Svansson, 1984) and reduced depth penetration in benthic macroalgae (Michanek, 1971). Further evidence has been cited above.

The initial phase of the changes described at the 5- and 10 m depth levels involved increased quantities of annual filamentous red algae, largely as epiphytes on the perennial phaeophycean *Halidrys siliquosa*. Increased quantities of annual filamentous algae, sometimes at the expense of perennial algae, have often been attributed to eutrophication (e.g. Kangas *et al.*, 1982; Larsson, 1984; Rueness, 1973 and Wallentinus, 1983). Most studies have been dealing with annual green- and brown algae in shallow waters, however. The mechanisms involved in such changes have been ascribed to changes in the competitive relationship between annual and perennial algae. When nutrients are highly limiting during the growth season, perennial algae often still have a capacity for growth by utilization of internal reserves of nitrogen (e.g. Chapman & Craigie, 1977). Annual algae, on the other hand, do not have the opportunity of storing nutrients during the winter period. Their capacity for growth is therefore dependent on the availability of nutrients during the growth season, and should increase with increasing nutrient levels. Relationships between annual and perennial seaweeds with respect to nutrient uptakes have been documented by Wallentinus (1984a, b).

The structural changes observed in our communities over the period 1970–76 (Figs. 2 and 3) could well be explained by the processes described above, under the influence of eutrophication. Further support for this hypothesis is given by a tendency of initial increase in *Halidrys*-cover, since nutrient enrichment during the growth season has been found to increase the capacity for growth also in perennial algae (Chapman & Craigie, 1977).

The development on our sites after 1976 could hardly be fully explained as direct effects of eutrophication, but possibly as a consequence. Part of the explanation is probably related to defence mechanisms in *Halidrys* against epibionts. Two different defense-mechanisms have been described in the literature: (1) Actively growing parts of *Halidrys* have been shown to continuously cast-off outer layers of cell walls, thereby also removing newly settled epibionts (Moss, 1982). (2) Extracts of *Halidrys* have shown antibiotic activity (Hornsey & Hide, 1974), and the likely chemical agents are polyphenols (e.g. Glombitza & Sattler, 1973; Glombitza *et al.*, 1980; Sattler *et al.*, 1977). The efficiency of these mechanisms, and especially the first, are probably related to the vitality of individual plants. If we assume that the increasing epiphytic growth of annuals on *Halidrys* (as a consequence of eutrophication) impaired the vitality of *Halidrys* due to shading, decreased rates of water exchange or direct competition for nutrients, this could explain why intense settling of *Mytilus* became possible and eventually led to the breakdown of *Halidrys*. This was further supported by our observations of discoloration of the parts of *Halidrys* where intense settling of *Mytilus* was observed.

Initial settling of *Mytilus* on *Halidrys* seems to be a common feature. We have examined *Halidrys* plants and found mean densities of newly-settled *Mytilus* ranging between 200–300 individuals per plant. Normally, however, these mussels seem to disappear before they reach sizes above 2–3 mm, possibly as a consequence of epidermis cast-off.

One puzzling detail is why *Mytilus*-settling appeared two years later at the 10 m-level than at the 5 m-level. A simple explanation could be that, due to hydrographic stratification, *Mytilus*-spat did not occur in sufficient numbers at 10 m in 1977. Another possibility is that the competitive advantage of annual red algae was less pronounced at the deeper site due to generally lower light levels and higher nutrient levels. This could lead to a slower break-down of defence mechanisms in *Halidrys*.

One feature of the development after the elimination of *Mytilus* in 1980 is hard to explain as a consequence of eutrophication. That is the partial recovery of *Halidrys* on the 5 m site. Only time will show if this was a temporary feature, perhaps related to a high proportion of young and actively growing shoots with efficient antifouling mechanisms during the initial period, or if a complete recovery will eventually be reached. Relatively high quantities of annual red algae during the same period still provide a partial support for the eutrophication hypothesis.

The variations in cover of filamentous red algae at the 15 m site can probably not be directly related to a possible eutrophication process. One reason is that nutrients are probably very seldom limiting at this depth. It is more probable that light is a limiting factor, and that the variations in cover indicate changes in water turbidity. This is in agreement with the fact that frequent plankton blooms were recorded over the period 1978–81 (Lindahl & Hernroth, 1983), when algal cover exhibited marked minima. Phytoplankton quantities have been considerably lower over the period 1982–85 (Lindahl, personal communication), when an increase in red-algal cover was recorded.

2. Changes in hydrographical properties

A possible alternative hypothesis for explaining the structural changes observed is related to long-term temperature anomalies. In Fig. 5, monthly temperature deviations at 5 m over the period 1965–83 are depicted. The data were obtained from daily recordings at the Bornö Hydrographical Station in the inner part of the Gullmar Fjord (courtesy Fishery Board of Sweden, Hydrographical Laboratory). A comparison of the temperature data with the changes observed in the studied sites reveal a few features of possible significance. (1) The first period of increase in red-algal cover (1972–76) coincided with a period of pronounced positive temperature anomalies. Positive anomalies were again recorded in 1982–83, when relatively high red-algal cover was recorded. Hydrographical data were not available after 1983, but it is likely that negative anomalies occurred in 1984–85, which is not in accordance with the relatively high figures of cover recorded in 1985. Alternatively, it could be noted that negative anomalies occurred over the period 1976–80, coinciding with the main decrease in *Halidrys*-cover.

We have no information about the significance of temperature for the considered algal populations. It could also be noted that there was not a very consistent relationship between temperatures and the algal populations, in spite of the observations noted above. We do not consider it a likely factor explaining the observed changes, but are presently not able to completely dismiss it as a factor of possible significance.

3. Population-dynamic processes

A third factor that should be considered is the possible significance of population structure in the studied *Halidrys*-populations. It is conceivable that a homogenous population of ageing *Halidrys*-plants could gradually loose vitality with a resulting weakening of defense-mechanisms against fouling. In the present case, we had no knowledge about the population structure in *Halidrys* at the start of the study. Although this was not studied specifically, there was no indication of intense recruitment during the studied period. The observed fact that most of the regeneration of *Halidrys*-cover over the period 1981–85 was obtained by regrowth of remaining old individuals contradicts this hypothesis. It is still a possibility that cannot be completely discarded.

A summary of the alternative, and possibly interacting, factors mentioned above that could explain the shift in algal composition is given in Fig. 6.

4. Predator-prey relationships

One important factor, influencing the community development in the studied sites, was the relationship between *Mytilus edulis* and its' principal predator, *Asterias rubens.* Our data demonstrated a close relationship between the occurrence of *Mytilus*-settling and a resulting gradual build-up of *Asterias*-densities. After the first intense settling of *Mytilus* at the 5 m level, it took relatively long (c. 2 yrs) before this build-up was complete. The process was much faster on the 10 m site, probably due to the fact that large numbers of *Asterias* were al-

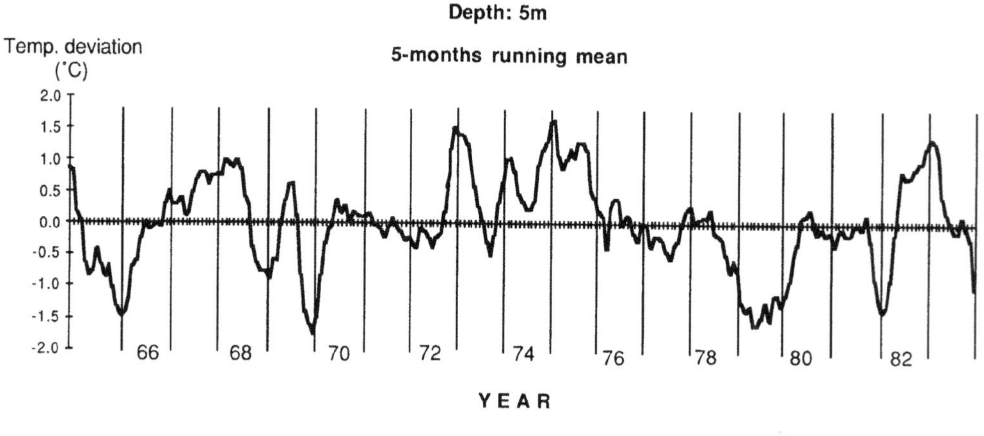

Fig. 5. Running means of temperature deviations over the period 1965–1983 at 5 m depth. Data from Bornö Hydrographical Station, inner part of the Gullmar Fjord.

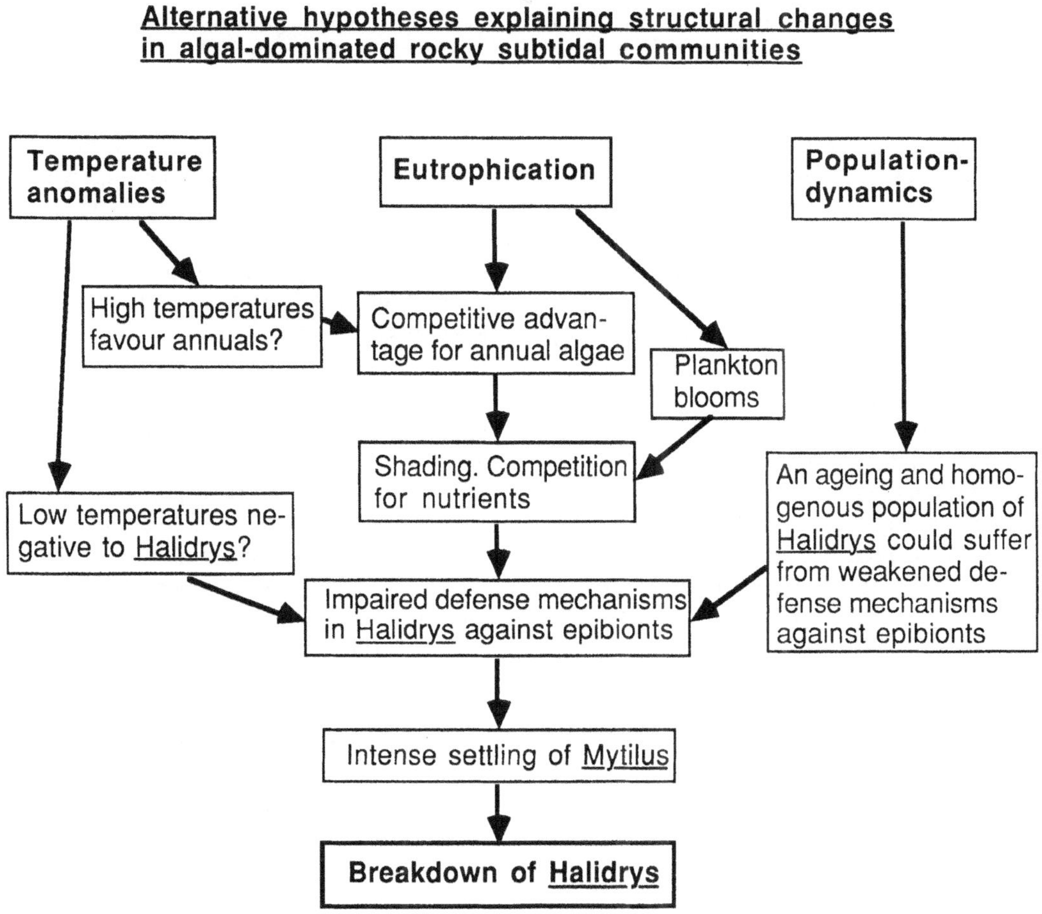

Fig. 6. Summary of a few interactions of possible significance for the observed structural changes in algal-dominated communities.

ready present close-by. The data also suggest a higher degree of seasonal variation in the occurrence of *Asterias* at the 5 m-level. It can be concluded that *Asterias*-predation prevented a long-term substrate monopolization by *Mytilus,* which is in accordance with findings elsewhere (e.g. Paine, 1966, 1969).

Possible repercussions on other parts of the coastal ecosystem

A critical factor when judging possible effects of the changes reported here on the coastal ecosystem at large, is the degree of generality of these changes. There was little direct evidence to judge this, but much circumstantial evidence supporting the idea that the changes reported here were of a general nature: (1) Lundälv (1985) showed that fixed sites of the type employed in this study were capable of reflecting largescale changes in a number of case-studies. Available data from sites in the Gullmar Fjord area, although situated in an environment very different from the one discussed here, also reflected partly similar long-term changes (c.f. Edler, 1984), including a minimum in algal cover towards the end of the 1980-ies and a co-inciding maximum in faunal opportunists (mainly *Ciona intestinalis*). (2) Long-term data on *Asterias*-densities in southern Norway correlated well with Swedish data (Lundälv & Christie, this volume). This could be viewed as in-direct evidence of similar changes going on in Norway, although direct observations on algal-dominated communities were not obtained. (3) Much 'subjective' evidence supported the generality of observed trends. This included much public concern over increasing quantities of filamentous annual algae along the entire northern Swedish west coast, and leading to rotting masses of algae in enclosed water bodies as well as deterioration of established bathing sites. Experienced marine botanists also noted declines in several species of perennial macroalgae (M. Pedersén, pers. comm.). Interviews with fishermen along the southern Swedish west coast also indicated changes in accordance with the ones reported here (Almer, 1979, 1982). During the same time-period (1980–85), inshore fishing also seems to have deteriorated markedly and many experienced divers have reported expanding areas of oxygen-depleted sediments (e.g. personal observations; H. Hallbäck & L.-O. Loo, pers. communication).

If we thus assume that the changes recorded here were of a general nature, it should be meaningful to make some rough calculations about possible consequences.

1. Effects of decreasing quantities of perennial macroalgae

There are almost no published data on normal quantities of perennial macroalgae along the northern Swedish west coast. Our own unpublished data, as well as other data (Ulmestrand, pers. communication), suggest that quantities varying between 1 500–2 000 g $d.wt \cdot m^{-2}$ are common within the depth range 1–10 m. It has been estimated that there is a total bottom area of approximately 100 km^2 in the depth range 0–3 m (the effect of small-scale topography to greatly increase the actual surface available for epibenthic organisms has not been taken into account) along the northern Swedish west coast between the city of Gothenburg and the Norwegian border (E. Degerman, pers. communication). It should be reasonable to assume that the bottom area in the depth range 0–10 m is then at least 200 km^2. If we assume that perennial benthic macroalgae occupy half of this area and that their biomass decreased by 1 000 g $d.wt \cdot m^{-2}$ over the period 1977–81, it could be interesting to speculate over the consequences for nutrient cycling in the coastal region. Perennial macroalgae are known to have a capacity for storing large quantities of nitrogen in the late autumn and winter period (e.g. Wallentinus, 1984a). Literature reports indicate that this storage could amount to between 2–4% (e.g. Chapman & Craigie, 1977) of the biomass (dry weight).

Under these assumptions, it could be calculated that the decrease in biomass of perennial algae would bring about a decrease in nitrogen uptake by perennial algae over the autumn- and winter period of about 2 000 tons N. This could be compared to uncertain estimates of nitrogen-inputs from land-drainage along the same coastal area varying between 1 400–19 000 tons yr^{-1} (Thorell, 1981; Berggren, 1982). If our calculations are in the right order of magnitude, the changes in algal community structure could thus mean a significant extra contribution to a possible eutrophication process, at least over part of the year. It is also interesting to note that ab-normal phytoplankton blooms were recorded exactly over the period when quantities of perennial algae were at a minimum (1978–81). The blooms were most pronounced relatively late in the autumn (October–November). This co-incides with the time of year when nitrogen storage by perennial macroalgae should have started. It is therefore conceivable that a reduction of these algae could have favoured the build-up of plankton blooms by increasing the pool of available nitrogen in the free water. Alternative hypotheses to explain the plankton blooms, involving shifts in pelagic food-web patterns, have been suggested (Lindahl & Hernroth, 1983). The latter hypothesis is not necessarily contradictory to the one given here. Instead, several factors, including a possible general eutrophication process (op. cit.; Rosenberg, 1985), could have contributed to the observed anomalies.

2. Effects of increasing quantities of annual algae

Our data, as well as a wealth of 'subjective' evidence partly mentioned above, indicated a marked and general increase in the production of annual, mostly filamentous, algae over the studied period. Annual algae differ from most perennial species in that there is a rapid overturn of algal biomass during the growth season. Parts of the algae constantly become dislodged or die off. By the end of the growth season almost all of these algae die off. Dislodged and dying algae largely sink to deeper bottoms where they decay under the consumption of oxygen and release of nutrients. Such decaying masses of algae can often be observed in bottom depressions while diving. This shift in algal composition from perennials to annuals should further increase the de-stabilization of annual nutrient cycles (c.f. Wallentinus, 1984a, b).

Parallel to what has been said about perennial algae, there is very little quantitative information about actual biomasses of annual algae along the Swedish west coast. However, drip-free wet weights ranging between a few kg up to 60 $kg \cdot m^{-2}$ were obtained in shallow estuaries along the northern Swedish west coast in the summer of 1984 (Lundälv & Ulmestrand, unpublished data). Production rates have not been studied at all.

The observed increase in the production of annual algae should lead to an increased transport of organic matter in the form of decaying algae to deeper bottoms. Let us assume that the annual production of ephemeral algae increased on the average by 1 000 g $d.wt \cdot m^{-2}$ over 100 km^2 along the northern Swedish west coast over the period studied. This would imply an increased transport of organic matter to deeper bottoms ranging perhaps between 50000 – 100000 tons $d.wt \cdot yr^{-1}$, depending on how much of the production that was consumed by herbivores or exported. The figures used are of course tentative and could be higher or lower, but should still be realistic enough to imply that shifts in macroalgal community structures could be quantitatively important in coastal areas of the type considered here. Further support for this notion, as mentioned above, is offered by numerous observations by divers of expanding areas of oxygen-depleted bottoms (covered by sulphuric bacteria) along the Swedish west coast over the last few years. This phenomenon could not be readily explained by increased phytoplankton production/sedimentation, since existing data indicated relatively low phytoplankton quantities after 1981 (O. Lindahl, pers. comm.). Increasing cover of red algae at our 15 m site during the same time period also indicated low turbidity.

Conclusions

1. A marked shift in the structure of algal-dominated rocky subtidal communities along the northern Swedish west coast occurred over the time period 1970 – 85 and especially 1977 – 82. The basic element of this shift was an increase in ephemeral, largely filamentous, algae at the expense of coarser perennial algae.

2. The reasons for the observed change are uncertain, but available evidence seems to indicate that a general eutrophication process has contributed. The changes recorded in this study could also possibly be partly related to long-term temperature fluctuations and/or internal population dynamical processes.

3. Hypothetical calculations concerning possible repercussions on other parts of the coastal ecosystem from shifts in the structure of macroalgal communities, together with circumstantial evidence from pelagic and benthic environments, seem to indicate that macroalgal communities could be quantitatively more important for the cycling of nutrients and organic matter than hitherto anticipated, at least in a coastal system of the type considered here. The sparse evidence presented seems to motivate further studies into these problems.

4. Long-term substrate monopolization by the blue mussel *Mytilus edulis,* after intense settling, was efficiently prevented by a rapid build-up of predation pressure from the common starfish *Asterias rubens.*

5. Photographic recording of sublittoral fixed sites seems to provide costefficient and sensitive means for long-term monitoring related to primary production.

Acknowledgements

We thank S. Mattsson for part of the picture analysis and B. Bergström, A. Holme, K. Lindahl and I, Svane for field work assistance over the years.

Financial support was obtained through many research contracts with the National Swedish Environment Protection Board and the Swedish Natural Science Research Council. Logistic support was also provided by the Kristineberg Marine Biological Station (Royal Swedish Academy of Sciences). All is gratefully acknowledged.

References

Almer, B., 1979. Av yrkes- och binäringsfiskare observerade förändringar vid Hallandskusten under 1970-talet. Report from the Fishery Board in county Halland (mimeo), 6 pp.

Almer, B., 1982. Förändringar i kustnära havsområden åren 1980–1981. Report from the Fishery Board in county Halland (mimeo), 8 pp.

Berggren, H., 1982. Föroreningsbelastning på Västerhavet. S 12 (mimeo).

Beukema, J. J., G. C. Cadée & H. Hummel, 1983. Differential variability in time and space of numbers in suspension and deposit feeding benthic species in a tidal flat area. Oceanol. Acta No. Sp.: 21–26 (Proc. 17th Eur. Mar. Biol. Symp.).

Bowman, R. S. & J. R. Lewis, 1977. Annual fluctuations in the recruitment of *Patella vulgata* L. J. Mar. Biol. Ass. U.K. 57: 793–815.

Buchanan, J. B., M. Sheader & P. F. Kingston, 1978. Sources of variability in the benthic macrofauna off the south Northumberland coast, 1971–1976. J. Mar. Biol. Ass. U.K. 58: 191–209.

Chapman, A. R. O. & J. S. Craigie, 1977. Seasonal growth in *Laminaria longicruris:* Relations with dissolved inorganic nutrients and internal reserves of nitrogen. Mar. Biol. 40: 197–205.

Colebrook, J. M., 1982. Continous plankton records: phytoplankton, zooplankton and environment, North East Atlantic and North Sea, 1958–1980. Oceanol. Acta 5: 473–480.

Connell, J. H., 1972. Community interactions on marine rocky intertidal shores. Ann. Rev. Ecol. Syst. 3: 169–192.

Connell, J. H., 1974. Ch. 2. Ecology: field experiments in marine ecology. In: R. Mariscal (ed.), Experimental Marine Biology. Academic Press, New York: 21–54.

Cushing, D. H., 1975. Marine ecology and fisheries. Cambridge Univ. Press, Cambridge, 278 pp.

Dayton, P. K., 1971. Competition, disturbance and community organization: the provision and subsequent utilization of space in a rocky intertidal community. Ecol. Monogr. 41: 351–389.

Edler, L., 1984. The West coast. In: R. Rosenberg (ed.), Eutrophication in marine waters surrounding Sweden – a review. Swedish National Environment Protection Board, Report 1808: 74–111.

Edler, L., G. Aertebjerg & E. Graneli, 1982. Exceptional plankton blooms in the entrance to the Baltic Sea – The Kattegatt and Belt Sea area. ICES, C.M. 1982/L:20: 6 pp.

Glombitza, K. W. & E. Sattler, 1973. Antibiotica aus Algen. 10. Trifuhalol, ein neuer Triphenyläther aus *Halidrys siliquosa.* Tetrahedron Lett. 43: 4277–4280.

Glombitza, K. W., C. Geisler & G. Eckhardt, 1980. Polyhydroxyphenyl ethers from Phaeophycea *Halidrys siliquosa* (III). Bot. mar. XXIII: 735–740.

Green, N., 1980. Underwater stereophotography applied in ecological monitoring. Niva Report No. OF-80613: 99 pp.

Hartnoll, R. G. & S. J. Hawkins, 1980. Monitoring rocky-shore communities: a critical look at spatial and temporal variation. Helgoländer wiss. Meeresunters. 33: 484–494.

Hawkins, S. J. & R. G. Hartnoll, 1983. Changes in a rocky shore community: an evaluation of monitoring. Mar. Envir. Res. 9: 131–181.

Hornsey, I. S. & D. Hide, 1974. The production of antimicrobial compounds by British marine algae. Br. Phycol. J. 9: 353–361.

Hällfors, G., P. Kangas & Å. Niemi, 1984. Recent changes in the phytal at the south coast of Finland. Ophelia 3: 51–59.

Josefson, A. B., 1981. Effekter av syrebrist och långsiktiga förändringar av bottensediment på den makrobentiska infaunan i Gullmarsfjordens djupbassäng: En preliminär rapport. Rep. to the Nat. Swedish Envir. Prot. Bd. (mimeo): 27 pp.

Josefson, A. B., 1985. Spatial and temporal structure of marine infauna in the Skagerrak (Eastern North Sea). Ph.D. thesis, University of Gothenburg: 70 pp.

Kangas, P., H. Autio, G. Hällfors, H. Luther, A. Niemi & H. Salemaa, 1982. A general model of the decline of *Fucus vesiculosus* at Tvärminne, south coast of Finland in 1977–81. Acta Bot. Fennica 118: 1–27.

Larsson, U., 1984. The Baltic. In: R. Rosenberg (ed.), Eutrophication in marine waters surrounding Sweden. Swedish National Environment Protection Board, Report 1808: 17–73.

Lewis, J. R., 1964. The ecology of rocky shores. English University Press, London: 323 pp.

Lewis, J. R., 1976. Long-term ecological surveillance: practical realities in the rocky littoral. Oceanogr. Mar. Biol. Ann. Rev. 14: 371–390.

Lewis, J. R., 1980. Options and problems in environmental management and evaluation. Helgoländer wiss. Meeresunters. 33: 452–466.

Lewis, J. R., 1984. Temporal and spatial changes in benthic communities: COST 47 approach. Mar. Poll. Bull. 15: 397–402.

Lindahl, O. & L. Hernroth, 1983. Phyto-zooplanktoncommunity in coastal waters of western Sweden – an ecosystem off balance? Mar. Ecol. Prog. Ser. 10: 119–126.

Lundälv, T., 1971. Quantitative studies on rocky-bottom biocoenoses by underwater photogrammetry. A methodological study. Thalassia Jugosl. 7: 201–208.

Lundälv, T., 1974. Underwater photogrammetry – a new device for biodynamical studies in the marine environment. In: B. Afzelius (ed.), Yearbook of the Swedish Nat. Sci. Res. Council 1974. Allmänna Förlaget, Stockholm: 222–229.

Lundälv, T., 1982. Effekter av planktonblomningar vid Bohuskusten. In: Anon (ed.), Algblomning. Liber Allmänna Förlaget, Stockholm: 11–19.

Lundälv, T., 1985. Detection of long-term trends in rocky sublittoral communities: Representativeness of fixed sites. In: P. G. Moore & R. Seed (eds), The ecology of rocky coasts. Hodder and Stoughton Educational, U.K.: 329–345.

Lundälv, T. & H. Christie, 1986. Comparative trends and ecological patterns of rocky subtidal communities in the Swedish and Norwegian Skagerrak area. Hydrobiologia 142: 71–80.

Michanek, G., 1971. A review of world seaweed resources. Proc. 7th int. Seaweed Symp., Sapporo, Japan. Univ. Tokyo Press: 248–250.

Moss, B. L., 1982. The control of epiphytes by *Halidrys siliquosa* (L.) Lyngb. (Phaeophyta, Cystoceiraceae). Phycologia 21: 185–191.

Paine, R. T., 1966. Food web complexity and species diversity. Am. Nat. 100: 65–75.

Paine, R. T., 1969. The *Pisaster – Tegula* interaction: prey patches, predator food preferences, and intertidal community structure. Ecology 50: 950–962.

Rosenberg, R., 1985. Eutrophication – the future marine coastal nuisance? Mar. Poll. Bull. 16: 227–231.

Rosenberg, R. & L. Edler, 1981. Laholmsbukten – oviss framtid för störd miljö. Forskning och Framsteg 3: 36–39.

Rueness, J., 1973. Pollution effects on littoral algal communities in the inner Oslofjord, with special reference to *Ascophyllum nodosum*. Helgoländer wiss. Meeresunters. 24: 446–454.

Sattier, E. K., W. Glombitza, F. W. Wehrli & G. Eckhardt, 1977. Antibiotica aus algen. 16. Polyhydroxyphenyläther aus der Phaeophycee *Halidrys siliquosa*. Tetrahedon 33: 1239–1244.

Sutherland, J. P., 1974. Mutliple stable points in natural communities. Am. Nat. 108: 859–873.

Svansson, A., 1984. Hydrography of the Gullmar Fjord. Inst. of Hydrogr. Res. Göteborg 23, 90 pp.

Thorell, L., 1981. Föroreningstillförsel från landområden. Föroreningssituationen i Skagerrak-Kattegatt. Nordiska Ministerrådets Sekretariat i Oslo. Gotab, Stockholm: 87–93.

Torlegård, A. K. I. & T. Lundälv, 1974. Under-water analytical system. Photogramm. Engng. 40: 287–293.

Wallentinus, I., 1983. Vegetationsklädda bottnar i Östersjön. Rep. to 'Eutrofiering i Marin Miljö', Nat. Swedish Envir. Prot. Bd., 56 pp.

Wallentinus, I., 1984a. Partitioning of nutrient uptake between annual and perennial seaweeds in a Baltic archipelago area. Hydrobiologia 116: 363–370.

Wallentinus, I., 1984b. Comparisons of nutrient uptake rates for Baltic macroalgae with different thallus morphologies. Mar. Biol. 80: 215–225.

Hydrobiologia 142: 97–111 (1986)

Rocky subtidal assemblages on the west coast of Ireland*

John Costelloe, Brendan F. Keegan & Gerd F. Könnecker
Department of Zoology, University College Galway, Galway, Ireland

Keywords: *Ciona intestinalis,* spawning, settlement, west coast of Ireland

Introduction

Prior to 1969, no study of marine epibenthic communities as such had been conducted in Irish coastal waters. Over the previous half century the sessile invertebrates of the Irish sublittoral had been singularly neglected.

As part of a continuing investigation, which seeks to determine the structure and functioning of 'clean' benthic communities in Greater Galway Bay (on the west coast of Ireland), Könnecker (1973, 1977) and Könnecker & Keegan (1983) have partly described the epifaunal associations of the area. Within this same programme, Shin (1981) used test panels to follow changes in the structure of a sessile community. More recently, Costelloe & Keegan (1984) and Costelloe (1985) have reported on the first of a number of autecological investigations on suspected 'key' (Lewis, 1978) epibenthic species. Within the broad objectives of some of this work, a long term programme was initiated to record the development of sessile assemblages on natural and artificial substrates. This was concentrated on a single site and was pursued against a background of synoptic data on the physical and chemical environment. With the advent of the COST-647 project, however, this was completely redressed to the extent of changing the study location and introducing the recommended 'core' protocol (COST 47 Activity Report) of photographic non-destructive sampling on vertical surfaces.

Because of the general paucity of information on the Irish epibenthos, and justifying the attention to ascidian dominated assemblages, a description of the methodology and some of the findings of the precursor programme are included hereunder with those of the COST-647 investigations.

Study areas

For the pre-COST-647 programme, artificial panels were placed on the Black Rock Reef in the inner Galway Bay area (Fig. 1; see also Ryland & Nelson-Smith, 1975 and Keegan *et al.,* 1976). The reef projects from the northern granitic shoreline and extends westward for some 2 Km. With a maximum depth of 15 m, the area is strewn with glacial debris, from small pebbles to boulders a few meters in diameter. For the most part, intermittent sandy patches otherwise blanket the bedrock. Salinity, temperature and dissolved oxygen values showed recurring annual patterns at the Black Rock Reef from February 1979 to March 1981. Peak surface and bottom temperature recordings were 17 °C, while minimum surface and bottom recordings were 6 and 5 °C, respectively. Surface salinity recordings ranged between 14.5‰ and 34‰ and fluctuated erratically. Bottom salinity values, however, were more stable, varying between 32 and 34‰. Dissolved oxygen values were lowest at both the surface and the bottom (5.8 and 4.9 ml/L, respectively) over the summer period. Maximum recorded winter values were 8 ml/L and 7 ml/L at the surface and the bottom, respectively.

A more suitable study location (Bird Rock) with vertical surfaces was chosen in Kilkieran Bay (Fig. 1; see also Könnecker & Keegan, 1983) to facilitate the COST-647 core programme. Könnecker (1977) and Könnecker & Keegan (1983) designated the epifaunal communities below 10 m at this location to,

*Contribution number 266 to the School of Marine Science

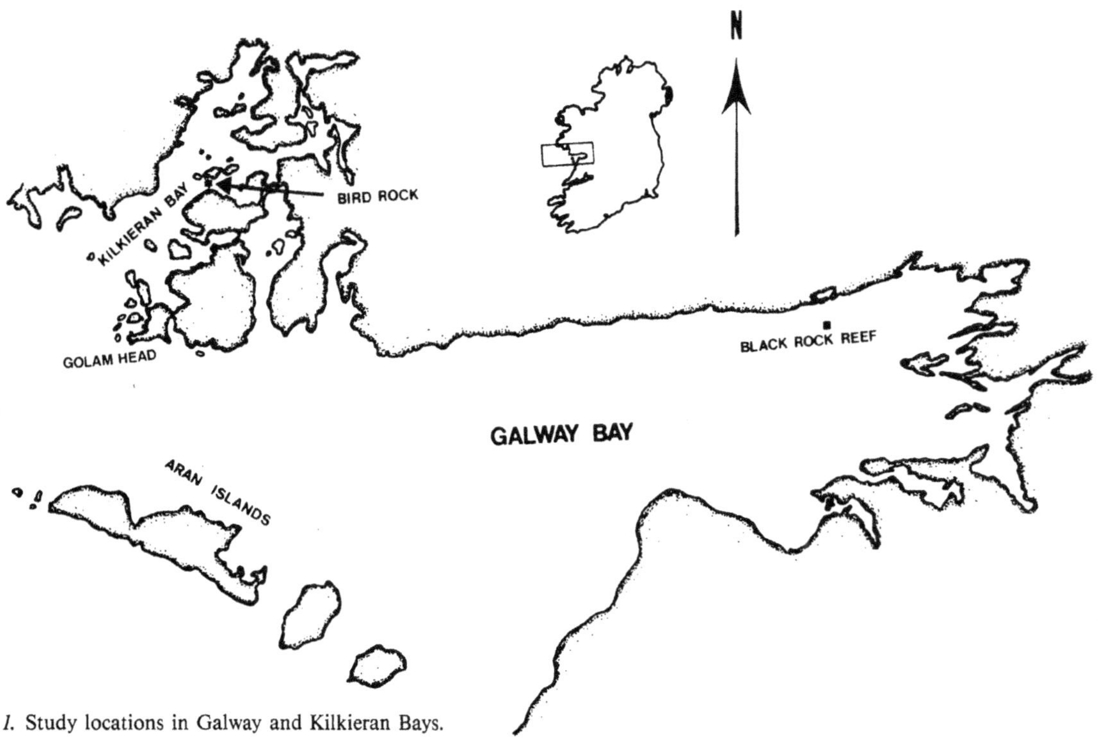

Fig. 1. Study locations in Galway and Kilkieran Bays.

what he termed, the *'Raspailia-Stelligera* Association'. This association sequentially follows the '*Laminaria saccharina* Association' with increasing water depth and is, in general, found below the level of the macrophytic algae. These associations are found in moderate to slack current flow and apparently thrive in conditions of high turbidity.

Since August 1983 to September 1985 the Cost-647 programme has been concentrated on near vertical granitic surfaces at the Bird Rock location. Here a shelf, covered in muddy sediment and with occasional boulders, slopes gradually to a depth of 8 – 12 m. The ground thereafter falls away as a series of granitic cliffs interspersed with boulder fields. This continues to a depth of 30 – 35 m, before giving away to deposits of biogenic gravel. Previous work has shown the current pattern, at this site, to be complex. Whilst the incoming tide runs along the cliff in a west-east direction, a reverse flow has not been observed. Instead a backeddy from the channel runs in the same direction (Fig. 2) during the ebb tide. Current strength varies but has not been found to exceed one knot.

Two positions, at 15 and 25 m depth, were selected for on-going study with near vertical surfaces. A shallower site, as specified in the COST-647 core programme, could not be established due to the lack of suitable vertical surfaces in this region.

Salinity and temperature recordings at Bird Rock showed little variation between those taken on the surface and on the bottom (25 m). This is probably due to the complex nature of the current pattern at this site which results in vertical mixing of the water column. Mean mid-summer and mid-winter temperatures at both the surface and the bottom ranged between 15 and 16.5 °C, and 6.4 and 6.9 °C, respectively. Similarly, mean mid-summer and mid-winter salinity values at both the surface and the bottom ranged between 34.3 and 34.7‰, and 30.02 and 32.4‰, respectively.

Materials and methods

Settlement/colonization experiments were carried out at the Black Rock Reef over a 14 month period, from February 1980 to April 1981. Roofing slates (30×30×0.5 cm), fabricated from a mixture of asbestos and cement, were employed as experimental surfaces. Supporting panels of hard

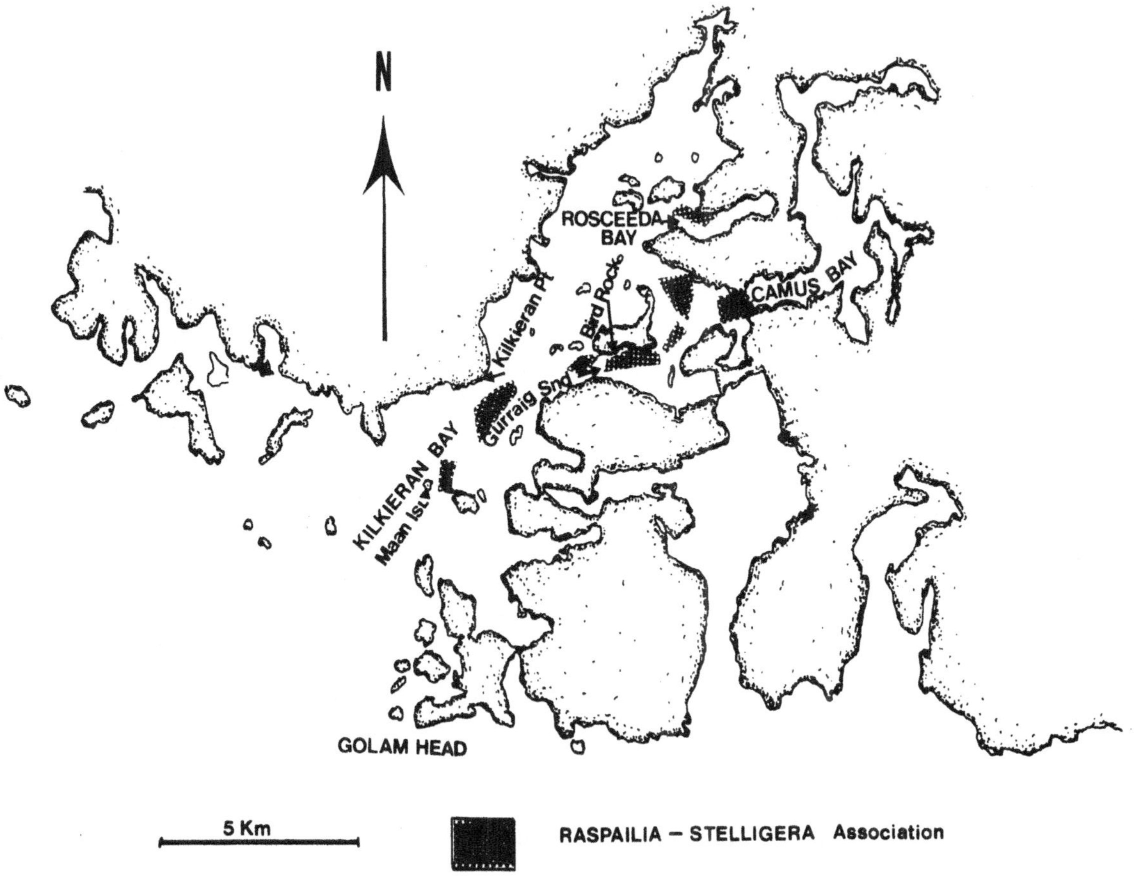

Fig. 2. Kilkieran Bay.

plastic or marine plywood were first sandwiched between pairs of slates, before being attached over the entire horizontal surface and along two of the vertical sides of metal frames (Fig. 3). One to three horizontal and vertical plates, exposed from the initial day of submergence, were lifted at regular sampling intervals. On each surface, the percentage cover of individual species was estimated using a random point system modified from that of Sutherland (1977). Each plate was divided into a 1×1 cm modular grid of 900 points. Two hundred computer generated random points were taken and the species under each of these location points were identified under a dissecting microscope. The percentage cover of each species was calculated from the formula:

$$\text{Percentage cover} = \frac{\text{The number of points a species occupies}}{\text{The number of points in the modular grid}} \times 100$$

Only those species which were attached directly to the slate surface were counted in the estimation.

Within the COST-647 core programme at Bird Rock two sampling locations (15 and 25 m) have been monitored since August 1983 to the September 1985. Vertical stakes have been secured 4 m apart, against vertical rock surfaces and on each sampling occasion a horizontal bar with 0.5 m intervals is first suspended from hooks on the vertical stakes. At each depth, 16×0.25 m^2 (8 above and 8 below a horizontal dividing bar) were stereophotographed. Stereophotographs are taken at each interval using a metal frame (50×50 cm), with flanking data plates similar to that described by Lundalv (1971) in Sweden and Rørslett *et al.*, (1978) in Norway. The frame extends from a camera assembly which incorporates two Nikonos II cameras with 15 mm lenses. Each camera is synchronised with its own electronic flash (Oceanic

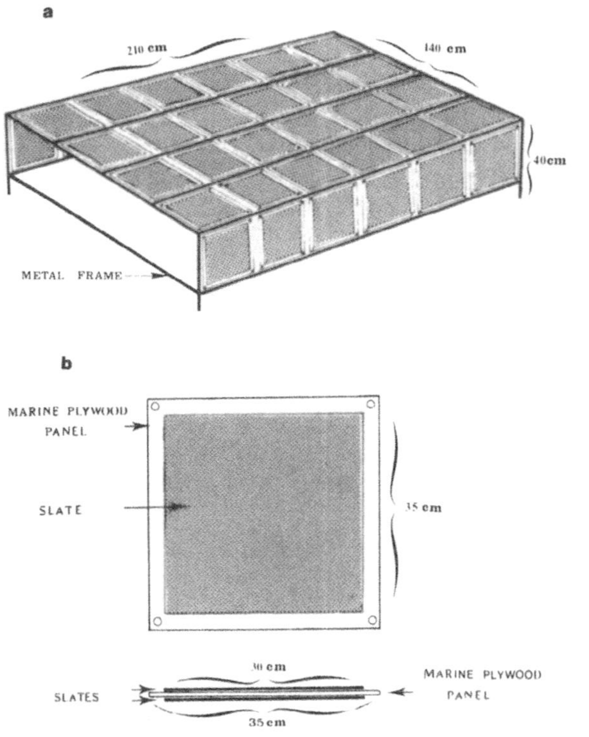

Fig. 3. (a) Frame and settlement panels. (b) Settlement panel with supporting sheet of marine plywood.

2002 and 2003). Mounted in parallel, with a distance of 20 cm between the centre of the lenses, the cameras are simultaneously fired by adjustable extensions of a common trigger bar. The distance between lens face and reference frame is 60 cm.

In practise, the camera frame is moved along both the upper and lower sides of the horizontal reference bar. Eight stereophotographs are taken on each side and the resulting transparencies are examined under a viewer constructed from two stereomicroscopes. Individual animals are counted, indentified to species level where possible and estimates taken of the percentage cover they occupy. Percentage cover estimates are calcualted using a systematic or random point quadrat technique. This is achieved by either using the intersects (110 points) on a grid graticule eyepiece or placing a small sheet of acetate with pin head random points (150 points) directly on the transparencies and observing directly the species occuring under each point. Only those individuals attaching directly to the underlying substrate are considered in this analysis. Percentage cover of each species is calculated as described above. The areas below the horizontal bar were denuded of their animal/plant life at the beginning of the study.

Augmenting the stereophotography, 3 panels ($30 \times 30 \times 0.5$ cm), as employed on the Black Rock Reef, are interchanged and suspended vertically at both depths on each sampling occasion. General epifaunal collections are also made at regular intervals. Animals are identified and qualitative estimates are made of their abundances.

Following preliminary observations it became apparent that the tunicate, *Ciona intestinalis*, was the dominant species at both the 15 and the 25 m site. Ten to fifteen individuals were subsequently taken at regular intervals for gonad analysis. Gonads and adjoining sections of the alimentary tract were fixed in Bouins fluid and stored in 70% alcohol. Tissues were dehydrated in graded alcohols and were subsequently cleared in xylene or toluene. Samples were then embedded in 56 °C mp paraffin wax, sectioned on a rotary microtome at 4–8 μm and stained with Ehrlich's haematoxylin and eosin.

Sections were examined under a light microscope to assess the stage of maturity, and to allow size frequency counts of, the oocytes of the female. Measurements of 100 or more oocytes were taken from each female and the results grouped into 20 μm size classes. Oocyte measurements were taken across the greatest axis and only from those cut across the nucleus.

Results

Settlement panels at the Black Rock Reef

The developmental sequence of epifaunal assemblages at the Black Rock Reef varied considerably between plates, depending on their angles of inclination. Assessment of the percentage cover occupied by the major taxa during the study period indicated an almost complete dominance by ascidian species on all but the horizontal upper surfaces and particularly the horizontal lower exposures (Fig. 4). The major contributors to this dominance were the species *Ascidiella aspersa, Ascidiella scabra* and *Ciona intestinalis* with lesser cover of *Aplidium pallidum, Aplidium punctum, Trididem-*

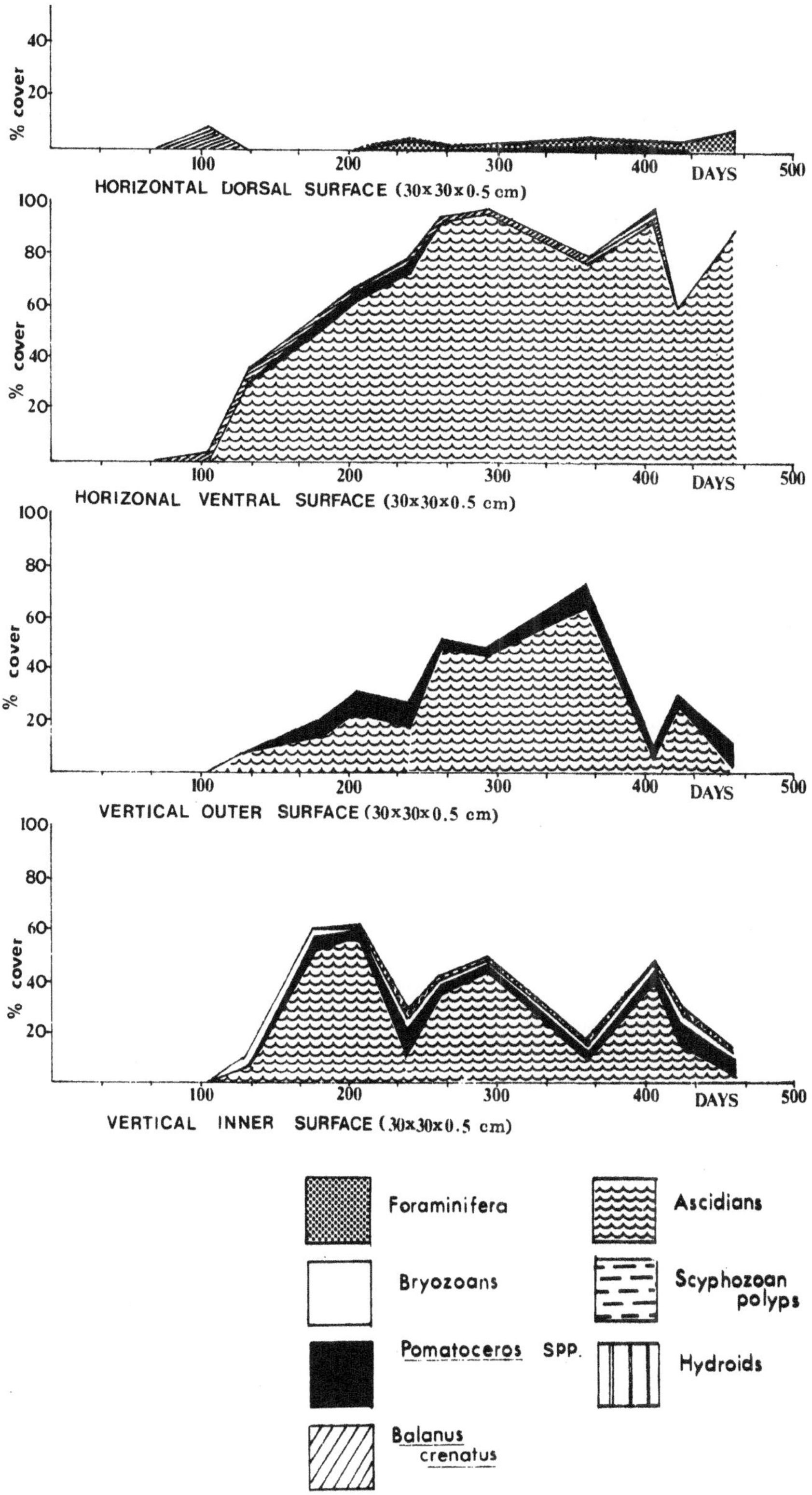

Fig. 4. Variations in the cumulative percentage cover of the dominant sessile species on 30×30 cm settlement panels, at the Black Rock Reef.

mum tenerum, Leptoclinides faeroensis, Diplosoma listerianum, Corella parallelogramma, Ascidia mentula Ascidia virginia, Polycarpa gracilis, Botryllus schlosseri, Botrylloides leachii, Pyura tessellata and *Molgula complanata.* In addition to this dominance of the occupied percentage cover, the ascidians also occupied high percentages of the total space available for settlement. This was particularly prevalent on the lower surfaces where the percentage cover occupied by the Ascidiacea was generally greater than 65% of the total cover. By comparison, their occupation of the upper surfaces was less than 1% while that of the vertical surfaces generally ranged in value between that of the horizontal upper and lower surfaces.

Species composition at Bird Rock

The sessile epifauna occurring at Bird Rock is both varied and rich. Records of the species present were made from general observations, collections and settlement panels throughout this study. A total of 131 species were recorded (Table 1) of which 3 belong to the Phylum Protozoa, 35 to the Porifera, 24 to the Coelenterata, 40 to the Ectoprocta, 2 to the Entoprocta, 1 to the Mollusca, 4 to the Annelida, 2 to the Crustacea and 20 to the Ascidiacea. *Spirorbidae* spp. and *Anomia* spp. were not identified to species level.

The epifaunal communities recorded below 10 m at the two study locations fall broadly into the '*Raspailia – Stelligera* Association' (Könnecker, 1973, 1976; Könnecker & Keegan, 1983). In Kilkieran Bay, this association dominates the rocky substrates of the Gurraig Sound and Camus bay and is also found below 20 m off Kilkieran Point, in the entrance of Rosceeda Bay and in the deeper part of the channel to the south-east of Maan island (Fig. 2).

The dominant members of this association are the poriferans, the bryozoans and the ascidians. Within the association, the species distribution manifests different habitat preferences in terms of substrate topography, particularly in the degree of inclination. While many species settle on and colonize all available surfaces, those which seemingly react to substrate inclination include *Raspailia ramosa* f. *pumila, Stelligera stuposa, Polymastia boletiformis, Ascidia virginea, Tethya aurantium* and *Corella parallelogramma* favouring upper sur-

Table 1. Species list compiled from general collections and settlement.

Phylum Protozoa

Folliculina ampulla (O. F. Muller)
Folliculina gigantea Dons
Haliphysema tumanowiczii Bowerbank

Phylum Porifera

Leucosolenia botryoides (Ellis & Solander)
Clathrina coriacea (Montagu)
Scypha ciliata (Fabricius)
Scypha compressa (Fabricius)
Leuconia barbata (Duchassaing & Michelotti)
Pachymatisma johnstoni (Bowerbank)
Poecilastra compressa (Bowerbank)
Stryphnus ponderosus (Bowerbank)
Stelletta grubei Schmidt
Dercitus bucklandi (Bowerbank)
Polymastia boletiformis (Lamarck)
Tethya aurantium (Pallas)
Prosuberites epiphytum (Lamarck)
Suberites carnosus (Johnston)
Cliona celata Grant
Cliona vastificata Hancock
Mycale rotalis (Bowerbank)
Mycale contarenii (Bowerbank)
Hemimycale columella (Bowerbank)
Jophon hyndmani (Bowerbank)
Stylopus dujardini (Bowerbank)
Hymedesmia paupertas (Bowerbank)
Hymedesmia sp.
Amphilectus fucorum (Esper)
Plocamia coriacea (Bowerbank)
Hymeraphia stellifera Bowerbank
Haliclona rosea (Bowerbank)
Haliclona oculata (Pallas)
Haliclona cinerea (Grant)
Halisarca dujardini Johnston
Aphysilla rosea Schulze
Aphysilla sulfurea Schulze
Dysidea fragilis (Montagu)
Stelligera stuposa (Montagu)
Raspailia ramosa (Montagu)

Phylum Coelenterata

Class Hydroidea

Coryne fruticosa Hincks
Tubularia indivisa Linnaeus
Bougainvillia ramosa (Van Beneden)
Clythia johnstoni (Alder)
Calycella syringa (Linnaeus)
Laomedae flexuosa (Hincks)
Obelia geniculata (Linnaeus)
Obelia dichotoma Linnaeus
Campanularia angulata Hincks
Halecium halecinum (Linnaeus)
Sertularella polyzonias (Linnaeus)
Sertularia operculata (Linnaeus)
Sertularia cupressina (Linnaeus)

Table 1. Continued.

Sertularia argentea (Linnaeus)
Nemertesia anteninna (Linnaeus)
Nemertesia ramosa (Linnaeus)
Plumularia catharina Johnston
Plumularia setacea (Ellis & Solander)
Antennella secundaria (Gmelin)
Hydrallmania falcata (Linnaeus)

Class Anthozoa

Alcyonium digitatum (Linnaeus)
Parerythropodium coralloides (Pallas)
Gonactinia prolifera (M. Sars)
Actinothoe spyrodeta (Gosse)

Phylum Bryozoa

Scruparia ambigua (d'Orbigny)
Scruparia chelata (Linnaeus)
Electra pilosa (Linnaeus)
Chartella papyracea (Ellis & Solander)
Callopora lineata (Linnaeus)
Callopora dumerilii (Audouin)
Alderina solidula (Hincks)
Amphiblestrum flemingii (Busk)
Scrupocellaria scruposa (Linnaeus)
Scrupocellaria reptans (Linnaeus)
Bicellariella ciliata (Linnaeus)
Beania mirabilis Johnston
Bugula plumosa (Pallas)
Bugula flabellata (Thompson in Gray)
Cribrilina radiata (Moll)
Escharoides coccinnea (Abildgard)
Parasmittina trispinosa (Johnston)
Escharella immersa (Fleming)
Escharella variolosa (Johnston)
Schizomavella linearis (Hassall)
Schizomavella auriculata (Hassall)
Microporella ciliata (Pallas)
Fenestrulina malusii (Audouin)
Hippothoa divaricata (Lamaroux)
Hippothoa flagellum (Manzoni)
Celleporella hyalina (Linnaeus)
Cellepora pumicosa (Pallas)
Schizotheca divisa (Norman)
Turbicellepora avicularis (Hincks)
Crisidia cornuta (Linnaeus)
Crisia ramosa (Lamarck)
Crisia denticulata (Lamarck)
Crisia eburnea (Linnaeus)
Berenicea patina (Lamarck)
Berenicea suborbicularis Hincks
Disporella hispida (Fleming)
Tubulipora liliacea (Pallas)
Nolella dilatata (Hincks)
Amathia lendigera (Linnaeus)
Alcyonidium gelatinosum (Linnaeus)

Phylum Entoprocta

Pedicellina cernua (Pallas)
Barentsia gracilis (Sars)

Phylum Mollusca

Class Bivalvia

Anomia sp.

Phylum Annelida

Class Polychaeta
Pomatoceros triqueter (Linnaeus)
Pomatoceros lamarcki (Rioja)
Hydroidea norvegica (Gunnerus)
Spirorbis spp.

Phylum Crustacea

Class Cirripedia

Balanus crenatus Brugiere
Verruca stroemia (O.F. Muller)

Phylum Ascidiacea

Clavelina lepadiformis (Muller)
Pycnoclavella aurilucens (Garstang)
Polyclinum aurantium Milne-Edwards
Sidnyum turbinatum Savigny
Diplosoma listerianum (Milne-Edwards)
Leptoclinides faeroensis Bjerkan
Ciona intestinalis (Linnaeus)
Perophora listeri Forbes
Corella parallelogramma (Muller)
Ascidiella aspersa (Muller)
Ascidiella scabra (Muller)
Ascidia virginea Muller
Ascidia mentula Muller
Dendrodoa grossularia (Van Benden)
Botryllus schlosseri (Pallas)
Botrylloides leachii (Savigny)
Microscosmus claudicans (Savigny)
Pyura tessellata (Forbes)
Pyura squamulosa (Alder)
Pyura microscosmus (Savigny)
Trididemnum tenerum (Verrill)

faces; *Ciona intestinalis, Prosuberites epiphytium,* and *Alcyonidium gelatinosum* favouring vertical surfaces; *Ascidia mentula, Scrupocellaria scruposa, Scrupocellaria reptans* and *Crisia* spp. favouring overhanging (i.e. horizontal lower) surfaces. In general, the substrate preferences stay constant for all depths. The distribution and intensity of colonization, however, does vary to some degree with depth. Figure 5 shows the topography, the location of the two stereophotographic sites and the distribution of *C. intestinalis* at Bird Rock. The most striking feature of this area is the high percentage cover of *Ciona intestinalis.* Below 25 m it gradually gives way to an *Ascidia virginea/Pyura* sp. dominance while above 8–10 m it yields to the

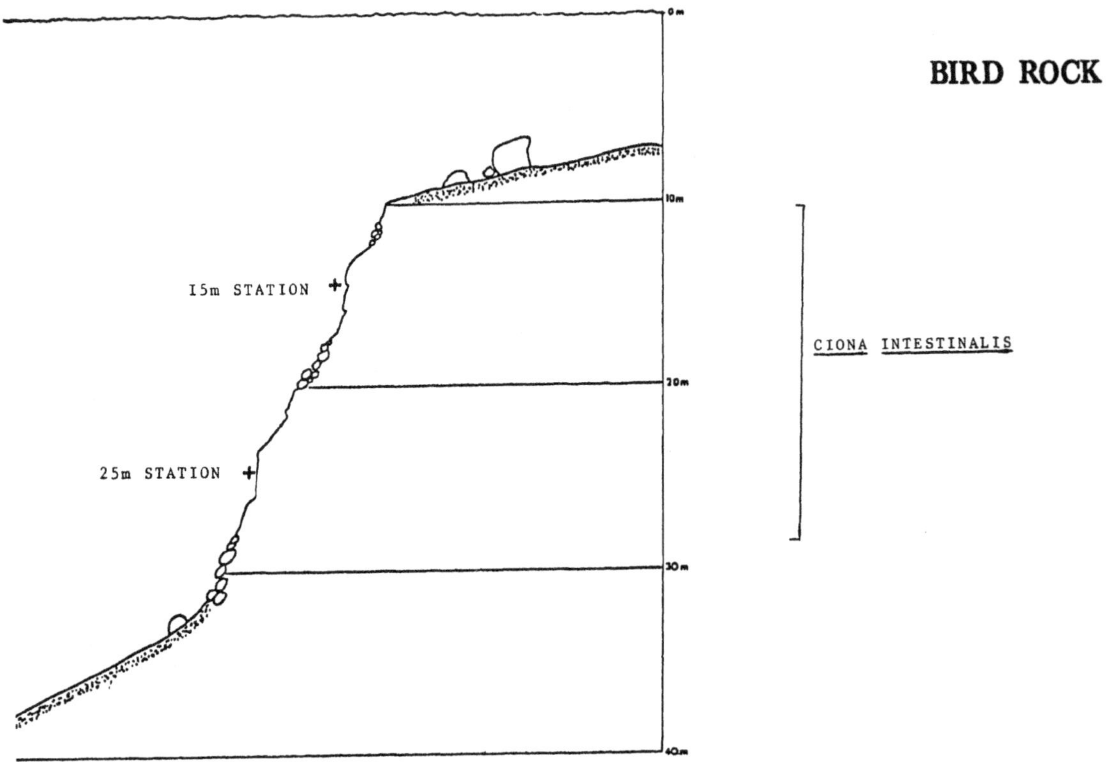

Fig. 5. The Bird Rock location.

'*Laminaria saccharina* Association' (Könnecker, 1973). Within which, heavy algal growth limits the epifauna to vertical or overhanging surfaces and to the algal surfaces themselves. Onces more the dominant faunal elements are the ascidians, particularly *Dendrodoa grossularia, Ascidia* and *Ascidiella* spp.

Stereophotographic results over a 25 month period have shown that the ascidian, *C. intestinalis,* was the dominant member, in terms of spatial distribution at both the 15 and 25 m sites (Fig. 10 and 11). Twenty one sessile and four vagile species have been identified photographically from both these sites (Table 2). The Porifera, are numerically dominant with the species *Haliclona oculata* and *Pachymatisma johnstoni* being the more conspicuous members of the group. However, due to slow growth of the sponge species over the relatively short study period and the frequent masking of other species due to overgrowth by the ascidians, the results are confined to *C. intestinalis.*

Reproductive cycle of Ciona intestinalis

Oocyte size frequency histograms (Fig. 6) taken from October (1983) through to June 1985 indicate a single annual spawning period in early summer for *C. intestinalis.* Primary oocytes are numerous and undergo little change in size over the winter months. During this period they range between 10 and 80 μm in diameter and, in general, are still attached to the germinal epithelium. In early spring, oocytes undergo a rapid increase in mean diameter. When they become detached from the germinal epithelium, at about the 60–80 μm stage, the ova acquire a complete cover of cuboidal inner follicular cells and a thin sheet of squamous outer follicular cells. At a late stage of development, many primary oocytes are still forming in the lumen of the gonad. This results in the rapid maturation of later developing oocytes and a wide range of oocyte size categories being present at the onset of spawning. The entire contents of the gonads are rarely shed at spawning and it is presumed that residual gametes

Table 2. Species recorded from stereophotographs.

Sessile species

Phylum Porifera

Haliclona oculata (Pallas)
Pachymatisma johnstonia (Bowerbank)
Tethya aurantium (Pallas)
Scypha ciliata (Fabricius)
Stelligera stuposa (Montagu)
Leucosolenia botryoides (Ellis & Solander)
Cliona sp.
Myxilla sp.
Sponge sp. 1(unidentified)
Sponge sp. 2(unidentified)

Phylum Coelenterata

Plumularia sp.
Caryophyllia smithi Stokes

Phylum ectoprocta

Crisia sp.

Phylum Annelida

Pomatocerossp.
Hydroidea norvegica (Gunnerus)
Spirorbid sp.

Phylum Ascidiacea

Ciona intestinalis (L.)
Clavelina lepadiformis (O.F. Muller)
Botryllus schlosseri (Pallas)
Pyura sp.

Phylum Rhodophyta

Lithothamnion sp.
Vagile species

Phylum Mollusca

Calliostoma zizyphinum (L.)

Phylum Echinodermata

Henricia oculata (Pennant)
Asterina gibbosa (Pennant)
Echinus esculentus L.

are resorbed subsequent to spawning. Mature oocytes typically range between 160–180 μm in diameter.

Throughout the reproductive cycle, varying quantities of sperm were found surrounding the alimentary tract of *C. intestinalis.* Spermatogenic activity greatly increases in early spring and by early summer dense quantities of sperm surround the alimentary tract.

Settlement of Ciona intestinalis

Evidence from stereophotography, settlement panels and histology indicate that settlement of *C. intestinalis* occurs over a single period between April and July each year. The greatest intensity of settlement was recorded during the months of May and June. At both depths settlement rate was intense with greater than 400 individuals settling per 30×30×0.5 cm settlement plate (Fig. 7). This rate of settlement would be equivalent to greater than 3500 individuals settling/m^2 on virgin surfaces. Settlement rates remained relatively consistent over the two year period at the different depths with the exception of the 25 m site during the summer of 1985. At this location, heavy settlement, with greater than 800 individuals per settlement panel was recorded between April and June.

Population dynamics of Ciona intestinalis

Preliminary stereophotography results from both depth locations at Bird Rock would seem to indicate that the total numbers and percentage cover of *C. intestinalis* follow a recurring annual pattern. Recruitment, as already noted on settlement panels, occurred in early summer at both depth locations. Due to the small size, overcrowded conditions and the tiering effect, the absolute numbers of *C. intestinalis* present at this time were impossible to discern. The highest intensity of settlement was generally observed between May and June with between 500 and 1000 per 0.25 m^2 on both cleared and undisturbed surfaces (Fig. 8 and 9). However, >1000 individuals per 0.25 m^2 were observed on cleared surfaces at the 25 m^2 site in June 1984. Following the early summer settlement, the numbers of *C. intestinalis* declined rapidly in late summer and autumn. This was mainly due to loss of individuals in overcrowded conditions and mortality among adults which generally seem to have a maximum life span of 15 to 16 months at Bird Rock. Animals which settle in early summer spawn the following summer and subsequently die in early autumn. The loss of individuals became greatly reduced over the winter and early spring, with the net result of between 50 and 150 individuals per 0.25 m^2 at the 15 and 25 m locations present prior to spawning (Fig. 8 and 9). In early autumn 1985,

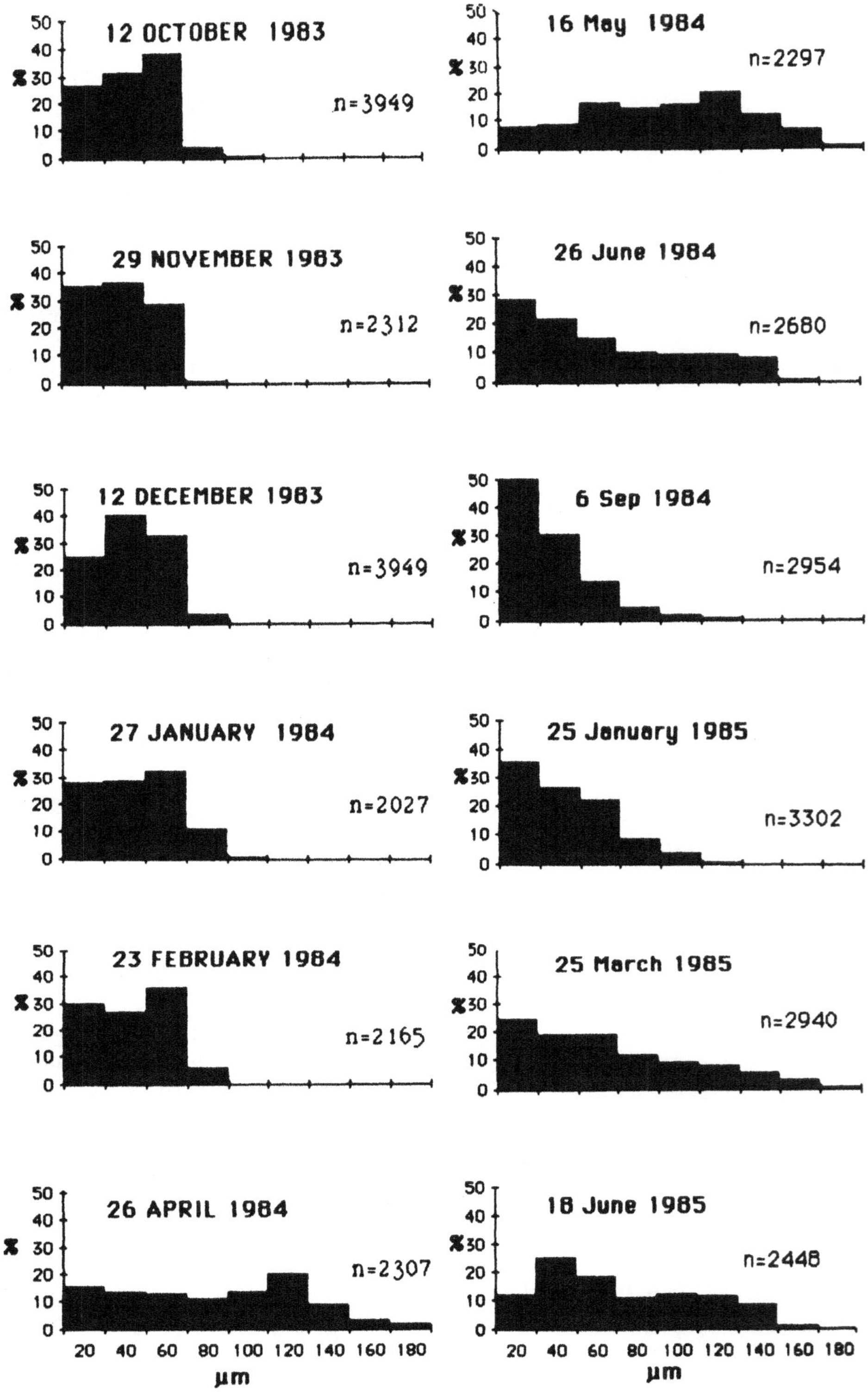

Fig. 6. Histograms of *Ciona intestinalis* oocyte size-frequencies grouped in 20 µm size-classes.

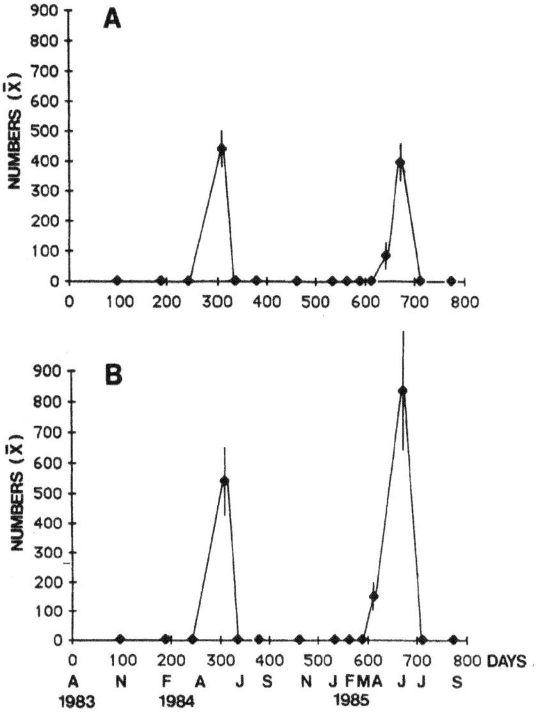

Fig. 7. Mean numbers of *Ciona intestinalis* per 30×30 cm settlement panel. (A), 15 m station; (B), 25 m station. Vertical bars represent standard deviations.

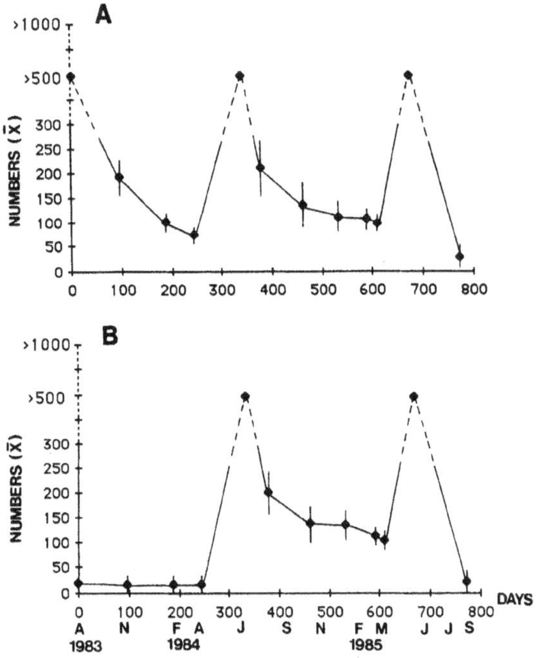

Fig. 8. Mean numbers of *Ciona intestinalis* per 0.25 m² at 15 m on (A) undisturbed surfaces (B) cleared (23 Aug. 1983) surfaces. Vertical bars represent standard deviations.

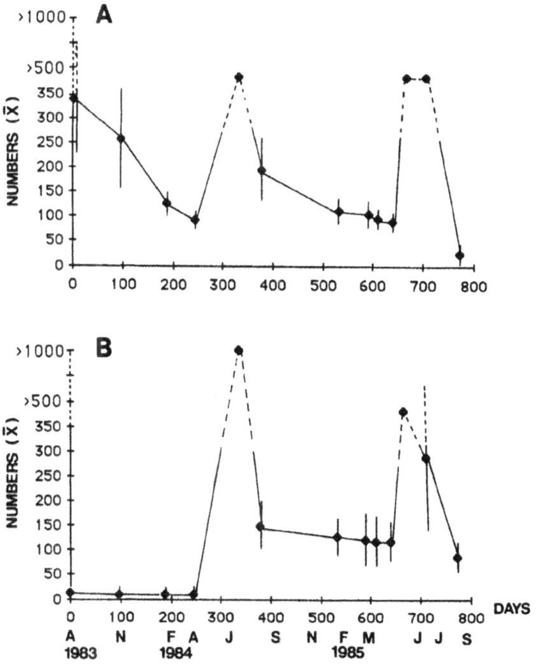

Fig. 9. Mean number of *Ciona intestinalis* per 0.25 m² at 25 m on (A) undisturbed surfaces (B) cleared (23 Aug. 1983) surfaces. Vertical bars represent standard deviations.

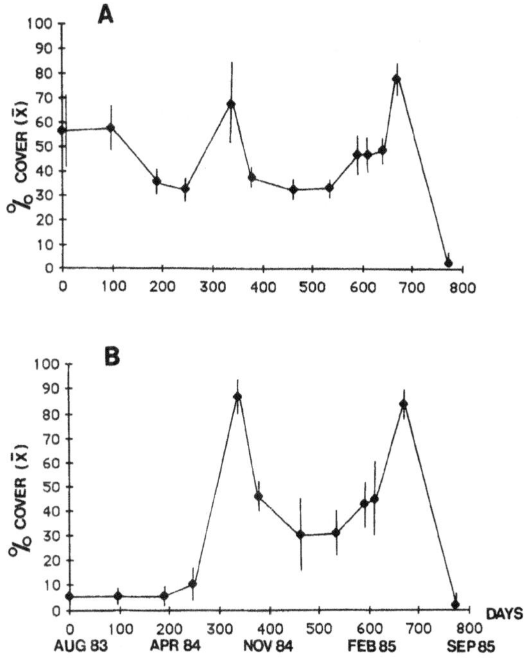

Fig. 10. The mean percentage cover of *Ciona intestinalis* per 0.25 m² at 15 m on (A) undisturbed surfaces (B) cleared (23 Aug. 1983) surfaces. Vertical bars represent standard deviations.

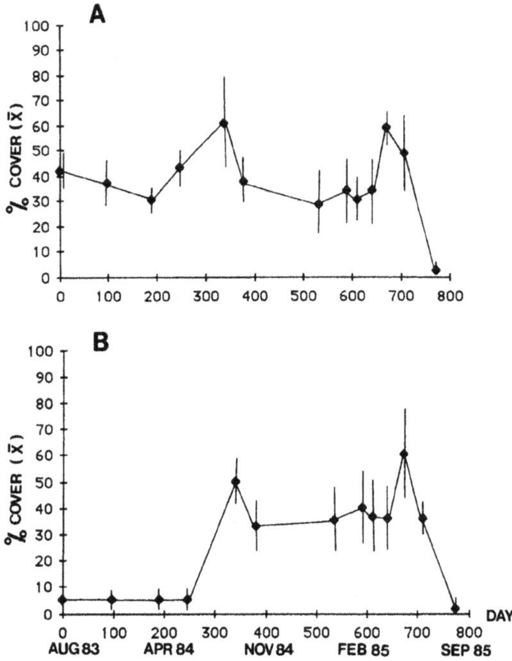

Fig. 11. The mean percentage cover of *Ciona intestinalis* per 0.25 m^2 at 25 m on (A) undisturbed surfaces (B) cleared (23 Aug. 1983) surfaces. Vertica bars represent standard deviations.

however, the population numbers declined rapidly subsequent to settlement and resulted <100 individuals per 0.25 m^2 on all surfaces. The mean percentage cover occupied by *C. intestinalis* at the 15 m and 25 m sites at Bird Rock followed a similar pattern to that of population numbers (Fig. 10 and 11). New recruitment in early summer may account for >80% of the total cover at the 15 m sites and >60% cover at the 25 m sites. In both 1983 and 1984 this was followed by an initial and substantial drop in percentage cover which then stabilized over the winter-spring period at 30–50% of the total cover at each location. As already outlined, when population numbers declined over this period, individual growth maintained and, in some instances, increased the percentage cover occupied prior to spawning. In 1985, however, population cover at both depths fell to <10% of the total cover following initial high recruitment. Settlement rates of *C. intestinalis* showed higher recruitment rates on cleared surfaces in the first year. However, population recruitment, numbers and percentage cover occupied on initially cleared surfaces subsequently varied little from those of undisturbed surfaces.

Discussion

Black Rock Reef

The investigation at the Black Rock Reef is included here through being the immediate precursor of the COST-647 programme and also serves to highlight the dominance of solitary ascidians at this location. The only comparable work in Irish coastal waters is that of Shin (1981) who found similar orientation preferences and dominance of the total percentage cover by solitary ascidians, especially *Ascidiella aspersa.* Both studies serve to highlight the dominant role of ascidians in the shallow sublittoral on the Irish west coast. It was unfortunate that the topography on the Black Rock Reef did not allow for the use of the stereophotogrammetric method. This apart, the results do throw light on the regulating effect of solitary ascidians with large annual variations in population densities and will be treated in a further paper.

Bird Rock

Recent work by Könnecker & Keegan (1983) has shown that the epibenthic assemblages of Kilkieran Bay can be allocated to seven discrete 'Associations' which are named for one or more of their faunal or floral elements. The '*Raspailia-Stelligera* Associations' found below 10 m in Camus Bay, Rosceeda Bay, off Kilkieran Point, south-east of Maan island and the Gurraig Sound generally belong to a euryhaline and eurythermal inshore regime. Sponges and ascidians feature prominently in this association which, as already outlined, displays a rich and diverse fauna below 10 m at Bird Rock. Stereophotographs of permanently marked test areas at 15 and 25 m have shown that the ascidian, *C. intestinalis,* is dominant in terms of spatial cover at this location.

C. intestinalis is a hermaphrodite and generally displays cross fertilization (Morgan, 1945). Evidence from settlement panels and histology would seem to indicate that spawning occurs over a single spawning period between April and early July at 15 and 25 m at Bird Rock in Kilkieran Bay. A variety of factors, including temperature (Runnstrom, 1927, 1936; Dybern, 1965; Gulliksen, 1972; Nomaguchi, 1974) and light (Berril, 1947; Lambert & Brandt, 1967; Whittingham, 1967; Georges, 1971) has been suggested to influence spawning. Dybern

(1965) suggests that the lower temperature limit for normal embryonic development seems to be about 8 °C and spawning generally takes place at temperatures between 8 and 12 °C. The temperatures recorded at Bird Rock at the time of spawning of *C. intestinalis* would seem to agree with these findings. The length and frequency of spawning varies considerably, depending on seasonal fluctuations, as well as regional variations of water temperature. In temperate regions spawning is generally restricted to the summer months. Both Runnstrom (1927) and Dybern (1965) found two distinct spawning periods, one in spring and the other in summer on the Swedish coast. Gulliksen (1972) found a similar spawning pattern to that found at Bird Rock but also recorded some *Ciona* larvae in the plankton over the winter period.

For warmer waters, Yamanouchi (1975) recorded continuous breeding of *C. intestinalis* throughout the year and that the number of eggs released at each spawning ranged between 2000 and 3000. The single spawning period of *C. intestinalis* at Bird Rock was preceeded by a gradual development and build up of gametes over the winter. In Meditterranean material, Peres (1952) has shown experimentally that after winter the ovary is first to develop and influences the development of the testicular follicles. This would also appear to be the case for the Bird Rock population which, at the onset of gametogenesis, shows a narrow range of oocyte diameters. However, as oogenesis and spermatogenesis proceed, mature oocytes, measuring 160–180 μm in diameter, coexist with new developing oocytes. Dybern (1965) found a similar range of oocytes in the gonad prior to spawning and also noted that mature oocytes measured 150 to 170 μm in diameter. The gonoducts are generally emptied and residual gametes resorbed subsequent to spawning of *C. intestinalis* at Bird Rock. Similar phagocytosis of the sex cells in the ovary and the testicular follicles have been reported by Peres (1952) and Dybern (1965).

The basic requirement for settling species is space which, is often one of the main limiting resources of fouling organisms, especially in physically stable environments (Dayton, 1971; Paine, 1974). The ability of epifaunal species to colonize free space depends on such factors as larval availability, competitiveness and subsequently the biotic and abiotic factors governing community development. Svane (1983) recognized four ascidian reproductive patterns within coexisting species with very similar ecological characters and larval types. He considered *C. intestinalis* as belonging to 'Type 1' which are essentially oviparous and annual producing large numbers ($>10^5$) of small eggs (200 μm) and undergoing semelparous reproduction in summer (June-July). 'Type 1' reproductive strategies typically show large annual variations in population densities, mainly caused by fluctuations in recruitment and adult mortality. At Bird Rock the mean density and percentage cover of *C. intestinalis* are highly variable although they show an annual recurring pattern. Dybern (1965) and Nomaguchi (1974) were of the opinion that growth and life span of *C. intestinalis* are closely related to temperature. Individuals which settle and survive at Bird Rock generally live up to 15–16 months. Dybern (1965) found that populations living close to the surface in the Gullmarfjord may live for two breeding seasons. However, in general, he noted that populations between 0–7 m and 15–30 m lived for 12 months. Millar (1952) found a similar life expectancy for Scottish populations but reported that a number of individuals attained an age of 18 months.

The loss of individuals subsequent to new settlement at Bird Rock may be attributed to a variety of factors including mortality of adults, dislodgement due to overcrowding, biological interference and possibly predation. Subsequent to spawning, adult populations generally die leaving new available space by late summer-early autumn. Svane (1983) suggested that *C. intestinalis* allocates most of its energy to reproduction leaving a less protective tunic and a less secure hold on the substrate. In addition, many newly settling larvae settle in high numbers on existing adults thereby increasing drag effect which results in their eventual dislodgement. Further to natural adult mortality, dislodgement of juveniles due to intraspecific competition for space results in reduction of the population densities and the percentage cover occupied at Bird Rock. Gulliksen & Skjaevaland (1973) and Yamaguchi (1975) noted that predation on *C. intestinalis* is an important population regulating factor. There was no evidence of predation at Bird Rock. However, small numbers (<1 per 4 m^2)) of *Echinus esculentus* were noted foraging over the test areas and these may dislodge settled ascidians. Lundalv (1985) has suggested that the foraging of *E. esculentus* may account for local scale variations in population densities. Following the heavy settlement and the subsequent drop in the high densities and percentage cover occupied, *C. intestinalis* stabilize at 30–50% of the total cover prior to spawning. However, in 1985 shortly after settlement, the population at Bird Rock fell to <10% of the total cover. Over this period hydrographic conditions (salinity and temperature) did not show any significant deviations from expected values. However, foul weather, causing turbulent water conditions with heavy suspended load, persisted throughout the summer of 1985 when *C. intestinalis* was patently overcrowded. Natural mortality of adult populations, intraspecific competition for space, turbulent conditions, biological dislodgement and possibly, in some instances, smothering of the juveniles would account for the substantial drop in populations by the end of the summer.

The settlement and survival of *C. intestinalis* on cleared and undisturbed surface differed little once the cleared surfaces had been colonized. The latter sites initially present low competition substrates for settlement; with time they rapidly stabilize and show similar *C. intestinalis* densities and percentage covers to undisturbed surfaces.

The short duration of the Bird Rock experiment to date has not afforded the authors previous experience of the reduction in numbers of *C. intestinalis* which has been such a feature of 1985. It remains to be seen if the situation is redressed with the next spawning or if the relatively low numbers endure for a longer period as has characterized *C. intestinalis* populations in other geographical locations.

References

Berrill, N. J., 1947. The development and growth of *Ciona intestinalis*. J. Mar. biol. Ass. U.K. 26: 616–625.

Cost 47 Activity Report, 1985. Commission of European Communities. XII/ENV/46/85.

Costelloe, J., 1985. The annual reproductive cycle of the holothurian *Aslia lefevrei* (Dendrochirota: Echinodermata). Mar. Biol. 88: 155–165.

Costelloe, J. & B. F. Keegan, 1984. Feeding and related morphological structures in the dendrochirote *Aslia lefevrei* (Holothuroidea: Echinodermata). Mar. Biol. 84: 135–142.

Dayton, P. K., 1971. Competition, disturbance and community organization: the provision and subsequent utilization of space in a rocky intertidal community. Ecological Monographs 41: 351–389.

Dybern, B. I., 1965. The life cycle of *Ciona intestinalis* L. *f. typica* in relation to the environmental temperature. Oikos 16: 109–131.

Georges, D., 1971. La lumiere et le declement de la ponte chez *Ciona intestinalis*. In D. J. Crisp (ed.), Fourth European marine biology symposium. Cambridge University Press, London: 561–569.

Gulliksen, B., 1972. Spawning, larval development, growth, biomass and distribution of *Ciona intestinalis*. Sarsia 51: 83–96.

Gulliksen, B. & S. H. Skjaeveland, 1973. The sea-star, *Asterias rubens* L. as predator on the ascidian, *Ciona intestinalis* (L.), in Borgenfjorden, North-Trondelag, Norway. Sarsia 52: 15–20.

Keegan, B. F., B. O'Conner, D. McGrath, & G. Könnecker, 1976. The *Amphiura filiformis-Amphiura chiajei* community in Galway Bay (west coast of Ireland)--A preliminary account. Thalass. Jugosl. 12: 189–198.

Könnecker, G., 1973. Littoral and Benthic Investigations on the west coast of Ireland – (Section A: Faunistic and Ecological studies). The sponge fauna of Kilkieran Bay and adjacent areas. Proc. R. Ir. Acad. 73: 451–472.

Könnecker, G., 1977. Epibenthic assemblages as indicators of environmental conditions. In B. F. Keegan, P. O'Ceidigh & P. J. S. Boaden (eds), Biology of benthic organisms. Pergamon Press, Oxford: 391–395.

Könnecker, G. F. & B. F. Keegan, 1983. Littoral and benthic investigations on the west coast of Ireland--XVII. The epibenthic animal associations of Kilkieran Bay. Proc. R. Ir. Acad. 83B: 309–324.

Lambert, C. C. & C. L. Brandt, 1967. The effect of light on the spawning of *Ciona intestinalis*. Biol. Bull. mar. biol. Lab., Woods Hole. 132: 222–228.

Lewis, J. R., 1978. Benthic baselines. – A case for international collaboration. Mar. Pollut. Bull. 9: 317–320.

Lundalv, T., 1971. Quantitative studies on rocky bottom biocoenoses by underwater photogrammetry. A methodological study. Thalass. Jugosl. 7: 201–208.

Lundalv, T., 1985. Detection of long-term trends in rocky sublittoral communities: Representativeness of fixed sites. In P. G. Moore & R. Seeds (eds), The ecology of rocky coasts. Hodder & Stoughton Educational, U.K.

Millar, R. H., 1952. The annual growth and reproductive cycle in four ascidians. J. mar. biol. Ass. U.K. 31: 41–61.

Morgan, T. H., 1945. The conditions that lead to normal or abnormal development of *Ciona*. Biol. Bull. LXXXIII: 50–62.

Nomaguchi, T. A., 1974. Seasonal variations in the life span of the ascidian *Ciona intestinalis*. Exp. Geront. 9: 231–234.

Paine, R. T., 1974. Intertidal community structure. Experimental studies on the relationship between a dominant competitor and its principal predator. Oecologia 15: 93–120.

Peres, J. S., 1952. Recherches sur le cycle sexuel de *Ciona intestinalis* (L.). Arch. Anat. Micr. Morph. Exp. 14.

Rørslett, B., N. W. Green & K. Kvalvagnaes, 1978. Stereophotography as a tool in aquatic biology. aquatic Botany. 4: 73–81.

Runnstrom, S., 1927. Über die thermopathie der fortpflanzung und entwicklung mariner tiere. Berg. Mus. Arb., naturv. Rekke 2: 1–67.

Runnstrom, S., 1936. Die anpassung de fortpflangzung und entwicklung mariner tiere an die temperaturverhaltnisse verschiedener verbreitungsgebiete. Ibid. 3: 1–36.

Ryland, J. S. & A. Nelson-Smith, 1975. Littoral and benthic investigations on the west coast of Ireland--IV. (Section A: Faunistic and ecological studies) Some shores in counties Clare and Galway. Proc. R. Ir. Acad. 75B: 245–266.

Shin, P. K. S., 1981. The development of sessile epifaunal communities in Kylesalia, Kilkieran bay (west coast of Ireland). J. exp. mar. Biol. Ecol. 54: 97–111.

Sutherland, J. P., 1977. Development and stability of the fouling communities at Beauford, North Carolina. Ecological Monographs. 47: 425–446.

Svane, I., 1983. Ascidian reproductive patterns related to long term population dynamics. Sarsia 68: 249–255.

Yamaguchi, M., 1975. Growth and reproductive cycles of the marine fouling ascidians *Ciona intestinalis, Styela plicata, Botrylloides violaceus* and *Leptoclinum mitsukurii* at Aburatsubomoroiso Inlet (Central Japan). Mar. Biol. 9: 231–234.

Whittingham, D. G., 1967. Light induction of shedding of gametes in *Ciona intestinalis* and *Molgula manhattensis.* Biol. Bull. mar. biol. Lab., Woods Hole 132: 292–298.

Hydrobiologia 142: 113–119 (1986)

Long term changes in the benthic communities of Loch Linnhe and Loch Eil (Scotland)

T. H. Pearson, G. Duncan & J. Nuttall
Scottish Marine Biological Association, Dunstaffnage Marine Research Laboratory, P.O. Box 3, Oban, Argyll, Scotland

Keywords: long-term changes, benthos, organic enrichment, temperature anomalies

Abstract

Population fluctuations over a twenty year period in the macrobenthic communities of Loch Linnhe and Loch Eil, on the western coast of Scotland, are related to changing organic inputs to the area and to long-term temperature changes. It is suggested that the carrying capacity of sedimentary benthos is dependent on organic input, but that the species composition of the communities may be modified by climatic fluctuations acting on the spawning success and subsequent recruitment of particular species. Confirmation of such hypotheses generated from field time-series data should be obtained by experimentation.

Introduction

Studies of the fluctuations in the benthic populations of Loch Linnhe and Loch Eil, a sea loch system on the west coast of Scotland, have been in progress since 1963. The distribution of populations throughout the system at that time have been described (Pearson, 1970, 1971a) and subsequent changes, attributable to the influence of the discharge of pulp and paper mill effluent to the system which first occurred in 1966, have been detailed (Pearson, 1971b, 1975). Discharge of effluent to the system was greatly reduced in 1980 following closure of the pulp mill. Throughout the period of effluent discharge carbon inputs to the benthic ecosystem were greatly enhanced (between 4× and 14× mean input from planktonic sources, Pearson, 1982) and the considerable fluctuations in populations and changes in species composition recorded during this period were largely explained as a consequence of this carbon enrichment. Following the cessation of pulp mill effluent input in 1980 (a low level of organic effluent input from the paper mill continues to be discharged to the loch) changes in the benthic populations have followed, to a great extent, the patterns predictable as a consequence of declining organic enrichment (Pearson & Rosenberg, 1978). However over the twenty year period covered by the observations there have been considerable fluctuations in other major environmental variables in the system in addition to organic input. The possibility that long-term variation in winter temperature minima and means could be responsible for subsequent variation in species abundance and succession in benthic communities off the Northumberland coast was suggested by Buchanan *et al.*, (1978). These authors suggested that a reversal in 1971 of the overall temperature regime from a period of cold winters in 1965 to 1970 to a warmer period of winter temperatures in 1971–76 might explain an observed switch in the benthic communities. In the later, warmer, years small polychaete species predominated over the larger, previously dominant, species. Since some of the species noted by Buchanan *et al.*, as becoming predominant on the Northumberland grounds in the warmer, post 1971, period, are the same as those noted as increasing in Loch Linnhe and Loch Eil in the period following effluent input, it seems prudent to examine the full data set to establish what, if any, role temperature variation may have played in mediating the faunal successions observed in the Linnhe/Eil system.

Study area

Loch Linnhe and Loch Eil form the inner end of the Firth of Lorne, an extensive fjordic complex on the west coast of Scotland (Fig. 1). The two lochs are separated by a narrow and shallow sill, a region of considerable tidal mixing into which effluent is discharged. Data from a single sampling station in each loch will be considered, Station 24 at 34 m depth near the head of Loch Eil, and Station 53 at 120 m depth in the centre of Loch Linnhe. A further station, LY1, in the central Firth of Lorne some 38 km SW of the sill dividing Loch Linnhe from the rest of the Firth, has been sampled in 1975–76 and from 1980 onwards. Since this area is populated by communities similar to those origi-

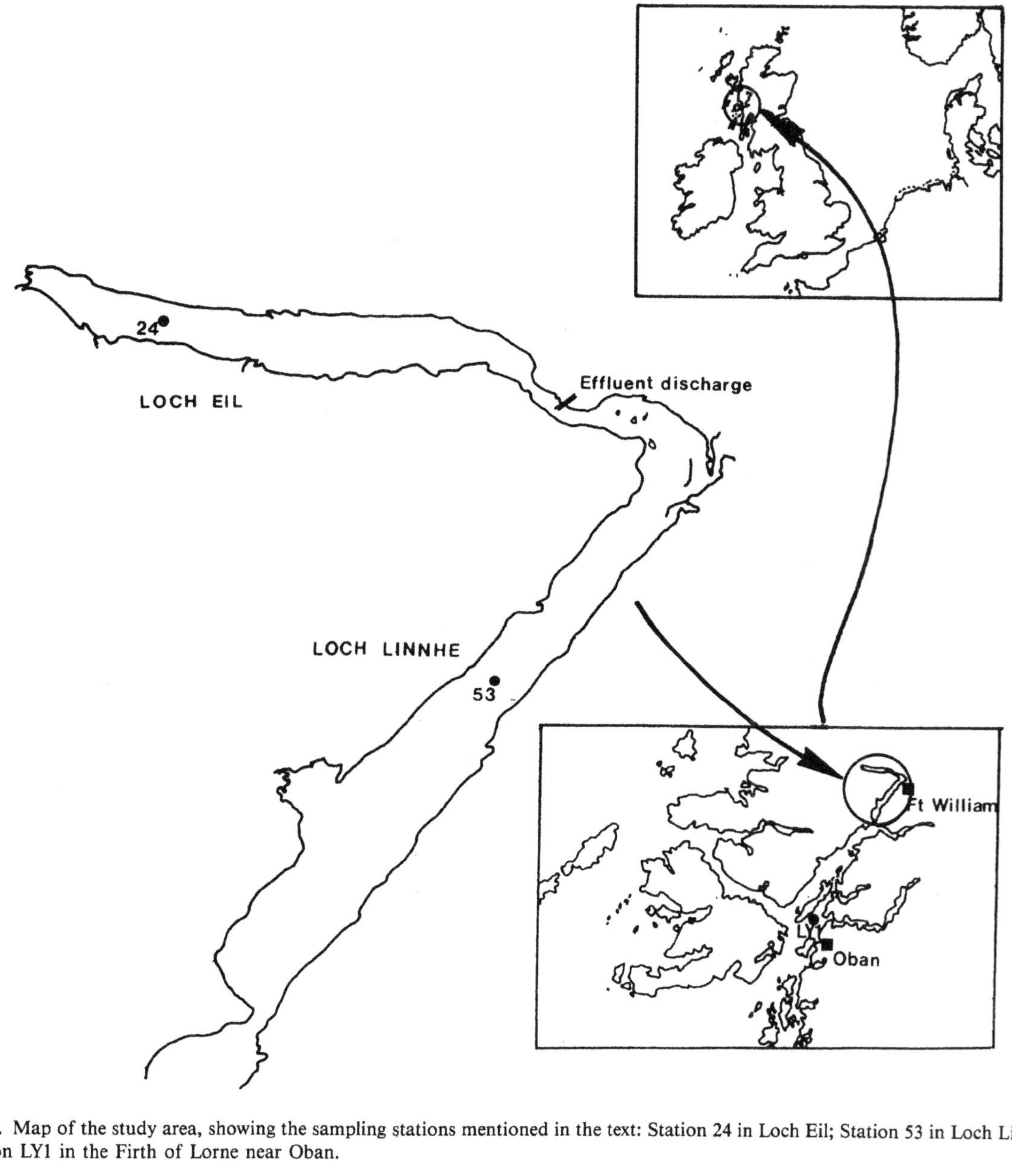

Fig. 1. Map of the study area, showing the sampling stations mentioned in the text: Station 24 in Loch Eil; Station 53 in Loch Linnhe; Station LY1 in the Firth of Lorne near Oban.

nally documented in the Loch Linnhe/Eil system and is well beyond any possible influence from the effluent discharged to that system this data is included for comparison.

Short-term changes in benthic populations in response to fluctuations in effluent input

Detailed studies of changes in many aspects of the benthic ecosystem of Loch Eil in response to fluctuations in effluent input were undertaken in 1975–76 (Pearson, 1981, 1982a & b). These suggested that the macrobenthic populations responded to changes in effluent input with a time lag of between two and six months (Fig. 2). Such essentially short-term changes were largely confined to populations of small annelid worms, whose opportunist life history strategies are suited to the rapid exploitation of ephemeral resources. Such populations were predominant at Station 2, situated closest to the source of effluent in the narrows between the two lochs. At Station 24 near the head of the loch, whilst not predominant, they formed a significant proportion of the populations.

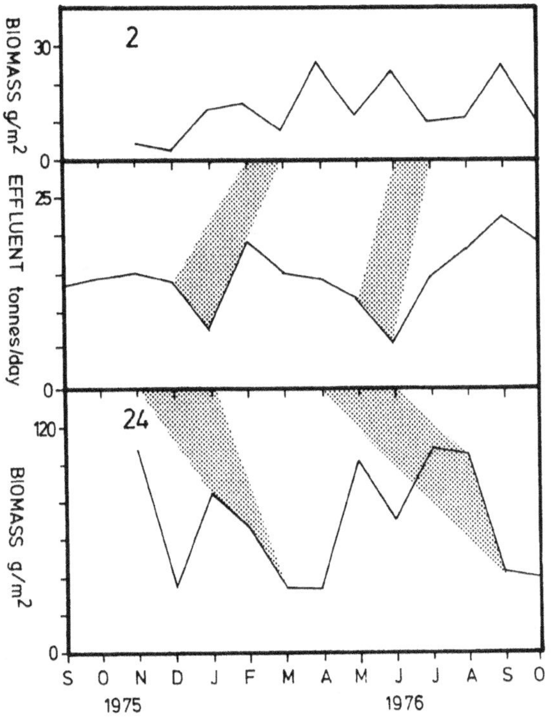

Fig. 2. Monthly variation in biomass (wet weight g/m²) at two sampling stations in Loch Eil during 1975 and 1976 compared with fluctuations in effluent input. Station E2 was situated in deep water 2.5 km from the effluent discharge point. Station 24 was situated at the head of the loch 8 km from the discharge point. At both stations a fall in effluent discharge levels was followed by a decline in benthic biomass after a period of between 6 and 8 weeks.

Annual fluctuations in benthic populations

Changes in the total biomass (wet weight) of macrofauna occurring at the two stations in Loch Linnhe and Loch Eil over the twenty year period 1964–83 in relation to the amount of effluent discharged and the prevailing temperature regime are shown in Fig. 3. Effluent input, recorded as the an-

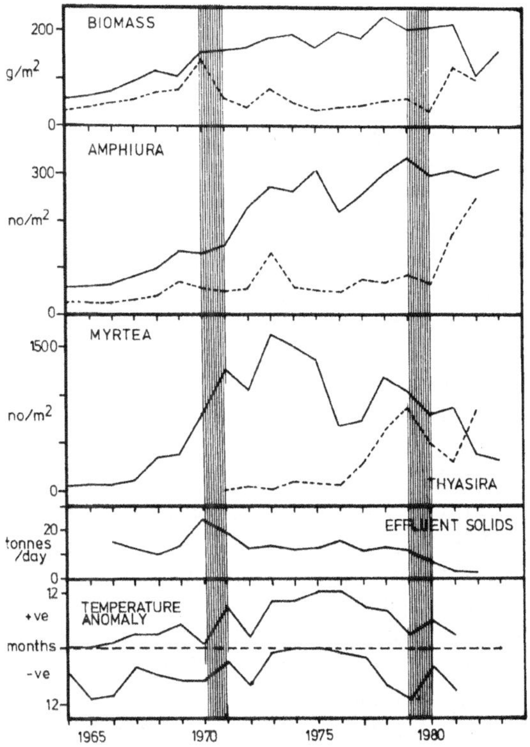

Fig. 3. Variation, over a 20 year period, in the total benthic biomass (annual mean g wet weight per m²) and the mean annual abundance of two dominant species (*Amphiura chiajei* and *Myrtea spinifera* at Station 53 [solid lines]; *A. chiajei* and *Thyasira flexuosa* at Station 24 [pecked lines]) compared with fluctuations in effluent inputs (annual mean tonnes suspended solids per day) and in temperature anomalies (sum of number of months in each year in which positive and negative deviations from the long-term mean were recorded). Note the predominantly negative anomalies prior to 1971 and the predominantly positive anomalies in the period 1972–78.

nual mean tonnes per day of suspended solids discharged, commenced in 1966 and between then and 1969 fluctuated between 10 and 15 tonnes per day. Between 1969 and 1971 the mean daily input fluctuated between 15 and 25 tonnes per day but thereafter declined to fluctuate between 10 and 15 tonnes until 1979. By 1982 it had declined to less than 5 tonnes per day. The trends in temperature over the period are shown as accumulative monthly temperature anomalies for each year i.e. the sum of the number of months in each year in which positive and negative deviations from the long-term mean were recorded. These detailed records are from an area in the Firth of Clyde some 110 km south of the study area (Barnett & Watson, in press and pers. comm.), but where the prevailing temperature regime is comparable to the Firth of Lorne area. The records demonstrate that in the period 1964 to 1970 negative temperature anomalies predominated. There was an abrupt switch in 1971 and for the subsequent eight years positive anomalies were most evident. Thus the two periods when the effluent inputs to the system were changing most rapidly, i.e. in 1970–72 when inputs increased substantially and 1979–81 when they decreased most markedly, the temperature regime was switching to and from a warmer sequence of years. The overall biomass recorded at Station 24 in Loch Eil (Fig. 3) showed an overall decline in both 1970–71 and in 1979–80, whereas at Station 53 in Loch Linnhe there was no deviation at either time from the steadily increasing trend recorded over almost the entire period. Abundance records over the period for three particularly important species are also illustrated in Fig. 3. The brittle star *Amphiura chiajei* is an important community member at both stations. At the head of Loch Eil numbers remained only slightly greater than the 1964 levels for much of the period, but rose substantially from 1980 onwards. At Station 53 numbers rose progressively to be six times their original level by 1979. Thereafter no further increases occurred. The bivalve mollusc *Myrtea spinifera* increased substantially in numbers at Station 53 following the introduction of effluent to the system, to reach 30 times its original density in 1972–73. Thereafter it declined progressively to be about 8 times its 1964 density in 1983. *Myrtea* did not occur in any numbers at Station 24, but another small bivalve, *Thyasira flexuosa,* which first appeared as a regular member of the community there in 1971, reached high numbers in 1979 and again in 1982, following a decline in 1980–81.

Changes in the principal population statistics at the two Linnhe/Eil stations and at the Firth of Lorne control station are compared with effluent and temperature fluctuations over the period 1971–1983 in Fig. 4. At the stations at the head of Loch Eil all three statistics fluctuated considerably throughout the period. These general changes appear to correspond more closely to the changes in effluent input than to temperature but there are no close correlations in either case. At Station 53 in Loch Linnhe biomass and species numbers fluctuated between relatively narrow limits during the period, but total abundance varied considerably

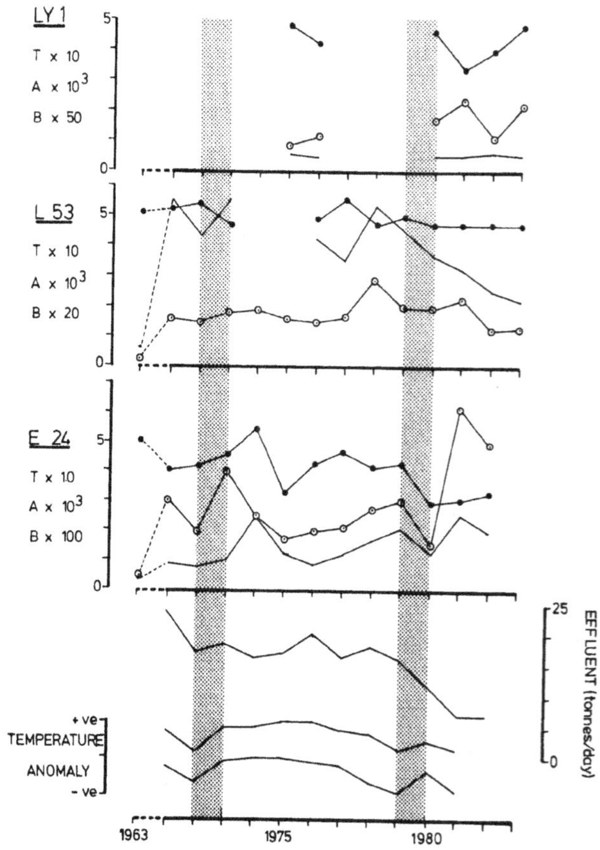

Fig. 4. Changes in the principal population statistics at the Stations in Loch Linnhe and Loch Eil over a twelve year period, and at the Station in the Firth of Lorne 1980–84, compared with annual mean fluctuations in effluent input and with temperature anomalies. T, total number of taxa recorded; A, total abundance (number per m²); B, total biomass (g wet weight per m²).

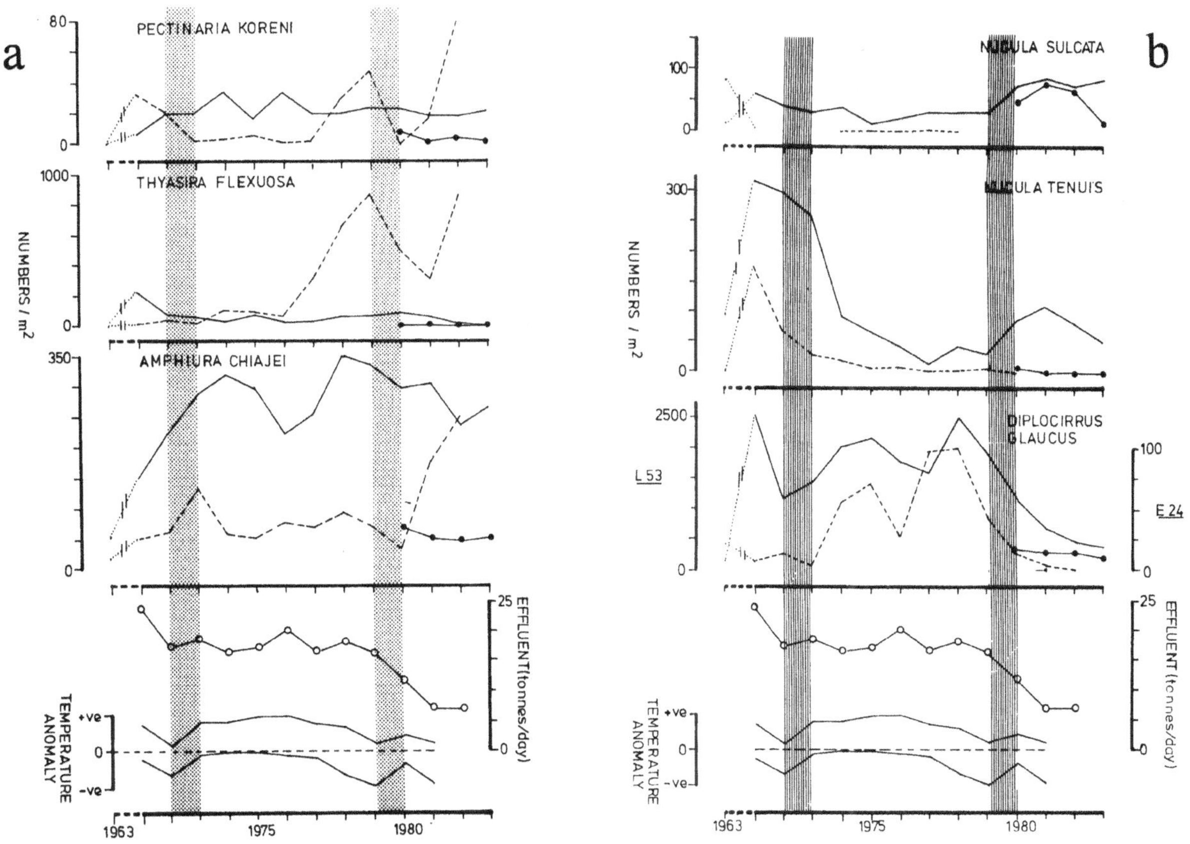

and followed the trend in effluent input relatively closely. At the reference station in the Firth of Lorne for the brief period when records are available total abundance varied little but biomass and species numbers changed markedly, although not obviously in response to temperature. Fig. 5a–d compares changes in the abundances of some dominant species in each area with fluctuations in effluent input and in temperature. These can be classified in three categories: (1) those whose fluctuations appear to be a response to changing effluent input; (2) those which apparently respond to both effluent input and temperature; and (3) those which appear to have no great response to either. In the first category include the sub-surface deposit feeders *Pectinaria koreni* and *Thyasira flexuosa,* the surface deposit feeders *Amphiura chiajei* and *Diplocirrus glaucus* and the sub-surface predators *Goniada maculata* and *Nephthys hystricis.* In the second group are the subsurface deposit feeders *Nucula sulcata, Nucula tenuis* and *Levensenia gracilis,* the surface deposit feeders *Prionospio cirrifera* and *Terebellides stroemi* and the surface predator *Pholoe minuta.* The feeding mode of the sub-surface dwelling *Ancistrosyllis groenlandica* is unknown. The only species showing little population response to either variable is the sub-surface carnivore *Glycera alba.*

Discussion

This synopsis of some of the gross population changes which have occurred in the Linnhe/Eil system over a twenty year period provides little evidence to support a major revision of the conclusions drawn previously as to the cause of those changes: namely that the benthic populations and communities of the area have fluctuated in response to gross changes in organic inputs to the system (Pearson, 1975, 1982b). Thus the major changes in species numbers, abundances and biomass at the stations in both lochs appeared to respond to changes in organic input and not to tem-

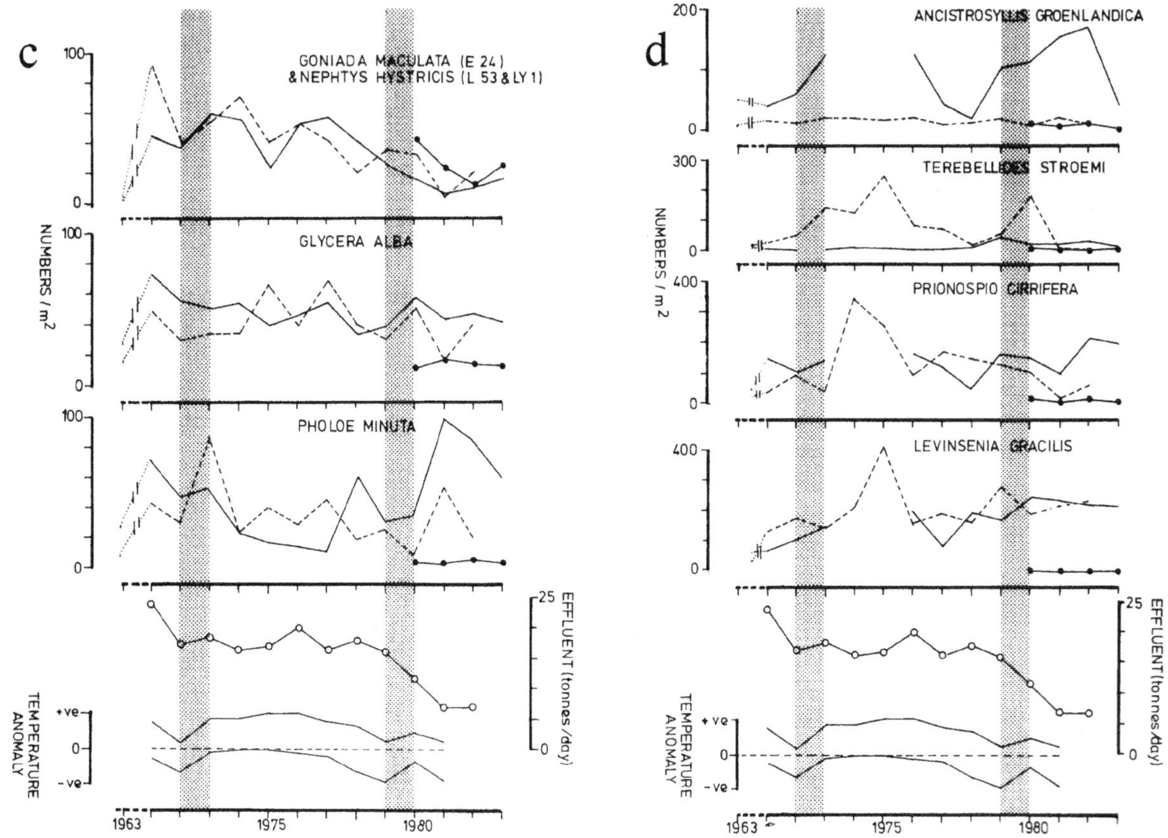

Fig. 5a–d. Changes in the abundance (annual mean numbers per m²) of some dominant species at Station 24 (broken line), Station 53 (continuous line) and Station LY1 (solid dots) over a twelve year period, compared with fluctuations in effluent input and with temperature anomalies (see Fig. 4 for units).

perature. Moreover the species showing the greatest changes are those known to be sensitive to changes in organic enrichment in a variety of different areas (Pearson & Rosenberg, 1978). However examination of the population trends for a number of such species suggested that some may be responsive to the temperature regime in addition to the organic enrichment. Thus the populations of *Terebellides, Prionospio* and *Levensenia* all peaked in 1974–75 at Station 24 during the period when the greatest positive and lowest negative temperature anomalies were being recorded, and when effluent inputs were relatively stable. Buchanan *et al.*, (1978) noted that during the same period *Levensenia* (recorded as *Paraonis gracilis*) and another *Prionospio* species, *P. malmgreni,* become predominant in the soft mud communities off the Northumberland coast. In that area, during the period of high winter temperature, a suite of species dominated by small polychaete worms replaced a suite in which larger polychaetes predominated. This succession was not dissimilar from that observed to be taking place in Loch Linnhe and Loch Eil under the impact of pulp mill effluent, and it is possible that the switch in the prevailing temperature regime may have had some influence on the species composition of the less radically impacted communities of the area. Despite the changes recorded in community composition the overall production of the Northumberland populations remained essentially unchanged (Buchanan *et al.,* 1974), suggesting that climatic changes were reflected, not in the carrying capacity of the habitat, but in the species composition. Similarly in the Linnhe/Eil system climatic change did not appear to influence the principal population parameters (species richness, biomass and total abundance) but may have been reflected to a certain extent in species composition. Thus it may be suggested that the carrying capacity of sedimentary benthos is dependent on the organic input, i.e. food availability, but that the species composition of the communities may be modified by climatic fluctuations. The mechanisms by which such modifications might be imposed are not immediately obvious, but it is probably through some temperature sensitive influence on spawning success and larval survival and hence recruitment. Barnett & Watson (in press) have demonstrated that recruitment of an infaunal bivalve species to the benthos of a shallow area in the Firth of Clyde was greatly enhanced by higher than normal temperatures in the period prior to spawning. Similarly spatfall failures were related to subnormal prespawning temperatures. Thus abnormally high or low temperatures during the spawning period may exert an influence on which species successfully recruit in any particular year. A complete assessment of the influence of temperature on spawning success and subsequent recruitment will necessitate more detailed observations. These could most usefully be made using manipulative experimentation, or by the use of mesocosms in which both organic inputs and temperatures can be controlled (c.f. Oviatt *et al.*, 1984) rather than by further necessarily extensive field observations. Nevertheless it is only by the use of extensive time series of benthic data such as those accumulated in Loch Eil/Linnhe that the potential impact of both short and long-term environmental change can be assessed. Moreover it is equally obvious that a more geographically widespread data series would have provided a better assessment of the relative impacts of organic enrichment and temperature fluctuations on the benthic communities.

Acknowledgements

The Loch Linnhe, Loch Eil surveys have been financed in part by Scottish Pulp and Paper Mills Ltd.

References

Barnett, P. R. O. & J. Watson, in press. Long-term changes in some benthic species in the Firth of Clyde. Proc. R. Soc. Edinb. B.

Buchanan, J. B., P. F. Kingston & M. Sheader, 1974. Long-term population trends of the benthic macrofauna in the off-shore mud of the Northumberland coast. J. mar. biol. Ass. U.K. 54: 785–795.

Buchanan, J. B., M. Sheader & P. F. Kingston, 1978. Sources of variability in the benthic macrofauna off the south Northumberland coast, 1971–76. J. mar. biol. Ass. U.K. 58: 191–209.

Oviatt, C. A., M. E. A. Pilson, S. W. Nixon, J. B. Frithsen, D. T. Rudnick, J. G. Kelly, J. F. Grassle & J. P. Grassle, 1984. Recovery of a polluted estuarine ecosystem: a mesocosm experiment. Mar. Ecol. Prog. Ser. 16: 203–217.

Pearson, T. H., 1970. The benthic ecology of Loch Linnhe and Loch Eil, a sea-loch system on the west coast of Scotland. I. The physical environment and distribution of the macrobenthic fauna. J. exp. mar. Biol. Ecol. 5: 1–34.

Pearson, T. H., 1971a. Studies on the ecology of the macrobenthic fauna of Lochs Linnhe and Eil, west coast of Scotland. II. Analysis of the macrobenthic fauna by comparison of feeding groups. Vie Milieu, Suppl. 22: 53–91.

Pearson, T. H., 1971b. The benthic ecology of Loch Linnhe and Loch Eil, a sea-loch system on the west coast of Scotland. III. The effect on the benthic fauna of the introduction of pulp mill effluent. J. exp. mar. Biol. Ecol. 6: 211–233.

Pearson, T. H., 1975. The benthic ecology of Loch Linnhe and Loch Eil, a sea-loch system on the west coast of Scotland. IV. Changes in the benthic fauna attributable to organic enrichment. J. exp. mar. Biol. Ecol. 20: 1–41.

Pearson, T. H., 1981. The Loch Eil project: introduction and rationale. J. exp. mar. Biol. Ecol. 55: 93–102.

Pearson, T. H., 1982a. The Loch Eil project: population fluctuations in the macrobenthos. J. exp. mar. Biol. Ecol. 56: 305–321.

Pearson, T. H., 1982b. The Loch Eil project: assessment and synthesis with a discussion of certain biological questions arising from a study of the organic pollution of sediments. J. exp. mar. Biol. Ecol. 57: 93–124.

Pearson, T. H. & R. Rosenberg, 1978. Macrobenthic succession in relation to organic enrichment and pollution of the marine environment. Oceanogr. Mar. Biol. Ann. Rev. 16: 229–311.

Hydrobiologia 142: 121–127 (1986)

Long-term studies at a benthic station off the coast of Northumberland

John B. Buchanan & J. Jonathan Moore
Dove Marine laboratory, Cullercoats, North Shields, Tyne and Wear, United Kingdom

Keywords: benthos, stability, density dependent mortality, temperature, organic enrichment, North Sea

Abstract

Biannual sampling (March and September) has been carried out over the period 1972 to 1985 at a muddy sand station, 55 m depth, with fauna belonging to the deeper offshore edge of Petersen's *Amphiura filiformis* community.

During the period 1974 to 1980 the community exhibited a high degree of persistence stability. This stability was lost between 1980 and 1983, with rising total numbers and biomass and changes in species ranking. There is some evidence of a downward reversal between 1984–85.

Evidence suggests that the principal stabilising process is density dependent mortality mediated by competition in a food limited environment. The principal destabilising process appears to be periodic fluctuations in the organic flux to the bottom. A secondary destabilising process is clearly concerned with fluctuating winter temperature. In competitive terms, cold winters favour increased survival in the dominant species at the expense of the lesser ranked species. This process is, however, more ephemeral and subject to adjustment within the time scale of a year.

Introduction

This paper summarises the temporal variability encountered between 1972 and 1985 at a silty sand station (M1) at a depth of 55 m, 6.5 miles offshore. In terms of its faunal constitution this station would be regarded in the classical Petersen sense as belonging to the deeper outside edge of the *Amphiura filiformis* community. In terms of production, however, the community is dominated by deposit feeding polychaetes, and *A. filiformis* is in fact the only significant suspension feeder. The monitoring has witnessed a comparitively prolonged period of persistence stability during the years 1974–80, followed by a period of change and instability during the years 1981–85. This latter period has involved changes in species ranking, total numbers and total biomass. Periods of change and periods of stability are clearly both of considerable biological interest.

The sampling station and sampling methods

The sampling regime and techniques have been fully described in Buchanan & Warwick (1974) and have been standardised throughout. Macrofauna are considered to be those animals held on a 0.5 mm sieve. The nature of the bottom sediment has remained stable throughout the exercise, and consists of a mixture of fine and medium sand (0.125–0.5 mm) together with 20% by weight of silt (<0.063 mm). Based on the experience of the earliest surveys the station was sampled in March and September corresponding to the low and high points respectively of the total numbers of individuals in the annual cycle. A quantitative assessment of recently settled macrofauna of meiofaunal size (<0.5 mm) was made by analysing ten subcores, five from each of two 'Haps' cores (Kanneworf & Nicolaisen, 1973).

Analytical methods

For the purpose of data analysis, the full data set was reduced to include only those taxa which occured with a density of at least 6 individuals per square metre in any one sample. This reduced the data set to 125 taxa.

The year to year similarities have been compared using Czekanowski Percentage Similarity (CPS) (Pielou, 1984). This is taken as –

$$CPS = 200 \times \sum \frac{\min (x_{i1}, x_{i2})}{(x_{i1} + x_{i2})}$$

where x_{i1} and x_{i2} are the amounts of species i in samples 1 and 2 respectively (i=1, 2, 3...., s).

The between sample (year) matrix derived from the CPS analysis was then subjected to two further analyses:

a) Linear Graphical Ordination – a graphical method devised to describe the similarity trends of individual or grouped years.

b) Principal Coordinate Analysis (Gower, 1967) – an eigenvector method of ordination that proceeds from any distance or dissimilarity matrix. Transformation of the CPS matrix into distance matrices was carried out with the Manhattan metric.

General review of variability 1972–85

The total numbers of individuals together with the ash free dry weight biomass are shown in Fig. 1

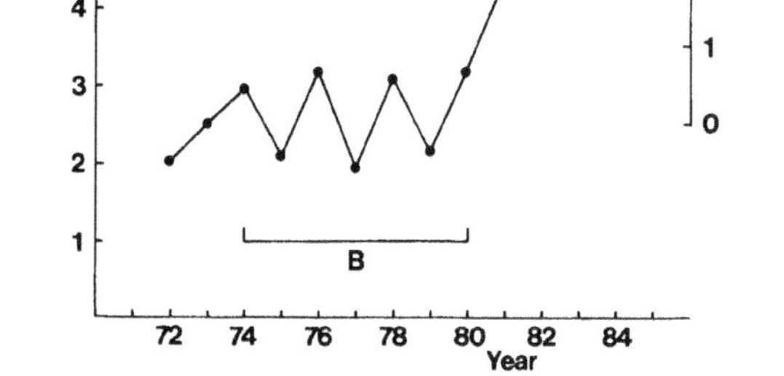

Fig. 1. Variation in the total number of individuals (N, joined), and the biomass (ash free dry weight), per square metre.

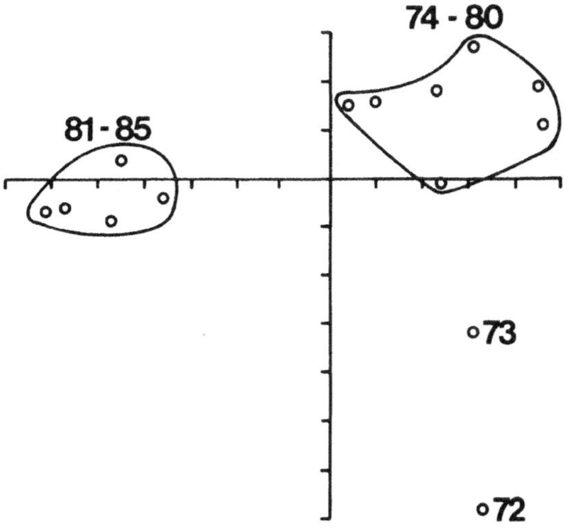

Fig. 2. Principal Coordinate Analysis (first two axes) of samples from station M1, using Czekanowski Percentage Similarity.

for March samples from 1972 to 1985. Total numbers are seen to oscillate between approximately 2000 and 3000 per square metre during the period 1972–80. Between 1980 and 1981 this cycle was broken and there is a sharp rise in numbers to 4348, and in 1982 to 5730 per square metre. Numbers then oscillate between 4000 and 6000 for the remaining years. Sample biomass may be seen to follow a similar trend, at least with respect to the marked rise between 1980 and 1981.

The ordination, Fig. 2, derived from Principal Coordinates Analysis of a Czekanowski distance matrix suggests that the monitoring period can be broken into three sub-periods. A. The starting years of the monitoring exercise (1972–73) appear to be the final stage of a period of faunal instability. B. 1974–80, a period of high persistence stability. C. 1981–85, a second shorter period of stability but fluctuating and at a higher level of biomass and total numbers. This rise in numbers and biomass was accompanied by changes in ranking as is shown in Table 1. The introduction of *Mysella bidentata* and *Heteromastus filiformis* into high ranking positions in the average top twenty species was clearly an important change.

In Fig. 3 a graphical method of comparing the two periods has been devised which preserves the chronological continuum and allows an assessment of the time scales involved in the period of the change. The similarities between one sample (year) and all the other years (x axis) are plotted from left

Table 1. Mean number of individuals per square metre for the twenty top ranked species in each of the two defined periods.

Period B 1974–80	N m^{-2}	Period C 1981–85	N m^{-2}
Prionospio malmgreni	333	Prionospio malmgreni	638
Paraonis gracilis	212	Mysella bidentata	331
Thyasira flexuosa	120	Amphiura filiformis	312
Magelona minuta	113	Heteromastus filiformis	305
Amphiura filiformis	107	Chaetozone setosa	269
Myriochele oculata	103	Paraonis gracilis	256
Chaetozone setosa	99	Pholoe minuta	256
Nephtys spp.	96	Myriochele oculata	223
Phoronis muelleri	85	Spiophanes bombyx	204
Harpinia antennaria	80	Tharyx sp.	201
Tharyx sp.	76	Phoronis muelleri	174
Owenia fusiformis	75	Nemertea spp.	157
Spiophanes bombyx	66	Nephtys spp.	127
Lucinoma borealis	64	Nucula tenuis	116
Ampharete baltica	53	Oligochaete spp.	100
Pholoe minuta	50	Thyasira flexuosa	79
Nemertea spp.	46	Montacuta ferruginosa	73
Abra nitida	39	Harpinia antennaria	70
Glycera rouxii	38	Ampharete baltica	68
Ampelisca tenuicornis	36	Owenia fusiformis	63

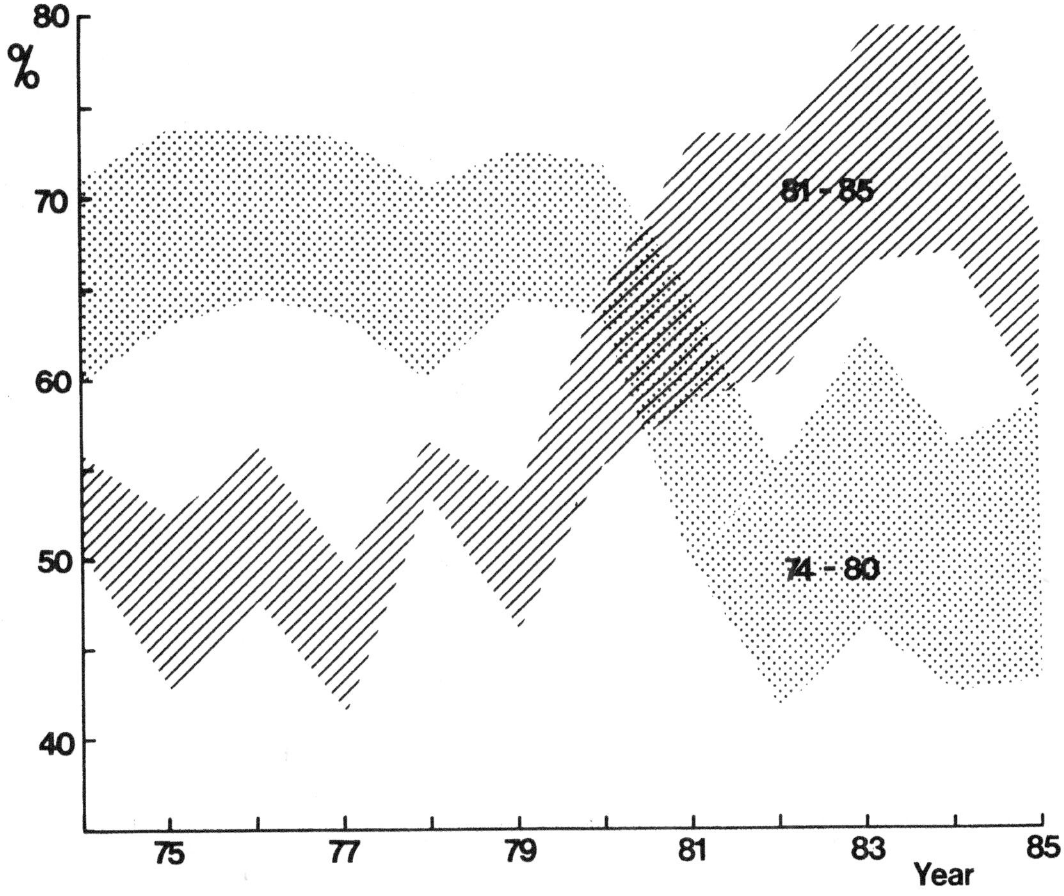

Fig. 3. Linear presentation of between sample similarities (Czekanowski Percentage Similarity). For construction details see text.

to right and joined with a continuous line. This is repeated for each year in turn, and these lines may then be grouped into bands of years that follow a similar path. In Fig. 3 these bands can be seen to correspond to the defined periods and it shows that the second 'stable' period is not as stable as it might seem from the ordination. In fact the 1985 sample shows a marked divergence from the rest of the period, suggesting that the community may be changing once more.

The effect of temperature

During an earlier stage in the monitoring exercise (Buchanan *et al.*, 1978) it was suggested that variability in winter temperatures (averaged between December and May) had an important effect on faunal variability resulting in 'cold winter states' and 'warm winter states'. This hypothesis is no longer supported by the data, but winter temperatures are shown to have a clear influence on ephemeral changes in the diversity of the March sample. In Fig. 4 the variability in the Shannon-Weiner index of community diversity is plotted alongside the variability of the winter temperature anomaly. Warm winters result in the March sample having a high diversity, whilst cold winters result in a reduced diversity. The relationship is very convincing ($r = 0.883$, $p = < 0.001$) and is almost wholly due to changes in equitability rather than species richness. Correlation of Shannon-Weiner diversity (H) against equitability (J) for March samples gives a coefficient (r) of 0.9064 whilst correlation of H against log2S (where S= total numbers of species) gives a coefficient of only 0.4558. This effect is short-lived and winter temperature does not appear to have any effect on diversity of the following September sample ($r = -0.2382$).

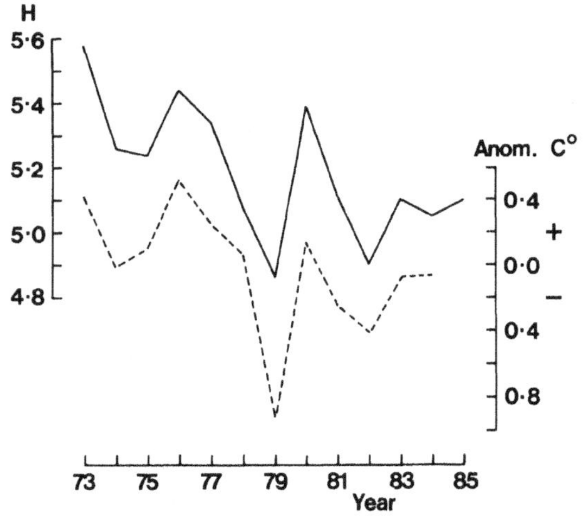

Fig. 4. Variation in Shannon-Weiner index (H(s)) for March samples (solid line), and temperature anomaly from mean winter sea-water temperature (December-May) (dashed line). Mean is 6.41 °C.

The effect of increased organic input

The marked rise in total numbers of individuals and biomass suggests that benthic production has risen and, assuming that food is limiting, this suggests that there has been an increase in organic input to the community. Without direct measurement of organic content of the bottom sediments or of organic input to the bottom it is not possible to prove or disprove this hypothesis. There is, however, circumstantial evidence to support the hypothesis from phytoplankton abundance data for the North Sea.

A crude but effective measure of phytoplankton abundance in the sea area adjacent to the Northumberland coast (area C2) is carried out by means of the Continuous Plankton Recorder. This measure is a 'Net colour' index and is carried out by a visual comparison of net greeness. A greeness scale is calibrated to give a numerical estimate of actual phytoplankton abundance by tests based on acetone extractions (Reid, 1978).

In Fig. 5 the net colour index has been summed

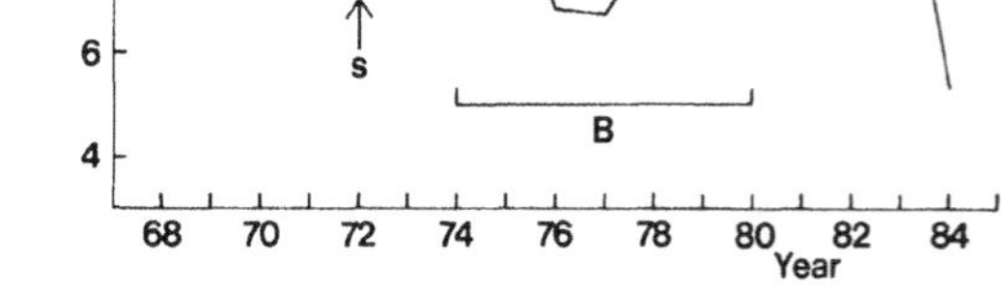

Fig. 5. Variation in the 'Net colour' index of phytoplankton abundance for the North Sea (area C2), actual (solid line) and three year running average (dashed line). The start of the benthos monitoring programme (s) and the stable period (B) are indicated.

for each year and plotted over the period 1968 to 1984. It shows that during the period 1973 to 1980 the net colour index of phytoplankton abundance was low in comparison to the five previous years (1968–72) and that in 1981 and 1982 there was an abrupt rise followed by an equally abrupt fall in 1983 and 1984. In the benthos, the stable period with low numbers and biomass ('B' in Fig. 5) can be seen to correspond almost exactly with the period of low phytoplankton abundance.

Density dependent mortality

The seven years 1974–80 represented a period of relatively high persistence stability with year to year faunal similarity (Czekanowski Percentage Similarity) constantly within the range of 60–70% and the total numbers of individuals in March within the range of 2000–3000 per square metre (Figs 1 and 3). Considering this period in more detail, the fluctuations in total numbers of individuals are shown in Fig. 6, taking into consideration the September as well as the March samples. These represent the high and low points respectively of the annual cycle. In Fig. 6 the twenty top-ranked species are shown separately (Fig. 6b) and account for between 70 and 80% of the total individuals in any one year. The lesser ranked species, accounting for 20–30% of the individuals are shown in Fig. 6a. In either case the curves are substantially similar suggesting that almost the entire community is synchronised in a distinct pattern of fluctuation. The annual cycle shows that in each year there is a period of sieve recruitment, March to September, followed by a period of mortality, September to March. It can be shown that this cycle is not dependent on the settlement times of the constituent species. Although each individual species has a well defined settlement period there are as many species settling in autumn and winter as in spring and summer, and indeed the settlements are spread fairly evenly throughout the year. Spring and summer, however, appears to be the time of maximum somatic growth in most species, leading to greater sieve recruitment from April to September.

In addition to the annual cycle, there is also clear evidence of a biennial cycle (Fig. 6) where a low March figure is followed by a high September and a high March figure by a correspondingly low September. This high-low-high sequence held steady for the entire period 1973–80. If the March figure is high, then sieve recruitment between March and September is proportionally low, but a low March figure is invariably followed by higher recruitment. This relationship is shown in Fig. 7 where sieve

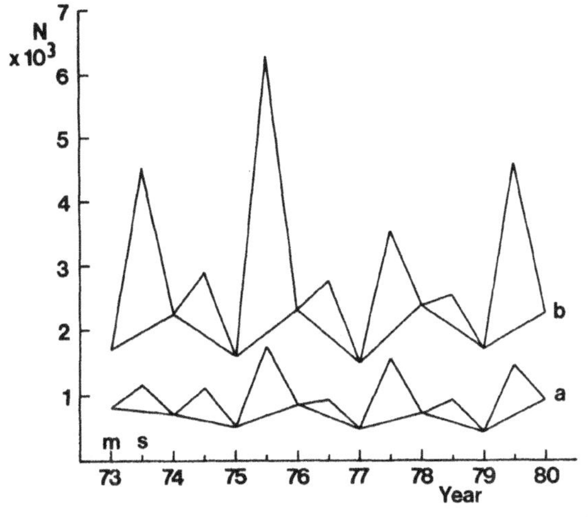

Fig. 6. Number of individuals represented by the twenty top ranked species (b), and the remaining species (a), in samples from March (m) and September (s).

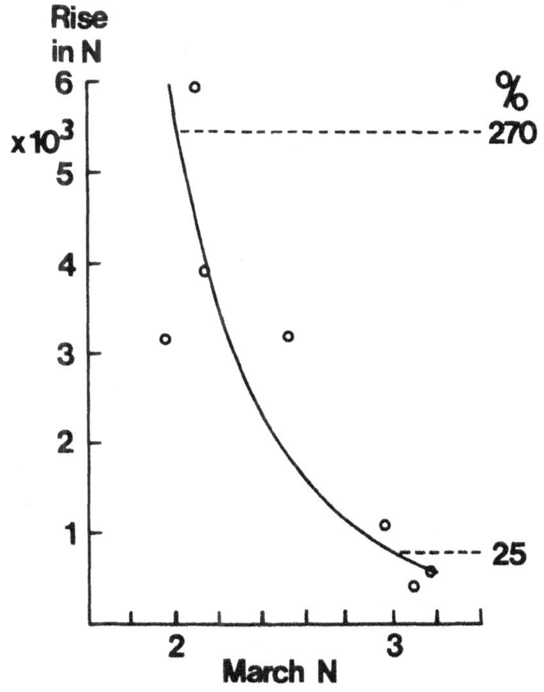

Fig. 7. Calculated regression of rise in numbers of individuals from March to September against numbers of individuals present in March. Percentage mortality at two points on line is indicated.

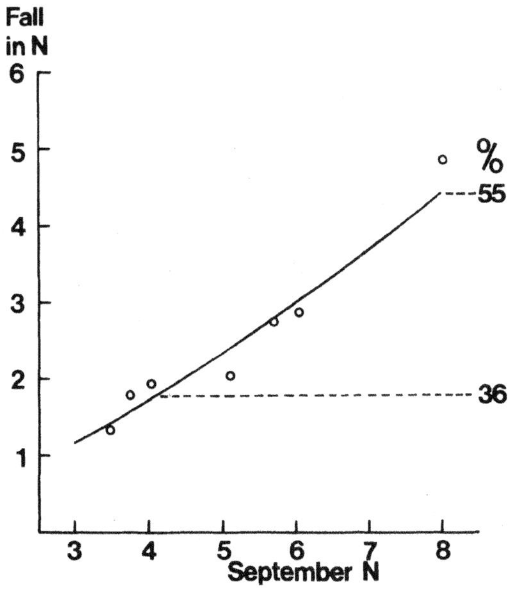

Fig. 8. Calculated regression of reduction in numbers of individuals from September to March against numbers of individuals present in September. Percentage mortality at two points on line indicated.

recruitment is graphed against the March total numbers. A power curve regression line has been fitted ($r=0.877$, $p=<0.02$). In addition, a higher proportional mortality invariably follows a high September figure for total individuals and a comparatively low mortality follows a low September. Fig. 8 shows the power curve relationship of this mortality ($r=0.975$, $p=<0.01$). Both curves represent strong evidence for density dependent relationships.

It is tempting, but in some ways misleading, to regard the two power curves as dealing with recruitment on one hand and mortality on the other. In fact only 'sieve recruitment' is represented by the curve in Fig. 7. Sieve recruitment may be far removed in time from the actual recruitment of a cohort to the bottom. Unless an impracticably small sieve mesh is employed, there is generally a time-lag between actual settlement of a cohort on the bottom and complete sieve recruitment. To illustrate the point, two examples of recruitment are shown in Figs 9 and 10. These deal with the polychaetes *Nephtys hombergi* (Fig. 9) and *Heteromastus filiformis* (Fig. 10). Data from meiofaunal cores have been combined with data from a 0.5 mm sieve

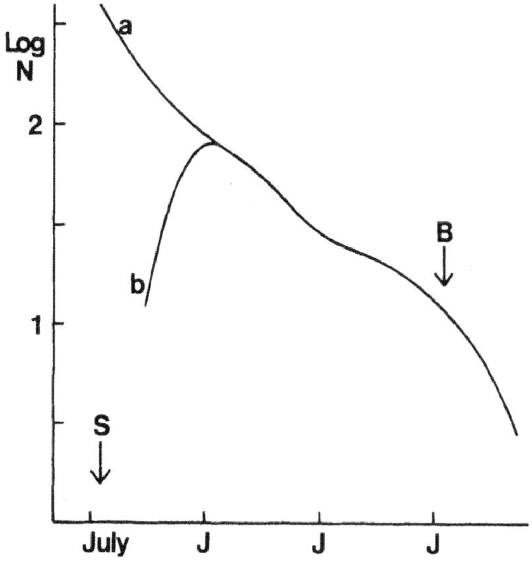

Fig. 9. Combined survivorship and sieve recruitment curve for *Nephtys hombergi.* Times of settlement (S) and breeding (B) are indicated.

in order to follow the progress of a cohort from settlement on the bottom to the time of breeding. In effect this provides a survivorship curve (curve a) together with a sieve recruitment curve (curve b). For both of the species considered, full sieve

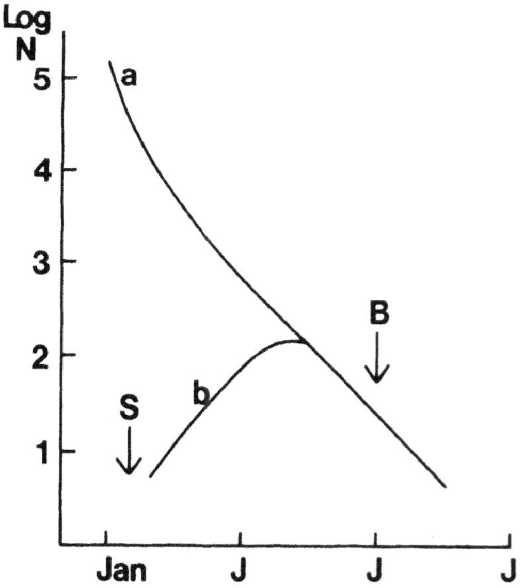

Fig. 10. Combined survivorship and sieve recruitment curve for *Heteromastus filiformis.* Times of settlement (S) and breeding (B) are indicated.

recruitment of the cohort is not achieved until a year or more after actual bottom settlement. These species are in no way unusual in this respect and many of the small-bodied polychaetes, which dominate the community, show quite similar patterns. The peak of sieve recruitment occurs where the curve b meets the survivorship curve and the actual height of this peak will in effect reflect the mortality which has already taken place in this cohort since the time of settlement. Returning to the power curves (Figs 7 and 8) it is clear that the rise in numbers from March to September and the fall in numbers from September to the following March are both in fact measures of mortality. This mortality apparently serves to adjust the survivorship of each species according to the total number of individuals present in the community.

Discussion

The community is dominated by deposit feeding polychaetes and the rapid response to the post 1980 organic enrichment suggests that food limitation plays an important role. The principal stabilising process appears to be density dependent mortality acting, with varying severity on different constituent species, through inter-specific and intra-specific competition for the finite food resource. In the stable period 1974–80, these processes maintained a stable ranking order together with stability in total numbers of individuals. After 1980, organic enrichment released the constraint of food limitation and almost all species exhibited a rising upward trend in numbers and the accurate density dependent control ceased to act. At the same time, it is clear that the period of enrichment altered the competitive balance within the community resulting in a quite different ranking, with a number of species, notably *Heteromastus filiformis,* reacting rapidly and with opportunism to the changed environmental circumstances. Between 1972 and 1980 this species had always maintained a modest presence at a density of 20–40 individuals per square metre, but after 1980 there was a tenfold increase to 300–500 individuals.

The expectation would be that if the higher organic input persisted, the fauna would stabilise at the new higher level. However, evidence suggests that the input has already fallen and that many of the species populations have shown a substantial fall in 1984–85 from their previous high levels. The linear presentation (Fig. 3) certainly indicates that the fauna has already embarked upon a further period of change.

The evidence from the monitoring suggests that the principal stabilising process is density dependent mortality mediated by competition in a food limited environment. The principal destabilising process appears to be periodic fluctuations in the organic flux to the bottom. A secondary destabilising process is clearly concerned with fluctuating winter temperature. In competitive terms, cold winters favour increased survival in the dominant species at the expense of the lesser ranked species. This process is, however, more ephemeral and subject to adjustment within the time scale of a year.

Acknowledgements

This research was financed by a series of grants from the Natural Environment Research Council.

We are grateful to the Institute of Marine Environmental Research for permission to use data collected by the Continuous Plankton Recorder in the North Sea.

References

Buchanan, J. B., M. Sheader & P. F. Kingston, 1978. Sources of variability in the benthic macrofauna off the south Northumberland coast, 1971–1976. J. mar. biol. Ass. U.K. 58: 191–209.

Buchanan, J. B. & R. M. Warwick, 1974. An estimate of benthic macrofaunal production in the offshore mud of the Northumberland coast. J. mar. biol. Ass. U.K. 54: 197–222.

Gower, J. C., 1967. Multivariate analysis and multidimensional geometry. The Statistician 17: 13–28.

Kanneworff, E. & W. Nicholaisen, 1973. The 'Haps', a frame-supported bottom corer. Ophelia 10: 119–128.

Reid, P. C., 1978. Continuous plankton records: large scale changes in the abundance of phytoplankton in the North Sea from 1958 to 1873. Rapp. P.-v. Réun. Cons. int. Explor. Mer 172: 384–389.

Pielou, E. C., 1984. The Interpretation of Ecological Data, A Primer on Classification and Ordination. J. Wiley & Sons, N.Y., 263 pp.

Hydrobiologia 142: 129–135 (1986)

Changes of the macrozoobenthos at 3 monitoring stations in the western Baltic Sea and the Sound

Kurt Jensen
Marine Pollution Laboratory, Jægersborg Alle 1 B, DK-2920 Charlottenlund, Denmark

Keywords: benthos, Baltic, salinity, eutrophication, *Amphiura filiformis*

Abstract

The soft bottom fauna of the western Baltic Sea and the Sound has been sampled and analysed every year since 1979 under the Baltic Monitoring Programme. Furthermore, benthos studies have been carried out in the area at intervals from as far back as 1871. In the area a distinct halocline exists between the overlying low saline Baltic water and the high saline North Sea water.

The variation in the species richness, abundance and biomass of the soft bottom fauna is mainly related to 3 abiotic factors.

First, many species live at the limit of their distribution. The low salinity of the Baltic Sea prevents their penetration into the Baltic proper. However, the marine species may be able to survive and grow but not to reproduce. Consequently, the population will depend on an influx of larvae for it's survival.

Second, the distinct halocline prevents the transport of oxygen to the deeper parts of the Baltic Sea. Oxygen will be supplied under special weather conditions where inflow of high-saline oxygen rich North Sea water occur. The incidences of salt water inflow have increases in the last four decades.

Third, an increasing load of the Baltic Sea with nutrients and organic matter has influenced the fauna. The result have been an increased biomass of the benthos above the halocline. Below the halocline the result has been a decrease in the biomass and a change in the species composition.

Introduction

The Baltic Sea is an inland sea consisting of two distinct water bodies, i.e. outflowing freshwater from the large rivers of East Europe and inflowing salt water from the North Sea. The entrance to the Baltic Sea, through the Danish Straits, the socalled Transition area, is narrow and shallow. The Baltic Sea may be regarded as a huge fjord with a stratified brackish water system (Melvasalo *et al.*, 1981).

The low saline Baltic water and the high saline North Sea water mix very slowly, and a distinct halocline is maintained, often supported by a thermocline. The halocline is at a depth of approximately 15 meters with distinct seasonal variation in the transition area decreasing to about 80 meters in the central Baltic Sea.

The freshwater input exceeds the evaporation, the result being a net transport out of the Baltic Sea through the Danish Straits. The total water transport through the transition area is 3500 km^3 per year (Ærtebjerg Nielsen *et al.*, 1981) or more than four times the volume of this area (Melvasalo *et al.*, 1981).

The inflow of high saline water is irregular and depends on weather conditions. Occasionally under westerly storms, large amounts of North Sea water are pressed through the transition area and into the deeper parts of the Baltic Sea. These intrusions stabilize the halocline and supply oxygen to the bottom layer in the deeps.

Also the outflow is greatly influenced by weather conditions. The tidal range is only some centimeters but the wind may press surface water back into the Baltic Sea and raise the sea level a couple of meters. This process may change the surface water current in the transition area from northgoing to southgoing.

The hydrographical conditions of the Baltic Sea have been studied since 1877 (Matthäus, 1979). From the beginning of this century the salinities of both the bottom and the surface waters have increased significantly (Matthäus, 1978 & 1979) presumably due to an increased frequency of salt water intrusions (Kullenberg & Jacobsen, 1981). In the bottom layer of the Gotland Deep the increase is approximately one per mille to $12.6^0/_{00}$.

Reliable measurements of nutrients – nitrate and phosphate – in the Baltic Sea exist from the last 20–25 years (Gundersen, 1981) as well as measurements of oxygen content of the water and salinity from a 30 year period (Launiainen *et al.,* 1986). From 1979 an intensive programme is proceeding in the Joint Baltic Monitoring Programme. In the Baltic Monitoring Programme macrozoobenthos samples from a series of stations in the Baltic Sea and the transition area have been analysed once a year. The results of these efforts may be compared to results from earlier investigations in the same area dating as far back as to 1871 (Zmudzinski *et al.,* 1986). Though the Baltic Monitoring Programme fixes a certain procedure for macrozoobenthos assessment (Anon, 1984a) the comparison with previous investigations is difficult due to changes in sampling procedures, taxonomic grouping, and problems in the judgement of the reliability of old data.

The changes in the composition and abundance of the soft bottom fauna in the western Baltic are under major influence of three factors of which the two are caused by the hydrographical structure of the Baltic area. The weather may change the current picture and make the halocline move up and down, especially in the transition area, and may occasionally lead to the salt water intrusions into the deeper parts. For the marine species this leads to irregular larval recruitment and occasional extermination of populations. The third factor is increasing supply of organic matter to the bottom fauna due to eutrophication. Other factors such as reduced predation by fish as proposed by Persson (1981) may also lead to long term changes in the benthic communities.

Recruitment

The salinity of the bottom water in the Baltic area decreases from about $30^0/_{00}$ in the Kattegat to about $10^0/_{00}$ in the central Baltic Sea. Thus, most of the true marine species of the benthos disappear in that area (Järvekulg, 1979). At the edge of their distribution these species live in a zone where they are able to survive as adult specimens but require an influx of settling larvae to maintain the population, which is often vulnerable to changes in the environmental conditions.

Amphiura filiformis lives in the transition area through the Sound and has its distributional limit off Copenhagen. The structure of populations of *A. filiformis* has been described from Galway Bay, Ireland (O'Connor, 1983) and in the northern Sound (Muus, 1981). Both found a slow and regular recruitment of the very long lived species resulting in one-peak size class histograms for the year classes older than the 0-generation, where the individual yearclasses cannot be discriminated.

Size class distributions in 1982 and 1983 of a population of *A. filiformis* from a sampling station in the Sound south of the island of Ven (Fig. 1) just north of the margin of *A. filiformis*' limit of penetration into the Baltic are depicted in Fig. 2. The histograms represent 249 specimens. The two years differ in the way that in the 1982 samples no supply of settlers from the previous years can be observed, while this is the case in 1983 illustrating the irregularity of larval recruitment. The phenomenon probably also contributes to the variations of the abundance of the species (Fig. 3).

Sampling of macrozoobenthos has been carried out at this station since 1979 following the procedure outlined in the Baltic Monitoring Programme (Anon, 1984a). Many of the true marine species disappear just south of this station. Also other species living at their salinity limit show this fluctuating abundance (Fig. 3). They are all long lived marine species, *A. filiformis, Thyasira flexuosa* and *Corbula gibba* have their limit of distribution just south of the station while *Cyprina islandica* is also found in Arkona Basin.

In 1984 *A. filiformis* has disappeared from the station. Measurements in the Baltic Monitoring Programme have shown that in the autumn 1983 low values of oxygen content followed by a winter with a relatively long period of low salinity and low

Fig. 1 The sampling stations in the Sound and the western Baltic Sea.

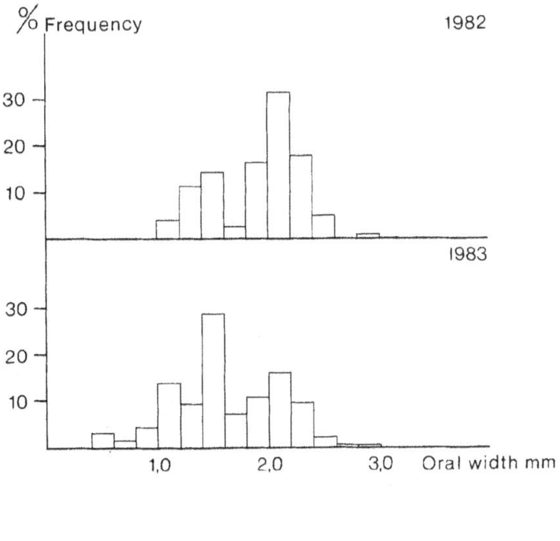

temperature of the bottom water were observed. These conditions may have contributed to the extermination of the population.

The average number of species sampled per year is 43.2 (sd = 7.1). Only three species have been sampled every year, i.e. *Cyprina islandica, Corbula gibba* and *Scoloplos armiger.* Six species have not been present in one of the years of sampling, and the remaining species have not been found two years or more.

Fig. 2. Size distribution of *Amphiura filiformis* in the Sound south of Ven 1982 and 1983.

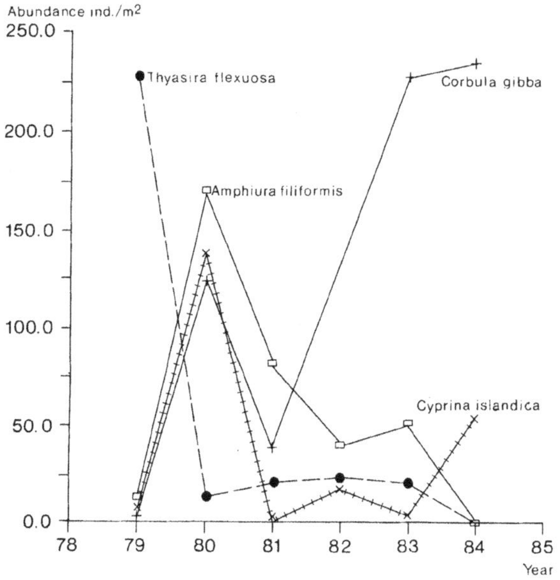

Fig. 3. Abundance of 4 marine species at the station in the Sound 1979–84.

The phenomenon of a population living in a place where it does not reproduce but depends on an influx of larvae has also been observed in the northern Sound where a stock of *Pecten opercularis* is maintained exclusively on imported larvae (K. W. Ockelmann, pers. comm.). In its extreme the invasion of larvae may only occur at rare occasions and one cohort of a long-lived species may be a significant element in a community for years without any reproduction. An example has been seen in the northern Sound where the sea-star *Luidia sarsi* suddenly appeared after one successful spatfall (Fenchel, 1965).

Salt water intrusions

Since the start of the Baltic Monitoring Programme in 1979 macrozoobenthos samples have been taken at 3 stations using the same procedure (Anon, 1984a), i.e. the Sound south of Ven, Arkona Basin and Bornholm Deep (Fig. 1).

The total abundance of the macrozoobenthos has fluctuated simultaneously at the three stations all with a distinct peak in 1980 (Fig. 4). This correlation is not likely to be explained by irregular larval recruitment as the stations contain different species spawning at different time of the year etc.

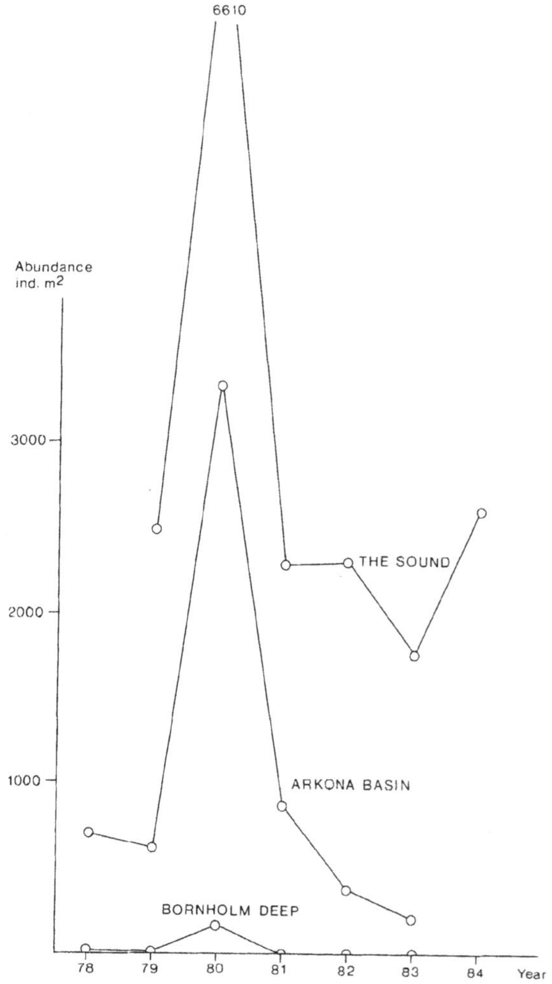

Fig. 4. Abundance of macrozoobenthos at 3 stations (figure 1) 1978–84 (from Zmudzinski *et al.,* 1986 and own results).

The reason for this fluctuation should be searched in quite large scale events in the Baltic Sea. In the winter 1980 stormy weather gave rise to a relatively large intrusion of salt water into the Baltic, which contributed to both an increase in salinity and in oxygen content of the near-bottom water (Fig. 5).

Comparisons with previous investigations of the soft bottom benthos in the Baltic Sea could give an idea of the influence of the salt water intrusions on the fauna. However, comparisons suffer several problems. Sampling and laboratory treatment have changed, the taxonomy has become more and more sophisticated, and the purpose of those investigations was different, often to estimate the amount of available fish food, which influenced the accuracy.

The very large fluctuations in the occurrence and abundance of the species from one year to another makes it difficult to discriminate the effect of salt water intrusions from other short term fluctuations.

However, if occasions of salt water instructions are compared to the species richness of macrozoobenthos samples from Arkona Bassin and Bornholm Deep over the period 1957–83 a pattern of similarity occur (Fig. 6). Periods of increased salinity lead to an increased number of taxa in the samples.

In 1951 an exceptionally great inflow of salt water was followed by a very long period of bottom water stagnation and low oxygen levels. At the Bornholm Deep station the percentual contribution of the major taxa to the total biomass changed drastically at that time. Bivalvia constituted about 90% of the total biomass up to the fifties. After that they were totally replaced by polychaetes, which now constituted more than 90% (Andersin *et al.*, 1978; Zmudzinski *et al.*, 1986).

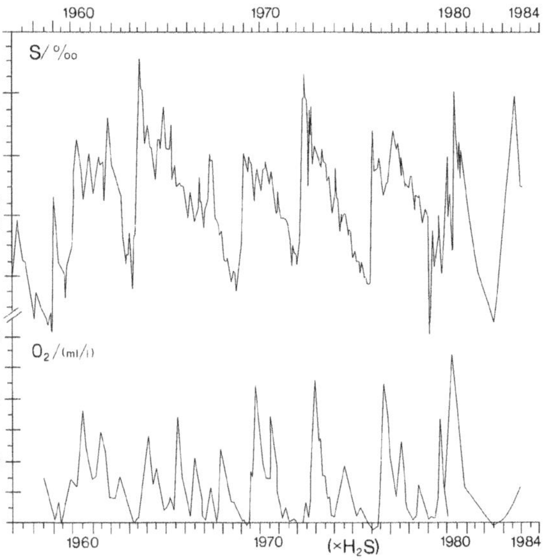

Fig. 5. Variations in salinity and oxygen content of near bottom water at Bornholm Deep 1957–84 (from Launiainen *et al.*, 1986).

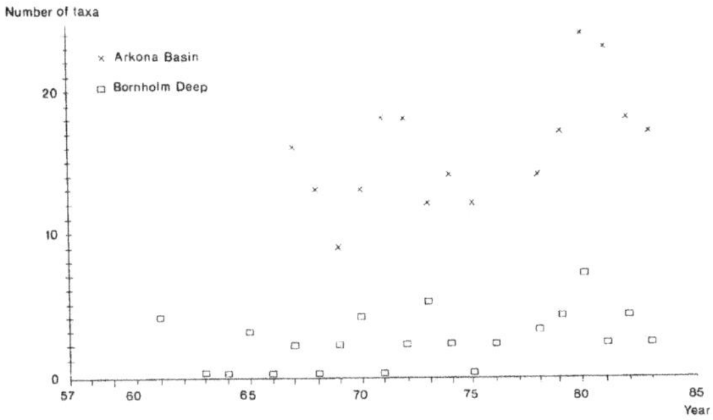

Fig. 6. Species richness at Arkona Basin (X) and Bornholm Deep (o) (from Zmudzinski *et al.*, 1986).

Eutrophication

The amount of nutrients leaking from agricultural and forest areas into the Baltic Sea and the amount of nutrient from sewage discharge have increased in the last decades (Pawlak, 1980; Anon, 1984b). Consequently the nutrient levels in Baltic water have increased. The concentrations of phosphate and nitrate in the surface layer in winter have increased in the whole Baltic area. In the surface water of the Gotland Deep the winter concentration of phosphate has increased from approx 0.2 to approx 0.7 umol/l and the nitrate from approx 2.0 to approx 4.5 umol/l from 1960 to 1983. The reason for this long-term increase is believed to be an interacting effect of human activities and the more frequent occurrence of salt water intrusions (Nehring, 1986).

In the central Sound the total content of nitrogen in the surface water in winter has increased about 40% in the period from 1975 to 1984, and a similar trend has been observed in the content of phosphate since the 1950's (Øresundskommissionen, 1984).

The accumulation of nutrients in the surface layer during winter results in a vernal bloom of phytoplankton. The size of this bloom is directly correlated to the available amount of nutrients. They are totally consumed and no further supply takes place from the bottom layer in that period, where the stratification between the two water bodies normally is distinct. Direct measurements of the size of the vernal bloom are difficult because the duration is short and it does not take place at the same time from year to year, and even more than one distinct peak may occur (Ærtebjerg, pers. comm.).

It has been shown that in the Baltic Sea the sedimentation of the spring bloom is the main contribution of organic matter to the bottom (Kanneworff & Christensen, 1986; Peinert *et al.*, 1982).

Hence, there is very good reason to believe that the amount of sedimentated organic matter has increased significantly.

The influence of the increasing eutrophication on the soft bottom fauna in the Baltic Sea and the transition area has been investigated by several authors (Andersin *et al.,* 1978; Cederwall & Elmgren, 1980; Kölmel, 1979; Pearson *et al.,* 1985).

The increased supply of organic material to the bottom due to the eutrophication may cause two different developments.

If the oxygen supply to the bottom fauna is still high, which is mostly the case above the halocline, the trend will be increasing biomass of the macrofauna. This has been demonstrated in the Baltic proper by Cederwall & Elmgren (1980). They found that the biomass had increased 4–7 times over the period from 1921 to 1976–77. It is difficult to exclude the possible role of the changes in the fish stocks, and this may also have contributed to the increase in benthos biomass. However, increases in benthic biomass related to eutrophication have also been demonstrated in other Baltic areas, i.e. in the Kiel Bay above the halocline (Brey, 1986) and along the Swedish west coast below the halocline where the oxygen situation probably is more favourable (Smith, 1985). Also Persson (1981) found an increase in the biomass along the Swedish coast. However, he contributes the increase to changes in the fish stock caused by commercial fishery.

In areas where the oxygen supply to the bottom is more slow and where oxygen depletion may occur more readily – below the halocline – the development of the soft bottom fauna has been almost reverse. At the station of the Baltic Monitoring Programme in the Bornholm Deep the macrozoobenthos has been sampled frequently since 1949, a dramatic decrease of the biomass has been observed from approx. 25 to below 1 g/m^2 (Zmudzinski *et al.*, 1986). At the station in the Sound south of Ven the biomass of the macrozoobenthos was 103 g/m^2 wet weight in average in 50 samples (Petersen, 1913). In the Baltic Monitoring Programme the sampling at this station was done with the more efficient Van Veen grab instead of the Petersen grab and sieving was done through a 1-mm screen instead of a 1.7-mm. Despite that the average biomass of 15 samples was only 70 g/m^2 wet weight. This indicates a decrease of the benthic biomass here, too.

A recent investigation in the northern Kattegat showed a significant decrease in the biomass and the specimens were smaller compared to Petersen's results from the beginning of this century (Pearson *et al.,* 1985). A similar trend towards smaller specimens has been reported by Smith (1985).

At the Bornholm Deep station the long lived bivalve species *Macoma calcarea* and *Astarte borealis* have been replaced by polychaetes amongst which the opportunistic species *Capitella capitata* and *Heteromastus filiformis* (Andersin *et al.,* 1978; Zmudzinski *et al.,* 1986). This change could very well be accelerated by an increasing load of organic material (Pearson & Rosenberg, 1978).

Stations in the Arkona Basin have been sampled since 1922 (Persson, 1981; Zmudzinski *et al.,* 1986). No trend is distinct here and a change in the composition of the percentual contribution of the different taxa to the biomass can not be observed. The halocline in the Arkona Basin is very close to the bottom and minor inflow of salt water often occur. This leads to great fluctuations in the oxygen content of the bottom layer through the year. The effect of this may veil the effect of eutrophication.

Conclusion

The macrozoobenthos of the soft bottom of the western Baltic Sea and the Sound has shown large changes in abundance, biomass and composition in the last decades. The increasing eutrophication of the area has resulted in an increasing supply of organic matter. The development follows the pattern as described by Pearson & Rosenberg (1978).

In areas with a poor oxygen supply, mainly below the halocline, the development has reached the point of a changed species composition towards opportunistic polychaete species and the total biomass has decreased.

In areas where the oxygen supply is better, mainly above the halocline, the species composition has not changed and the biomass has increased.

Simultaneous with this development the fauna is influenced by incidents of salt water intrusions into the deep layer of the Baltic Sea, which has occurred more frequently in the last decades and resulted in increased salinity.

Also the short-term, days or hours, changes in currents and salinity picture influence the macrofauna leading to irregular larval recruitment of the marine species in the area.

The short-term changes and the salt water intrusions result in year-scale fluctuations in species composition, abundance and biomass of the soft bottom macrozoobenthos, which veil the long-term trends in the development caused by e.g. increasing eutrophication.

References

Andersin, A.-B., J. Lassig, L. Parkkonen & H. Sandler, 1978. The decline of macrofauna in the deeper parts of the Baltic proper and the Gulf of Finland. Kieler Meeresforsch., Sonderheft 4: 23–52.

Anon, 1984a. Guidelines for the Baltic Monitoring Programme for the Second Stage. Baltic Sea Environ. Proc., 12: 1–251.

Anon, 1984b. Iltsvind og Fiskedød i 1981, Omfang og årsager. Miljøstyrelsen, Copenhagen, 247 pp.

Brey, T., 1985. Increase in macrozoobenthos above the halocline in Kiel Bay comparing the 1960s with the 1980s. Mar. Ecol. Progr. Ser., 28: 277–292.

Cederwall, H. & R. Elmgren, 1980. Biomass increase of benthic macrofauna demonstrates eutrophication of the Baltic Sea. Ophelia, Suppl. 1: 287–304.

Fenchel, T., 1965. Feeding biology of the sea-star *Luidia sarsi* Duben & Koren. Ophelia: 2: 223–236.

Gundersen, K., 1981. The Distribution and Biological Transformations of Nitrogen in the Baltic Sea. Mar. Pollut. Bull., 12: 199–205.

Järvekulg, A., 1979. Donnaj Fauna Vostochnoi Chasti Baltiiskogo Morj. Tallin, Valgys, 382 pp.

Kanneworff, E. & H. Christensen, 1986. The oxygen consumption of the sediment in relation to sedimentation of phytoplankton (Presented at the 20th European Marine Biology Symposium, Hirtshals, 1985). Ophelia 26: (in press).

Kullenberg, G. & T. S. Jacobsen, 1981. The Baltic Sea: an Outline of its Physical Oceanography. Mar. Pollut. Bull. 12: 183–186.

Kölmel, R., 1979. The Annual Cycle of Macrozoobenthos: its Community Structure under the influence of Oxygen Deficiency in the Western Baltic. In: E. Naylor & R. G. Hartnoll (eds); Cyclic Phenomena in Marine Plants and Animals. Pergamon Press, Oxford: 19–28.

Launiainen, J., W. Matthäus, S. Fonselius & E. Francke, 1986. First Periodic Assessment of the State of the Baltic Sea. 1. Hydrography. Baltic Sea Envir. Proc. 16 (in press).

Matthäus, W., 1979. Long-term variation of the primary halocline in the Gotland Basin. ICES C.M. 1979/C:22, 14 pp.

Matthäus, W., 1978. Zur mittleren jahreszeitlichen Veränderlichkeit des Oberflachensalzgehaltes in der Ostsee. Gerlands Beitr. Geophysik 87: 369–376.

Melvasalo, T., J. Pawlak, K. Grasshoff, L. Thorell & A. Tsiban, 1981. Assessment of the effect of pollution of the natural resources of the Baltic Sea, 1980. Baltic Sea Envir. Proc. 5B: 1–426.

Muus, K., 1981. Density and Growth of Juvenile *Amphiura filiformis* (Ophiuroidea) in the Øresund. Ophelia 20: 153–168.

Nehring, D., 1986. First Periodic Assessment of the State of the Baltic Sea. 2. Nutrients. Baltic Sea Envir. Proc. 16 (in press).

O'Connor, B., T. Bowmer & A. Grehan, 1981. Long-term assessment of the population dynamics of *Amphiura filiformis* (Echinodermata: Ophiuroidea) in Galway Bay (west coast of Ireland). Mar. Biol. 75: 279–286.

Ockelmann, K. W. (pers. comm.) Marine Biological Laboratory, Strandpromenaden 5, DK-3000 Helsingør, Denmark.

Pawlak, J., 1980. Land-based inputs of some major pollutants to the Baltic Sea. Ambio 9: 163–167.

Pearson, T. H., A. Josefson & R. Rosenberg, 1985. Petersen's Benthic Stations Revisited. I. Is the Kattegat Becoming Eutrophicated? J. Exp. Mar. Biol. Ecol. 92: 157–206.

Pearson, T. H. & R. Rosenberg, 1978. Macrobenthic succession in relation to organic enrichment and pollution of the marine environment. Oceanogr. mar. Biol. A. Rev. 16: 229–311.

Peinert, R., A. Saure, P. Stegmann, C. Stienen, H. Haardt & V. Smetacek, 1982. Dynamics of primary production and sedimentation in a coastal ecosystem. Neth. J. Sea Res. 16: 276–289.

Persson, L.-E., 1981. Were macrobenthic changes induced by thinning out of flatfish stocks in the Baltic Proper? Ophelia 20: 137–152.

Petersen, C. G. J., 1913. Havets Bonitering II. Om havbundens dyresamfund og om disses betydning for den marine zoogeografi. Beretning Danm. Biol. Stn. 21: 1–42.

Smith, S., 1985. Dramatic changes during 10 years of monitoring the bottomfauna off the Swedish coast in the Kattegat. Poster presented at the 20th European Marine Biology Symposium in Hirtshals, Denmark, 1985.

Zmudzinski, L., F. Gosselck, H. Cederwall, K. Jensen & H. Rumohr, 1986. First Periodic Assessment of the State of the Baltic Sea. 5. Zoobenthos. Baltic Sea Envir. Proc. 16 (in press).

Ærtebjerg, G. (pers. comm.). Marine Pollution Laboratory, Jægersborg Alle 1 B, DK-2920 Charlottenlund, Denmark.

Ærtebjerg Nielsen, G., T. S. Jacobsen, E. Gargas & E. Buch, 1981. Evaluation of the Physical, Chemical and Biological Measurements. The Belt Project. The National Agency of Environmental Protection, Denmark, 122 pp.

Øresundskommissionen, 1984. Øresund, tillstånd – effekter av närsalter. Naturvårdsverket Rapport 3008, 93 pp.

Hydrobiologia 142: 137–150 (1986)

Temporal changes of community structure and biomass in two subtidal macroinfaunal assemblages in La Coruña bay, NW Spain

E. López-Jamar, G. González & J. Mejuto
Instituto Español de Oceanografía, Centro Costero de La Coruña, Ap. 130, 15080 La Coruña, Spain

Keywords: macroinfauna, temporal variation, community structure, disturbance

Abstract

Benthic samples were collected from 1982 to 1985 in two stations in La Coruña Bay in order to study the temporal changes of two subtidal benthic assemblages: (1) a *Tellina fabula* – *Paradoneis armata* community, and (2) a *Thyasira fluxuosa* facies of an *Abra alba* community. The *Tellina* station is located in an area with little human disturbance, whereas the *Thyasira* station is located inside the organically polluted harbour area, where dredging operations took place at the beginning of this study. Temporal changes in both communities follow quite different patterns. The sedimentary environment of the *Tellina* station remains fairly constant through time: sediment size ranges from 2.80 to 3.85 Φ, and organic matter content (loss by ignition) varies between 3.15 and 4.53%. In the *Thyasira* station these sedimentary variables cover a much wider range: 3.10 to 5.75 Φ mean diameter and 5.19 to 17.05% organic matter.

Community structure is quite stable in the *Tellina* station. The mean similarity of every sample with the following ones is relatively high and stable, and diversity (H′) and evenness (J′) fluctuations are small. However, in the *Thyasira* station the community structure has changed drastically after dredging operations ended as a result of the recolonization and further stabilization of the community.

At the *Tellina* station biomass varies between 10 and 32 g ashfree DW m^{-2}, and it does not show a discernible temporal pattern. However, at the *Thyasira* station biomass increased from less than 1 g ash-free DW m^{-2} during the dredging period up to 16 g ash-free DW m^{-2} 18 months after the end of dredging. In conclusion, the *Tellina fabula* community seems very stable through time, whereas the *Thyasira flexuosa* facies has wider fluctuations due to human disturbances.

Introduction

The study of the natural spatial and temporal variability of benthic communities is necessary in order to evaluate the changes that may occur due to catastrophic or man induced disturbances of the environment and the subsequent recovery of the community. In the framework of the COST 647 Project, benthic studies were undertaken in the Galician Rías Bajas (López-Jamar, 1981, 1982; López-Jamar & Mejuto, 1986). The infaunal benthos of the Rías Bajas is greatly affected by human activities, such as mussel culture on rafts (López-Jamar, 1982; Tenore *et al.*, 1982), paper mill discharges (López-Jamar, 1978; Mora *et al.*, 1982), urban sewage and fish trawling. These disturbances usually cause the benthos of the Rías Bajas to be very unstable in relation to time. The Rías Altas, located farther North, are smaller bays which are less affected by human activities, so the long-term study of the benthic communities should be more meaningful to understand the natural patterns of variation through time.

The spatial distribution of the infaunal communities of La Coruña Bay has been described earlier (López-Jamar & Mejuto, 1985). The two main benthic assemblages are: (1) a *Tellina fabula* – *Paradoneis armata* community, inhabiting hard-packed fine sand in a large area of the Bay; and (2) a *Thyasira flexuosa* facies, inhabiting organic-rich mud inside the polluted harbour area. Temperature and salinity conditions of the bottom water in both sites vary very little throughout the year: temperature ranges between 11 and 13 °C, and salinity is always higher than $33^0/_{00}$. The place selected for the study of the *Thyasira* assemblage is located in an area where harbour dredging was carried out, and it was almost completely defaunated during the dredging operations (López-Jamar & Mejuto, in press c). Moreover, in this site many human disturbances take place because of its proximity to an important dock, whereas the *Tellina* community is located in a shallower area where the effect of human activities is much smaller.

This paper presents the results of three years of study of the temporal variations of these two faunal assemblages, comparing community structure, biomass, diversity and sedimentary environment.

Material and methods

Benthic samples were collected from July 1982 to September 1985 in two stations in La Coruña Bay using a Bouma box corer (0.0175 m^2 surface are and 10 to 20 cm deep sediment samples. The *Tellina* station (9 m deep) is located outside the harbour area, whereas the *Thyasira* station (16 m deep) is placed inside the harbour (Fig. 1). Five samples were taken at each station for community structure and biomass determinations, although at the *Tellina* station ten samples were collected to follow the *Tellina fabula* population dynamics. This sample size was decided after estimating the minimum area required to have good estimates of the temporal variation of the common species with a reasonable effort. Although the rare species could be not adequately collected with this sample size, the sampling frequency and the high abundance of the infaunal dominants made nearly impossible to sample a larger area. A subsample was also taken

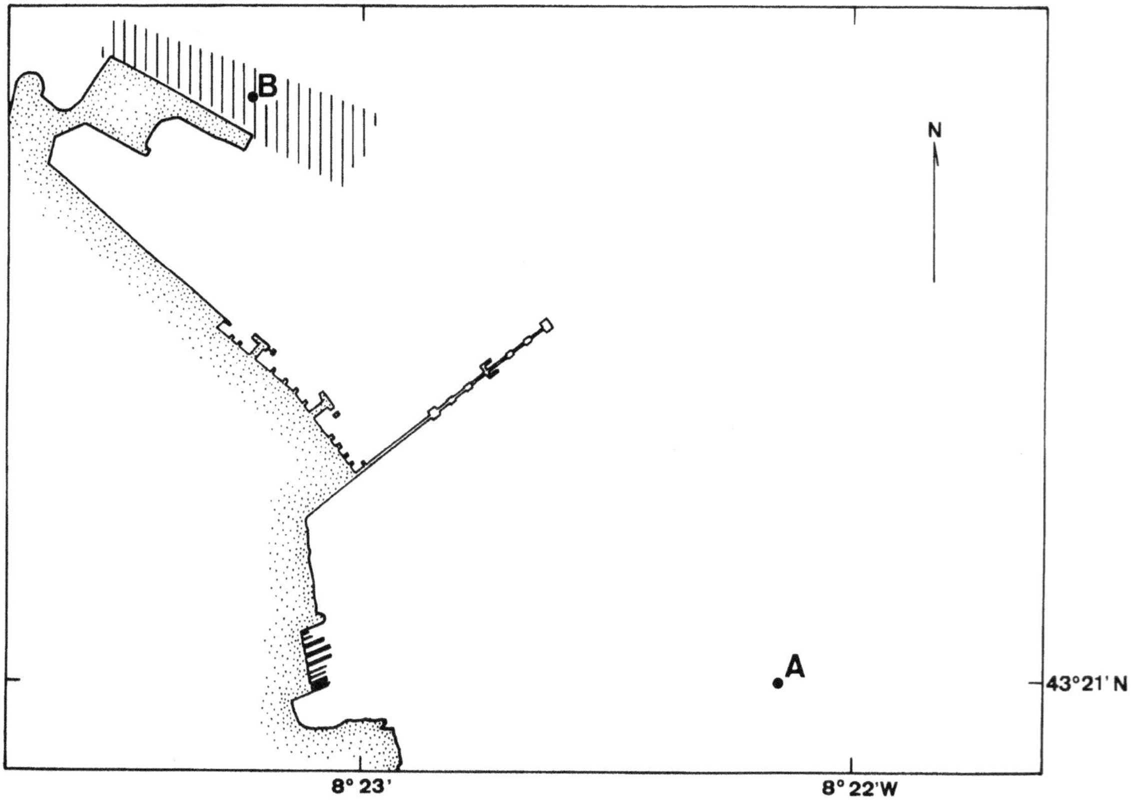

Fig. 1. Situation of the sampling stations in the study area. A: *Tellina* station; B: *Thyasira* station. The shaded area indicates the zone that was dredged.

for sediment analysis and organic matter content determinations.

Particle size analysis was performed by a combination of dry sieving and sedimentation techniques (Buchanan, 1984). Organic content of the sediment was estimated as the loss in weight of dried samples (100 °C, 24 h) after combustion (500 °C, 24 h). The faunal samples were sieved through a 0.5 mm mesh, anaesthetized with a menthol solution, and then preserved in 5% buffered formaldehyde previously containing Rose Bengal as a staining agent to facilitate the sorting of organisms (Mason & Yevich, 1967). Most benthic ecologists recommend using the 0.5 mm mesh instead of the 1 mm for successional studies of infaunal benthos. Moreover, a previous study of the benthos of La Coruña Bay indicated high abundances of small polychaetes (Spionidae, Cirratulidae), that could be underestimated bv using the 1 mm sieve. After each cruise, the preserved samples were sorted and wet weight of each species recorded. Wet weight to ash-free dry weight (AFDW) correlations had been previously determined and were used to estimate biomass. The Morisita index (Morisita, 1959) modified by Horn (1966) was used to calculate faunal similarity among samples. Diversity was calculated using abundance data with the Shannon function (Shannon & Weaver, 1963). Evenness was also estimated as defined by Pielou (1966).

Results

Sedimentary environment

At the *Tellina* station, temporal variations of sediment size are relatively small; mean diameter ranges from 2.80 Φ (143 μm) to 3.60 Φ (82 μm), and there is no discernible temporal pattern (Fig. 2). At this station sediment is composed of

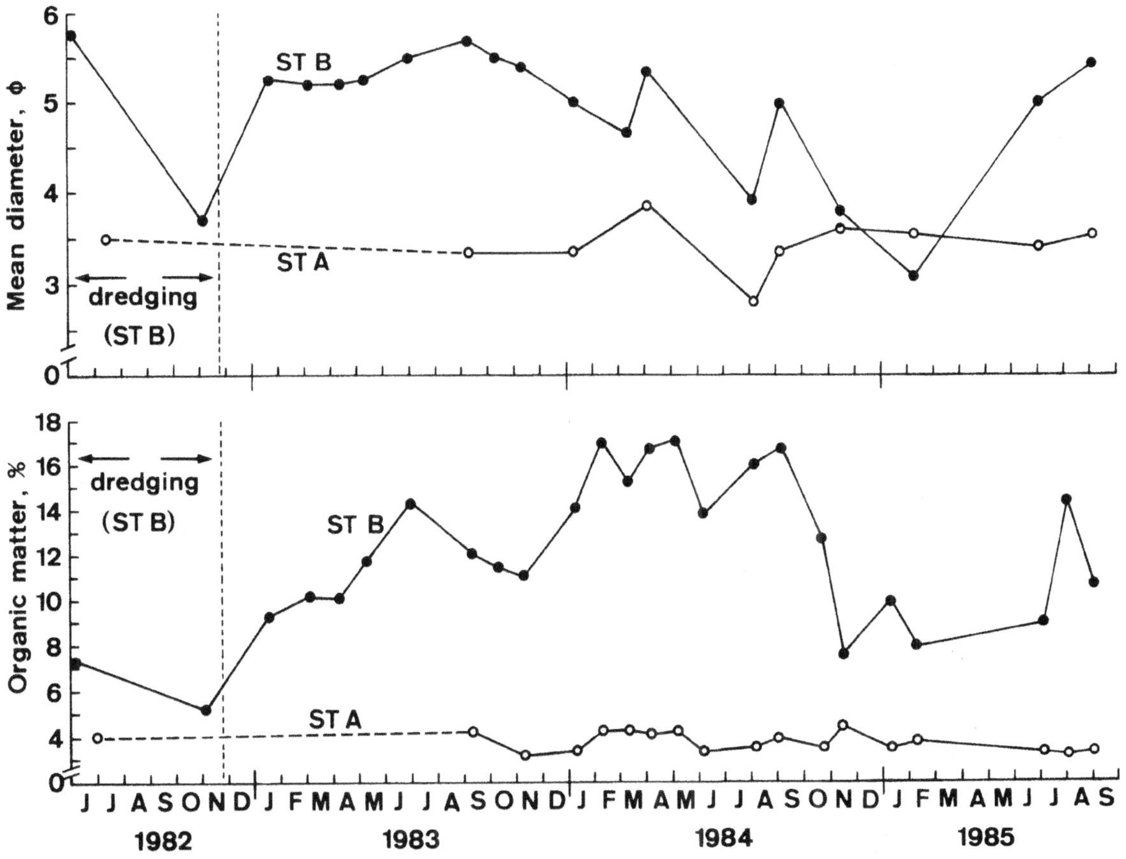

Fig. 2. Temporal variation of organic matter content and sediment size in both stations. A: *Tellina* station; B: *Thyasira* station. Arrows indicate the dredging period.

very fine to fine sand. The organic matter content is low to moderate (3.35 to 4.53% DW) and it varies very little through time. However, at the *Thyasira* station both sediment size and organic content have a much wider variation. Mean diameter ranges from 3.10 Φ (116 μm) to 5.75 Φ (19 μm). At the end of the dredging period the mean diameter was relatively large (3.70 Φ) because dredging removed preferentially the finer fractions. After dredging ended (November 1982), mean diameter decreased gradually until September (5.70 Φ), but increased thereafter (3.10 Φ in February 1985) (Fig. 2). During the final period of this study (summer 1985) sediment size tends to decrease again (5.00 and 5.45 Φ in July and September 1985, respectively). Sediment at this station is very heterogeneous, consisting of black mud mixed with a variable fraction of sand and gravel. Dead shells of *Thyasira flexuosa* constitute an important fraction of the sediment. Organic content is moderate during the dredging period (5.19 to 7.38% DW), but increased regularly after dredging concluded, reaching values higher than 16% DW 15 months later. However, organic content decreased sharply starting in September 1984, and values lower than 10% DW were recorded in several sampling dates (Fig. 2). The sulphide smell in most of the sediment samples of this station was evident, suggesting that anoxic conditions prevail.

Community structure

At the *Tellina* station species richness ranges from 29 to 59 species. The highest values occur in late summer, and they decreased in the winter months (Fig. 3). This pattern of variation has also been noted by Dauvin (1979) in a similar community from the Bay of Morlaix (Atlantic coast of France). Total number of species recorded at this station during the study period was 93. Polychaetes are the major group (57 species), followed by molluscs (22 species). The dominant organisms are the polychaetes *Paradoneis armata* (2389 to 7321 individuals m^{-2}), *Spio filicornis* (16 to 8595 individuals m^{-2}), *Spiophanes bombyx* (251 to 2012 individuals m^{-2}), *Pseudopolydora kempi* (0 to 3349 individuals m^{-2}), *Capitella capitata* (0 to 1554 individuals m^{-2}), and the bivalve *Tellina fabula* (932 to 1871 individuals m^{-2}). Table 1 shows the maximum and average values of abundance and biomass of the dominant species in this station. This benthic assemblage was described in more detail by López-Jamar & Mejuto (1985), and it can be included in the boreo-lusitanian *Tellina* community described by Stephen (1930) from a Scottish shore. A very similar community was described by Shin *et al.,* (1982) from the subtidal area of Galway Bay (West coast of Ireland).

The *Thyasira* station has a lower number of species, varying from 13 to 38. Obviously, the lower values occurred during the dredging period (13 to 15 species), and species richness increased thereafter (Fig. 3). Total number of species recorded over the whole study period was 80. Again, polychaetes are the dominant group (47 species), followed by molluscs (19 species). The most abundant

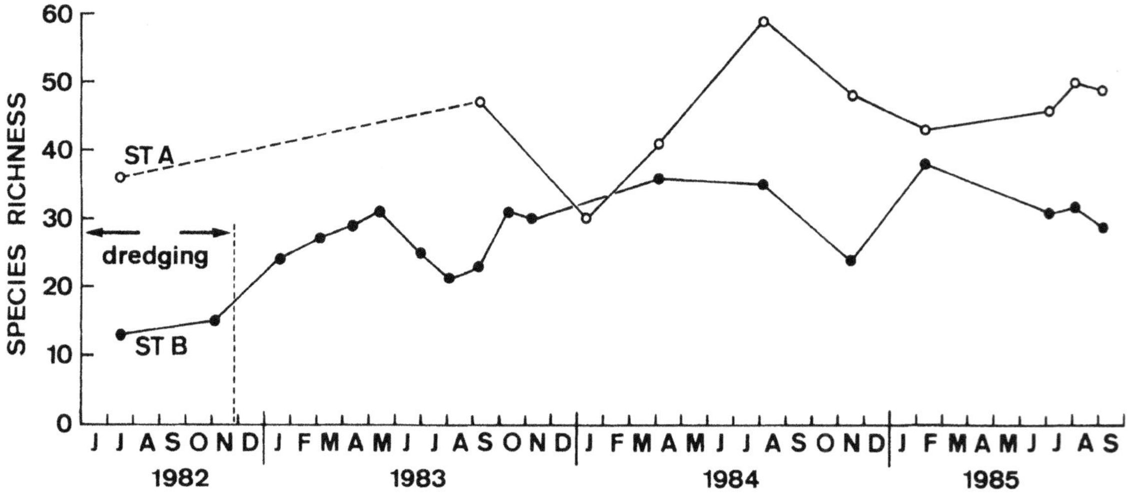

Fig. 3. Temporal variation of species richness in both stations. Symbols as in Fig. 2.

Table 1. Maximum and average abundance (number of individuals m^{-2}) and biomass (mg m^{-2} AFDW) of the dominant species of the *Tellina* station. Only species occuring at least in 50% of the samples are listed.

Taxa	Abundance		Biomass	
	Max.	Aver.	Max.	Aver.
Polychaetes				
Sigalion mathildae Aud. & M. Edwards	57	34	200	62
Phyllodoce laminosa Savigny	194	48	458	97
Eteone sp.	240	48	77	20
Exogone sp.	69	16	1	1
Nephtys hombergi Savigny	103	26	1123	207
Glycera rouxii Audoin & M. Edwards	212	93	368	119
Hyalinoecia bilineata Baird	149	55	1483	475
Lumbrineris gracilis (Ehlers)	149	115	3309	966
Schistomeringos caeca (Webster & Ben.)	240	69	3	1
Ophryotrocha sp.	46	14	2	1
Spio filicornis (Müller)	8595	3208	438	218
Prionospio malmgreni Claparède	309	151	16	8
Prionospio cirrifera Wiren	69	15	2	1
Pseudopolydora kempi (Southern)	3349	414	1521	156
Spiophanes bombyx (Claparède)	2012	1157	836	314
Magelona sp.	522	229	169	57
Chaetozone sp.	375	85	109	23
Paradoneis armata Glémarec	7321	4416	1630	929
Capitella capitata (Fabricius)	1554	619	34	9
Notomastus latericeus Sars	171	67	846	307
Mediomastus sp.	1749	856	351	144
Owenia fusiformis delle Chiaje	114	32	112	16
Myriochele heeri Malmgren	217	55	7	2
Brada villosa (Rathke)	457	164	1273	373
Polycirrus sp.	23	9	100	12
Molluscs				
Tellina fabula Gmelin	1871	1255	11128	540
Venus striatula (da Costa)	594	151	5126	765
Abra alba (Wood)	103	48	1142	225
Thracia phaseolina (Lamarck)	331	91	2083	458
Mysella bidentata (Montagu)	34	11	2	1
Thyasira flexusa (Montagu)	251	53	567	85
Pharus legumen (L.)	80	36	6674	1883
Mactra sp.	217	86	2595	415
Bivalve indet.	160	45	4	1
Nassarius reticulatus (Ström)	46	16	7248	2783
Cylichna cylindracea (Pennant)	69	23	70	18
Echinoderms				
Leptosynapta inhaerens (O. F. Müller)	11	5	1072	167
Leptosynapta bergensis (Östergren)	67	28	592	135
Crustaceans				
Cumacea indet.	514	194	8	5
Ostracoda indet.	354	98	23	4
Others				
Peloscolex sp.	229	42	3	1
Nemertina indet.	389	230	7083	1378

Table 2. Maximum and average abundance (number of individuals m^{-2}) and biomass (mg m^{-2} AFDW) of the dominant species at the *Thyasira* station. Only species occuring at least in 50% of the samples are listed.

Taxa	Abundance		Biomass	
	Max.	Aver.	Max.	Aver.
Polychaetes				
Pholoe minuta (Fabricius)	183	42	6	1
Phyllodoce laminosa Savigny	263	77	653	61
Ophiodromus flexuosus (d. Chiaje)	514	167	138	69
Glycera rouxii Audouin & M. Edwards	160	38	53	11
Lumbrineris gracilis (Ehlers)	606	180	213	32
Ophryotrocha sp.	1657	440	17	3
Spio filicornis (Müller)	1474	116	7	2
Prionospio malmgreni Claparède	343	55	54	8
Pseudopolydora kempi (Southern)	1371	375	652	105
Scolelepis fuliginosa (Claparède)	800	88	213	43
Chaetozone sp.	3326	1032	2310	322
Cirriformia sp.	343	95	108	24
Capitella capitata (Fabricius)	2343	756	226	35
Notomastus latericeus Sars	411	135	1013	223
Mediomastus cf. *fragilis* Rasmussen	411	62	41	7
Brada villosa (Rathke)	617	158	286	41
Ampharete acutifrons (Grube)	389	65	119	15
Molluscs				
Thyasira flexuosa (Montagu)	22071	7528	7121	2734
Mysella bidentata (Montagu)	57	14	3	1
Abra alba (Wood)	926	236	6473	1552
Abra nitida (Müller)	208	87	1213	337
Tellina fabula Gmelin	69	21	4	1
Nassarius incrassatus (Ström)	160	37	2894	819
Philine aperta (L.)	229	40	1962	138
Echinoderms				
Leptosynapta inhaerens (O. F. Müller)	91	28	786	84
Crustaceans				
Cumacea indet.	69	16	4	1
Others				
Peloscolex sp.	240	52	10	2
Cerianthus sp.	69	24	1418	365
Nemertina indet.	80	18	26	4

organism is the bivalve *Thyasira flexuosa* (8 to 22071 individuals m^{-2}), followed by the polychaetes *Chaetozone* sp. (103 to 3326 individuals m^{-2}), *C. capitata* (183 to 2343 individuals m^{-2}), *Ophryotrocha* sp. (0 to 1657 individuals m^{-2}), *S. filicornis* (0 to 1474 individuals m^{-2}), and *P. kempi* (0 to 1371 individuals m^{-2}). The maximum and average abundance and biomass values of the dominant species are listed in Table 2.

A similarity analysis was performed in order to evaluate temporal changes in the community structure. In both stations, faunal similarity of each sample with the following ones was calculated, and the results were plotted against time (Fig. 4). The patterns obtained were very different at each station. In the *Tellina* station, the similarity of each sample to the next ones is usually higher than 60%, and a temporal pattern cannot be distinguished. However, at the *Thyasira* station, these values are very low at the beginning of the study, and they start to be high and relatively constant six months after the dredging concluded. Thus, the community structure at the *Tellina* station is quite stable through time, whereas in the *Thyasira* station the

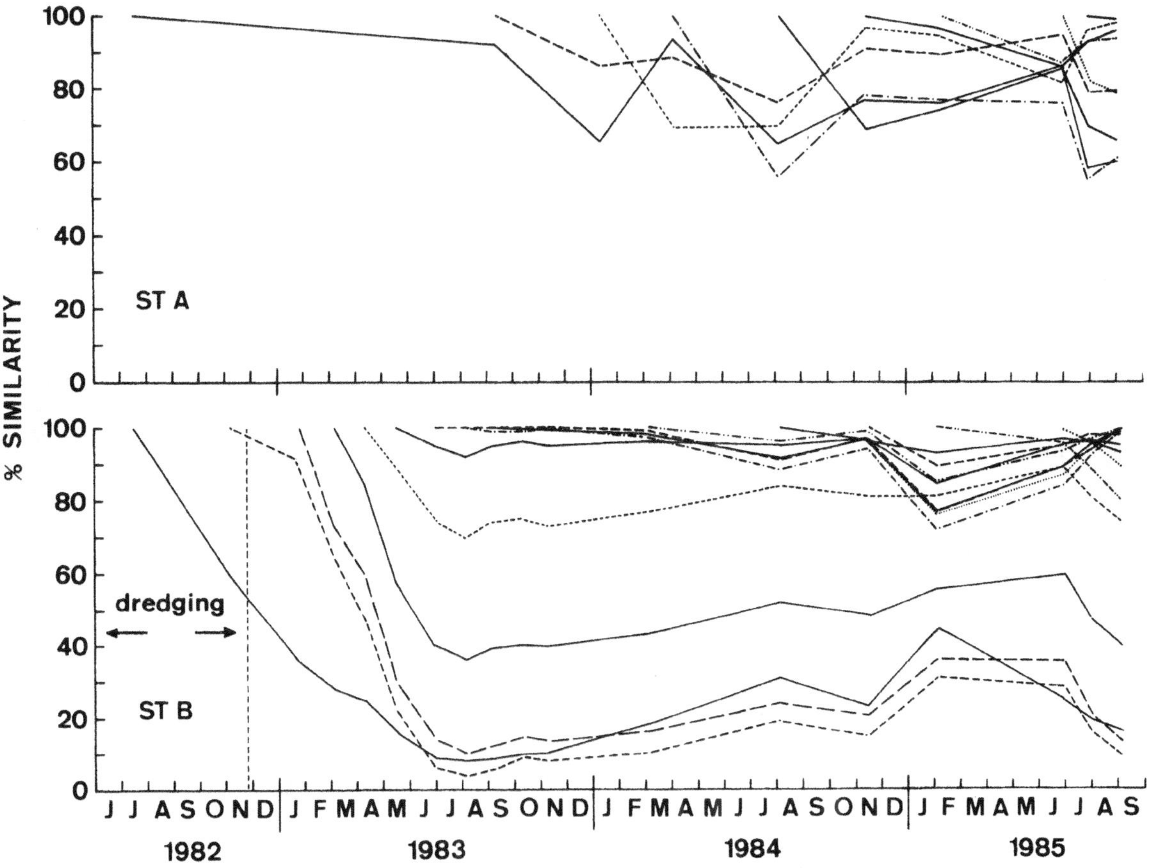

Fig. 4. Faunal successive similarity in both stations. Each line corresponds to the similarity of one sample with the following ones. Symbols as in Fig. 2.

community changed rapidly after dredging and it reached a certain degree of equilibrium six months later. The oscillations of the similarity values since the *Thyasira* station recovered are smaller than those of the *Tellina* station. This fact could suggest a greater stability at the *Thyasira* community, but it is rather related to the high dominance of *T. flexuosa,* which causes the similarity of successive samples to be very high once the community has been recovered.

Diversity and evenness

Diversity follows different temporal patterns in each station. In the *Tellina* community relative maxima occur in late summer coinciding with the highest number of species. Minimum values occur in the winter months due to a lower species richness as well as to a slightly reduced evenness. In this community diversity ranges from 2.92 to 4.13, and its variations through time are relatively smooth and display a cyclic pattern. Evenness varies from 0.51 to 0.72, and its variations are approximately parallel to those of diversity (Fig. 5).

In the *Thyasira* station diversity and evenness increased initially at the beginning of the postdredging period, but then decreased sharply to much lower values, reaching the minimum in August 1983 ($H' = 1.23$; $J' = 0.28$). From this date on, diversity and evenness increased regularly until February 1985, when they started to decrease again. This temporal pattern of diversity is mainly caused by the variation in the abundance of *T. flexuosa.* Just after dredging ended, the input of new species to the area causes a rise in diversity and evenness values; however, *T. flexuosa* soon becomes over-

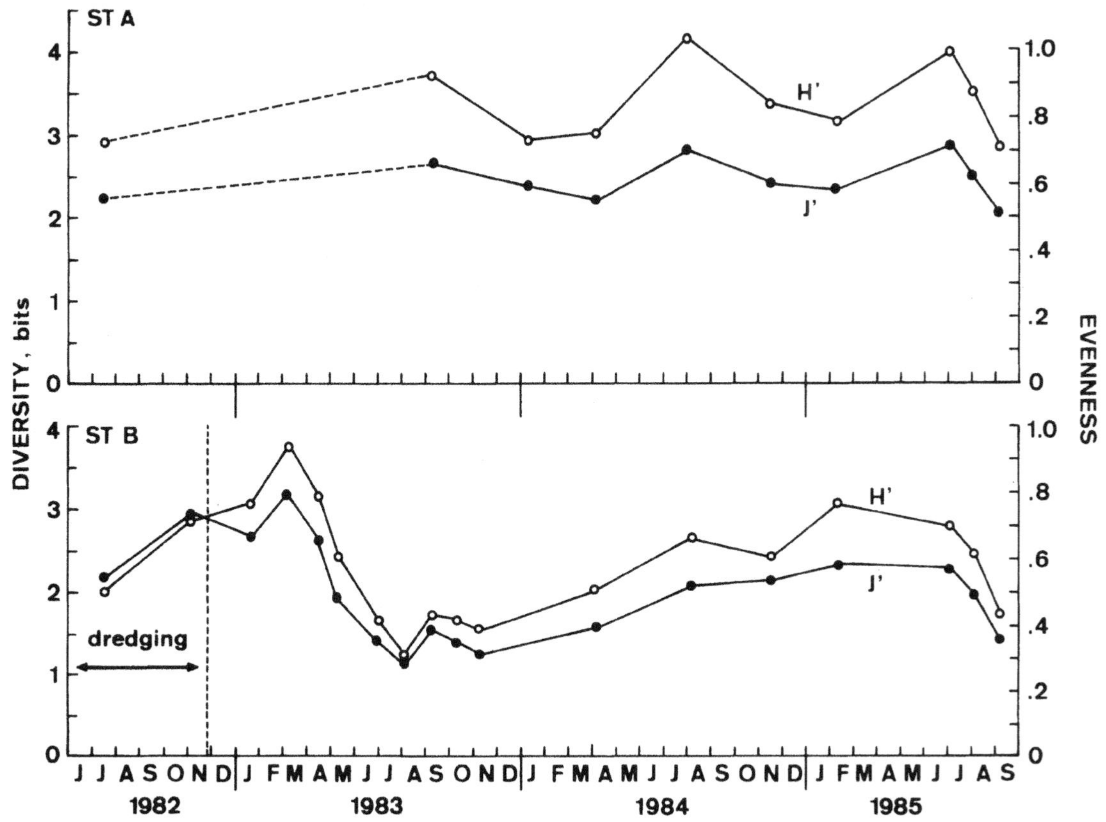

Fig. 5. Temporal variation of diversity (H′) and evenness (J′) in both stations. Symbols as in Fig. 2.

whelmingly dominant, causing a sudden drop in diversity. As recolonization goes on, other species reach their natural densities while *T. flexuosa* abundance decreases. This fact causes a gradual increase in diversity and evenness. Nevertheless, in 1985 the recruitment of *T. flexuosa* was very successful and this species reached very high densities; consequently, diversity and evenness decreased again. At this station the range of variation of diversity and evenness is wider than that of the *Tellina* community ($1.23 \leq H' \leq 3.76$; $0.28 \leq J' \leq 0.79$) (Fig. 5).

Temporal variations of the dominant species

Tellina *station*

The polychaete *P. armata* is the dominant species in terms of abundance in most of the samples. Its density was always higher than 2000 individuals m^{-2} (Fig. 6). The highest abundance occurred in June (7321 individuals m^{-2}) and November 1984 (6938 individuals m^{-2}), but its temporal variation does not follow a distinct pattern. *S. filicornis* is another very abundant polychaete in this station. Two relative maxima occur in both winters 1983–84 and 1984–85, but the highest densities were recorded during late summer in 1985 (6378 and 8595 individuals m^{-2} in August and September, respectively) (Fig. 6). Dauvin (1979) pointed out that the highest density of this species also occurred in late summer in the Bay of Morlaix. Although density of *S. filicornis* in La Coruña Bay is very variable, it shows a tendency to increase since this study started.

The dominant bivalve in this community is *T. fabula*. Its density is quite constant throughout the study period, usually varying between 1000 and 2000 individuals m^{-2}. In La Coruña Bay this species shows lower abundance oscillations than those of a similar community in the German Bight (Salzwedel, 1979). The polychaetes *Mediomastus* sp.

and *C. capitata* also vary little through time, although *C. capitata* densities are slightly higher in summer. The abundance of *S. bombyx* is also very constant, except during the winter of 1983–84, when it has much lower values. The opportunistic species *P. kempi* is usually present at very low densities, but in August 1984 it reached a high abundance (3349 individuals m^{-2}) (Fig. 6).

Thyasira *station*

The dominant species in this station is the bivalve *T. flexuosa,* whose density usually accounts for more than 50% of the total at every sample. During the dredging period this species was present at very low densities, but soon after dredging ended, its abundance increased very rapidly, reaching 10000 individuals m^{-2} 5 months later. The abundance remained very high during 1983 and increased even more in the first half of 1984 due to a successful new recruitment, reaching densities higher than 20000 individuals m^{-2}. However, in summer 1984 its abundance started to decrease down to ca. 3000 individuals m^{-2}. In July 1985 a new recruitment occurred and high densities (up to 12000 individuals m^{-2}) were present again (Fig. 7).

In this station there are several opportunistic species that reached high densities during the first stages of the recolonization, but whose abundance usually remained lower once the community has recovered. *C. capitata* densities were relatively high after dredging ended (>2000 individuals m^{-2}), but decreased later to much lower values. *P. kempi* also showed a sharp increase during the first months of succession (>1300 individuals m^{-2})

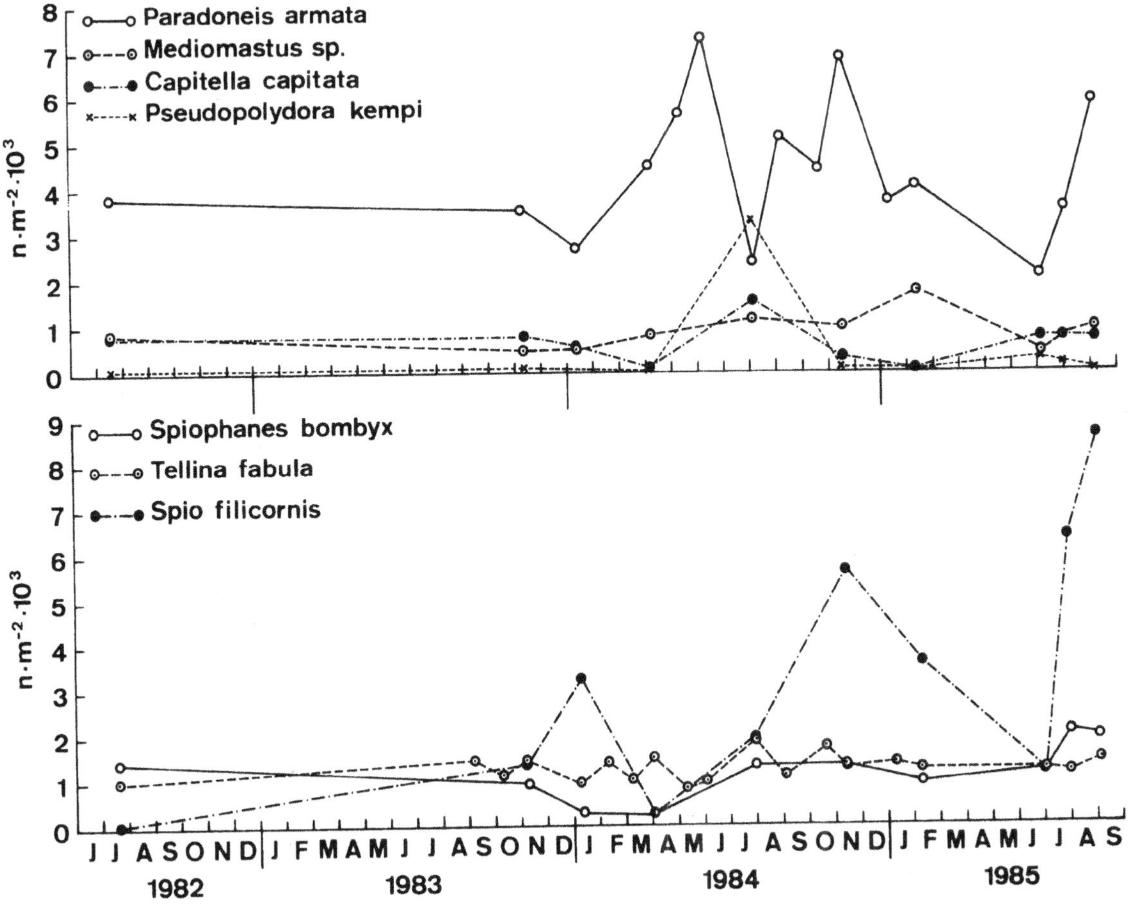

Fig. 6. Temporal variation of population density of the dominant species in the *Tellina* station.

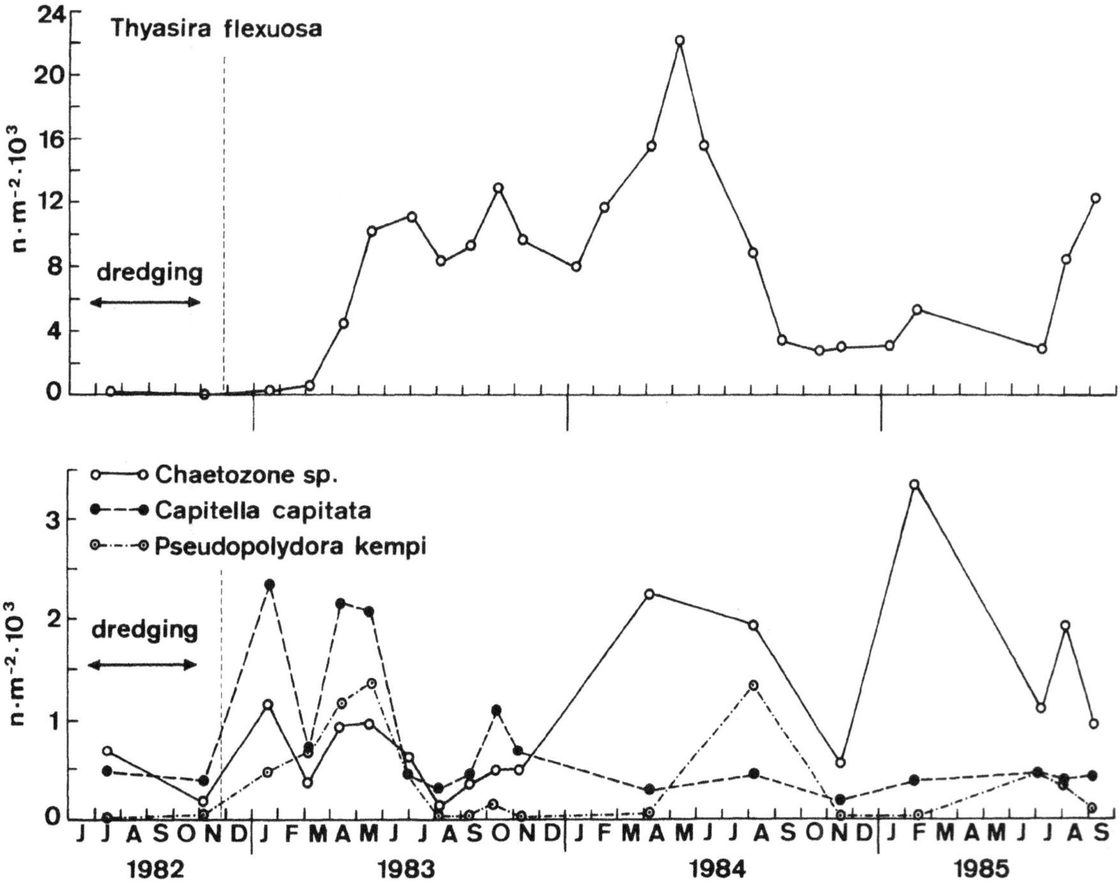

Fig. 7. Temporal variation of population density of the dominant species in the *Thyasira* station.

and decreased thereafter. This species shows a marked peak again in summer 1984, and another smaller one in summer 1985. The polychaete *Chaetozone* sp. displays strong oscillations in its abundance, although it seems to have a tendency to increase with time (Fig. 7).

Biomass

In the *Tellina* station biomass ranges from 11.9 to 32.0 g m^{-2} AFDW. Every year the maximum values occurred in summer, whereas the winter biomass is usually lower (Fig. 8). Molluscs are the dominant group in terms of biomass, and they usually account for more than 50% of total biomass in every sample. *T. fabula* (1.8 to 11.1 g m^{-2} AFDW) and *Nassarius reticulatus* (0 to 7.2 g m^{-2} AFDW) are the two species that constitute the highest proportion of the total. Polychaetes are the second most important group. Their contribution to total biomass varies from 12.1 to 48.6%. The dominant species in terms of biomass are *P. armata* (0.3 to 1.6 g m^{-2} AFDW), *Lumbrineris gracilis* (0.3 to 3.3 g m^{-2} AFDW), *Brada villosa* (0.1 to 1.3 g m^{-2} AFDW), *Hyalinoecia bilineata* (0 to 1.5 g m^{-2} AFDW) and *S. bombyx* (0.1 to 0.8 g m^{-2} AFDW). The rest of the taxonomic groups usually form a small fraction of total biomass.

In the *Thyasira* station total biomass varies between 0.7 and 16.7 g m^{-2} AFDW. Biomass values are very low during and immediately after dredging, but later they steadily increase, reaching the maximum values in summer 1984, 20 months after dredging ended. A slight tendency to decrease can be noticed after that time.

During the dredging period polychaetes are the

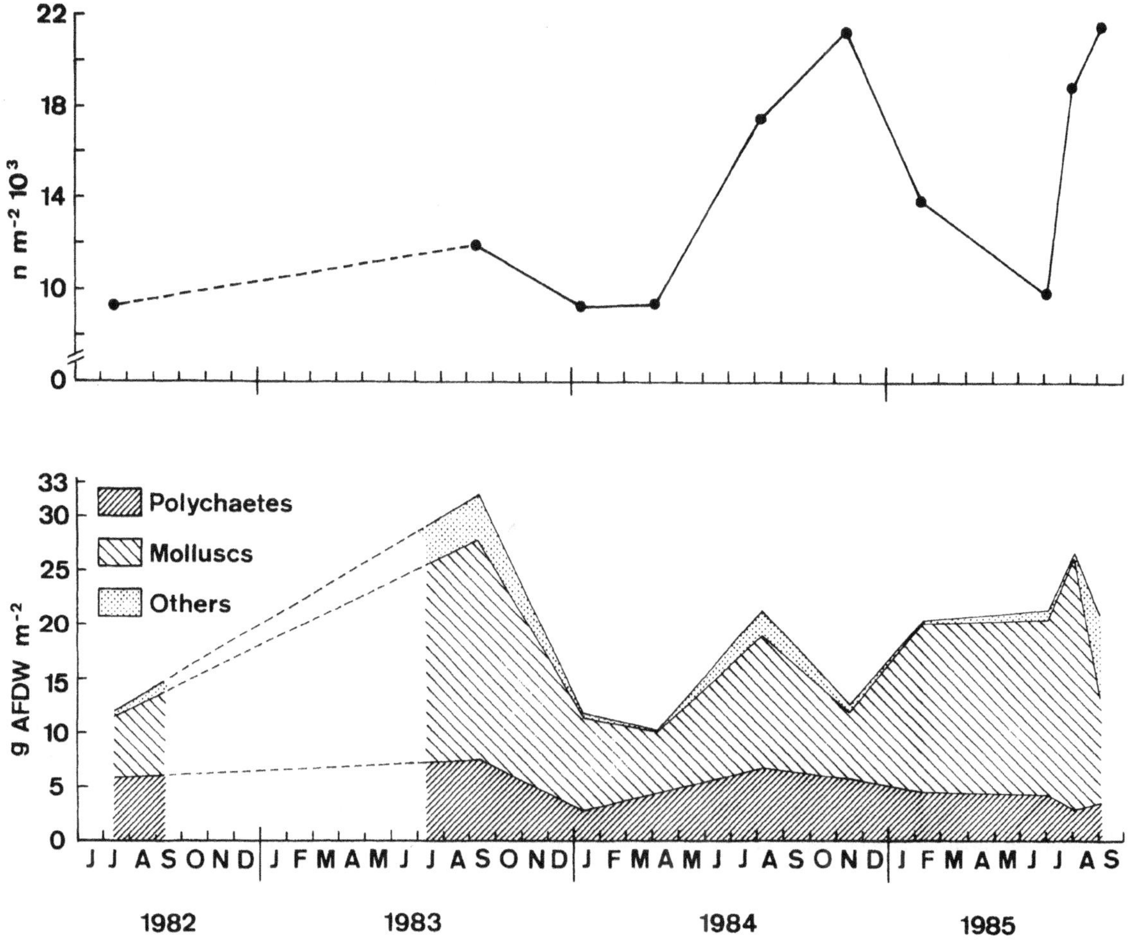

Fig. 8. Temporal variation of total faunal density, total biomass, and proportion of the total biomass of the different groups in the *Tellina* station.

most important group, but molluscs soon start to dominate, mainly due to the contribution of *T. flexuosa* (Fig. 9). Six months after dredging ended, molluscs always constitute more than 50% of the total biomass. *T. flexuosa* is the dominant species (0.1 to 7.1 g m^{-2} AFDW). *Abra alba* (0 to 6.5 g m^{-2} AFDW) and *Nassarius incrassatus* (0 to 2.9 g m^{-2} AFDW) also constitute an important fraction of the total, although their biomass values are very irregular with time. Among the polychaetes, the dominant species are *Chaetozone* sp. (0.1 to 2.3 g m^{-2} AFDW), *B. villosa* (0 to 0.3 g m^{-2} AFDW), and *Notomastus latericeus* (0 to 1.0 g m^{-2} AFDW). The anthozoan *Cerianthus* sp. can occasionally be important in biomass.

Discussion

Most coastal soft-bottom benthic communities exhibit seasonal and long-term variability. In temperate latitudes, population density may change substantially due to the seasonal patterns of reproduction. This fact can be noticed in the *Tellina* community, where diversity and evenness display a seasonal pattern due to a higher abundance of several species (mainly *S. filicornis*) in winter. Thus, in this community diversity and evenness are lower in these months (Fig. 5).

The great fluctuations of the sediment features at the *Thyasira* station may be a result of man induced disturbances, such as dredging at the begin-

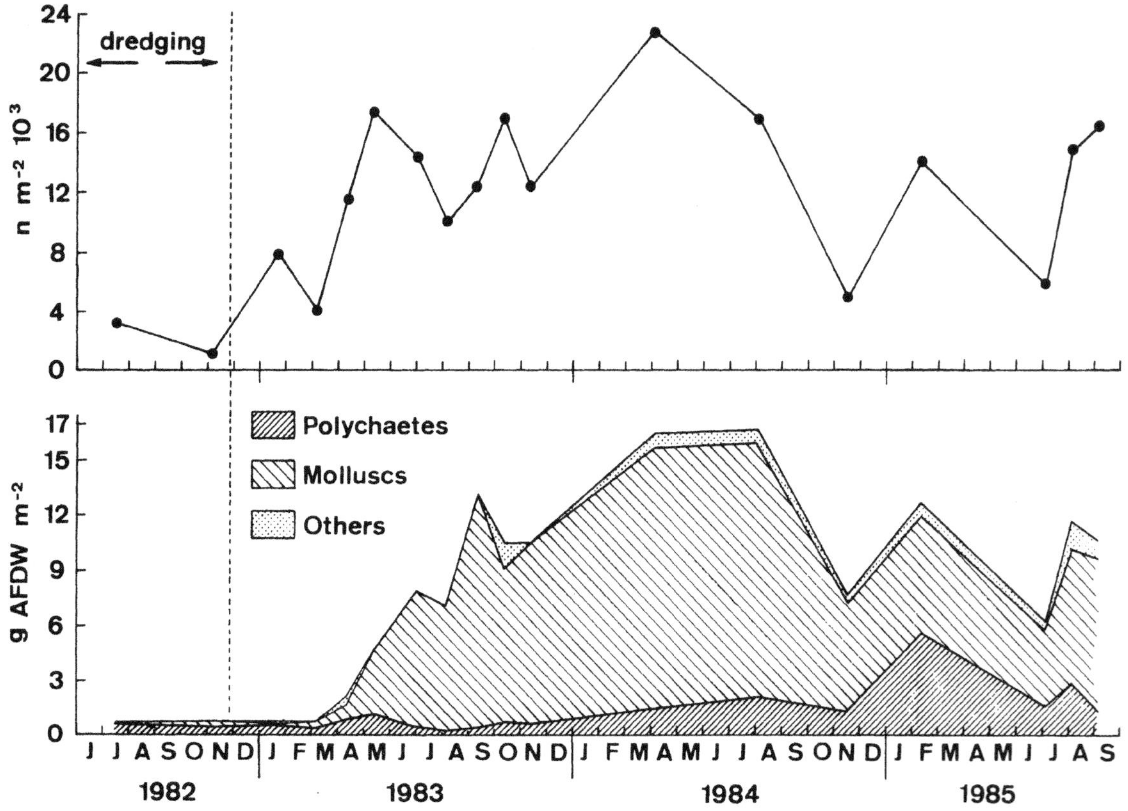

Fig. 9 Temporal variation of total faunal density, total biomass, and proportion of the total biomass of the different groups in the *Thyasira* station.

ning of the study, as well as minor and more frequent disturbances associated with large ship traffic. By contrast, the sediment stability at the *Tellina* station is related to the lack of natural or induced disturbances of the environment during the study period.

In the *Tellina* community of La Coruña Bay, the present study has not detected major changes in community structure over a three-year period. Species composition remained very stable, and the population density of the dominant species (excluding *S. filicornis* and *P. kempi*) varied very little through time. Dauvin (1979) reported that *S. filicornis* population density exhibits wide temporal fluctuations in the Bay of Morlaix. *P. kempi* is an opportunistic species whose population size can dramatically increase under certain conditions, i.e., after an oxygen defficiency during summer. In the *Tellina* station, *P. kempi* population density showed a marked peak in summer 1984, probably as a response to low oxygen conditions in the bottom water. Nevertheless, the abundance of other infaunal dominants in this community, such as *T. fabula* and *S. bombyx,* remains fairly stable during the study period. However, Ziegelmeier (1963, 1970) found population irruptions of *S. bombyx* and temporary extinctions of *T. fabula* in a very similar community of the German Bight during a 17-year study. Thus, the apparent stability of the *Tellina* community in La Coruña Bay may be related to the lack of natural or man induced disturbances during the relatively short time of study, and an extended time-series would probably reveal a long-term variability.

The *Thyasira* station exhibits a much wider variation of community structure, biomass and diversity. This fact is the result of: (1) a short-time, major disturbance, such as the dredging operations carried out during the second half of 1982; and (2) a series of minor and more frequent disturbances related to the harbour activities: large ship traffic, accidental dumping, etc. As Flint & Younk (1983) pointed out, large ship traffic can cause bottom sediment disruption and high turbidity, thus affecting the infaunal community. This fact can explain the wide-range variations of the sediment features at the *Thyasira* station. These authors indicated that in an area influenced by ship traffic, the benthic community exhibited lower species number, densities, and diversity than the adjacent areas not affected by the large ship traffic, and these findings agree with the results of this study.

The occurrence of the bivalve *T. flexuosa* as the first macroinfaunal colonist after dredging is related to the fact that the end of dredging coincided with the main peak of the settlement of this species in La Coruña Bay. Without this synchronism, the first colonizers probably would have been small polychaetes. Nevertheless, *T. flexuosa* remained as the dominant organism during the whole period studied. This species is known to increase its abundance after drastic disturbances such as oil spills (Dauvin, 1982). In La Coruña Bay there is an oil terminal close to the *Thyasira* station, and heavy metal and hydrocarbon concentrations in the sediment are very high (Cabanas, personal communication). Thus it is likely that *T. flexuosa* is not affected very much by this type of pollution, and then its recovery after dredging can be faster than that of other organisms. This species has an opportunistic behaviour, reaching very high densities in a short time after dredging ended.

In the *Thyasira* station, the community composition and structure remains as one that is typical of a pioneering assemblage during the three years of study. The continual minor disturbances, both in time and space, of the sediments in this area can probably keep this community in a state of continual disruption. This allows the opportunists to persist more successfully than non-opportunistic species (Flint & Younk, 1983). Under conditions of frequent disturbance, only those species with a high population growth rate can be successful.

In La Coruña Bay, as well as in other estuaries affected by industrial and urban activities, the occurrence of frequent and small-scale disturbances can impose different patterns of community structure than those found in areas where these events do not accur. The results of this study agree with the model of benthos response suggested by Boesch & Rosenberg (1981): (1) communities in less constant environments are more resistant to disturbance; and (2) colonists in inconstant environments affected by disturbance are usually species already dominant in the community rather than alien opportunistic species. In La Coruña Bay, the benthos that recolonized the dredged area is composed of the same species that were dominant in the surrounding areas (López-Jamar & Mejuto, in press), and they seem to be very successful after a major disturbance (dredging).

In summary, this study of the temporal variation of the benthos of La Coruña Bay indicates that: (1) a community composed mainly of opportunistic species can persist under continuous disturbance conditions (the *Thyasira* community); and (2) the study period was not long enough to detect temporal trends in an undisturbed benthic assemblage (the *Tellina* community). Consequently, longer term studies are needed to understand adequately the natural patterns of variability in coastal benthic communities, and observations of the benthos of La Coruña Bay will continue in subsequent years.

Acknowledgements

The authors are indebted to Dr. T. H. Pearson for his ideas on the research planning. Thanks also to Dr. Brendan O'Connor for the identification of some polychaetes, and to Concepción Fernández and Alicia Alonso for their help in sorting and sediment analysis. This research was conducted within the COST 647 Project.

References

Boesch, D. F. & R. Rosenberg, 1981. Response to stress in marine benthic communities. In G. M. Barret & R. Rosenberg (eds), Stress effect on natural ecosystems. John Wiley & Sons, New York: 179–200.

Buchanan, J. B., 1984. Sediment analysis. In N. A. Holme & A. D. McIntyre (eds), Methods for the study of marine benthos. Blackwell Scientific Publications, Oxford: 41–65.

Dauvin, J. C., 1979. Recherches quantitatives sur le peuplement des sables fins de la Pierre Noire, Baie de Morlaix, et sur sa perturbation par les hydrocarbures de l' 'Amoco-Cadiz'. PhD. Thesis. Université Pierre et Marie Curie, Paris.

Dauvin, J. C., 1982. Impact of Amoco Cadiz oil spill on the muddy fine sand *Abra alba* and *Melinna palmata* community from the Bay of Morlaix. Estuar. coast. mar. Sci. 2: 329–348.

Flint, R. W. & J. A. Younk, 1983. Estuarine benthos: long-term community structure variations, Corpus Christi Bay, Texas. Estuaries 6: 126–141.
Horn, H. S., 1966. Mesures of 'overlap' in comparative ecological studies. Am. Nat. 100: 419–424.
López-Jamar, E., 1978. Macrobentos infaunal de la Ría de Pontevedra. Boln Inst. esp. Oceanogr. 4: 111–130.
López-Jamar, E., 1981. Spatial distribution of the infaunal benthic communities of the Ría de Muros, North-West Spain. Mar. Biol. 63: 29–37.
López-Jamar, E., 1982. Distribución espacial de las comunidades bentónical infaunales de la Ría de Arosa. Boln Inst. esp. Oceanogr. 7: 255–268.
López-Jamar, E. & J. Mejuto, 1985. Benthos infaunal en la zona submareal de la Ría de La Coruña. 1. Estructura y distribución espacial de las comunidades. Boln Inst. esp. Oceanogr. 2: 99–109.
López-Jamar, E. & J. Mejuto, 1986. Evolución temporal de cuatro comunidades infaunales submareales de las Rías de Arosa y Muros. Resultados preliminares. Boln Inst. esp. Oceanogr. 3: 95–110.
López-Jamar, E. & J. Mejuto, in press. Benthic recolonization after dredging operations in La Coruña Bay, NW Spain. Oceanis.
Mason, W. T. & P. P. Yevich, 1967. The use of Phloxine B and Rose-bengal stains to facilitate sorting benthic samples. Trans. am. microsc. Soc. 86: 221–223.
Mora, J., M. A. García & R. Acuña, 1982. Contribución al conocimiento de las poblaciones de la macrofauna bentónica de la Ría de Pontevedra. Oecol. aquat. 6: 51–56.
Morisita, M., 1959. Measurement of interespecific association and similarity between communities. Mem. Fac. Sci. Kyushu Univ. (Ser. E) 3: 65–80.
Pielou, E. C., 1966. The measurement of diversity in different types of biological collections. J. theor. Biol. 13: 131–144.
Salzwedel, H., 1979. Reproduction, growth, mortality, and variations in abundance and biomass of *Tellina fabula* (Bivalvia) in the German Bight in 1975/76. Veröff. Inst. Meeresforsch. Bremerh. 18: 111–202.
Shannon, C. E. & W. Weaver, 1963. The mathematical theory of communication. Univ. Illinois Press, Urbana, 111 pp.
Shin, P. K. S., M. E. Conneely & B. F. Keegan, 1982. Littoral and benthic investigacions on the West coast of Ireland – XV. The macrobenthic communities of North Bay (Galway Bay). Proc. r. irish Acad. 82(B): 133–154.
Stephen, A. C., 1930. Studies on the Scottish marine fauna. Additional observations on the fauna of the sandy and muddy areas of the tidal zone. Trans. r. Soc. Edinb. 56: 521–535.
Tenore, K. R., L. F. Boyer, R. M. Cal, J. Corral, C. Garcia Fernández, N. González, E. González-Gurriarán, R. B. Hanson, J. Iglesias, M. Krom, E. López-Jamar, J. McClain, M. M. Pamatmat, A. Pérez, D. C. Rhoads, G. de Santiago, J. Tietjen, J. Westrich & H. L. Windom, 1982. Coastal upwelling in the Rias Bajas, NW Spain: contrasting the benthic regimes of the Rias de Arosa and the Muros. J. mar. Res 40: 701–772.
Ziegelmeier, E., 1963. Das Makrobenthos im Ostteil der Deutschen Bucht nach qualitativen und quantitativen Bodengreiferuntersuchungen in der Zeit von 1949–1960. Veröff Inst. Meeresforsch. Bremerh. Sonderbd 1: 101–114.
Ziegelmeier, E., 1970. Über Massenvorkommen verschiedener makrobenthaler Wirbelloser während der Wiederbesiedlungsphase nach Störungen durch 'katastrophale' Umwelteinflüsse. Helgoländer wiss. Meeresunters. 21: 9–20.

Hydrobiologia 142: 151–158 (1986)

Demographic equilibrium: The case of an *Amphiura filiformis* assemblage on the west coast of Ireland

Brendan D. S. O'Connor, David McGrath & Brendan F. Keegan
Department of Zoology, School of Marine Sciences, University College, Galway, Ireland

Keywords: *Amphiura* community, long-term study, west coast of Ireland

Introduction

Amphiura filiformis dominated muddy-sand assemblages have been described at a number of localities from shallow European waters (see O'Connor *et al.*, 1983 for review). One location in Galway Bay, with consistently high densities of *A. filiformis*, has been under examination since 1974 (except for the period October 1976–October 1978) and has been reported upon several times in the literature: Keegan *et al.* (1976) outline the distribution of macrofaunal assemblages in Inner Galway Bay and list the species which are prominent at the study site; O'Connor & McGrath (1980), Bowmer (1982), Bowmer & Keegan (1983), O'Connor *et al.* (1983) and O'Connor *et al.* (1986) describe various aspects of the ecology and biology of *A. filiformis*. The two other numerical dominants in the assemblage i.e. the polychaete *Pholoe minuta* Fabricius and the bivalve *Mysella bidentata* (Montagu) have also been studied (Heffernan *et al.*, 1983; O'Foighil *et al.*, 1984; Heffernan, 1985).

In 1980, the study methods were adapted to conform to the protocol of the COST647 project (Keegan *et al.*, 1982). Data are presented here from July 1980–October 1985 on community structure and dynamics at the monitoring site. These data are discussed in the light of monitored environmental factors and are compared with similar studies in other geographical localities.

Study area and methods

Station description

For the present work, all samples were taken at the Margaretta station, Inner Galway Bay (53°13.16′N, 9°6.30′W) at a depth of 18 m. The sediment is a poorly sorted fine-sand (*ca.* 80% fine sand) with an organic component of *ca.* 7%. Temperature and salinity data, collected since 1973, show bottom temperatures to remain within the range 5.8–16.5 °C and salinities stay above 33.5. O'Connor *et al.* (1983) demonstrated that during the years 1972–1979, there was a two degree drop in bottom temperature. Since then there has been a recovery of 2.1 °C. Oxygen levels are always saturated and bottom current speeds are in the region of 0.25 $m \cdot sec^{-1}$. Harte *et al.* (1982) have demonstrated the presence of a gyre in Inner Galway Bay based on residual flow patterns. This anticlockwise-turning gyre is at, or close to, the centre of the highest *A. filiformis* densities. (For further site description see Keegan *et al.*, 1976).

Sampling procedure

Five replicate faunal samples and one sediment sample were taken four times a year with a modified 0.1 m^2 van Veen grab (0.5 mm mesh doors fitted to the upper surface). The grab performed consistently well on the sediment at the monitoring station taking a sample of between 7–9 L to an average depth of 11–13 cm. Faunal samples were placed in 0.5 mm mesh sieves and the finer sediment washed out by 'puddling' in a large tray filled with seawater. The material remaining on the sieve after washing was stained with eosin red and preserved in either 70% alcohol or neutral buffered formalin. Samples were split into 1 and 0.5 mm fractions in the laboratory and the 1 mm fraction was directly sorted under a binocular microscope (× 10 magnification) or after elutriation in a fluidised sand bath (P. Barnett, unpublished MS). (The 0.5 mm fraction was retained for examination of post-larval settlement). All the major taxa were identified (with the exception of the Nemertea and Tanaidacea) and counted. Only those determined to species level were used for analysis.

Data analysis

The fauna from the five replicates were amalgamated to form one sample, representing 0.5 m^2, for each sampling session from 1980 to 1985. Species were given unique 6 character codes based on the first three letters of their generic and specific epithets e.g. AMPFIL = *Amphiura filiformis.* These codes and species counts were entered on a data file for analysis of community structure, dynamics and diversity. Rank (Boesch, 1977) and Inverse analyses were performed to determine community dominants, with the former being run on the complete data set and the latter on a transformed (log n(x+1)) and reduced set (i.e. omission of species which occurred less than twice). Group Average sorting (Clifford & Stephenson, 1975) was employed to cluster the values generated using the Bray-Curtis dissimilarity index (Bray & Curtis, 1957).

Normal analysis was used to elucidate community dynamics on the transformed and reduced data set. The pseudo-F test (Boesch, 1977) was used to determine those species which characterise sample clusters. Diversity as Shannon-Weiner (Pielou, 1975) and Brillouin (Washington, 1984) indices, evenness (Pielou, 1975), richness (Margalef, 1958) and numbers of individuals were used to describe the assemblage.

Results

Of the 231 species of benthic macrofauna identified, 120 remained after reduction. This latter group was used in the statistical analyses.

Community structure

The dendrogramme resulting from Inverse analysis (Fig. 1) on the data shows a cluster which fuses at a level of 0.345 and contains 46 species (Table 1) which are deemed to be most typical of the assemblage at large. Based on present knowledge, an examination of the feeding types within these taxa shows that 28 are deposit feeders, 12 are carnivores/scavengers and 6 are suspension feeders. The results from the Rank analysis are listed in Table 2.

Community dynamics

Normal analysis of the time series data resulted in the dendrogramme presented in Fig. 2. Whilst the level of fusion of all clusters is below that which is considered to be statistically significant i.e. *ca.* 0.6, four groups are clearly visible. Reading from left to right, the first contains ten samples from August 1980–December 1982 and fuses at 0.27; the second group, fusing at 0.3 and with five samples includes the months May 1983–August 1984. Group three has the lowest fusion coefficient of 0.21 and comprises four samples, November 1984–May 1985, while the last group has the highest coefficient of 0.37 and contains the February samples from 1983 and 1984. These four clusters will be referred to as A, B, C and D respectively.

In order to explain this pattern in terms of faunal change, a pseudo-F test was run on the clusters. This test identifies those species which characterise these groups and these are listed under different headings in Table 3. 'Rare species' are those which occur in densities of less than 10 per sample while those listed under 'Sporadic Occurrence' appear irregularly in high numbers during the study. Certain species are known commensals and are so listed while those under the heading 'Small Species' are either small, surface dwelling polychaetes or juvenile bivalves whose adults are rarely if ever recorded. The fifth category includes those species which show upward, downward or other trends in densities. The first four categories are not considered further here as their densities are too low, their occurrence too sporadic or dependant on the presence of known hosts, or, in the case of small species, the sampling technique is likely to be inadequate to collect them efficiently. Within the fifth group, the following variations have been noted: *Pholoe, Mysella* and *Eulima* show clear downward trends over the duration of the study, *Lumbrineris* and *Nephtys* show a downward trend in the early part of the study, and an upward trend latterly, while *Echinocardium* shows an increase in numbers in the

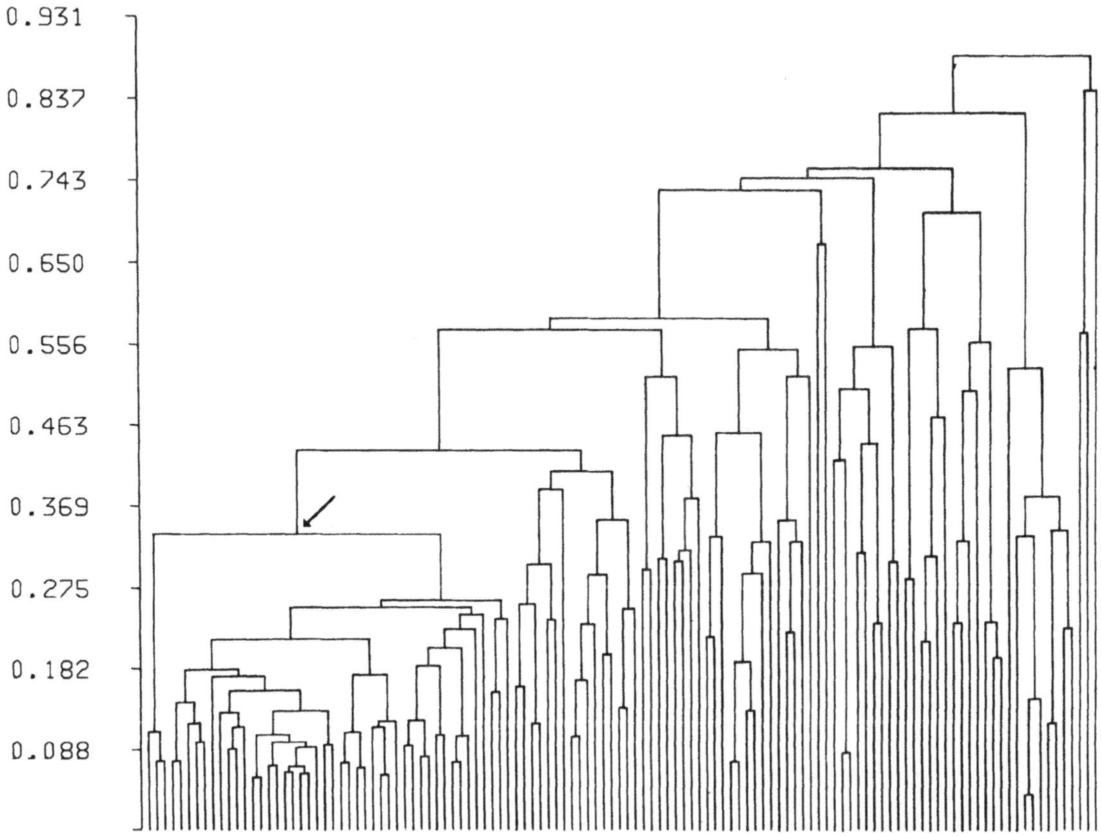

Fig. 1. Dendrogramme of Inverse analysis on the reduced data set from the Margaretta station, Galway Bay 1980–1985.

Table 1. Species contained in arrowed cluster of Fig. 1. (Order as in dendrogramme).

Pholoe minuta	Ophiodromus flexuosus
Amphiura filiformis	Montacuta ferruginosa
Mysella bidentata	Cerianthus llyodii
Prionospio malmgreni	Magelona alleni
Nucula turgida	Spio martinensis
Minuspio cirrifera	Harpinia pectinata
Chaetozone setosa	Paranaitis kosterensis
Magelona minuta	Eulima glabra
Aricidea catherinae	Venus striatula
Cylichna cylindracea	Melinna palmata
Thracia phaseolina	Spiophanes bombyx
Lumbrineris gracilis	Harpinia crenulata
Nephtys hombergii	Mediomastus fragilis
Ampelisca tenuicornis	Goniada maculata
Owenia fusiformis	Cultellus pellucidus
Phthisica marina	Harpinia antennaria
Diplocirrus glaucus	Gyptis capensis
Leptosynapta bergensis	Echinocardium cordatum
Thyasira flexuosus	Notomastus latericeus
Onoba vitrea	Leptopentacta elongata
Harmothoe andreapolis	Sthenelais limicola
Leucothoe lilljeborgi	Eudorella truncatula
Spisula subtruncata	Corbula gibba

Table 2. Result of Rank Analysis.

Rank analysis (maximum score 210)		
Amphiura filiformis	:	201
Mysella bidentata	:	189
Pholoe minuta	:	172
Lumbrineris gracilis	:	110
Nucula turgida	:	58
Nephtys hombergi	:	52
Ampelisca tenuicornis	:	48
Prionospio malmgreni	:	45
Minuspio cirrifera	:	36
Chaetozone setosa	:	24

second half of the study (Fig. 3). Although not listed in Table 2, *Amphiura* densities, shown in Fig. 4, remain relatively stable.

The computed community statistics, i.e. diversity, indices, evenness, richness and numbers of individuals per sample are shown in Fig. 5. As can be seen from the last index, the samples from February 1983 and 1984 differ from the others in having

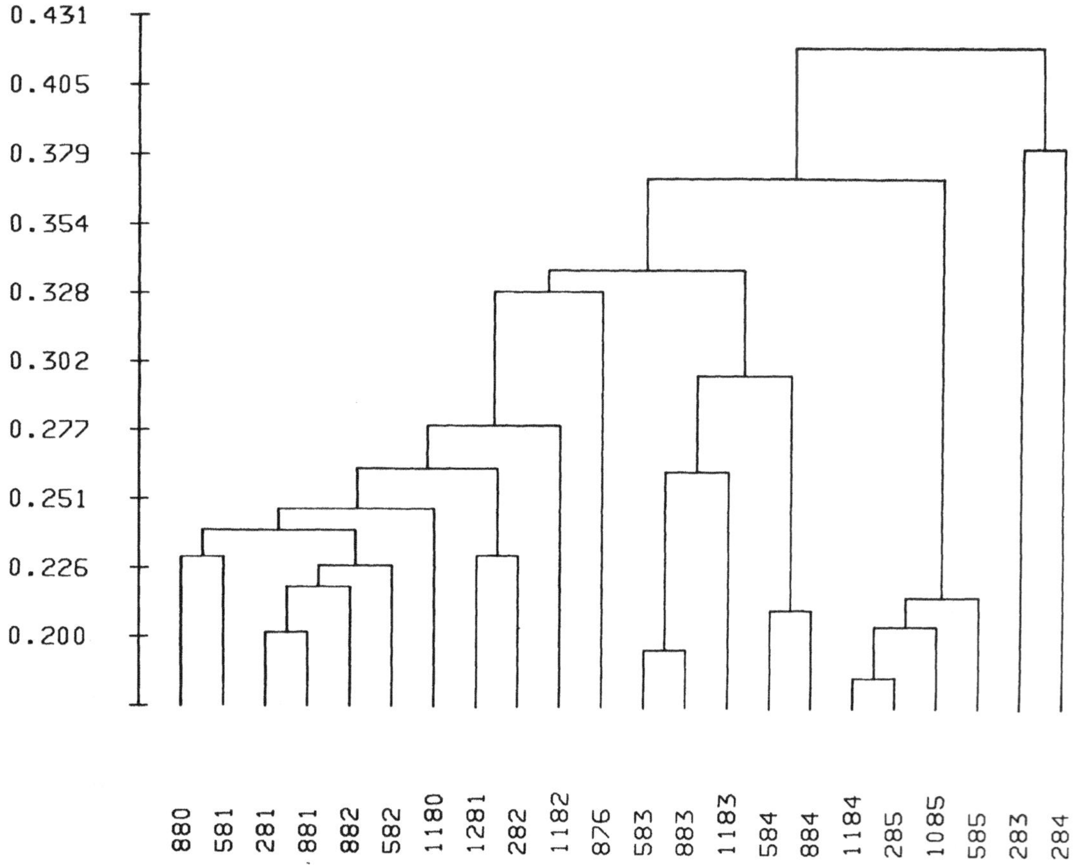

Fig. 2. Dendrogramme of Normal analysis on the reduced data set from the Margaretta station, 1980–1985.

less than 1000 individuals. This difference is reflected in the other indices only for the 1983 sample because of the low number of species (45) recorded on that date compared to 61 for the same month in the following year. In both samples, small species such as *Minuspio* and *Prionospio* were either absent or represented by single individuals and both *Pholoe* and *Mysella* had diminished densities. Fig. 2 (Cluster D) highlights the similarity and overall discreteness of these two samples. They were taken during severe sea conditions and both grab performance and on-board processing of the samples suffered as a result. Further examination of Fig. 5 shows that the evenness, Shannon-Weiner & Brillouin indices demonstrate little change. The number of individuals per 0.5 m^2 and species richness values show a downward trend, especially in the latter half of the study.

Discussion

To the authors' knowledge, the lack of seasonal cycles and significant variation in the long-term quantitative composition of the *A. filiformis* assemblage has not been previously recorded for other soft-bottom benthic communities. Buchanan *et al.* (1978) working on a somewhat similar community off the English East coast do, however, comment that '...the faunal groupings and the delineation of associations are entirely similar to those found by Buchanan (1963), and it must be assumed that the broad features of faunal distribution have remained stable for somewhat more than a decade'. They nonetheless note a clear annual cycle in terms of numbers of species and biomass and also show a long-term change in faunal composition. 'Gaining', 'losing' and 'neutral' species, i.e. those with

Table 3. Categories of species showing variable trends in density over the study period.

Rare species	*Downward trend*
Paranaitis kosterensis	Pholoe minuta
Eteone longa	Eulima glabra
Gyptis capensis	Mysella bidentata
Goniada maculata	
Phthisica marina	*Variable trend*
Diastylis laevis	Lumbrineris gracilis
Harpinia antennaria	Nephtys hombergi
Harpinia crenulata	
Leucothoe lilljeborgi	*Upward trend*
Ophiura affinis	Echinocardium cordatum
Commensals	
Harmothoe andreapolis	*Sporadic occurrence*
Harmothoe lunulata	Scalibregma inflatum
Devonia perrieri	
Montacuta ferruginosa	
Small species	
Aricidea catherinae	
Minuspio cirrifera	
Prionospio malmgreni	
Chaetozone setosa	
Magelona minuta	
Mediomastus fragilis	
Nucula turgida	
Spisula subtruncata	
Corbula gibba	
Thyasira flexuosa	

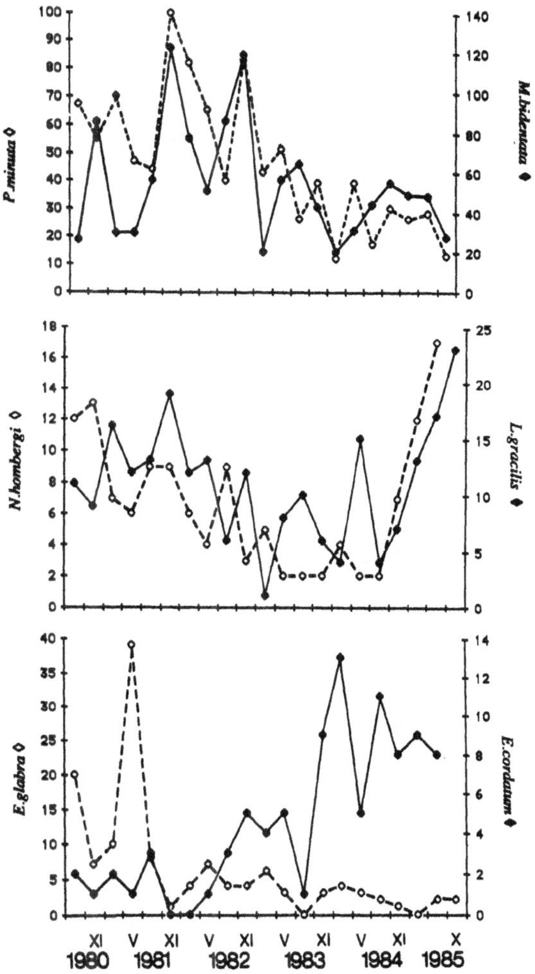

Fig. 3. Mean densities per 0.1 m^2 of species pairs *P. minuta/M. bidentata, N. hombergi/L. gracilis* and *E. glabra/E. cordatum* from the Margaretta station, 1980–1985.

increasing, decreasing and stable densities respectively, are recorded over the six year period of their study. Other long-term data in the literature e.g. Pearson & Rosenberg (1978), Rachor (1980) and Reid (1979), report changes in benthic assemblages which are attributed to aperiodic stresses such as the impact of industrial wastes or major, atypical climatic events.

Buchanan *et al.* (1978) attribute such changes as they observed to a gradual increase in sea temperature. However, Buchanan & Moore (1986) now suggest that the more likely cause of the observed changes is increased organic load due to sewage dumping. Conversely, while Pearson & Rosenberg (1978) pointed to the waste from a pump mill as the most likely cause for changes in species abundances at their study site, Pearson (1986) points out the correlation between such fluctuations and a concommitant change in sea temperature. Temperature variation, in terms of air temperature, has also been used by Glémarec (1979) and Princz *et al.* (1984) to explain the temporal evolution of what is again a very similar assemblage to that reported here. As noted above, long-term change in bottom temperature has been recorded at the monitoring site in Galway Bay. This temperature change cannot be correlated with such density fluctuations as have been observed at the site. Moreover, the extent to which a 2 °C variation in temperature can effect species (e.g. *Pholoe, Nephtys, Lumbrineris* and *Mysella*) which are well within their geographic ranges has yet to be demonstrated. *Amphiura* itself, with no significant change over time, does not appear to have responded to this change in tempera-

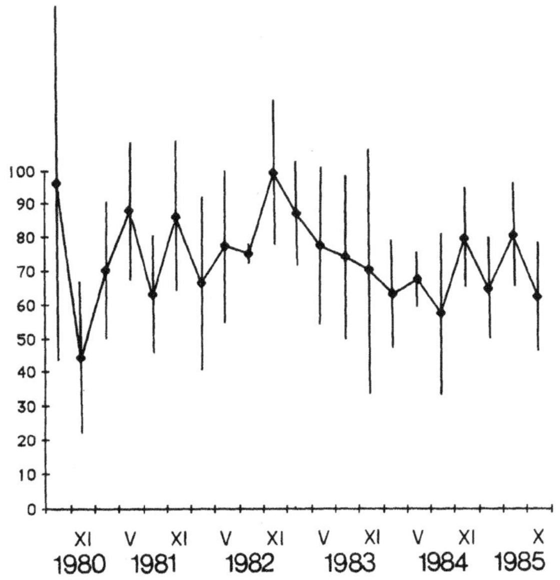

Fig. 4. Mean densities per 0.1 m^2 of *A. filiformis* from the Margaretta station, 1980–1985.

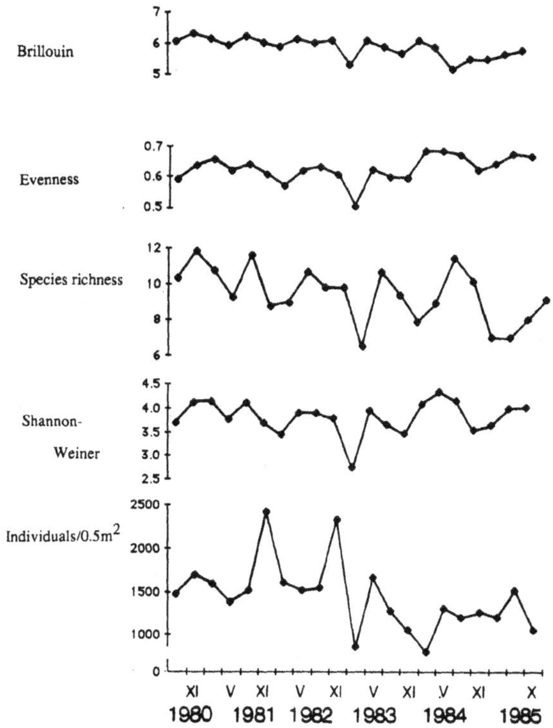

Fig. 5. Diversity Indices for Margaretta Station, Galway Bay, 1980–1985.

ture. Density changes in some species may represent parts of long-term cyclic patterns in abundance such as have been documented by Gray (1981).

The data presented here, and in O'Connor *et al.* (1983), show that *A. filiformis* densities have remained stable since at least 1969. The latter workers and Muus (1981) suggested a life-span of at least 20 years for this ophiuroid. In terms of both numbers and biomass, it is the macrofaunal dominant and is presumed to have a key role in structuring the assemblage (Ockelmann & Muus, 1978; O'Connor *et al.*, 1983). It is interesting to speculate as to whether it is the very long-term constancy of the *A. filiformis* population which has maintained the compositional character of the assemblage at large.

Apart from the density fluctuations in the species referred to above, the assemblage has remained largely unchanged in macrofaunal composition over the period considered here. Indeed there are data to suggest that this situation has persisted since 1976. O'Connor *et al.* (1983) present data from the October months of 1978–1980 inclusive which show that it was stable over that time period. Samples collected in August 1976 and processed in the same way as described above, were added to the data set for July 1980–October 1985 and are included in Fig. 1. It suggests little or no difference in the assemblage between 1976 and 1980.

Given its constancy over this 10 year period, the *A. filiformis* assemblage in Galway Bay can be seen as an equilibrium community *sensu* Rhoads & Boyer (1982). It displays a number of characteristics attributed to mature or 'Stage III' phases in soft-bottom marine assemblages by Rhoads *et al.* (1978) but differs in being dominated by the largely suspension/surface deposit feeding ophiuroid, *A. filiformis* rather than strictly infaunal deposit feeders. Nonetheless, infaunal deposit feeders, though low in numbers, are represented e.g. *Echinocardium cordatum* (Pennant), *Notomastus latericeus* Sars and *Leptosynapta bergensis* (Ostergren). A characteristic feature of equilibrium stages is the presence of large deep-burrowing species which, in addition to those listed above, include *Upogebia spp, Calianassa subterranea* (Montagu) and *Thracia pubescens* (Montagu), (Keegan *et al.*, 1976). Their burrowing range precludes them from being collected routinely by the van Veen grab. Just how long the equilibrium has been maintained is not known but the earliest samples collected in 1969 showed that *A. filiformis* was numerically dominant even then. Thus the assemblage appears to have persisted for at least 17 years.

Just how long this assemblage will endure is a matter of speculation. Glémarec (1979) discussing the causes of change in benthic communities, highlights seasonality, aperiodic stress-induced change and periodic stresses, cyclic in nature, which he relates to sunspot cycles. No such changes have been noted in the *A. filiformis* population, but the latter two might explain density fluctuations in other species of the assemblage. Equilibrium communities in other ecosystems persist for longer time scales than are currently known for sublittoral benthic systems and it may be that *Amphiura* communities, if not upset by anthropogenic activity, may persist for comparable periods. However, it is also possible that natural aperiodic perturbations, such as severe storms, might disrupt the Galway Bay assemblage. Such effects have been noted by Rees *et al.* (1977) and Rachor & Gerlach (1978) *inter alios.*

The persistent character of stable assemblages may simply be due to the survival of longevous species (Connell, 1985). Connell & Sousa (1983) have suggested that, to avoid this difficulty, numbers be compared over an interval in which complete turnover of all individuals has occurred. It is not known whether the *Amphiura* population consists of one or few successful recruitments which have persisted through the period of investigation or whether it is composed of many year classes with continuous low-level recruitment offsetting mortality. To date, answering this question has been frustrated by the inability to age *Amphiura* (O'Connor *et al.*, 1983).

A. filiformis assemblages with low densities of the brittle star are wide-spread in Galway Bay (O'Connor & McGrath, 1981). This present study however, has been carried out on a site which exhibits the highest density of *A. filiformis* in the area (O'Connor *et al.*, 1983) and is at or near to the centre of a gyre demonstrated by Harte *et al.* (1982). Whether the results presented here are representative of the *A. filiformis* assemblage throughout the Bay or whether the gyre may influence the assemblage remains to be ascertained.

Acknowledgements

This work was carried out under an National Board for Science and Technology Grant to Dr B. F. Keegan provided within the COST 647 Project on Coastal Ecology. Special thanks to technicians Albert Lawless and John Galvin, and Davy Burke, Skipper r.v. Ona III, for the many years of toil. Their help and humour made the long hours short.

Contribution No. 267 from the School of Marine Sciences, University College, Galway.

References

Boesch, D. F., 1977. Applications of numerical classification in ecological investigations of water pollution. Special Scientific Report No. 77, Virginia Institute of Marine Science.

Bowmer, T., 1982. Reproduction in *Amphiura filiformis* (Echinodermata: Ophiuroidea): seasonality in gonad development. Mar. Biol. 69: 281–290.

Bowmer, T. & B. F. Keegan, 1983. Field survey of the occurrence and significance of regeneration in *Amphiura filiformis* (Echinodermata: Ophiuroidea) from Galway Bay, west coast of Ireland. Mar. Biol. 74: 65–71.

Bray, J. R. & J. T. Curtis, 1957. An ordination of the upland forest communities of Southern Wisconsin. Ecol. Mono. 27: 325–349.

Buchanan, J. B., 1963. The bottom fauna communities and their sediment relationships of Northumberland. Oikos 14: 154–175.

Buchanan, J. B., M. Sheader & P. F. Kingston, 1978. Sources of variability in the benthic macrofauna off the south Northumberland coast, 1971–1976. J. mar. biol. Ass. U.K. 58: 191–209.

Buchanan, J. B. & J. J. Moore, 1987. Long-term studies at a benthic station off the coast of Northumberland. Hydrobiologia, this volume.

Clifford, H. T. & W. Stephenson, 1975. An introduction to numerical classification. Academic Press, New York, 299 pp.

Connell, J. H., 1985. Variation and persistance of rocky shore populations. In P. Moore and R. Seed (eds). The ecology of rocky coasts. Hodder and Stoughton, London, 57–70 pp.

Connell, J. H. & W. P. Sousa, 1983. On the evidence needed to judge ecological stability or persistence. Am. Nat. 121: 729–824.

Glémarec, M., 1979. Les fluctuations temporelles des peuplements benthique liées aux fluctuations climatique. Oceanol. Acta 2: 365–371.

Gray, J., 1981. The ecology of marine sediments. Cambridge University Press, Cambridge, 185 pp.

Harte, A. M., J. P. Gilroy & S. F. McNamara, 1982. A computer simulation of water circulation in Galway Bay. Eng. J. 35: 6–8.

Heffernan, P., B. O'Connor & B. F. Keegan, 1983. Population dynamics and reproductive cycle of *Pholoe minuta* (Fabricius) (Polychaeta: Sigalionidae) in Galway Bay, west coast of Ireland. Mar. Biol. 73: 285–291.

Heffernan, P., 1985. Demography of *Pholoe minuta* (Polychaeta: Sigalionidae) in Galway Bay, west coast of Ireland, with special reference to settlement and recruitment patterns. Mar. Biol. 84: 312–329.

Keegan, B. F., O'Connor, D. McGrath & G. Könnecker, 1976. The *Amphiura filiformis-Amphiura chiajei* community in Galway Bay (west coast of Ireland) – a preliminary account. Thallas. Jugo. 12: 189–198.

Keegan, B. F., B. O'Connor, D. McGrath & M. Lyes, 1982. COST 47 Newsletter Number 2. N.B.S.T., Dublin, 14 pp.

Margalef, D. R., 1958. Information theory in ecology. Gen. Sys. 3: 36–71.

Muus, K., 1981. Density and growth of juvenile *Amphiura filiformis* (Ophiuroidea) in the Oresund. Ophelia 20: 153–168.

O'Connor, B. & D. McGrath, 1980. The population dynamics of *Amphiura filiformis* (O. F. Müller) in Galway Bay, west coast of Ireland. In M. Jangoux (ed.), Echinoderms: past and present. A. A. Balkema, Rotterdam: 219–222.

O'Connor, B. & D. McGrath, 1981. Benthic Macrofaunal studies in the Galway Bay area. Vol. 1. The Macrobenthic Faunal Assemblages of Galway Bay. Ph.D. thesis (unpubl.), N.U.I.

O'Connor, B., T. Bowmer & A. Grehan, 1983. Long-term assessment of the population dynamics of *Amphiura filiformis* (Echinodermata: Ophiuroidea) in Galway Bay, west coast of Ireland. Mar. Biol. 82: 222–223.

O'Connor, B., T. Bowmer, D. McGrath & R. Raine, (In press). Energy flow through an *Amphiura filiformis* (Ophiuroidea: Echinodermata) population in Galway Bay, West Coast of Ireland: a preliminary investigation. In K. Jensen (ed.), Processes in the Sediment. Ophelia, special supplement.

Ockelmann, K. W. & K. Muus, 1978. The biology, ecology and behaviour of the bivalve *Mysella bidentata* (Montagu). Ophelia 17: 1–93.

O'Foighil, D., D. McGrath, M. E. Conneely, B. F. Keegan & M. Costelloe, 1984. Population dynamics and reproduction of *Mysella bidentata* (Bivalvia: Galeommatacea) in Galway Bay, Irish west coast. Mar. Biol. 81: 283–291.

Pearson, T., 1987. Long-term changes in the benthic communities of Loch Linnhe and Loch Eil (Scotland). Hydrobiologia, this volume.

Pearson, T. & R. Rosenberg, 1978. Macrobenthic succession in relation to organic enrichment and pollution of the marine environment. Oceanogr. Mar. Biol. Ann. Rev. 16: 229–311.

Pielou, E. C., 1975. Ecological Diversity. J. Wiley and Sons N.Y., 165 pp.

Princz, D., A. Menesguen & M. Glemarec, 1983. Temporal evolution over ten years in macrobenthos of muddy sands in the Bay of Concarneau (France). In L. Cabioch, M. Glemarec & J. F. Samain (eds), Fluctuation et succession dans les ecosystemes marins. Oceanol. Acta, special volume: 159–164.

Rachor, E., 1980. The inner German Bight – an ecologically sensitive area as indicated by the bottom fauna. Helgölander wiss. Meeresunters. 33: 522–530.

Rachor, E. & S. Gerlach, 1978. Changes in macrobenthos in a sublittoral sand area of the German Bight, 1967 to 1975. Rapp. P.–v. Cons. int. Explor. Mer, 172: 418–431.

Rees, E. I. S., A. Nicholaidou & P. Laskaridou, 1977. The effects of storms on the dynamics of shallow water benthic associations. In B. F. Keegan, P. O'Céidigh & P. J. S. Boaden (eds), The Biology of Benthic Organisms. Pergamon Press, Oxford: 465–474.

Reid, R. N., 1979. Long-term fluctuations in the mud-bottom macrofauna of Long Island Sound, 1972–1978. M.A. thesis (unpubl.), Boston University.

Rhoads, D. C., P. L. McCall & J. Y. Yingst, 1978. Disturbance and production on the estuarine seafloor. Am. Sci. 66: 577–586.

Rhoads, D. C. & L. F. Boyer, 1982. The effects of marine benthos on physical properties of sediments. In P. McCall and M. Tevesz (eds), Animal-Sediment Relations: The biogenic alteration of sediments. Plenum Press, N.Y.: 3–50.

Washington, H. G., 1984. Diversity, biotic and similarity indices: a review with special reference to aquatic ecosystems. Wat. Res. 18: 653–694.

Hydrobiologia 142: 159–170 (1986)

Modifications des écosystèmes des vasières cotières du sud-Bretagne

M. Glémarec, H. Le Bris & C. Le Guellec
Université de Bretagne Occidentale, Laboratoire d'Océanographie Biologique, UA 711 du C.N.R.S., 6, Avenue Le Gorgeu, 29287 – BREST Cédex

Keywords: soft benthos, successional models, long-term monitoring, Bay of Biscaye

Summary

The northern part of the Bay of Biscaye consists of muddy areas which were the subject of many surveys conducted from the early sixties revealing important modifications. Communities living in these muddy areas are very rich in quality as well as quantity and with high interactions between the different species. With their trophic ethology, their biogenic actions, these species modify largely the sediment quality, its compactness for example is related to the tubicolous species presence. The disappearance of certain dominant species is due to some climatic anomalies causing an important disturbance of the ecosystem. In the infralittoral muddy areas, *Zostera* beds had been replaced momently by the abundant populations of *Melinna*. In the coastal muddy sites the recent disappearance of the *Maldane* cause a momentary imbalance, and for many long years, an equilibrium based on a new dominant species may appear. The intervention of an allogenic factor may in this case, opposite autogenic succession of the community or call the equilibrium which seems established in question again.

Introduction

Depuis quelques années, un effort particulier est consacré aux théories de la succession écologique afin de comprendre les phénomènes dynamiques intervenant au sein des communautés benthiques. Sans que soit encore clairement définie une réelle théorie de la perturbation, il est sûr que celle-ci est devenue un véritable outil expérimental, simulée parfois au laboratoire, étudiée par ailleurs en vraie grandeur dans le cas de défaunation à la suite d'événements naturels (marées rouges) ou accidentelles (marées noires), ainsi ont été définis des modèles de dynamique successionnelle. En Bretagne, les méthodes d'évaluation quantitative à l'aide de bennes ne sont utilisées que depuis moins de 20 années, mais peut-être est-il possible aujourd'hui, sans beaucoup de recul, d'aborder cette réflexion sur ce que sont les modifications à long terme des écosystèmes sédimentaires marins. Celles qui sont liées aux fluctuations climatiques à petite échelle sont mieux connues aujourd'hui, ainsi que les fluctuations pluriannuelles. Au cours des vingt dernières années, les études des vasières de Bretagne sud révèlent des changements importants dont la première cause semble être d'ordre climatique, liée à une tendance séculaire qui s'est inversée après 1965. Encore faut-il pour identifier un tel phénomène, que soient clairement distinguées les fluctuations qui relèvent, soit de l'activité propre des organismes, soit du développement autogénique des écosystèmes. D'un autre côté, il ne faut pas exclure les facteurs allogéniques non climatiques. Si les actions anthropiques de type polluant semblent limitées dans les secteurs concernés, à moins qu'elles soient diffuses, les indices d'un envasement progressif, lié à une forte érosion des sols, les modifications des apports telluriques, sont à prendre en compte. La Bretagne méridionale offre différents sites d'étude, dont trois sont retenus par leur richesse et par les modifications importantes de leurs peuplements depuis 1962 (Fig. 1).

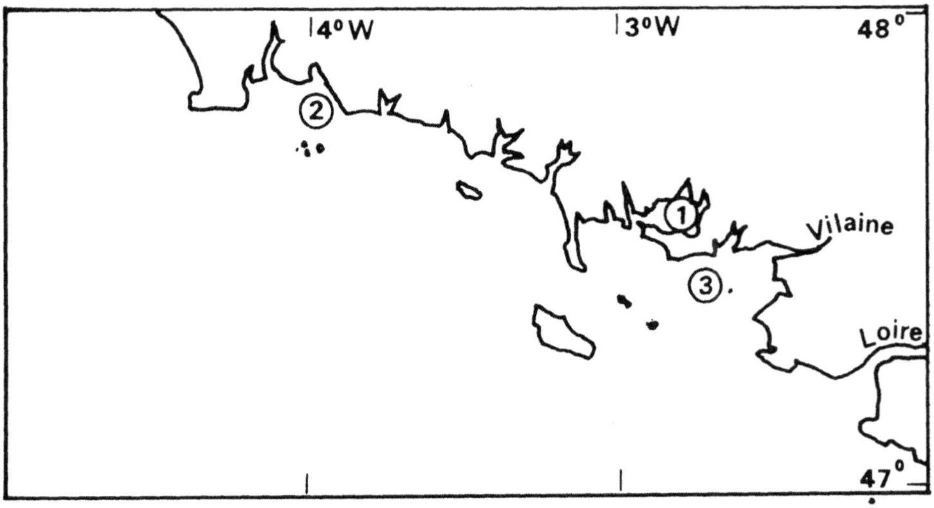

Fig. 1. Localisation des sites d'étude de Bretagne méridionale. **1**– Golfe du Morbihan, **2**– Baie de Concarneau, **3**– Baie de Vilaine.

Les vasières du golfe du Morbihan

La partie orientale de ce golfe est très peu profonde, tout y est envasé et les platiers largement étendus peuvent héberger des herbiers de Zostères. Comme dans l'ensemble de l'Atlantique nord-est, ces herbiers ont subi, à partir des années 1930, une destruction presque totale. Dès la fin des années 1960, il y a renaissance des herbiers sur l'ensemble du secteur touché, celle-ci pouvant apparaître de façon relativement brutale en Bretagne. La carte biosédimentaire (Fig. 2), établie en 1961 (Glémarec, 1964), montre les platiers subtidaux, sans herbiers, fortement érodés et constitués de vase compacte hébergeant un peuplement très dense de *Melinna palmata* (Polychète Ampharetidé). Ce peuplement n'avait jamais été décrit auparavant, depuis, Hily (1984) en rade de Brest a montré que cette espèce tubicole a de fortes potentialités pour recoloniser les vasières placées sous un flux particulaire important. Seul à l'abri de l'île d'Ars, un herbier restreint s'est maintenu. Le phénomène d'érosion, consécutif à la disparition des herbiers, est illustré par la présence de sables hétérogènes envasés, habités par de riches populations de bivalves suspensivores, *Tapes aureus, Solen marginatus*... ceci dans le souffle du chenal passant entre les îles d'Ars et d'Ilur.

Le même travail de bionomie, renouvelé en 1975, met en évidence la recolonisation des platiers par les herbiers au nord et à l'ouest de l'île Tascon.

Le faciès d'érosion est considérablement réduit, car le retour des herbiers a favorisé une sédimentation active. Ce suivi permet de situer la reprise des Zostères après 1965, comme à Roscoff ou dans l'archipel de Glénan. Nous ne pouvons qu'être frappés par l'anologie existant entre le déclin des herbiers entre 1930 et 1965 et le 'cycle de Russell' décrit par les planctonologistes (Southward *et al.*, 1975). Sans reprendre ici l'hypothèse d'un phénomène cyclique lié aux cycles solaires (Glémarec, 1979), nous restons convaincus que seul un phénomène climatique est responsable d'une dynamique successionelle d'une telle ampleur sur un secteur aussi vaste. Billiet et Servain (1979), en analysant les séries mensuelles de température de surface de 1960 à 1970 dans un secteur proche de la Bretagne (Carreau Mardsen 145) montrent 'que sur la période couvrant 110 ans, de 1860 à 1970, les années les plus froides eurent lieu de 1922 à 1924, tandis que 1957 et 1959 correspondent aux années les plus chaudes'. Les courbes d'anomalies de température (Fig. 3) épousent relativement bien l'évolution de la moyenne annuelle du nombre de WOLF R, représentatif de l'activité solaire, avec une période évidente de 11 ans.

Cayan (1985) développe une approche similaire sur les côtes est-américaines et met en évidence, dans le secteur concerné par les herbiers de Zostères (40°N-70°W), entre 1949 et 1984, les plus fortes températures dans les années 1950 et les plus faibles autour de 1965.

Il est frappant de constater que le déclin des her-

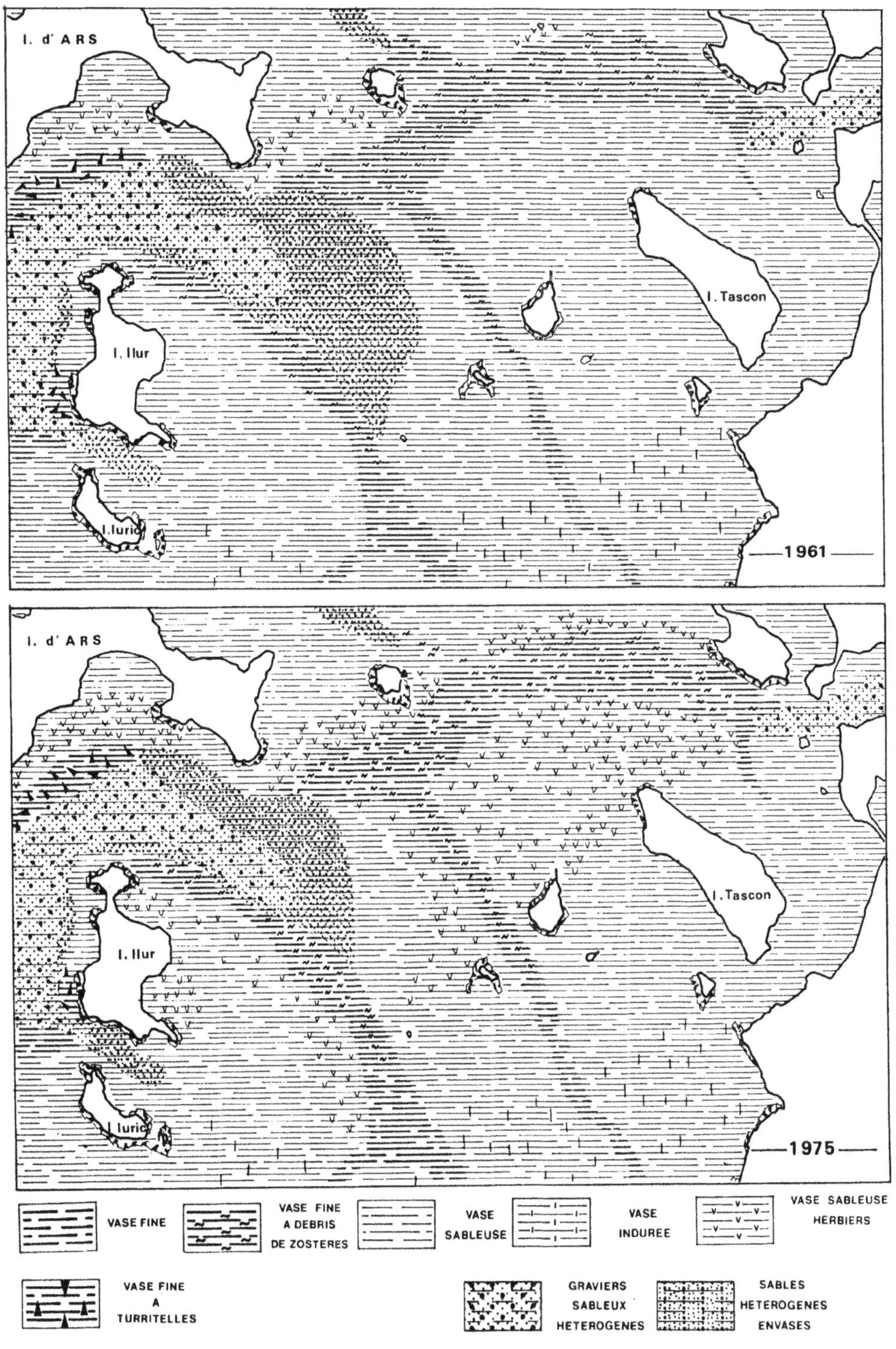

Fig. 2. Distribution des herbiers subtidaux et des différentes entités biosédimentaires, en 1961 et en 1975. On notera entre ces dates la colonisation des platiers par les herbiers et la réduction du faciès d'érosion, les sables hétérogènes envasés.

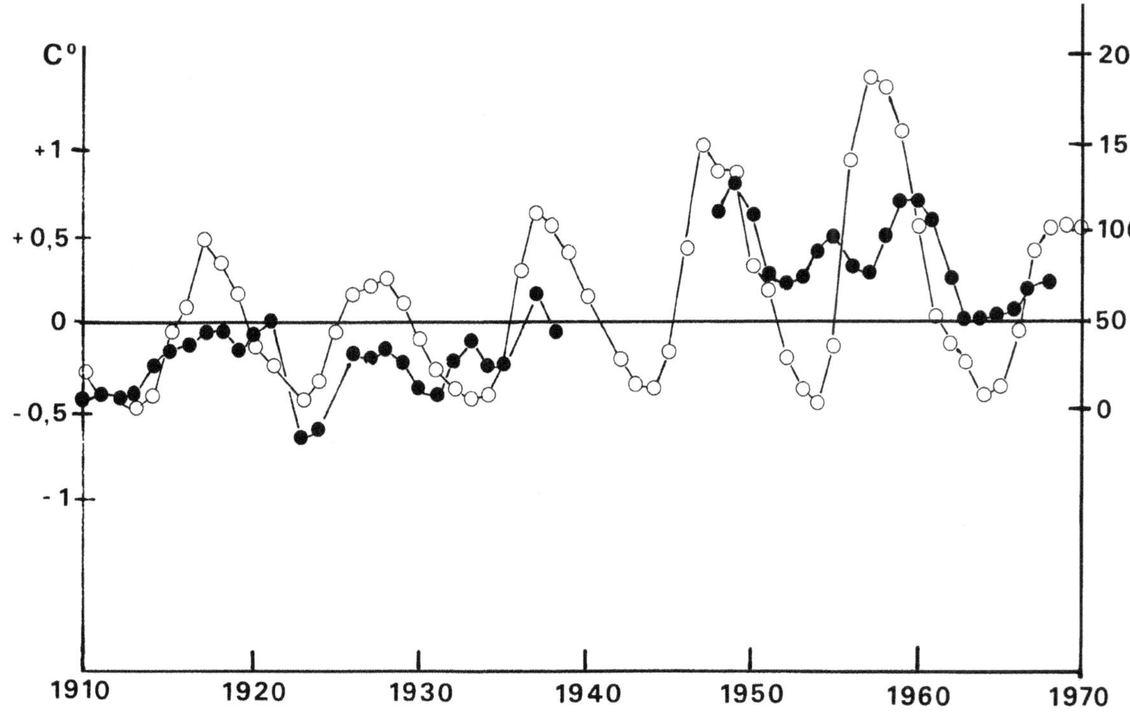

Fig. 3. Moyennes courantes sur trois ans des anomalies annuelles de la température de la mer en surface (•) et valeur annuelle du nombre de WOLF (o) au niveau du carreau Marsden n° 145, d'après Billiet & Servain.

biers correspond au maximum de la tendance séculaire du réchauffement, décrite par les climatologues. La cause initiale doit donc être d'ordre climatique comme l'ont prétendu dès 1954 Martin, puis Rasmussen en 1973. Den Hartog, dans une revue (sous presse) des divers facteurs impliqués, n'est pas aussi convaincu par cette hypothèse. A notre avis, un tel déclin n'est pas dû à une cause accidentelle, il n'aurait pas intéressé un secteur aussi vaste et avec un tel synchronisme. Le réchauffement des eaux a pu constituer, pour cette espèce végétale, des conditions adverses, puisqu'elle est d'affinité holarctique. Elle sera alors prédisposée à l'attaque des agents pathogènes ('mysterious wasting disease') et ensuite l'érosion sera une conséquence de cet affaiblissement, mais ces facteurs ne peuvent constituer les causes premières du déclin. La reprise des herbiers ayant été brutale, dans le cas des sites bretons par exemple, il y a nécessairement ensuite stabilisation, régulation du phénomène, puis les atteintes anthropiques croissantes que subit le milieu littoral, pourraient être les responsables d'un nouveau déclin vers la fin des années 1970.

Les vasières de la baie de Concarneau

Cette baie constitue l'extrémité occidentale de la dépression prélittorale sud-américaine (Pinot, 1974) et son fond plat est tapissé entre 15 et 35 mètres de profondeur par des vases et des sables envasés qui s'y maintiennent grâce à l'abri des houles dominantes d'ouest, formé par les îles comme les Glénan, les Moutons (Fig. 4). La première couverture bionomique de la baie en 1964 (Glémarec, 1969) a permis d'établir une carte biosédimentaire. D'ouest en est, et faisant suite aux sables fins situés au pied des massifs rocheux, ce sont d'abord les sables fins envasés (15 à 30% de pélites) qui hébergent la communauté à *Amphiura filiformis*, puis les vases sableuses (30 à 80% de pélites) à *Maldane glebifex*. Au centre de ces vasières, dans le secteur le plus abrité, des populations très denses de l'Amphipode *Haploops tubicola*, de l'ordre de 5000/m^2, se surimposent au peuplement à *Maldane*. Dix années plus tard, en 1974 et 1975, Ménesguen (1980), à partir de nouvelles campagnes de prélèvements et à l'aide d'analyses factorielles, met

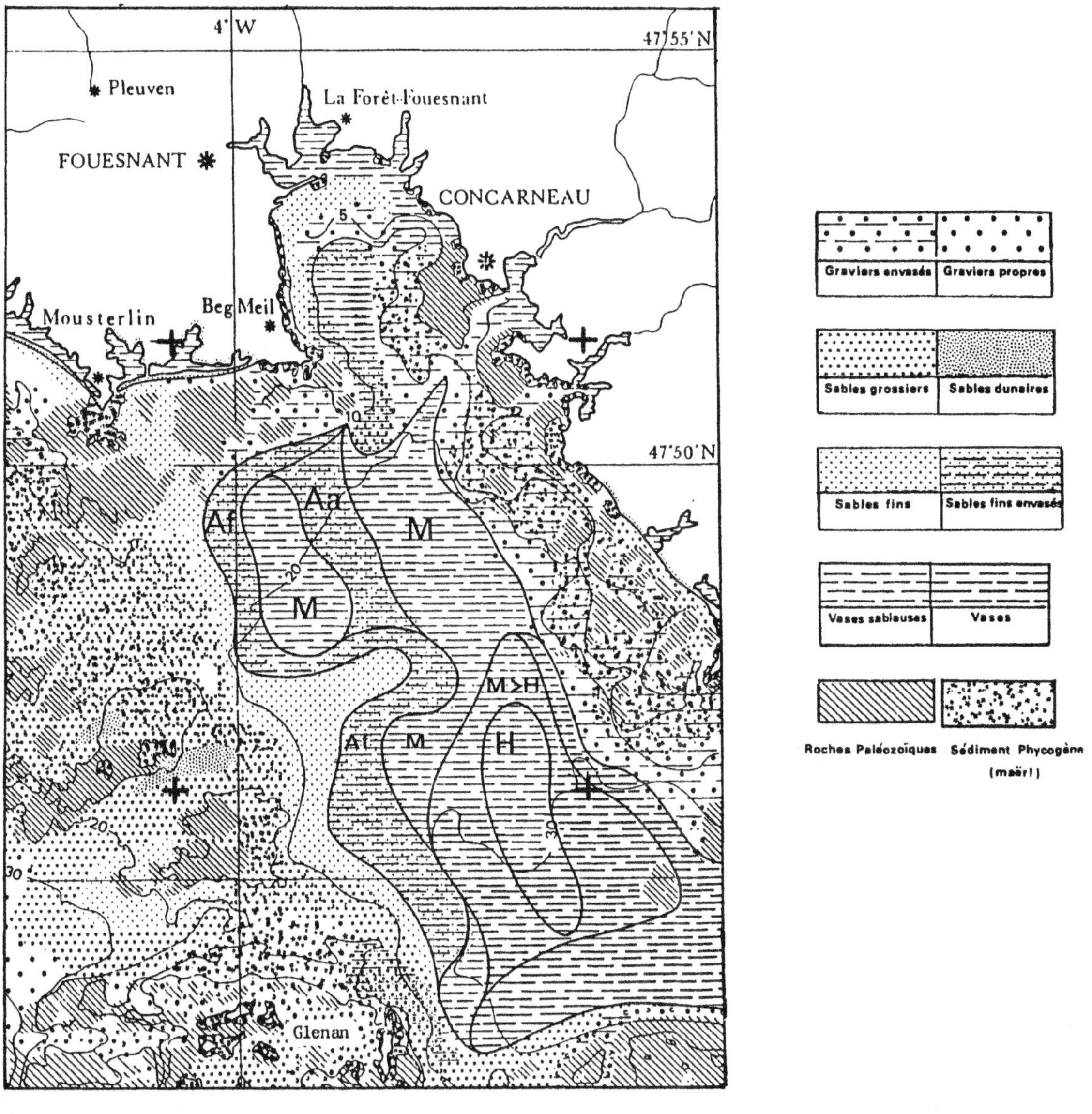

Fig. 4. Les entités biosédimentaires des sédiments envasés de la baie de Concarneau en 1964, désignées par l'espèce leader. **Aa:** *Abra alba,* **Af:** *Amphiura filiformis,* **M:** *Maldane glebifex,* **H:** *Haploops tubicola.*

en évidence la parfaite coïncidence de l'ordination des stations représentatives de ces peuplements selon le premier axe factoriel et l'augmentation de la teneur en pélites. Deux faits nouveaux apparaissent, l'essentiel du peuplement à *Maldane glebifex* a fait la place à une peuplement à *Nucula turgida* et *Abra alba* caractérisant toujours des vases sableuses mais de bien moindre consistance. D'autre part, les populations d'*Haploops* ont envahi, vers le nord, une large part des vases à *Maldane.* En 1977, une nouvelle couverture de la baie est réalisée et ces faits évolutifs majeurs se confirment; le peuplement à *Nucula turgida* et *Abra alba* (**Aa** sur la figure 5), à peine représenté en 1964 selon un axe nord-sud entre deux entités de vase compacte, occupe en 1977 la plupart de la superficie occupée précédemment par les *Maldane* dans cette partie nord de la baie. Les *Haploops* ont migré

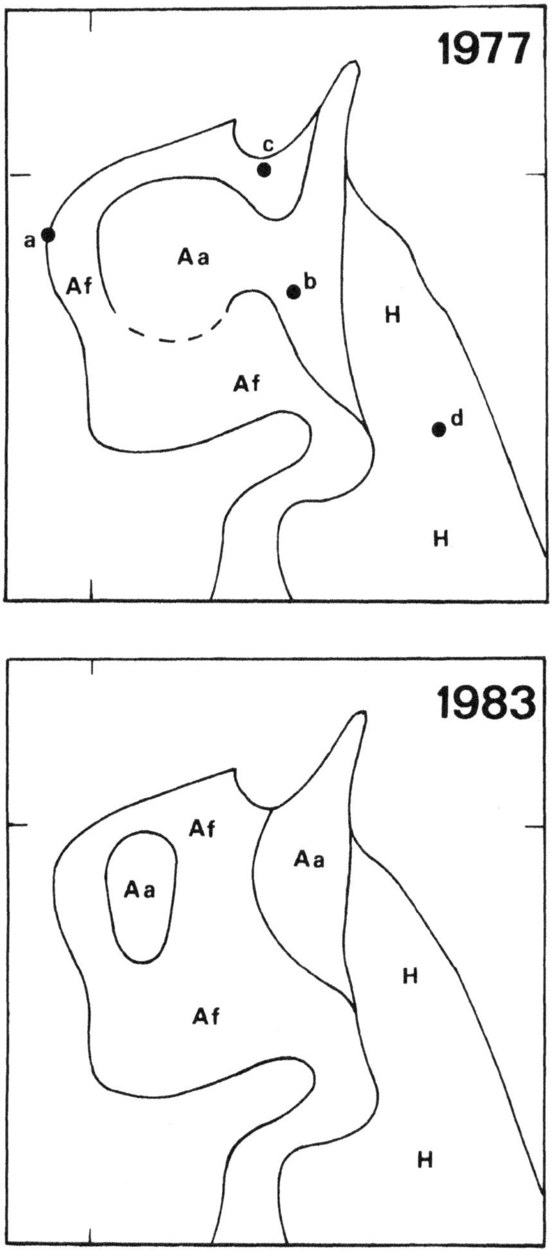

Fig. 5. Evolution temporelle des entités de peuplement en 1977 et en 1983. Parmi les nombreuses stations, celles utilisées pour la Figure 6 sont désignées **a, b, c, d.** Le périmètre concerné est indiqué sur la Figure 4 à l'aide de quatre croix.

vers le nord jusqu'au 47°50′N et le peuplement à *Amphiura filiformis* gagne aussi des sédiments plus envasés sous forme de faciès d'appauvrissement.

Une étude similaire effectuée en 1983 montre la stagnation des *Haploops* tandis que les vases sableuses molles à *Nucula turgida* et *Abra alba* se restreignent à deux taches dont l'une à la confluence du chenal d'entrée vers le port de Concarneau et de la sortie de la baie de la Forêt (Le Guellec, 1984).

Une analyse de l'évolution des peuplements et de leur composition, à quatre stations particulièrement choisies, est résumée par la figure 6. En 1964, les méthodes d'échantillonnage n'étaient pas réellement quantitatives (drague au lieu de benne), aussi cet examen ne peut-il se faire que sur les dominances qui révèlent clairement l'importance relative des principales espèces.

La station **a**, située à la limite occidentale de la vasière, a fait l'objet d'une étude suivie (Princz *et al.*, 1983). D'importantes fluctuations internes, liées notamment à la démographie de l'espèce dominante *Amphiura filiformis*, ne remettent pas en cause la structure du peuplement. Le déclin de cette espèce est évident en 1982 et 1983, mais il n'est pas possible de confirmer l'hypothèse d'un envasement croissant de ce secteur sur plus de dix années, comme l'ont suspecté Delanoë et Pinot (pers. commun.).

Les stations **b** et **c** étaient représentatives des vases sableuses compactes à *Maldane* en 1964. Cette espèce ayant considérablement régressé, en 1974 appraît une destructuration du peuplement sans aucune dominance avec des espèces ubiquistes comme *Notomastus latericeus, Abra alba*.., et illustrée par un modèle similaire de dominances relatives des espèces. Il faut attendre 1977 et 1983 pour assister à une évolution avec la prédominance d'*Amphiura filiformis* dans le cas de la station **b** et celle d'*Abra alba* à la station **c**. A l'aide des prélèvements réalisés à la benne Aberdeen, il est possible de chiffrer le déclin des *Maldane* depuis 1969. En effet, à cette date, la densité de cette espèce variait entre 250 et 450/m^2. En 1974, Ménesguen indique des valeurs oscillant entre 100 et 120/m^2, enfin en 1983 et 1985 cette densité ne dépasse nulle part 30/m^2. Dans le cas de la station **d**, on retrouve également ce déclin des *Maldane* qui en 1964 dominaient nettement les *Haploops*, les premières disparaissent presque complètement ensuite.

Pour essayer d'interpréter ces phénomènes temporels, il est nécessaire de revenir à l'éthologie alimentaire des espèces dominantes car, dans ces vasières, les densités sont suffisamment importan-

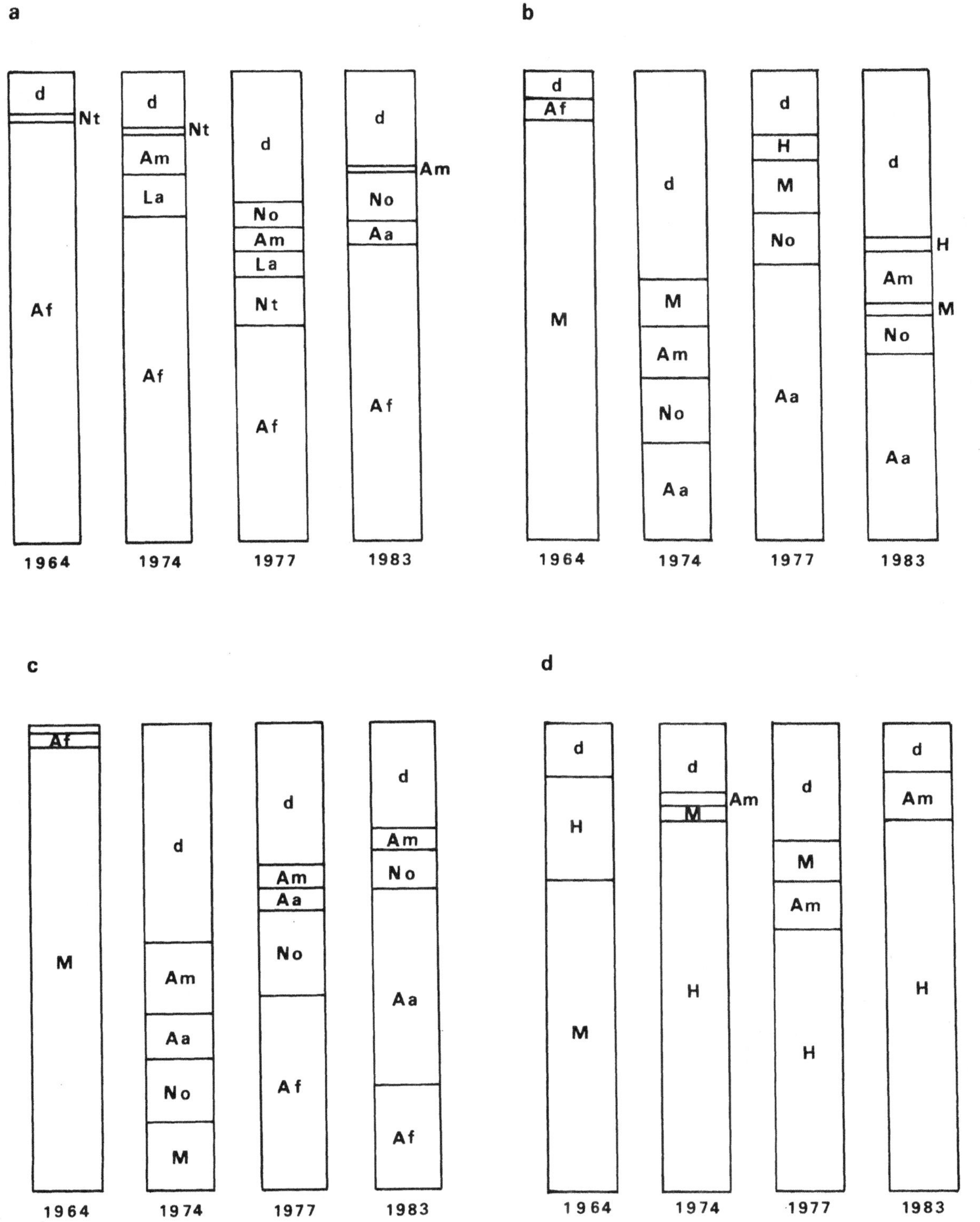

Fig. 6. Evolution temporelle des dominances des principales espèces dans les quatre stations **a, b, c, d. Aa:** *Abra alba,* **Am:** *Ampelisca sp.,* **Af:** *Amphiura filiformis,* **d:** divers, **H:** *Haploops tubicola,* **M:** *Maldane glebifex,* **No:** *Notomastus latericeus,* **Nt:** *Nucula turgida.*

tes pour que de fortes interactions entre espèces interviennent, pouvant aller jusqu'à l'exclusion compétitive (Woodin, 1978). En effet, les biomasses peuvent atteindre plus de 30 g de POS/m^2. *Maldane glebifex*, Polychète de la famille des Maldanidés, construit un manchon de vase de 2 à 3 cm de diamètre, pouvant dépasser 10 cm de longueur. Le ver dont la partie céphalique est dépourvue de tout appendice capte les particules à l'aide de sa trompe à la surface du sédiment, il est déposivore et construit son tube en les disposant de façon concentrique. Ce sont de véritables boudins de vase durcie qui ne se désagrègent que très lentement au sein du sédiment une fois l'animal mort. Leur compacité est suffisante pour supporter les remaniements à l'échelle géologique (Babin *et al.*, 1971).

Haploops tubicola construit aussi un tube, long de quelques centimètres mais aplati, dépassant largement du sédiment dans lequel il est planté verticalement, les Amphipodes suspensivores se nourrissent en se tenant à la partie supérieure du tube. L'ensemble des tubes dressés est si dense, qu'il constitue une véritable forêt.

Amphiura filiformis est une espèce déposivore et suspensivore qui vit enfouie à quelques centimètres dans le sédiment envasé, que ce soit des sables envasés ou des vases à condition qu'elles restent molles et ne soient pas consolidées.

Les bioperturbateurs, par leur activité alimentaire maintiennent une instabilité de la couche sédimentaire et créent ainsi un facteur limitant pour les suspensivores tubicoles pour lesquels, une certaine stabilité et une compacité restent favorables à l'installation des larves. Cette théorie de l'amensalisme entre groupes trophiques de Rhoads & Young (1970) peut être évoquée dans le cas des phénomènes décrits en baie de Concarneau. Dans le cas des *Maldane* vis à vis des *Haploops*, il est clair que leur action se traduit plus par la construction du tube que par le remaniement de la surface à des fins alimentaires. La présence nombreuse de tubes occupés par des animaux vivants ou inoccupés transforme le sédiment original et favorise l'implantation des *Haploops*. C'est un exemple parfait de ce que peut être un modèle de facilitation parmi les modèles de succession définis par Connell & Slatyer, 1977.

Par la création de tubes, *Maldane* favorise l'installation des *Haploops* et inhibe le propre développement de sa population. De même, le rôle des *Haploops* est d'accroître la consolidation du sédiment aussi leur développement n'apparait dans un premier temps, qu'à l'abri des remaniements crées par les houles lors des tempêtes. Une faune associée aux *Haploops* apparaît, c'est une épifaune de grandes *Ophiothrix fragilis*, de crabes, de pagures qui migrent des terrasses de maërl voisines et parmi l'endofaune, de palourdes caractéristiques de graviers hétérogènes (*Venerupis rhomboides*). Tout ceci peut se rapporter au modèle de facilitation. Les *Maldane* ayant régressé, les *Amphiura filiformis* migrent sur ces vases sableuses dont la compacité est amoindrie par l'activité de déposivores de surface comme les Bivalves *Nucula turgida* et *Abra alba* et d'autres plus profonds, *Notomastus latericeus* par exemple.

Par contre, la théorie de l'amensalisme entre groupes trophiques prend tout sa valeur dans le cas de ces vases sableuses molles à *Nucula turgida* et *Abra alba* que les *Haploops* ne peuvent coloniser et l'on est surpris de voir les Amphipodes stoppés dans leur extension géographique tant que les Bivalves déposivores ou les *Amphiura* maintiennent une certaine fluidité de la surface sédimentaire. Ceci rappelle le modèle d'inhibition de Connell & Slatyer.

Les apports sédimentaires actuels sont relativement faibles dans cette baie, car les rivières sont peu importantes, les vases fluides sans cesse alimentées par des apports telluriques n'existent pas. Les évolutions décrites peuvent donc s'expliquer uniquement par les interactions entre organismes sans que soit nécessairement recherchée une cause allogènique. Les constructeurs de tube fournissent une explication 'mécanistique' aux modèles de succession qui peuvent être résumés dans la figure 7.

Le déclin des *Maldane* reste à expliquer. Le recours aux modèles de succession fournit une première hypothèse; en effet, le modèle de facilitation peut expliquer la quasi disparition des *Maldane* au

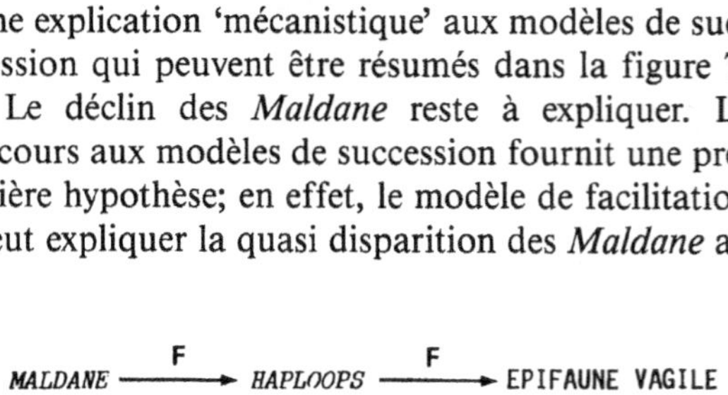

Fig. 7. Séquence possible de modèles de succession sur les sédiments envasés: I modèle d'inhibition, F modèle de facilitation.

profit des *Haploops* puisque le sédiment devient moins favorable pour les *Maldane*. Par contre, leur déclin dans les zones qui ne sont pas colonisées par les *Haploops*, et où s'installent des bivalves déposivores de surface et des espèces ubiquistes qui laissent supposer qu'il y a de l'espace disponible ou destructuration du peuplement, ne peut être compris qu'en faisant appel à une cause allogénique. L'envasement croissant suggéré par certains (*op. cit.*) ne peut être évoqué, car il est nécessairement favorable aux *Maldane*. L'examen de la distribution géographique fait apparaître que cette espèce tempérée chaude et tropicale atteint sa limite septentrionale dans le nord du Golfe de Gascogne, elle est très peu abondante en rade de Brest (Hily, 1984). Elle est connue du Golfe de Guinée en profondeur, de la Méditerranée, de l'Adriatique, mais ne semble pas pénetrer en Manche tandis qu'en province tempérée froide, elle est remplacée par une espèce vicariante *Maldane sarsi* Malmgren. Pour *Maldane gle-*

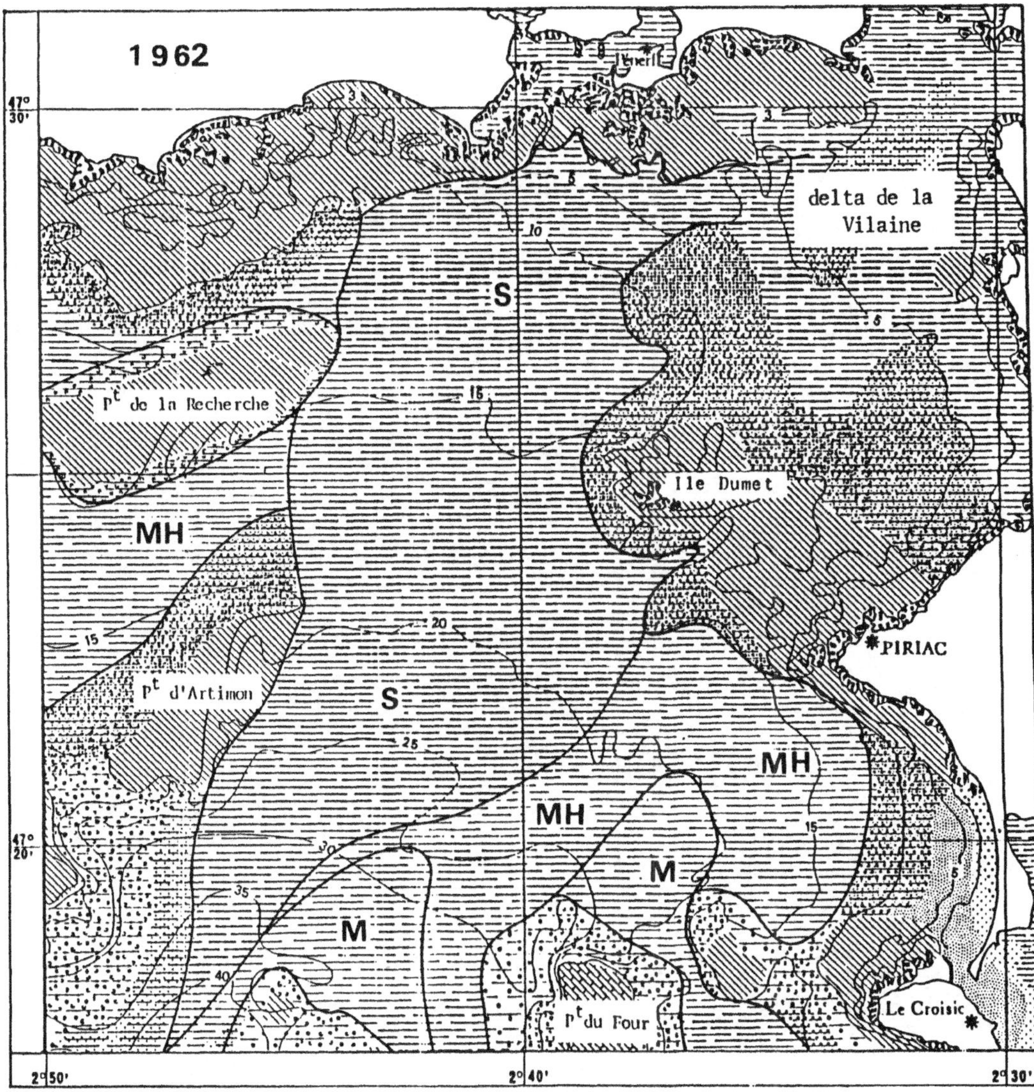

Fig. 8. Distribution des peuplements en baie de Vilaine en 1962. Les *Maldane* et les *Haploops* (**M-H**) n'occupent pas le centre de la baie, mais seulement des vases consolidées à l'ouest ou au sud-est à l'abri des massifs rocheux. **S:** peuplement des vases molles à *Sternaspis scutata, Virgularia tuberculata.*

bifex espèce à affinité chaude, son extension maximale en 1964, par rapport à ce qu'elle est depuis 1974, peut être comparée mais de façon inverse au déclin et au retour des Zostères dans notre région. Le réchauffement maximal de la fin des années 50 a pu favoriser le développement des *Maldane* tandis que le refroidissement intervenu après 1964 a pu restreindre l'extension des *Maldane* sans gêner d'autres espèces à affinité plus froide, sans exclure totalement les *Maldane*. Ces dernières sont limitées par la compétition qui est toujours forte sur ces vasières côtières.

La Baie de Vilaine

En 1962, cette partie orientale de la dépression prélittorale est composée d'une vasière très étendue où le taux de pélites est toujours voisin de 90%. Partout la vase est molle, habitée par le peuplement à *Sternaspis scutata, Virgularia tuberculata* et

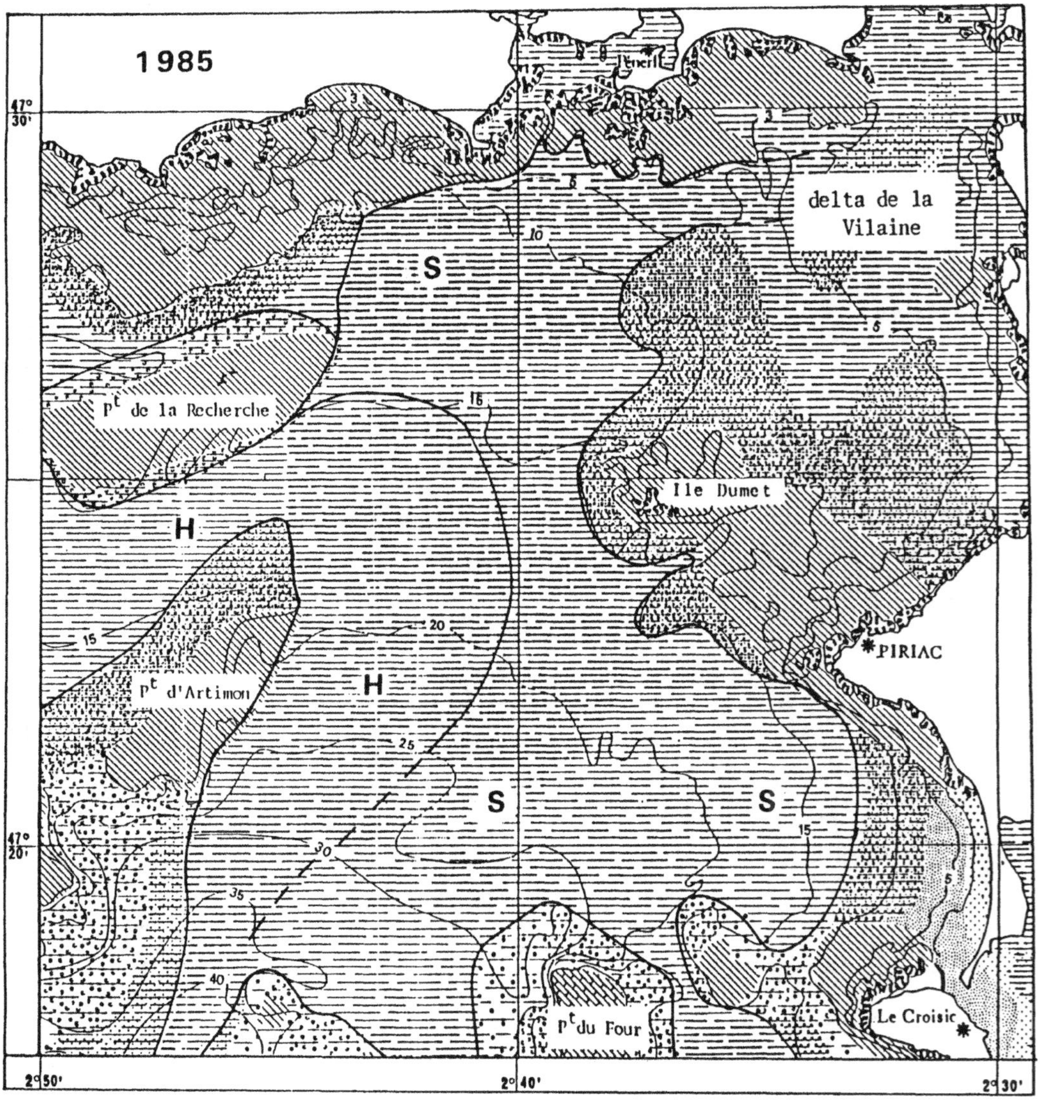

Fig. 9. Distribution des peuplements en baie de Vilaine en 1985. Les *Haploops* ont envahi la vasière vers l'est, les *Sternaspis* ont gagné la rade du Croisic, tandis que partout les *Maldane* ont regressé.

Amphiura filiformis. Elle est en communication directe avec l'estuaire de La Vilaine par le biais d'un delta sous-marin (Fig. 8). A l'ouest elle est limitée par des massifs rocheux (le plateau de la Recherche, le plateau d'Artimon) à l'abri desquels la vase sableuse héberge un peuplement à *Maldane* et *Haploops.* Ce type de peuplement existe aussi au sud-est, en rade du Croisic, à l'abri de plateaux rocheux qui limitent cette baie au sud.

Un barrage est construit sur La Vilaine à partir de 1965, et une nouvelle couverture bionomique, réalisée en 1985, révèle de profondes modifications au niveau du delta, le barrage ayant diminué l'action érosive des forts courants liés aux apports d'eau douce, lors des crues notamment. Au-delà de ce delta, au centre de la baie des changements sont aussi intervenus (Fig. 9). Les *Haploops* ont colonisé la vasière centrale à partir des zones abritées occidentales, ce qui suppose une consolidation du sédiment favorisée par l'activité propre de ces crustacés. De façon inverse, les vases sableuses compactes du sud de la baie hébergent le peuplement caractéristique des vases molles. Aujourd'hui le flux particulaire de la Vilaine ne parvient que partiellement au centre de la baie, ce qui a pu favoriser l'implantation progressive des *Haploops* sur la vase molle, atteignant des densités de 2 à 3000/m^2, toujours à l'abri des houles d'ouest. L'évolution sédimentologique et biologique au sud de la baie ne trouve pas aujourd'hui d'explication claire, elle suppose un acroissement de l'envasement puisque les vases sableuses à *Maldane* et *Haploops* deviennent des vases franches fluides comme l'indiquent les *Sternaspis scutata* par exemple. Si la Vilaine semble peu impliquée à la latitude du Croisic pour expliquer cette évolution, la partie sud de cette baie peut être influencée par le chenal nord de la Loire et l'on sait que les équilibres hydrodynamiques entre Vilaine et Loire ont été modifiés à la suite de grands travaux d'aménagement sur les deux estuaires.

Selon les secteurs, les *Haploops* se sont développés ou ont disparu à la suite de causes allogéniques liées à l'apport des flux de particules sédimentaires. Partout les *Maldane* sont en diminution ce qui renforce l'hypothèse climatique développée en baie de Concarneau, puisque le changement climatique évoqué ne peut intéresser qu'une vaste région comme le golfe de Gascogne et pas seulement une partie de la baie de Concarneau.

A l'abri de l'île Dumet, dans une station de vase sableuse, la densité des *Maldane* est encore de 170/m^2 en absence d'*Haploops,* ce qui laisse supposer aussi que les rapports de compétition restent prédominants. Lorsque les *Haploops* envahissent les vases molles en l'absence de *Maldane* ceci montre également que dans une telle succession, les *Maldane* ne constituent pas une étape obligatoire, elle est facultative mais peut alors faciliter le développement des *Haploops.*

Conclusion

La distribution des espèces en fonction du gradient d'envasement constitue un écocline ou un continuum où la place de chaque espèce est très liée aux rapports de compétition entre espèces. Par ailleurs, au facteur envasement, se surimpose le caractère fluidité du sédiment, aussi voit-on deux séquences possibles selon l'envasement croissant; la première sur vases consolidées:

– *Amphiura filiformis* – *Maldane glebifex* – *Haploops tubicola*;

la deuxième sur vases molles:

– *Amphiura filiformis* – *Abra alba* – *Nucula turgida* – *Sternaspis scutata, Virgularia tuberculata, Amphiura filiformis.*

Amphiura filiformis limitée vers 30% de pélites par les *Maldane glebifex* et par la compacité du sédiment voit son spectre de distribution étendu lorsque la vase reste fluide; elle côtoie alors les *Virgularia* et les *Sternaspis,* ce qui montre bien que l'action des *Maldane* se traduit par une compacité accrue du sédiment lors d'une dynamique temporelle, la succession autogénique peut expliquer le passage des vases à *Maldane* à celles à *Haploops,* puis à une épifaune. Le rôle des tubicoles est donc essentiel dans la consolidation des sédiments, par contre les dépositivores non tubicoles ont un rôle inverse en maintenant la fluidité du sédiment de surface. Cette évolution peut être modifiée par une perturbation allogénique et les trois sites étudiés apportent des éléments de réflexion complémentaires. La disparition des herbiers et leur remplacement dans les vasières infralittorales du Golfe du Morbihan par un peuplement de tubicoles (*Melinna palmata*) met en évidence le rôle des herbiers dans la sédimentation, sur un site où les apports telluriques sont relativement élevés.

Dans la baie de Vilaine, les flux particulaires ayant été modifiés, les *Haploops tubicola* peuvent coloniser les vases molles, leur rôle est bien de consolider le substrat. Dans cette baie et en baie de Concarneau, le déclin des abondantes populations de *Maldane* a des conséquences importantes et modifie les rapports d'équilibre entre les différents peuplements. L'année 1974 témoigne des déséquilibres importants où les espèces ubiquistes sont toutes d'égale importance, ensuite de nouvelles dominances apparaîtront. Le rôle des facteurs allogéniques est de perturber le cours normal d'une évolution autogénique, dans le cas des vasières cette dernière est essentiellement liée au rôle bioperturbateur de certaines espèces et au rôle inverse des tubicoles, qui ne sont pas tous des suspensivores. Cette dynamique successionnelle est d'autant plus évidente que les peuplements sont très riches, sans doute à leur maximum de capacité de charge, que ce soit dans l'infralittoral du golfe du Morbihan, ou dans les vasières côtières de la dépression prélittorale du nord du golfe de Gascogne. Indépendamment de tous les critères habituels de variabilité temporelle, il existe donc sur ces vasières une évolution permanente et une succession liée à l'action des espèces sur le sédiment.

Remerciements

Les auteurs tiennent à exprimer toute leur gratitude envers leurs collaborateurs qui nous ont aidés, tant en mer qu'au laboratoire, M. R. Le Liart, C. Michel, C. Tartu, R. Marc et E. Hussenot. Sont associés également à ce travail, M. Seaman, D. Aranda et D. Princz qui ont assuré une grande partie des dépouillements.

References

Babin, C., M. Glémarec, H. Termier & G. Termier, 1971. Rôle des Maldanes (Annélides Polychètes) dans certains types de bioturbation. Ann. Soc. Géol. Nord. 91: 203–206.

Billiet, P. & S. Servain, 1979. Eléments de climatologie sur le proche atlantique. Met. Mar. 103: 21–24.

Cayan, D. R., 1985. North atlantic seasonal sea surface temperature anomalies and associated statistics, 1949–1984. S IO Référence séries n° 85–19.

Connell, J. H. & R. D. Slatyer, 1977. Mechanisms of succession in natural communities and their role in community stability and organization. Am. Nat. 111: 1119–1144.

Den Hartog, C., 1986. Wasting 'disease' and other dynamic phenomena in Zostera-beds. (sous presse).

Glémarec, M., 1964. Bionomie benthique de la partie orientale du Golfe du Morbihan. Cah. Biol. Mar. V: 33–96.

Glémarec, M., 1969. Les peuplements benthiques du plateau continental Nord-Gascogne. Thèse Doctorat d'Etat, Paris, p. 1–167.

Glémarec, M., 1979. Les fluctuations temporelles des peuplements benthiques liées aux fluctuations climatiques. Oceanol. Acta II, 3: 365–372.

Hily, C., 1984. Variabilité de la macrofaune benthique dans les milieux hyper-trophiques de la rade de Brest. Thèse d'Etat. Université de Bretagne Occidentale, 359 pp.

Le Guellec, C., 1984. Evolution à long terme des peuplements benthiques de la baie de Concarneau. Rapp. Int. Université de Bretagne Occidentale: 1–28.

Martin, A. C., 1954. A clue to the eelgrass mystery. Trans. 19th North Amer. Wildlife Confer. Washington, D.C.: 441–449.

Menesguen, A., 1980. La macrofaune benthique de la baie de Concarneau. Thèse de 3ème cycle. Université de Bretagne Occidentale, 127 pp.

Pinot, J. P., 1974. Le précontinent breton entre Penmarc'h, Belle-Ile et l'escarpement continental, étude géomorphologique. Lannion. Imfram. 256 pp.

Princz, D., A. Menesguen & M. Glémarec, 1983. Temporal evolution over ten years in the macrobenthos of muddy sands in the bay of Concarneau (France). Oceanol. Acta. Actes du 17e Symposium Européen de Biologie Marine: 159–163.

Rasmussen, E., 1973. Systematics and ecology of the Isefjord marine fauna (Denmark). Ophelia 11: 1–430.

Rhoads, D. C. & D. K. Young, 1970. The influence of deposit-feeding organisms on sediment stability and community trophic structure. J. Mar. Res. 28: 150–178.

Southward, A. J., E. I. Butler & L. Pennycuick, 1975. Recent cyclic changes in climate and abundance of marine life. Nature 253: 714–717.

Woodin, S. A., 1978. Refuges, disturbance and community structure. Ecology 59: 274–284.

Hydrobiologia 142: 171–186 (1986)

Variations à long-terme (1977–1985) du peuplement des sables fins de la Pierre Noire (baie de Morlaix, Manche occidentale): analyse statistique de l'évolution structurale

Jean-Claude Dauvin* & Fréderic Ibanez**
**Station Biologique de Roscoff, CNRS-LP 4601 et Université P. & M. Curie (Paris VI), 29211 Roscoff*
***C.E.R.O.V., Station Zoologique, B.P. 28, 06230 Villefranche-Sur-Mer*

Keywords: macrobenthos, subtidal communities, long-term evolution, statistical analysis, West English Channel

Abstract

Long term monitoring for eight years of the fine sand community from the bay of Morlaix has allowed the determination of the principal stages of its structural evolution.

Principal component analysis on the general trends of 43 species has permitted the division of the chronological series into six successive periods. The rapid evolution of the community during the first four years after the Amoco Cadiz oil spill indicates a period of perturbation which is followed by a stabilisation of the macrobenthos. This evolution is a function of changes in direction in gradients within the populations.

The resemblance metric and the dissimilarity correlograms were used after regrouping the species into six trophic groups. The Mantel test allowed measurement of the contrasts between equilibrium and restructuration periods. The extended variogram function showed different behaviour of trophic groups related to demographic strategies. The generalization of these methods in benthic studies should allow the comparison in space and time of the scale of the physical factors responsible for the observed phenomena.

Introduction

En Manche occidentale, on observe dans les baies un gradient granulométrique et biocénotique très marqué depuis les cailloutis du large jusqu'aux vases sableuses du fond des baies; les peuplements de sédiments fins, les plus productifs de cette zone, présentent une répartition discontinue en taches isolées séparées par de vastes étendues de nature différente. L'évolution à long-terme de la macrofaune de l'une de ces unités insulaires a été suivie pendant huit années (1977–1985). Cette unité correspond au peuplement des sables fins peu envasés à *Abra alba – Hyalinoecia bilineata* de la Pierre Noire dans la partie orientale de la baie de Morlaix.

Survenant après une année d'observation, l'arrivée des hydrocarbures de l'Amoco Cadiz au niveau du fond, au printemps 1978, a profondément modifié la structure quantitative du peuplement, notamment par la destruction ou la réduction de populations d'Amphipodes dominantes (Dauvin, 1984). Evoluant dans un milieu aux caractéristiques hydrosédimentaires ne présentant que de faibles variations annuelles et interannuelles, la pollution a été l'élément perturbateur essentiel du système au cours de la période d'échantillonnage.

Bien que les méthodes d'analyses multivariées, telle l'analyse en composantes principales, soient appliquées couramment pour la cartographie des peuplements macrobenthiques (Lie & Kelley, 1970; Souplet & Dewarumez, 1980; Elkaïm *et al.*, 1982), dans la plupart des séries pluriannuelles d'observation du macrobenthos les phénomènes sont simplement décrits à partir de l'évolution des 'S.A.B.' (richesse en espèces, abondances, biomasses) et de celles des principales espèces (Pearson *et al.*, 1982; Rees & Walker, 1983; Bonsdorff & Osterman, 1984). Aussi dans cette publication nous avons essayé, à partir d'analyses en composantes principales, de déterminer objectivement les phases caractériques de l'évolution pluriannuelle du peuplement; l'interprétation de la segmentation de la série temporelle a été recherchée en relation avec les évolutions d'abondances des 43 espèces sélectionnées. Ensuite, la variation temporelle des ressemblances entre les observations a été étudiée en tenant compte du classement des espèces par groupes trophiques. Cette approche, inusitée dans le domaine benthique où l'évolution temporelle des groupes trophiques est généralement étudiée à partir des variations des dominances numériques des différents groupes (Maurer *et al.*, 1979; Bachelet, 1981; Pearson *et al.*, 1982), a ainsi permis de mettre en évidence des périodes de plus grande hétérogénéité temporelle révélatrice de restructuration du peuplement. Enfin, l'étude scalaire de l'évolution des groupes trophiques a mis en évidence leurs différences adaptatives.

Ces résultats s'inscrivent parmi ceux du projet COST 647 d'étude de la variabilité temporelle des populations et des peuplements macrobenthiques des mers Nord-Européennes; ils fournissent un exemple d'évolution du macrozoobenthos subtidal dans une région avec de faibles variations annuelles et pluriannuelles des paramètres hydrologiques.

Matériels et méthodes

Site et échantillonnage

Le peuplement des sables fins peu envasés à *Abra alba – Hyalinoecia bilineata* de la Pierre Noire (Cabioch, 1968) est localisé dans la partie orientale de la baie de Morlaix où il forme une tache de sable fin thixotropique par 15 à 20 m de profondeur. Les mécanismes de sédimentation (Auffret & Douvillé, 1974), les caractéristiques physico-chimiques et biologiques (Boucher *et al.*, 1984; Dauvin, 1984) et les conditions hydrologiques de ce peuplement sont maintenant bien connus.

Le point de prélèvement 'Pierre Noire' se situe à 0,28 mille au sud-est de la balise de la Pierre Noire, par 17 m de profondeur par rapport au zéro des cartes dans la partie la plus riche du peuplement. La médiane du sédiment, comprise entre 148 μm et 184 μm, varie peu au cours du temps; la fraction fine inférieure à 63 μm dépasse légèrement 1%. Le milieu correspond à une zone relativement oligotrophe avec une teneur en carbone organique de 0,47% du poids sec du sédiment (Beslier, 1981) et une teneur en azote organique de 0,035% (Chamroux & Mével, sous presse).

Après un premier échantillonnage en avril 1977 les prélèvements ont été réalisés mensuellement d'août 1977 à mars 1982 puis à raison de cinq relevés annuels de juin 1982 à mars 1985, soit un total de 71 observations. A l'exception des trois premiers relevés (5 prélèvements de 0,28 m^2 à la benne Hamon), l'échantillonnage a été réalisé au moyen d'une benne Smith Mc Intyre à raison de dix prélèvements unitaires de 0,1 m^2, soit une surface totale de 1 m^2. Tous les échantillons ont été lavés sur un tamis à ouverture circulaire de 1 mm de diamètre et triés deux fois après coloration au rose bengale. On désignera ci-dessous par N le nombre d'individus sur 1 m^2 (ou 'densité') d'une espèce.

Choix des descripteurs et lissages des données

388 espèces macrobenthiques ont été recensées à partir des 71 relevés, mais seules les espèces bien représentées en fréquence et en nombre d'individus, soit 43 (Tableau I), ont été conservées dans les analyses, à l'exclusion des espèces trop temporairement abondantes ou accidentellement disparues. En effet, certaines espèces principales du peuplement, comme *Ampelisca armoricana* Bellan-Santini & Dauvin et *Ampelisca tenuicornis* (Liljeborg), ont été temporairement ou totalement éliminées du peuplement au printemps 1978 lors de la pollution des fonds par les hydrocarbures de l'Amoco Cadiz (Dauvin, 1984); d'autres espèces, comme *Polydora antennata* Claparède, ne présentent que des proliférations estivales passagères (Dauvin, 1984). En définitive, 43 descripteurs ont été sélectionnés (Tableau I).

Les séries présentent un grand nombre d'observations manquantes (71 valeurs mensuelles échantillonnées sur 96). En se référant aux profils d'abondance, on a pu vérifier que les tendances locales étaient presque toujours exponentiellement croissantes ou décroissantes. Pour régulariser le pas d'échantillonnage, les données ont donc été transformées en logarithmes puis les valeurs manquantes interpolées linéairement entre deux valeurs successives. Les valeurs arithmétiques sont ensuite déduites par transformation réciproque. Cette interpolation n'a, en fait, servi qu'à la représentation graphique des abondances et pour les traite-

Tableau 1. Caractéristiques statistiques (densités (nombre d'individus sur 1 m^2): données brutes) des 43 espèces sélectionnées.

	Numéro dans les analyses numériques	Densité Moyenne	Ecart-type	Densité Minimale	Densité Maximale	Nombre de valeurs nulles
Abra alba (Wood)	1	135,2	114,1	7	575	0
Abra prismatica (Montagu)	2	39,2	36,8	1	202	0
Ampelisca brevicornis (Costa)	3	92,8	78,6	0	501	29
Ampelisca sarsi Chevreux	4	990,6	1594,9	1	6641	0
Ampelisca typica (Bate)	5	113,6	115,1	0	737	26
Ampharete acutifrons Malmgren	6	43,0	42,2	0	295	2
Apherusa ovalipes Norman & Scott	7	8,5	12,8	0	113	37
Apseudes latreillei (Milne-Edwards)	8	23,2	23,1	0	173	46
Aricidea cerrutii Laubier	9	49,9	27,2	0	122	1
Aricidea fragilis Webster	10	122,2	43,6	15	327	0
Aricidea minuta Southward	11	11,5	8,7	0	51	9
Bathyporeia elegans Watkin	12	53,9	42,1	0	342	10
Bathyporeia tenuipes Meinert	13	19,4	13,2	0	78	1
Chaetozone setosa Malmgren	14	226,4	140,8	0	725	1
Clymene oerstedii Claparède	15	34,1	22,1	3	174	0
Cultellus pellucidus (Pennant)	16	7,7	6,0	0	31	10
Diplocirrus glaucus (Malmgren)	17	6,7	4,1	0	31	9
Eteone longa Fabricius	18	7,5	3,6	0	29	1
Eulalia sanguinea Oersted	19	35,0	25,0	0	150	7
Exogone hebes (Webster & Benedict)	20	17,3	25,1	0	167	3
Heterocirrus alatus (Southern)	21	59,6	100,0	0	668	7
Hyalinoecia bilineata Baird	22	22,9	9,4	2	69	0
Lanice conchilega (Pallas)	23	60,1	49,4	0	482	9
Marphysa bellii (Audouin & M. Edwards)	24	168,2	92,6	15	634	0
Mediomastus fragilis Rasmussen	25	38,4	22,9	0	190	2
Nephtys hombergii Savigny	26	87,5	74,0	4	446	0
Notomastus latericeus Sars	27	42,2	34,9	0	208	4
Nucula hanleyi Winckworth	28	6,0	2,6	0	20	5
Nucula turgida Leckenby & Marshall	29	7,7	3,7	1	26	0
Odontosyllis gibba Claparède	30	38,7	27,5	0	161	4
Ophiura albida Forbes	31	14,9	8,8	1	51	0
Paradoneis armata Glémarec	32	410,0	134,9	26	850	0
Pariambus typicus (Kröyer)	33	18,0	17,8	0	154	28
Perioculodes longimanus (Bate & W. Wood)	34	36,5	20,5	0	195	1
Phtisica marina (Bate)	35	44,1	37,9	0	248	45
Phyllodoce groënlandica Oersted	36	14,9	16,2	0	94	24
Scoloplos armiger (Müller)	37	123,8	84,4	11	405	0
Spio filicornis (Müller)	38	525,2	380,7	37	3101	0
Spiophanes bombyx (Claparède)	39	31,6	24,7	7	144	0
Tellina fabula Gmelin	40	3,9	2,9	0	18	12
Thyasira flexuosa Montagu	41	70,3	81,1	0	364	8
Urothoë pulchella (Costa)	42	130,0	103,2	1	426	0
Venus ovata Pennant	43	37,5	31,8	0	135	2

ments multivariables nous n'avons tenu compte que des données réelles (voir plus loin). Dans de précédents travaux (Ibanez, 1983, 1984), nous avons souligné les propriétés du lissage des points moyens. Cet ajustement consiste à définir la tendance générale d'un processus temporel, comme le lieu des points situés à égale distance de son enveloppe inférieure et supérieure (Figures 4, 5 et 6). Cette technique présente moins d'arbitraire que les lissages polynômiaux ou par moyenne mobile: il n'y a pas de choix arbitraire à faire sur l'ordre du polynôme (variable suivant le descripteur utilisé) et les valeurs aux extrémités du signal ne sont pas éliminées. Les phénomènes de basse fréquence sont intégralement conservés et on observe que seules les oscillations de période inférieure à cinq prélèvements sont supprimées. Mais la propriété la plus remarquable de ce lissage est de rendre l'analyse en composantes principales tout à fait adaptée à l'étude des structures temporelles multivariables: l'ordination des observations dans l'espace factoriel conserve leur connexité temporelle, et permet un découpage de la série en périodes homogènes par rapport aux corrélations entre les descripteurs (Ibanez, 1983, 1984). En outre, il rend inutile une transformation préalable des données compte tenu de la stabilisation des variances par lissage.

Analyses en composantes principales

Après avoir ainsi estimé les tendances générales des 43 descripteurs à partir des 96 valeurs observées ou interpolées, nous avons appliqué une ACP sur les résultats du lissage afin de réaliser une segmentation de la série temporelle: nous l'appellerons 'ACP sur les tendances'. Toutefois comme nous ne disposions à l'origine que de 71 valeurs effectivement échantillonnées, il nous a semblé plus rigoureux d'effectuer la méthode d'ACP avec données manquantes: chaque valeur lissée correspondant à une interpolation préalable est donc éliminée du traitement. Legendre *et al.*, (1984) proposent de calculer d'abord la matrice de corrélation basée sur un nombre variable de degré de liberté pour chaque couple de descripteurs (en enlevant les couples d'observations contenant au moins une donnée manquante) puis d'extraire les axes principaux. Comme, dans ce cas, la matrice de corrélation n'est pas nécessairement définie positive, nous avons utilisé la méthode itérative de Wold (Bouvier, 1977) qui part directement du tableau des données initiales.

Les corrélations entre les descripteurs et les premiers axes principaux (les saturations) permettent une représentation graphique des proximités entre les espèces sans pour autant indiquer objectivement des groupes bien individualisés. Nous avons donc choisi de procéder à une méthode de classification des descripteurs. Par souci d'homogénéité avec l'ACP précédente sur les tendances, nous avons considéré dans un premier temps les corrélations de Bravais-Pearson entre les profils lissés des espèces. Ces corrélations sont ensuite changées en mesures de distance par la transformation arc cosinus et normées entre 0 et 1 en divisant par π (Blumenthal, 1953). Le modèle général de groupement par agglomération hiérarchique de Lance & Williams (1967) a été utilisé pour définir des classes de descripteurs (méthode de groupe flexible avec $\beta = -0{,}30$), qui ont été reportées sur la projection sur les plans principaux.

Analyse de l'évolution des groupes trophiques

Nous avons recherché, dans un deuxième temps, à définir d'autres analyses pour étudier la variabilité temporelle des organismes ayant les mêmes besoins trophiques. Les espèces ont été regroupées en six catégories trophiques (Tableau 2). Les espèces à la fois suspensivores et déposivores de surface

Tableau 2. Répartition des 43 espèces en six principaux groupes trophiques.

Groupe 1: S (suspensivores et mixtes)	Groupe 2: DS (déposivores de surface)
16 *Cultellus pellucidus*	1 *Abra alba*
43 *Venus ovata*	2 *Abra prismatica*
3 *Ampelisca brevicornis*	6 *Ampharete acutifrons*
4 *Ampelisca sarsi*	7 *Apherusa ovalipes*
5 *Ampelisca typica*	9 *Aricidea cerrutii*
23 *Lanice conchilega*	10 *Aricidea fragilis*
38 *Spio filicornis*	11 *Aricidea minuta*
41 *Thyasira flexuosa*	14 *Chaetozone setosa*
	17 *Diplocirrus glaucus*
	25 *Mediomastus fragilis*
	32 *Paradoneis armata*
	28 *Nucula hanleyi*
	29 *Nucula turgida*
	39 *Spiophanes bombyx*
	40 *Tellina fabula*
Groupe 3: DSS (déposivores de subsurface)	**Groupe 4: C (carnivores)**
8 *Apseudes latreillei*	18 *Eteone longa*
12 *Bathyporeia elegans*	22 *Hyalinoecia bilineata*
13 *Bathyporeia tenuipes*	24 *Marphysa bellii*
15 *Clymene oerstedii*	26 *Nephtys hombergii*
21 *Heterocirrus alatus*	36 *Phyllodoce groëndlandica*
27 *Notomastus latericeus*	
37 *Scoloplos armiger*	
42 *Urothoë pulchella*	
Groupe 5: H (herbivores)	**Groupe 6: O (omnivores)**
20 *Exogone hebes*	19 *Eulalia sanguinea*
30 *Odontosyllis gibba*	31 *Ophiura albida*
	33 *Pariambus typicus*
	34 *Perioculodes longimanus*
	35 *Phtisica marina*

(mixtes) ont été réunies avec les suspensivores strictes; les déposivores de surface, dont le regroupement avec les suspensivores a été proposé par certains auteurs, ont cependant été considérées comme un groupe trophique indépendant.

Variation des ressemblances entre les observations selon les groupes trophiques

La mesure de la ressemblance entre les échantillons est basée sur leur proximité dans l'espace des espèces (R^n). La longueur du vecteur liant un point échantillon de cet hyperespace à l'origine des axes représente l'abondance totale d'organismes en ce point, alors que sa direction figure les relations de dominance entre les espèces. Nous voulions comparer les changements de communauté indépendamment des variations liées à l'abondance des espèces considérées. La mesure de corde (Orloci, 1978), est une mesure de distance (ou de dissemblance), qui a déjà été utilisée par Mackas & Sefton (1982) et Mackas (1985) pour un raisonnement voisin en planctologie. Elle correspond à une distance euclidienne calculée sur les abondances normées à 1 pour chaque échantillon. Si on appelle x la densité de l'espèce i à la station j, on doit avoir (après normalisation):

$$\sum_{i=1}^{n} x_{ij}^2 = 1; j = 1, 2 \ldots, p.$$

Tous les vecteurs échantillons j ont alors un module unitaire bien que leur direction reste inchangée dans l'espace R^n. La distance de corde s'écrit:

$$D_{jk}^2 = \sum_{i=1}^{n} (x_{ij} - x_{ik})^2$$

Elle est nulle si deux échantillons possèdent les mêmes espèces dans les mêmes proportions, elle est maximale et égale à deux quand les espèces diffèrent totalement d'un prélèvement à l'autre. Elle permet donc de comparer la composition faunistique d'un prélèvement à un autre sans subir l'influence des variations temporelles de forte amplitude dues à une minorité d'espèces.

Pour un groupe trophique particulier, la ressemblance entre p stations s'exprimera par une matrice des carrés des distances, d'ordre p. Le test de Mantel (1967), repris par Sokal (1979), permet de tester la similitude de deux matrices de distances d'ordre p. Nous l'avons utilisé pour savoir si la classification des prélèvements à partir d'un groupe trophique donné était différente de celle obtenue avec chacun des autres groupes. Le test consiste à effectuer le produit vectoriel des deux matrices puis de comparer le nombre z ainsi calculé avec sa moyenne, paramètre que l'on estime ainsi que son écart-type par simulations en permutant au hasard les lignes et les colonnes de l'une quelconque des deux matrices. L'écart entre z et sa moyenne, normé par l'écart type, suit une loi de t de Student, le nombre de degrés de liberté correspondant au nombre de simulations effectuées.

L'échelle temporelle de la ressemblance entre les échantillons: le distogramme

De même que la fonction Auto-D^2 (Ibanez, 1984) est une extension au cas multivariable de la fonction d'autocorrélation, nous appellerons 'distogramme' une extension de la fonction de structure ou variogramme (Matheron, 1970). Cette dernière donne une estimation de la variance d'un processus selon différentes échelles de temps (ou d'espace). Si on considère la distance de corde moyenne $\bar{d}$ entre des stations séparées successivement par des intervalles de temps (h) de plus en plus grands, on définit une fonction de dissemblance $\bar{d}(h)$ (le distogramme), qui exprime la modification scalaire de la communauté statistique. D'après Mackas (1984) le distogramme théorique (dans un milieu turbulent) s'ajuste à une courbe d'allure parabolique: pour des échelles très vastes, il n'existe plus de corrélation entre les échantillons et la dissemblance tend vers sa valeur maximale possible. Ne disposant pas d'échantillons prélevés simultanément à la même station, la distance moyenne $\bar{d}$ ne peut être calculée pour un décalage nul ($h=0$). On peut l'estimer en extrapolant la fonction jusqu'a son intersection avec l'axe des ordonnées. Dans notre cas, cette distance minimale reste toujours légèrement supérieure à 0, traduisant un effet de la variabilité aléatoire locale (comparable à l'effet 'de pépite' des variogrammes).

Résultats

Analyses en composantes principales

Ordination des prélèvements

L'ordination par l'analyse en composantes principales des 71 prélèvements selon les axes 1 et 2 et les axes 2 et 3 est représentée sur les figures 1 et 2; l'inertie des trois premiers axes forme plus de 53% de l'inertie totale (ler axe: 20,86%; 2ème axe: 19,67% et 3ème axe: 12,86%). Dans cette analyse sur les données lissées, comme la dépendance entre les prélèvements successifs est amplifiée, les prélèvements proches dans le temps sont voisins dans l'espace factoriel (Ibanez, 1983).

Ainsi en joignant les observations successives dans l'espace des deux premiers axes, on obtient une ligne brisée relativement lissée d'allure parabolique ce qui indique que le premier axe est une fonction quadratique du second (effet Guttman: Benzecri et coll., 1973). Il convient donc d'interpréter globalement les projections des observations dans le premier plan propre plutôt que de considérer séparément les axes 1 et 2.

La branche supérieure de la parabole correspond à la dérive temporelle progressive du peuplement à la suite de sa perturbation par la pollution de l'Amoco Cadiz: entre 1978 et 1980. La composition spécifique atteint en septembre 1980 un maximum de divergence par rapport à celle du début des observations.

La branche inférieure de la parabole marque un retour progressif à la situation initiale, comme si après s'être écarté au maximum d'une position d'équilibre en 1980, l'écosystème tendait à retrouver la même structure interactive entre les espèces. Cependant, l'opposition entre les observations au début et à la fin de l'échantillonnage est évidente: elle rend compte de la réduction des densités au cours de trois dernières années des espèces précisément abondantes au début de l'observation (Paraonidés) et, inversement, de l'augmentation d'abondance d'espèces rares au début des relevés (*Urothoë pulchella*).

Hill (1973) préconise l'examen du troisième axe factoriel chaque fois que l'on est en présence de l'effet Guttman. Cet axe oppose les prélèvements avec une forte densité totale et ceux avec une faible densité totale. Ainsi, on observe une décroissance régulière de la densité depuis août 1977 (13164

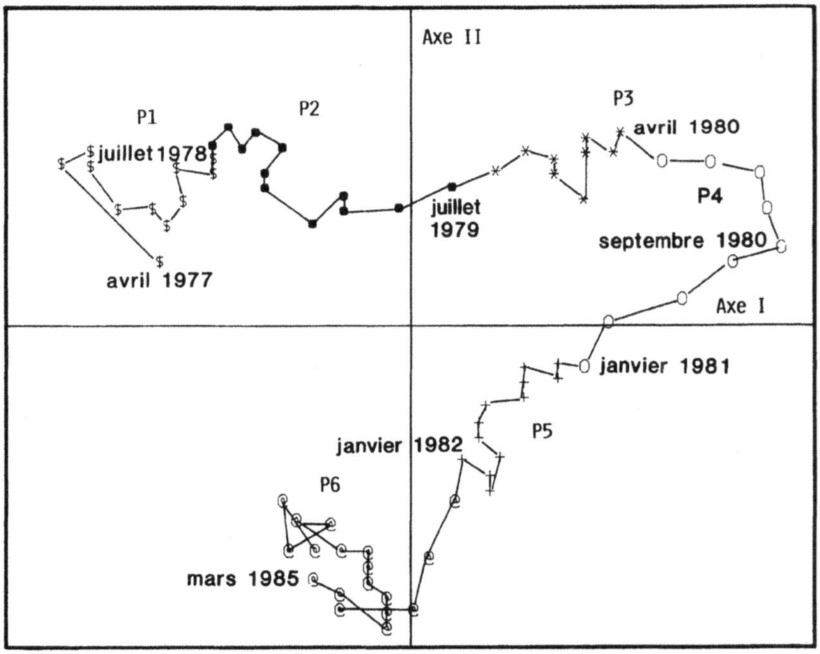

Fig. 1. Ordination des 71 échantillons dans le plan des deux premiers axes de l'ACP sur les tendances avec identification des groupes obtenus par agglomération hiérarchique (pl: période l, ...; l: échantillon l...).

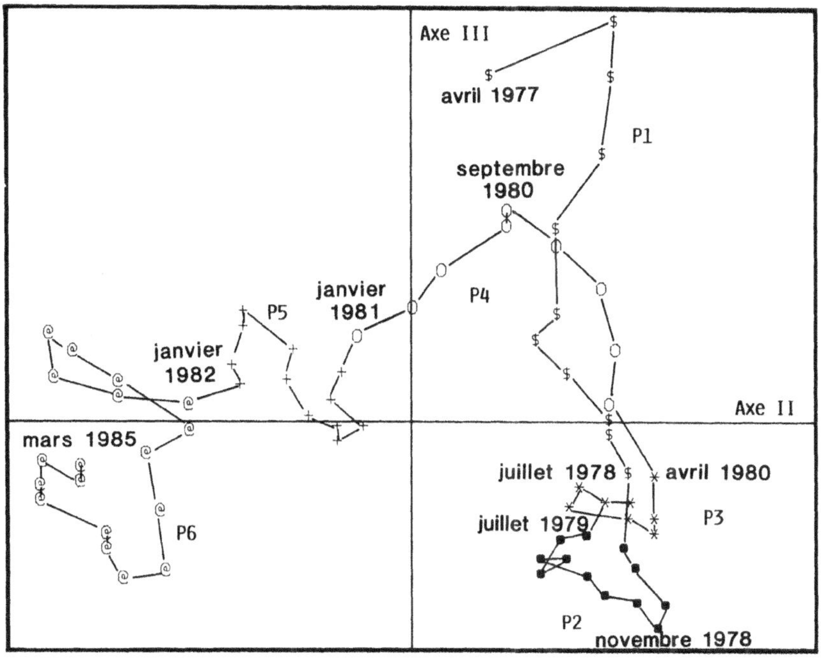

Fig. 2. Ordination des 71 échantillons dans le plan des deuxième et troisième axes de l'ACP sur les tendances.

ind·m^{-2}) jusqu'à novembre 1978 (2887 ind·m^{-2}) puis elle atteint un maximum en septembre 1980 (5069 ind·m^{-2}) et montre par la suite une tendance décroissante à l'exception des périodes estivales de 1981 et de 1982. En résumé, finalement, les axes 1 et 2 traduisent des changements de la composition spécifique et l'axe 3 l'évolution de la densité.

Le regroupement des prélèvements par agglomération hiérarchique de Lance & Williams (1967) à partir des trois premières composantes de l'analyse factorielle permet à un seuil de similarité de 0,80 de séparer la série en 6 groupes de prélèvements, lesquels se suivent de façon chronologique sans insertion de relevés éloignés dans le temps. Les six périodes sont: p 1, avril 1977-juillet 1978; p 2, août 1978-juillet 1979; p 3, août 1979-avril 1980; p 4, mai 1980-janvier 1981; p 5, février 1981-janvier 1982 et p 6, février 1982-mars 1985. Les cinq premiers groupes correspondent à des périodes de 9 à 16 mois, la sixième période regroupe les prélèvements effectués au cours des trois dernières années du suivi. L'évolution rapide de la structure numérique du peuplement au cours des quatre premières années après la pollution (été 1978-hiver 1982) indique ainsi une période de perturbation laquelle est suivie d'une période de stabilisation de l'écosystème benthique.

Ordination des espèces

L'ordination des 43 descripteurs biologiques a seulement été représentée selon les deux premiers axes (Fig. 3). Cependant, le regroupement des espèces par agglomération hiérarchique a été effectué à partir des trois premières composantes de l'analyse factorielle. Trois principaux groupes bien individualisés sont mis en évidence (au niveau de similarité de 0,9); le groupe A, lequel est subdivisé en trois sous-groupes, correspond aux espèces montrant au moins un maximum d'abondance entre avril 1979 et avril 1981; le groupe B comportant deux sous-groupes, réunit les espèces qui présentent des abondances plus soutenues à partir d'avril 1981; à l'opposé le groupe C, comportant deux sous-groupes, renferme les espèces plus abondantes au début du suivi. Dans l'espace factoriel des deux premiers axes, l'axe 1 est caractérisé par l'opposition du groupe A (notamment du sous-groupe A_1) et du groupe C_2; l'axe 2 contribue à séparer les espèces du groupe C1 de celles du groupe B2. Les deux premiers axes différencient les groupes d'espèces en fonction de leurs successions quantitatives. L'axe 1 reflète les successions d'abondance en relation avec la perturbation: dans un premier temps, forte régression des populations du groupe C2 puis stimulation des populations du groupe A1 et dans un

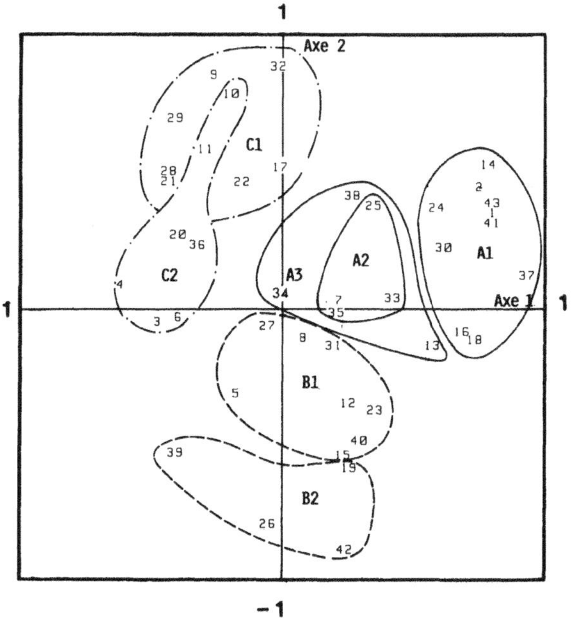

Fig. 3. Ordination des 43 espèces dans le plan des deux premiers axes de l'ACP sur les tendances avec identification des groupes obtenus par agglomération hiérarchique (1, 2, ...: numéros d'ordre des espèces: Tableau 1).

deuxième temps, reconstitution des populations du groupe C2 et déclin des populations du groupe A1. En ce qui concerne l'axe II, il traduit le remplacement des espèces abondantes au début de l'étude (C1) par celles abondantes à la fin de l'étude (B2). On rejoint donc les conclusions précédentes, tirées de l'effet Guttman sur les prélèvements: les différentes phases de l'évolution de l'écosystème benthique mises en évidence par l'ACP sur les prélèvements sont le reflet des modifications d'abondance des principales espèces groupées comme il vient d'être fait.

Le Tableau 3 donnant la dominance des dix premières espèces à chacune des six périodes p 1 à p 6 fournit également un élément d'interprétation de la dynamique à long terme du peuplement:

Période p 1: avril 1977 à juillet 1978 (avant la pollution): dominance des espèces du groupe C notamment d'*Ampelisca sarsi* (Fig. 4a) et de *Paradoneis armata* (Fig. 4b); fortes densités: 5362 ind $\cdot$ m^{-2} en moyenne.

Période p 2: août 1978 à juillet 1979 (première année après la pollution): très forte réduction des Ampeliscidés et de la densité moyenne totale du

Tableau 3. Dominance des dix premières espèces (données lissées) rangées par ordre de dominance décroissante au cours des six périodes distinguées par l'analyse en composantes principales.

p1 avril 1977 – juillet 1978	%	p2 août 1978 – juillet 1979	%	p3 août 1979 – avril 1980	%
Ampelisca sarsi	45,3	*Paradoneis armata*	24,5	*Spio filicornis*	28,4
Paradoneis armata	16,7	*Spio filicornis*	21,6	*Paradoneis armata*	13,1
Spio filicornis	6,0	*Chaetozone setosa*	9,0	*Chaetozone setosa*	11,8
Aricidea fragilis	4,5	*Heterocirrus alatus*	7,2	*Abra alba*	5,9
Chaetozone setosa	4,1	*Aricidea fragilis*	6,1	*Scoloplos armiger*	5,8
Marphysa bellii	4,0	*Abra alba*	5,6	*Marphysa bellii*	4,6
Aricidea cerrutii	2,5	*Marphysa bellii*	4,0	*Thyasira flexuosa*	4,4
Nephtys hombergii	1,7	*Aricidea cerrutii*	3,8	*Aricidea fragilis*	3,7
Hyalinoecia bilineata	1,4	*Perioculodes longimanus*	2,6	*Venus ovata*	2,3
Mediomastus fragilis	1,2	*Abra prismatica*	1,6	*Urothoë pulchella*	1,8

p4 mai 1980 – janvier 1981	%	p5 février 1981 – janvier 1982	%	p6 février 1982 – mars 1985	%
Paradoneis armata	11,6	*Ampelisca sarsi*	16,2	*Ampelisca sarsi*	24,6
Spio filicornis	10,7	*Paradoneis armata*	13,5	*Spio filicornis*	13,3
Chaetozone setosa	9,8	*Chaetozone setosa*	8,3	*Urothoë pulchella*	9,4
Marphysa bellii	9,7	*Urothoë pulchella*	7,3	*Paradoneis armata*	8,4
Abra alba	8,4	*Marphysa bellii*	6,6	*Nephtys hombergii*	5,9
Ampelisca sarsi	6,6	*Spio filicornis*	6,5	*Scoloplos armiger*	3,4
Scoloplos armiger	6,1	*Scoloplos armiger*	5,5	*Marphysa bellii*	3,2
Thyasira flexuosa	4,8	*Abra alba*	3,6	*Chaetozone setosa*	3,0
Urothoë pulchella	3,6	*Aricidea fragilis*	3,2	*Aricidea fragilis*	2,8
Aricidea fragilis	3,3	*Nephtys hombergii*	2,3	*Ampelisca typica*	2,6

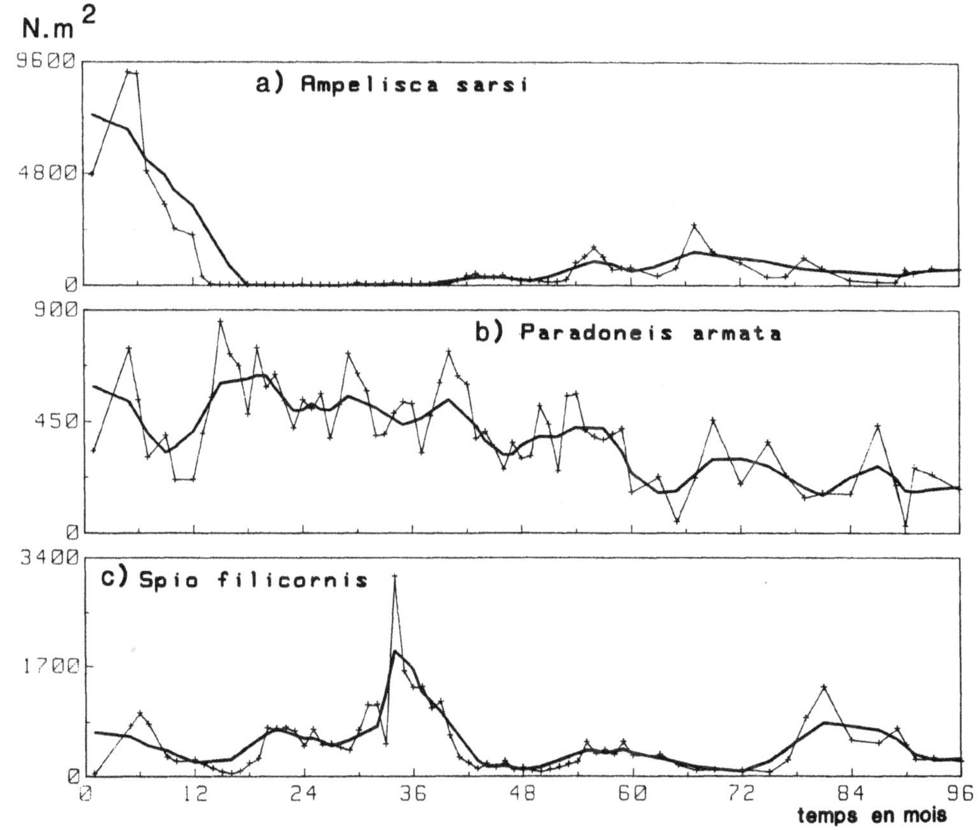

Fig. 4. Profils d'abondance d'*Ampelisca sarsi* (a), de *Paradoneis armata* (b) et de *Spio filicornis* (c) (trait fin) avec tracé de la tendance générale (trait gras) de t 1 (avril 1977) à t 96 (mars 1985).

peuplement: 2298 ind·m^{-2}. La dominance revient aux espèces du groupe C1 et à celles du groupe A3 parmi lesquelles on rencontre *Spio filicornis* (Fig. 4c). Parmi le groupe C1, la Polychète opportuniste de petite taille *Heterocirrus alatus* s'installe de façon modérée sur le peuplement (Fig. 5a).

Période p 3: août 1979 à avril 1980 (deuxième année après pollution): augmentation des densités moyennes totales: 3613 ind·m^{-2}. La dominance reste aux espèces des groupes C1 et A3, début des biostimulations des espèces du groupe A1 comme *Abra alba* (Fig. 5b): période de restructuration du peuplement.

Période p 4: mai 1980 à janvier 1981 (troisième année après pollution): légère augmentation des densités moyennes totales: 4140 ind·m^{-2} et faible disparité numérique entre les espèces (Tableau 3). On observe une stimulation d'espèces du groupe A1 comme *Marphysa bellii* (Fig. 5c) et du groupe A2 comme *Mediomastus fragilis* (Fig. 6a), le début des reconstitutions des espèces du groupe C2 (*Ampelisca sarsi*) et l'augmentation du pourcentage des espèces du groupe B2 comme *Urothoë pulchella* (Fig. 6b).

Période p 5: février 1981 à janvier 1982 (quatrième année après la pollution): période de transition entre le régime perturbé du peuplement et la période de stabilisation. Réduction des densités moyennes totales: 3122 ind·m^{-2}. *Ampelisca sarsi* redevient l'espèce la plus abondante; *Chaetozone setosa* et *Scoloplos armiger* montrent encore des densités élevées. On assiste à la fin des biostimulations des espèces des groupes A1 et A2, à la poursuite de l'installation des espèces du groupe B2 et à l'installation modérée des espèces du groupe B1 comme *Lanice conchilega* (Fig. 6c).

Période p 6: février 1982 à mars 1985: maintien des densités moyennes totales à de faibles valeurs: 2938 ind·m^{-2} (inférieures de moitié à celles de la période p 1); ralentissement des processus de

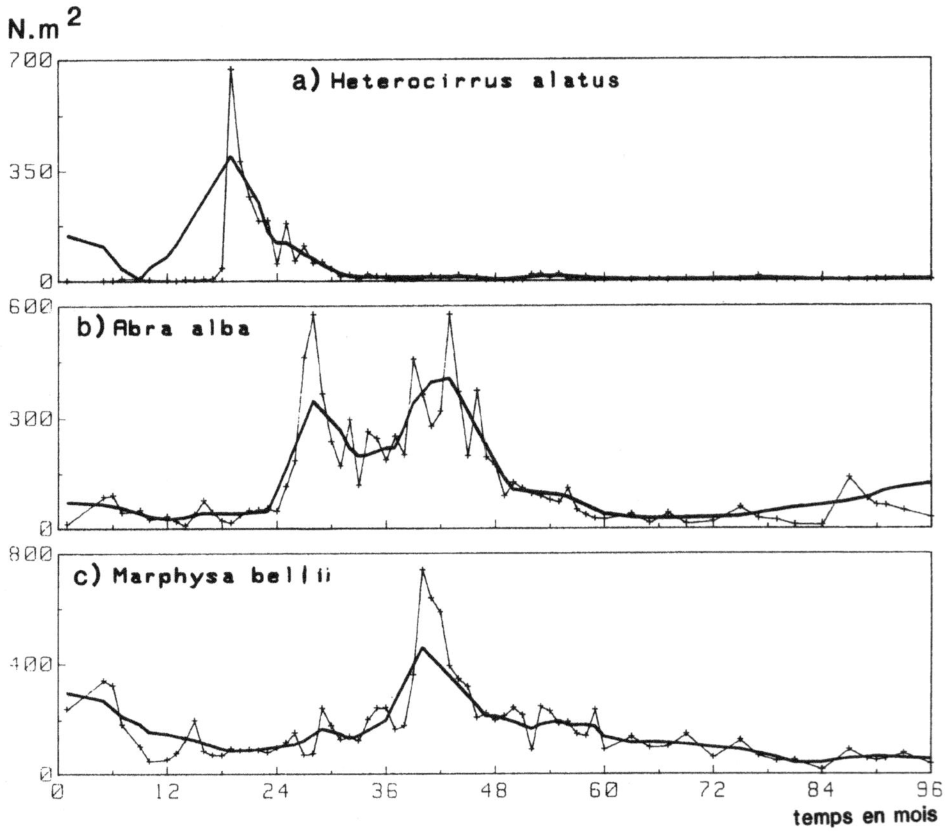

Fig. 5. Profils d'abondance d'*Heterocirrus alatus* (a), d'*Abra alba* (b) et de *Marphysa bellii* (c) (trait fin) avec tracé de la tendance générale (trait gras) de t 1 (avril 1977) à t 96 (mars 1985).

reconstitutions et stabilisation de la structure du peuplement. Les espèces des groupes C2, B2 et A3 deviennent dominantes, à l'inverse la part des espèces des groups C1, B1 et A1 se réduit.

Analyse de l'évolution des groupes trophiques

Comparaison des distances entre échantillons en fonction des groupes trophiques

La distance de corde exprime les différences entre les observations du point de vue de la dominance spécifique, indépendamment des valeurs absolues des abondances. L'algorithme de Mantel sert à tester l'identité de deux matrices de distances entre deux groupes de prélèvements. En considérant dans un même groupe d'observations deux types de catégories trophiques, ce test permet de comparer les resemblances entre les observations en fonction des abondances relatives des espèces appartenant à chaque groupe trophique. Si l'hypothèse nulle est vérifiée cela signifie que la dominance des espèces de chaque groupe est identique pour la période considérée, en d'autres termes, qu'il n'y a aucune compétition entre ces populations.

Le test de Mantel indique, par example, si les distances entre p prélèvements dans l'espace des huit espèces déposivores de subsurface sont statistiquement identiques à celles des mêmes p stations considérées dans l'espace des carnivores. Avec six catégories trophiques (Tableau 2), nous pouvons effectuer 15 comparaisons différentes. Les simulations pour le test ont été effectuées 20 ou 40 fois compte tenu de l'ordre des matrices de distances considérées.

Les tests effectués en prenant l'ensemble des 71 observations sont tous non significatifs – résultat prévisible puisque la classification globale sur les espèces précédentes n'a pas discriminé les groupes trophiques. Nous avons voulu ensuite décrire la variabilité faunistique pour des échelles de temps

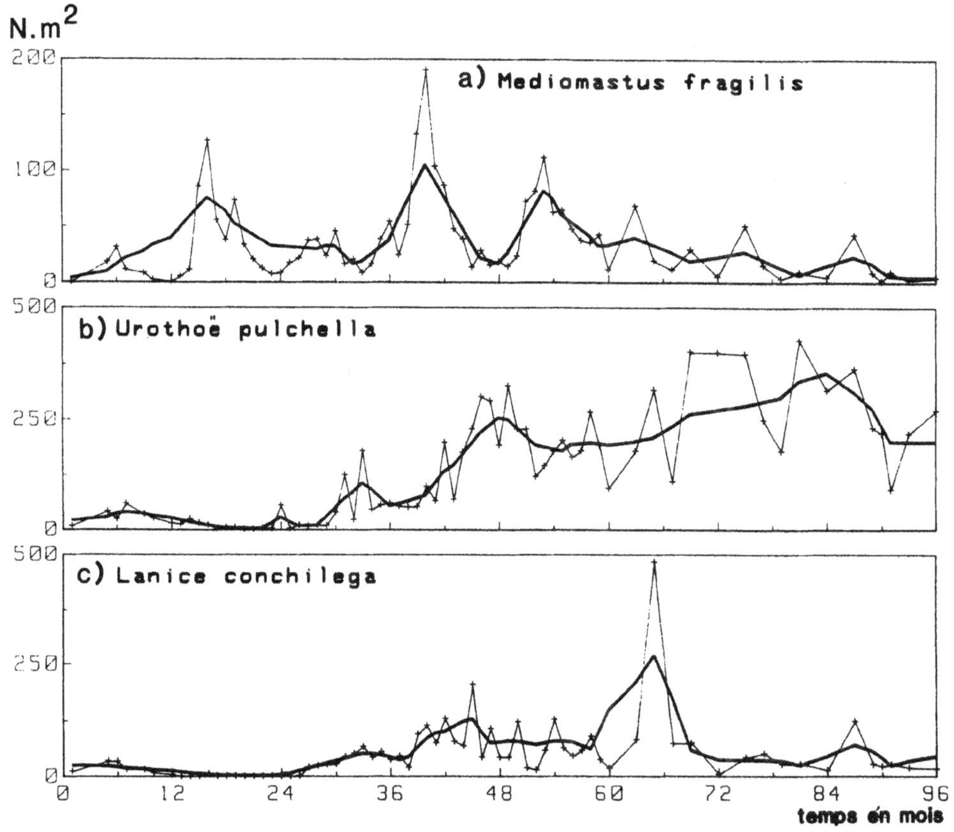

Fig. 6. Profils d'abondance de *Mediomastus fragilis* (a), d'*Urothoë pulchella* (b) et de *Lanice conchilega* (c) (trait fin) avec tracé de la tendance générale (trait gras) de t 1 (avril 1977) à t 96 (mars 1985).

plus courtes et pour lesquelles il peut exister une forte autocorrélation entre les prélèvements. Le test de Mantel a donc été appliqué aux six périodes décelées par l'ACP.

Les moyennes des 15 valeurs de t pour les six périodes sont consignées dans le Tableau 4. L'hétérogénéité maximale se situe au cours de la période p 3 (août 1979 à avril 1980) correspondant à une période de restructuration du peuplement. Cette période est caractérisée par des recrutements importants d'espèces appartenant à plusieurs groupes trophiques: suspensivores: installation de *Venus ovata,* début des reconstitutions des populations d'*Ampelisca;* biostimulations de déposivores de surface: *Abra alba, Abra prismatica* et de déposivores de subsurface, à l'inverse les carnivores et les omnivores présentent des populations en déclin. On assiste par conséquent au cours de la péride p 3 à des tentatives de colonisation du milieu par des espèces de plusieurs groupes trophiques.

Dans le détail des comparaisons on remarque un certain nombre de valeurs de t significatives, seules portées dans le Tableau 4. La première période (p 1) est très homogène et seules les distances entre les prélèvements dans l'espace des déposivores de surface (DS) et de subsurface (DSS) sont significativement différentes. La période suivante p 2 est totalement homogène. D'août 1979 à avril 1980 (p 3) on note une forte variabilité de la composition faunistique: les suspensivores (S) et les carnivores (C) sont les catégories au comportement le plus singulier. Seuls les herbivores (H), il est vrai représentés par trop peu d'espèces, semblent suivre une évolution moyenne peu remarquable. La période p 4 exprime les différences entre trois groupes, les déposivores de surface (DS), les carnivores (C) et les omnivores (O). L'avant dernière période p 5 montre une opposition entre déposivores de surface (DS) et trois catégories, suspensivores (S), déposivores de subsurface (DSS), omnivores (O). Les car-

Tableau 4. 'Test de Mantel'. $\bar{t}$: moyenne sur 15 comparaisons du t de student. t: valeurs du t de student significatives (seuil 5%). p 1: avril 1977 à juillet 1978; p 2: août 1978 à juillet 1979; p 3: août 1979 à avril 1980; p 4: mai 1980 à janvier 1981; p 5: février 1981 à janvier 1982 et p 6: février 1982 à mars 1985. DS: déposivores de surface; DSS: déposivores de subsurface; S: suspensivores; C: carnivores; O: omnivores.

	p1	p2	p3	p4	p5	p6
$\bar{t}$	0,97	0,96	1,77	1,42	1,50	1,11
t	2,12 (DS/DSS)	–	2,17 (S/DS)	2,13 (DS/C)	2,17 (S/DS)	2,29 (S/DSS)
	–	–	2,79 (S/DSS)	2,67 (DS/O)	2,64 (DS/DSS)	–
	–	–	2,31 (S/C)	2,15 (C/O)	1,95 (DS/O)	–
	–	–	2,27 (DS/O)	–	–	–
	–	–	2,11 (DSS/C)	–	–	–
	–	–	2,24 (C/O)	–	–	–

nivores ne se distinguent plus par leur comportement. La sixième période (p 6) est relativement plus homogène avec un contraste dans leur évolution temporelle pour les suspensivores et les déposivores de subsurface.

Etude scalaire de l'évolution des groupes trophiques

La fonction du $\bar{d}^2$ (distogramme: Figure 7) est calculée pour des observations séparées de un à quarante mois successivement, avec un pas constant de un mois. Un point commun à toutes ces fonctions est que leur valeur extrapolée pour un décalage nul, donnerait une valeur de $\bar{d}^2$ très voisine de 0: en d'autres termes, la variabilité ponctuelle de l'échantillonnage est faible et constante quel que soit le groupe trophique (ce que l'on n'observe pas avec des données planctoniques: Mackas, 1984). On distingue quatre types de variations bien caractéristiques:

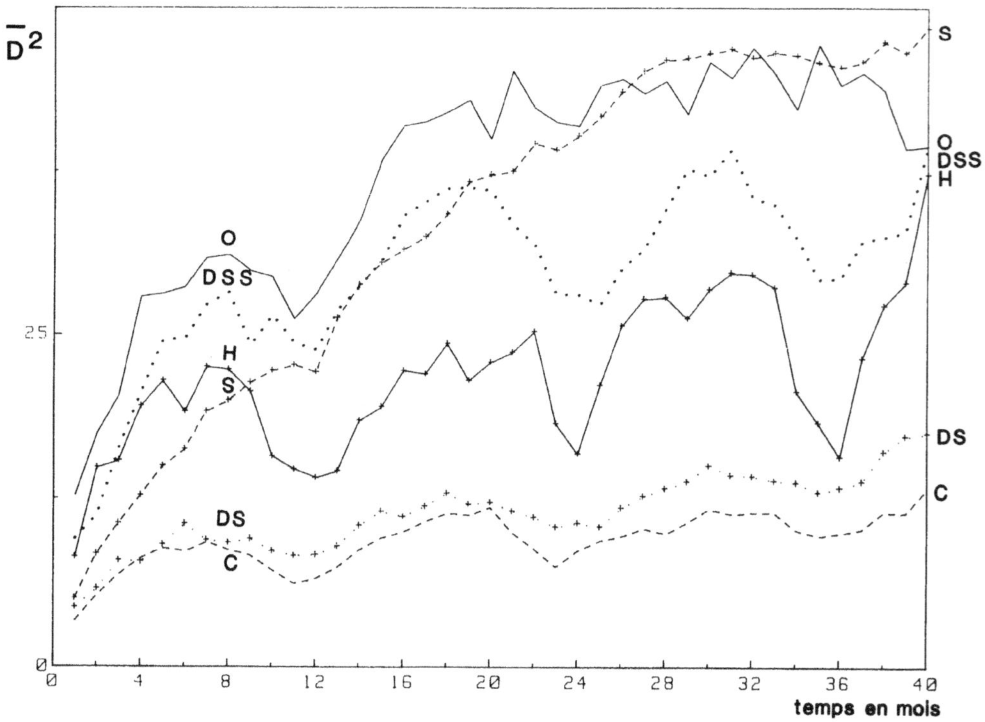

Fig. 7 Distogrammes pour les six catégories trophiques. En abscisse, l'échelle en mois, en ordonnée la distance au carré moyenne entre les observations (distance euclidienne). S: suspensivores et mixtes; DS: déposivores de surface; DSS: déposivores de subsurface; C: carnivores; H: herbivores et O: omnivores.

– Le type 1 (les carnivores (C) et les déposivores de surface (DS)) présente un cycle annuel reconnaissable mais de faible amplitude, surimposé à une très faible tendance croissante. La proportion des espèces dans ces deux groupes varie donc très peu au cours du temps. La répartition des organismes est homogène et traduit une évolution des populations marquée par une évolution saisonnière.

– Le type 2 (les herbivores (H) et les déposivores de subsurface (DSS)) montre une oscillation annuelle de période douze mois très prononcée. Compte tenu d'une assez forte tendance croissante générale les communautés sont plus comparables d'une année successive à une autre que séparées par 24 ou 36 mois notamment dans le cas des déposivores de subsurface. Pour ces deux catégories trophiques, l'assemblage spécifique varie très rapidement pour des intervalles de temps très courts de deux ou trois mois.

– Le troisième type (les omnivores (0)) amplifie le résultat précédent. On note une brutale tendance au changement de communauté dès les premiers décalages. On distingue une seule oscillation annuelle ce qui signifie que les structures évoluent également à plus long terme même si le processus de recrutement reste saisonnier. Au bout d'environ 18 mois la fonction prend l'allure d'un plateau relativement stable qui indique un seuil d'hétérogénéité maximale pour ce groupe trophique. Les omnivores ont donc un comportement plus sensible aux variations de l'environnement même si leur cycle biologique reste assez stable.

Le type 4 (les suspensivores (S)) est nettement particulier: le cycle saisonnier est pratiquement absent et la fonction prend une allure parabolique telle que la décrivait Mackas (1984) pour des organismes planctoniques en régime de turbulence. Tout changement d'échelle induit une modification des proportions des différentes espèces. L'aspect très lissé de la fonction laisse supposer un gradient particulier, une tendance croissante au changement à l'échelle de l'ensemble de la série des observations.

Discussion et conclusions

Evolution du peuplement macrobenthique

L'ACP sur les tendances générales des descripteurs a montré que les observations formant des groupes dans l'espace des espèces sont elles-mêmes voisines dans le temps: les communautés benthiques présentent une forte autocorrélation temporelle, leurs changements s'effectuant de façon graduelle. L'ordination des observations et la classification des espèces soulignent l'altération progressive des relations quantitatives entre les organismes, succédant à la pollution par les hydrocarbures de l'Amoco Cadiz. Deux ans et demi après cet événement, l'assemblage faunistique se trouve le plus profondément remanié. Par la suite on note l'indice d'un retour vers la situation avant la pollution, sans pour autant que le phénomène soit strictement réversible, car l'abondance relative des différents descripteurs est nettement modifiée par rapport à 1977–1978.

Ces phénomènes de successions se rapprochent ainsi de ceux décrits par Pearson & Rosenberg (1978) lors de pollutions de nature organique et de ceux décrits par Sanders *et al.*, (1980) et Glémarec & Hussenot (1982) lors de pollutions pétrolières. A partir du printemps 1982 la stabilisation des processus de reconstitution du peuplement, avec une tendance au retour à l'ordre hiérarchique des espèces observé avant pollution, semble indiquer que l'écosystème s'organise dans un nouvel état d'équilibre, avec cependant des densités demeurant très inférieures à celles mesurées avant pollution. Une tendance analogue vers un nouvel état d'équilibre se manifeste pendant le même temps chez le peuplement de Nématodes méiobenthiques (Boucher, 1985).

Néanmoins, les résultats de l'ACP ne permettent pas de conclure si l'écosystème est en train de rejoindre le point d'équilibre qui correspondrait à la phase antérieure à la pollution, ou si l'impact de celle-ci l'a fait dériver définitivement vers un nouveau type de structure.

Esai d'interprétation de la segmentation par des paramètres abiotiques

Outre l'évolution des abondances des principales espèces nous disposions de cinq paramètres abiotiques: température et salinité de l'eau au niveau du fond, hydrodynamisme calculé à partir de l'indice d'agitation de la mer, azote total dans le sédiment mesuré à partir d'avril 1978 (données de Chamroux & Mével, sous presse) et les teneurs en hydrocarbures (mesurées en spectrophotométrie aux infra-rouges) enregistrées d'avril 1978 à février 1981. Excepté l'azote et les hydrocarbures, les trois autres paramètres présentent des cycles annuels très marqués sans tendance générale ni accident notable permettant d'interpréter la segmentation de la série. Les écarts saisonniers des paramètres hydrologiques sont faibles: la température de l'eau varie ordinairement de 9 °C en hiver à un peu plus de 15 °C en été, la salinité passe de 34,5‰ en hiver à 35,3‰ au début de l'automne. L'hydrodynamisme est élevé en novembre-décembre puis février-mars et relativement faible en été.

La réserve organique dans le sédiment montre une augmentation plus ou moins régulière jusqu'à l'automne 1981 puis une décroissance jusqu'à l'automne 1984. Les plus fortes valeurs d'azote ont été enregistrées au cours du printemps 1979 et de l'été 1981. La période de concentration de l'azote dans le sédiment correspond aux périodes des biostimulations et de persistances de densités élevées chez les espèces des groupes A1 et A2 parmi lesquelles on rencontre de nombreuses déposivores. Il semble par conséquent exister une relation directe entre les installations de ces espèces et la réserve organique du sédiment. Après une phase de pollution primaire au cours du printemps et de l'été 1978 (plus de 700 $\mu g \cdot g^{-1}$ de sédiment sec) et une phase de pollution secondaire au printemps 1979 (460 $\mu g \cdot g^{-1}$) les teneurs en hydrocarbures demeurent inférieures à 50 $\mu g \cdot g^{-1}$ dans l'ensemble des relevés de juillet 1979 à février 1981. On relève ainsi que l'inversion de la dérive du peuplement en 1980 coïncide avec le fait que pour la première fois après la marée noire les recrutements printaniers et estivaux ont eu lieu sur des sédiments dont le degré de pollution est définitivement tombé en dessous du seuil significatif. (50 $\mu g \cdot g^{-1}$). Toutefois, l'évolution temporelle de l'azote ou des teneurs en hydrocarbures ne nous permet pas d'expliciter plus en détail les découpages de la série chronologique.

Il semble, par conséquent, que le peuplement évolue en outre en fonction des relations intra- et interspécifiques, dont il est difficile d'apprécier les responsabilités respectives.

Evolution des groupes trophiques

Le test de Mantel a montré les compétitions entre les différents groupes trophiques. Il a également mis en évidence le contraste entre des périodes d'équilibre et des périodes de restructuration du peuplement. Ainsi, deux points essentiels peuvent être soulignés. En premier lieu, ce sont les déposivores de surface (DS) qui présentent les évolutions les plus notables. En effet, ces espèces, qui se nourrissent à l'interface eau-sédiment, sont en compétition à la fois avec les suspensivores strictes, les mixtes (espèces à la fois suspensivores et déposivores de surface), les omnivores et à un moindre degré avec les déposivores de subsurface lors de leur phase juvénile. En second lieu, on observe une évolution temporelle dans l'opposition des groupes; les périodes p 1 (avant pollution) puis p 2 (très faibles recrutements) et p 6 (période de stabilisation) sont homogènes. Pendant ces phases d'équilibre, le peuplement présente peu de compétition entre les groupes trophiques; elles sont séparées par des périodes de forte compétition entre les espèces: périodes p 3 à p 5 correspondant à une restructuration du peuplement avec des évolutions rapides des proportions de chaque groupe trophique.

Le distogramme traduit des comportements très différents selon les six groupes trophiques. Il faut cependant noter que pour les échelles de temps supérieure à 12 mois, les conséquences de la perturbation due à la pollution de l'Amoco Cadiz prend une importance considérable dans les dissimilitudes temporelles. Ainsi, la tendance croissante générale rend compte essentiellement des reconstitutions des populations d'espèces au cours du temps, lesquelles interviennent à l'échelle de l'année. Les groupes trophiques les moins affectés par les mortalités dues à la pollution (carnivores, déposivores de surface et de subsurface, les herbivores) ont la tendance croissante la plus modérée. A l'opposé les omnivores et les suspensivores (parmi lesquels se classent les Amphipodes, *Phtisica marina, Pariambus typicus* et les *Ampelisca,* populations très fortement réduites lors du stress) présentent la tendance croissante la plus considérable.

La saisonnalité affecte tous les groupes trophiques à l'exception des suspensivores. On observe, effectivement, une concentration des périodes de recrutement de la majorité des espèces du peuplement du printemps au début de l'automne (maximale en juin-juillet). Ces recrutements, plus ou moins abondants selon les années, surviennent au moment des meilleures conditions trophiques de l'environnement (Dauvin, 1984): abondance de la production primaire et secondaire du plancton, développement d'une microflore estivale. A partir de l'automne, on assiste à de très fortes mortalités pouvant affecter, chez certaines espèces, la quasi-totalité des juvéniles recrutés en été. Ces évolutions saisonnières marquées sont assez régulièrement reproduites d'année en année: par conséquent on observe de faibles dissimilitudes inter-relevés pour des échelles de temps voisines de un an. L'évolution saisonnière est faiblement prononcée chez les carnivores et les déposivores de surface. Parmi ceux-xi se rangent surtout des espèces à stratégie démographique de type 'k' (Pianka, 1970). Dans le cas particulier des carnivores les espèces *Hyalinoecia bilineata, Marphysa bellii* et *Nephtys hombergii* vivent plusiers années et présentent une très grande stabilité numérique de leurs populations adultes et un rapport recrutement/stock faible, ordinairement compris entre 5 et 10. Seule l'espèce déposivore de surface *Mediomastus fragilis* se distingue par une stratégie de type 'r' (opportuniste). On observe ainsi une très forte régulation annuelle des populations dans ces deux groupes trophiques. Dans les cas des déposivores de subsurface, la majeure partie des espèces développent une stratégie démographique de type 'r': espèces annuelles à forte fécondité présentant des rapports R/S élevés (jusqu'à plus de 500 dans le cas de la Cirratulidae de petite taille *Heterocirrus alatus* qui a proliféré au cours de l'automne 1978). De même, les herbivores, représentés par des Polychètes Syllidiens, montrent de grandes variations quantitatives annuelles.

Appréciée à partir des suspensivores, la dissimilitude entre les relevés ne cesse de croître. On rencontre en effet dans ce groupe les cas de recolonisations les plus significatives notamment celle *d'Ampelisca sarsi,* espèce dominant numériquement le peuplement (Tableau 3). Les populations réduites d'Amphipodes ne se reconstituent qu'à une cadence limitée, à la fois à cause de leur distribution insulaire et de leurs caractéristiques biologiques: reproduction directe, faible fécondité, vie brève (Cabioch *et al.,* 1982). La reconstitution s'effectue cependant de façon progressive année après année; la dissimilitude inter-relevés devient par conséquent de plus en plus élevée en fonction de l'échelle temporelle.

Prospectives

Au cours de cette étude, deux types d'analyses quantitatives nouvelles en benthologie ont été utilisées, le test de Mantel et la fonction du distogramme. Ces techniques recèlent des possibilités d'application bien plus vastes que celles que nous avons présentées.

Ainsi, le test de Mantel, pourrait être élargi à d'autres types de comparaison:

- entre deux ensembles d'observations distincts dans le temps et/ou l'espace vis-à-vis d'un même groupe d'espèces.
- pour un même ensemble d'observations entre les distances obtenues dans l'espace des espèces et l'espace de paramètres du milieu.
- pour un même ensemble d'observations entre les distances dans l'espace des espèces et l'espace géographique, afin de tester s'il existe une autocorrélation spatiale (et/ou temporelle), et au besoin jusqu'à quelle échelle.

Le distogramme a servi à quantifier les modifications des relations de dominance en fonction de l'échelle temporelle pour six groupes trophiques. A l'exception des suspensivores tous les groupes ont en commun des variations de composition faunistique saisonnières, bien que leur intensité ou leur déterminisme soient très particuliers. Le distogramme permet une interprétation détaillée des stratégies démograhiques de chaque groupe et de l'impact des changements du milieu. Par cette fonction, on peut estimer, au moins qualitativement, l'importance relative de la dynamique des populations et des changements physiques, chimiques, sédimentologiques sur la variabilité temporelle des populations et des peuplements macrobenthiques.

Le distogramme a la propriété de pouvoir s'appliquer aussi à des études spatiales (Mackas, 1984). Non seulement, il pourrait permettre de connaître l'échelle spatiale des déformations de structure, donnant ainsi une estimation de la texture des peuplements, mais si on tient compte de différentes directions géographiques, il mettrait en évidence l'anisotropie structurale qu'il serait utile de relier aux gradients des facteurs externes: direction de la houle, des courants, éloignement à la côte, gradients sédimentaires, hydrologiques, etc...

Finalement le distogramme pourrait bientôt se révéler comme l'outil essentiel pour l'étude de l'échelle des changements dans l'espace géographique ou dans l'espace des facteurs externes. Ces applications en dehors de l'aspect descriptif auraient des conséquences fondamentales tant au niveau de la planification de l'échantillonnage que de la modélisation des écosystèmes marins.

Remerciements

Ce travail a été effectué dans le cadre du GRECO Manche (programme de recherches 'Amoco Cadiz' et 'Veille écologique': financement NOAA-CNEXO, Ministère de l'Environnement et région Bretagne). Les auteurs remerciement M. Joncourt pour son assistance technique et N. Guyard pour la dactylographie.

Références

Auffret, G. A. & J. L. Douvillé, 1974. Observations sur la dynamique des sables de la Pierre Noire (Baie de Morlaix, Finistère, France). Bull. B.R.G.M. 2: 5–18.

Bachelet, G., 1981. Données préliminaires sur l'organisation trophique d'un peuplement benthique marin. Vie Milieu 31: 205–213.

Benzecri, J. P. et coll., 1973. L'analyse des données, 2. L'analyse des correspondances. Dunod, Paris, 619 pp.

Beslier, A., 1981. Les hydrocarbures de l'Amoco Cadiz dans les sédiments subtidaux au nord de la Bretagne. Distribution et évolution. Thèse Doct. 3ème cycle, Caen, 204 pp.

Blumenthal, L. M., 1953. Theory and applications of distance geometry. Oxford University Press, London, 347 pp.

Bonsdorff, E. & C. S. Osterman, 1984. The establishment, succession and dynamics of a zoobenthic community – an experimental study. In P. E. Gibbs (ed.), Proceedings of the nineteenth European Marine Biology Symposium. Cambridge University Press: 287–297.

Boucher, G., 1985. Long-term monitoring of meiofauna densities after the Amoco Cadiz oil spill. Mar. Poll. Bull. 16: 328–333.

Boucher, G., S. Chamroux & C. Riaux, 1984. Modifications des caractéristiques physico-chimiques et biologiques d'un sable sublittoral pollué par hydrocarbures. Mar. Env. Res. 12: 1–24.

Bouvier, A., 1977. Programme ACPM. Analyses des composantes principales avec données manquantes. CNRS, Lab. Biométrie de Jouy-en-Josas. Documn. 77/17: 1–34.

Cabioch, L., 1968. Contribution à la connaissance des peuplements benthiques de la Manche occidentale. Cah. Biol. Mar. 9: 493–720.

Cabioch, L., J. C. Dauvin, C. Retière, V. Rivain & D. Archambault, 1982. Evolution à long terme (1978–1981) de peuplements benthiques des fonds sédimentaires de la région de Roscoff, perturbés par les hydrocarbures de l'Amoco Cadiz. Neth. J. Sea Res. 16: 491–501.

Chamroux, S. & G. Mével (sous presse). Conséquences et enseignements d'une pollution par hydrocarbures (Amoco Cadiz) sur la microflore bactérienne et la chimie du sédiment. Public. CNEXO, Actes colloq.

Dauvin, J. C., 1984. Dynamique d'écosystèmes macrobenthiques des fonds sédimentaires de la baie de Morlaix et leur perturbation par les hydrocarbures de l'Amoco Cadiz. Thèse Doct. Sci. Nat. Paris, 468 pp et annexes, 193 pp.

Elkaïm, B., F. Ibanez & F. Proniewski, 1982. Le benthos subtidal de l'estuaire de la Seine: analyse statistique des peuplements. J. exp. mar. Biol. Ecol. 64: 231–252.

Glémarec, M. & E. Hussenot, 1982. A three-year ecological survey in benoit and wrach Abers following the Amoco Cadiz oil spill. Neth. J. sea Res. 16: 483–490.

Hill, M. O., 1979. Reciprocal averaging: an eigenvector method for ordination. J. Ecol. 61: 237–249.

Ibanez, F., 1983. Optimisation de la représentation des séries chronologiques multivariables. Rapp. Comm. Int. mer Médit. 28: 113–115.

Ibanez, F., 1984. Sur la segmentation des séries chronologiques planctoniques multivariables. Oceanol. Acta 7: 481–492.

Lance, G. N. & W. T. Williams, 1967. A general theory of classificatory sorting strategies. I. Hierarchical systems. Computer J. 9: 373–380.

Legendre, P., D. Planas & M. J. Auclair, 1984. Succession des communautés de gastéropodes dans deux milieux différant par leur degré d'eutrophication. Can. J. Zool. 62: 2317–2327.

Lie, U. & J. Kelley, 1970. Benthic fauna communities of the coast of Washington and in Puget Sound: identification and distribution of the communities. J. Fish. Res. Board. Can. 27: 621–651.

Mackas, D. L., 1984. Spatial autocorrelation of plankton community composition in a continental shelf ecosystem. Limnol Oceanogr. 29: 451–471.

Mackas, D. L. & H. L. Sefton, 1982. Plankton species assemblages off southern Vancouver Island: geographic pattern and temporal variability. J. mar. Res. 40: 1173–1200.

Mantel, N., 1967. The detection of decrease clustering and a generalized regression approach. Cancer Res. 27: 209–220.

Màtheron, G., 1970. Application de la théorie des variables régionalisées. Cah. C.M.M. Fontainebleau 5, 212 pp.

Maurer, D., L. Watling, W. Leathem & P. Kinner, 1979. Seasonal changes in feeding types of estuarine invertebrates from Delaware Bay. J. exp. mar. Biol. Ecol. 36: 125–155.

Orloci, L., 1978. Multivariate analysis in vegetation research. Junk.

Pearson, T. H., G. Duncan & J. Nuttall, 1982. The Loch Eil project: population fluctuations in the macrobenthos. J. exp. mar. Biol. Ecol. 56: 305–325.

Pianka E. R., 1970. On r and k selection. Am. Nat. 104: 592–597.

Rees, E. I. S. & J. M. Walker, 1983. Annual and spatial variation in the *Abra* community in Liverpool Bay. Oceanol. Acta 6, suppl.: 165–169.

Sokal, R. R., 1979. Testing statistical significance of geographic variation patterns. Syst. Zool. 28: 227–232.

Souplet, A. & J. M. Dewarumez, 1980. Les peuplements benthiques du littoral de la région de Dunkerque. Cah. Biol. Mar. 21: 23–29.

Hydrobiologia 142: 187–197 (1986)

Suivi pluriannuel du peuplement à *Abra alba* dans la partie méridionale de la Mer du Nord (région de Dunkerque-France)

J. M. Dewarumez, C. Quisthoudt & A. Richard
Station Marine de Wimereux – B.P. 41 – F 62930 Wimereux

Keywords: *Abra* community, annual and pluriannual cycles, hydrologicals parameters, silting up, competitions

Abstract

The *Abra alba* community is considered as a faunistical unity well established on coastal areas of North Sea. This community shows, since several years, very important recruitments of certain species. Since 1978, recruitments of *Abra alba, Tellina fabula, Phyllodoce (Anaïtides) mucosa, Lagis koreni* or *Lanice conchilega* have been observed. However these demographic phenomena does not affect the structure of the community, which shows, through rank-frequency diagrams, the characteristic of a mature population, principally constituted of 7 species.

An annual cycle study shows a succession of recruitments of several species. A correspondance factorial analysis was realised from grain size datas as well as abundances of 18 main species. This analysis shows an annual cycle which seems to be linked to the clay part. Maximal values of clay part appears in summer and minimal values at the end of winter or early springtime.

The hydrological parameters studied, allowed to make the difference between external factors (continental contributions, phytoplanctonic activity etc...) and benthic activity.

Introduction

Le domaine benthique de la partie méridionale française de la Mer du Nord est constitué de cinq unités biosédimentaires majeures: le peuplement à cailloutis à épibiose sessile, le peuplement des graviers propres à *Amphioxus lanceolatus,* le peuplement des sables moyens propres à *Ophelia borealis,* le peuplement de l'hétérogène envasé et le peuplement des sables fin envasés à *Abra alba* (Souplet *et al.,* 1980). Le principal facteur régissant l'organisation et la distribution de ces peuplements est, sans nul doute, l'intensité des courants. Ceux-ci sont dus essentiellement au phénomène des marées. Le courant est amplifié au niveau du Pas de Calais (Cap Gris Nez et Cap Blanc Nez): il atteint dans ces zones 3 noeuds en vive eau moyenne. Cette intensité décroît ensuite vers l'est pour n'être que de 1,5 noeud à Dunkerque. Cette atténuation entraîne un granuloclassement et permet l'installation d'un gradient biosédimentaire dont l'aboutissement est le peuplement à *Abra alba.* Ce peuplement occupe sur la totalité du littoral une zone côtière dépassant rarement l'isobathe −10 m. Il fait l'objet d'une étude depuis 1973 (Dewarumez, 1976 et 1979; Souplet & Dewarumez, 1980).

Zone d'étude

Le peuplement à *Abra alba* est étudié en une station située entre Calais et Dunkerque, à 1,5 mille de l'extrémité du chenal de l'Aa (Fig. 1). Cette station a été déterminée en fonction des différentes études déjà citées. Elle représente les aspects les plus typiques du peuplement dans la zone considérée.

Ce peuplement est installé sur un sédiment constitué en majeure partie de sable fin: 94% pondéraux en moyenne des particules ont une taille comprise entre 0,1 et 0,5 mm. La fraction fine (inférieure à 0,1 mm) est néanmoins abondante:

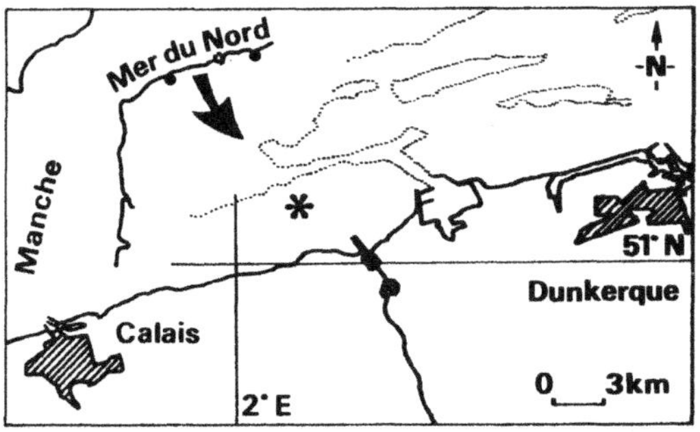

Fig. 1. Zone d'étude. * Station échantillonnée.

5% pondéraux en moyenne avec un taux de pélites pouvant atteindre 9,7%.

Depuis le début de nos observations (1974) la salinité de l'eau varie sur un cycle annuel entre 32 et 35$^{0}/_{00}$, la température de l'eau est minimale en février (4 °C) et maximale en septembre (19 °C).

Le milieu marin local reçoit de nombreux apports émanant du continent (Richard *et al.*, 1978). Ces apports sont amenés par des émissaires dans le port de Calais, des wateringues dans le port de Dunkerque, les eaux du petit fleuve l'Aa et de nombreux rejets industriels. Ces effluents enrichissent en particules en suspension les masses d'eaux déjà bien chargées après leur passage devant les estuaires picards et les grands ports de la Manche orientale. L'hydrodynamisme intense assure une dilution efficace mais il subsiste une zone littorale: véritable 'fleuve marin côtier' très chargé en matières en suspension sur une largeur pouvant atteindre 2 à 3 milles selon les conditions de vent (une étude sur la détermination de son extension réelle est en cours). Ce 'fleuve côtier' couvre le peuplement à *Abra alba* et l'alimente en particules fines: le taux de matières en suspension peut atteindre 50 mg/l au moment du maximum printanier.

Matériel et méthodes

Prélèvements hydrologiques

Les campagnes ont été réalisées à bord du N/O Sepia II. Les prélèvements d'eau ont été effectués à l'aide de bouteilles de type Niskin. Pour cette étude, seule la masse d'eau proche du fond a fait l'objet d'analyses multiparamétriques: température (thermomètre à renversement Richter & Wiese), salinité (salinomètre à induction Beckman), turbidité (mesure néphélométrique), taux de matières en suspension (MES). Les pigments chlorophylliens (Chla) sont déterminés par dosage spectrophotométrique suivant l'équation établie par Richards & Thomson (1952). Les teneurs en carbone et en azote particulaire (CTP, COP, NTP, NOP) sont obtenues avec un analyseur C.H.N. (Carlo Erba).

Prélèvements benthiques

Le prélèvement consiste en une série de prises de sédiment à l'aide d'une benne Smith-McIntyre d'une surface de 0,1 m^2. Dans le peuplement étudié 7 réplicats sont nécessaires pour appréhender correctement des données telles que la biomasse et la diversité spécifique (Dewarumez, 1979), pour des raisons de commodités de calcul 10 replicats sont effectués pour chaque prélèvement, la surface échantillonnée est donc de 1 m^2. Un onzième prélèvement est utilisé pour les analyses granulométriques.

Les prélèvements sont tamisés sur une maille circulaire de 1 mm de diamètre puis fixés au formol salé à 8%, et colorés au rose Bengal.

Nous avons utilisé les catégories granulométriques suivantes:

	taille des particules
Pélites	inférieures à 50 μm
Sablons	de 50 μm à 0,1 mm
Sable fin I	de 0,1 mm à 0,2 mm
Sable fin II	de 0,2 mm à 0,5 mm
Sable moyen	de 0,5 mm à 1 mm
Sable grossier	de 1 mm à 2 mm
Gravier	supérieures à 2 mm

Le traitement des données est effectué par la méthode de l'analyse factorielle des correspondances sur les données non transformées (densités des 18 espèces principales en nombre d'individus par mètre carré et proportions pondérales de l'échantillon granulométrique total). Ce type d'analyse permet de projeter sur un même plan les points variables et les points observations. Cette projection est, certes, un abus de présentation mais elle facilite l'interprétation en permettant de situer les observations par rapport à toutes les variables et les variables par rapport aux observations.

D'autre part, l'évolution et la structure du peuplement sont appréhendées à l'aide de diagrammes rang-fréquence (Frontier, 1976).

Résultats

Etude hydrologique

La courbe de variation annuelle de la température présente un maximum en été et un minimum en hiver. Les salinités sont généralement plus élevées en période estivale du fait de l'évaporation, des précipitations moins intenses et des apports telluriques plus restreints (Fig. 2).

Cette zone côtière est soumise à un brassage important; elle reçoit des apports continentaux non négligeables: les taux de matières en suspension y sont importants. L'observation microscopique d'un échantillon filtré sur acétate de cellulose (0,45 μm de porosité) montre la part importante du matériel détritique par rapport au matériel vivant. Les valeurs de MES et NTU sont bien corrélées (coefficient de corrélation: r = 0,92). Il est difficile d'établir le cycle annuel de la turbidité qui varie en fonction de facteurs externes affectant la masse d'eau littorale (vents, courants, apports terrestres...)

Ces valeurs ont notamment tendance à augmenter avec le coefficient de marée. Les plus fortes valeurs sont relevées en période hivernale (en décembre 84: MES = 87,26 $mg \cdot l^{-1}$).

L'analyse des deux profils (MES et NTU) met en évidence un premier pic au mois d'avril, correspondant à la reprise de l'activité biologique (Fig. 2). En été, les taux sont plus constants. Ils augmentent au début de la période hivernale. En effet, la fréquence des tempêtes diminue en été, tout comme le flux des apports continentaux (Fig. 3).

D'un point de vue de l'activité biologique, le premier bloom phytoplanctonique a lieu en avril alors que la température commence à augmenter (Chla, Fig. 2). Dès le mois de mai, les valeurs redeviennent faibles et stables jusqu'en août où apparaît la seconde poussée. Toutefois, des populations phytoplanctoniques subsistent tout au long de l'année, ce qui représente une source alimentaire disponible pour les communautés benthiques.

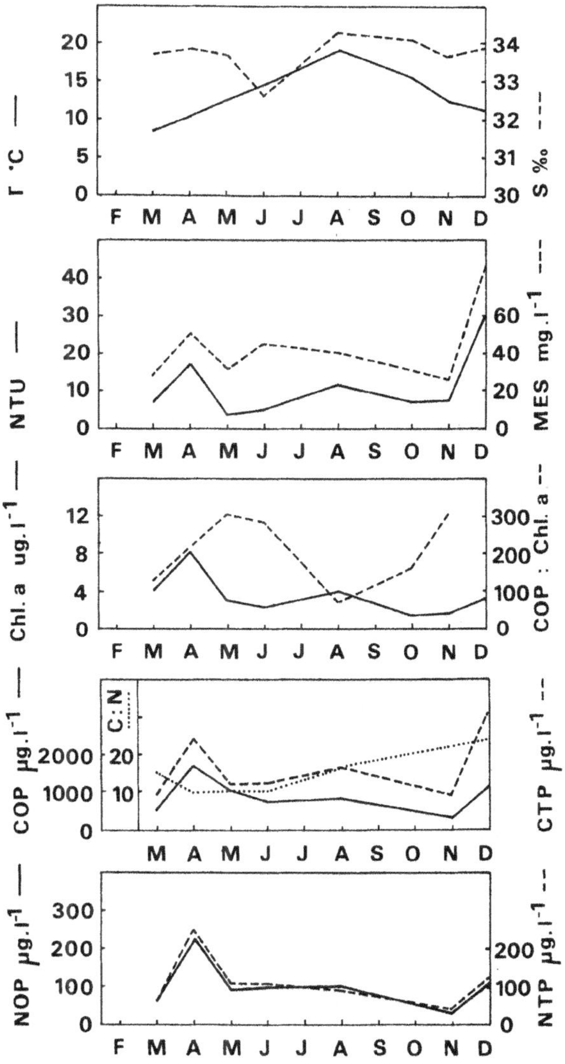

Fig. 2. Evolution des paramètres hydrologiques pendant l'année 1984.

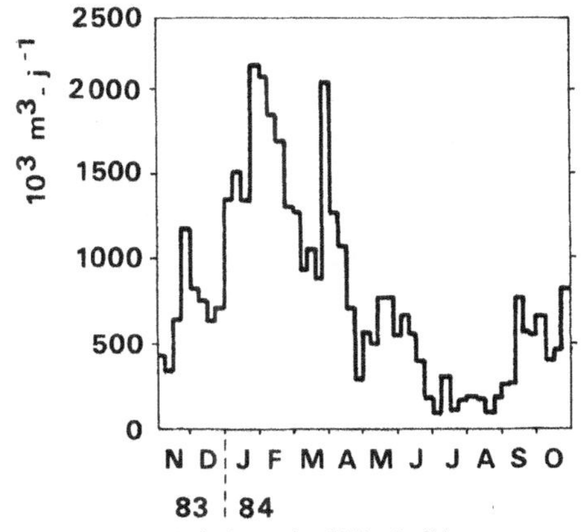

Fig. 3. Variations du débit de l'Aa.

Ce site côtier est fortement influencé par les apports telluriques (urbains, agricoles, industriels) notamment en période hivernale où les précipitations accrues font augmenter le débit des fleuves (Fig. 3). La fraction particulaire importante est alors constituée de détritus organiques (Fig. 2).

Les MES sont corrélées au carbone organique particulaire (r = 0,91). Les profils mettent en évidence un pic en avril confirmant la reprise de l'activité biologique.

Les rapports C/Chla, augmentent de mars à mai, alors que les teneurs chlorophylliennes diminuent; le COP serait d'origine autre que phytoplanctonique (apparition du zooplancton par exemple). Ce rapport diminue jusqu'en août moment auquel la deuxième poussée algale a lieu. Le COP serait alors d'origine chlorophyllienne, notons que le COP est bien corrélé pendant l'année à la Chla (r = 0,92). Puis ce rapport C/Chla reprend des valeurs élevées pendant la période hivernale. Par contre, les teneurs en azote particulaire sont plus constantes au cours de l'année, hormis une augmentation lors du premier bloom planctonique.

L'étude des variations du rapport C/N donne des indications quant à l'évolution de la matière organique au cours de l'année (Fleming, 1942 et Vinogradov, 1935 in Sverdrup *et al.*, 1970). La charge organique est très importante sur ce site côtier puisque les rapports C/N sont toujours supérieurs à 10 (Fig. 2) contrairement aux valeurs trouvées au large (Quisthoudt, 1983). Ainsi, les populations benthiques disposent d'une source trophique non négligeable. Nous observons un cycle annuel du rapport C/N: la période hivernale se caractérise par des valeurs élevées (Nov. 84: C/N = 22,15; Dec. 84: C/N = 24,29). Puis les taux diminuent brusquement en avril (C/N = 10,04) lors du développement des populations planctoniques. Il reste voisin de 10 jusqu'en juin (Fig. 2) pour réaugmenter progressivement. Ceci correspond à la succession des différentes populations phytoplanctoniques, zooplanctoniques et bactériennes. Cette dernière constitue une source alimentaire dont il faut signaler l'importance.

Evolution pluriannuelle du peuplement à Abra alba

Les données enregistrées depuis 1978 ont fait l'objet d'une analyse factorielle des correspondances. Cette analyse porte sur les effectifs des 18 espèces les plus abondantes du peuplement. Ces espèces représentent plus de 97% des individus récoltés. Les prélèvements considérés sont des prélèvements printaniers (fin mai-début juin) donc antérieurs au recrutement des espèces prises en compte. Les résultats rendent ainsi compte de l'efficacité *a posteriori* des recrutements de l'année précédente.

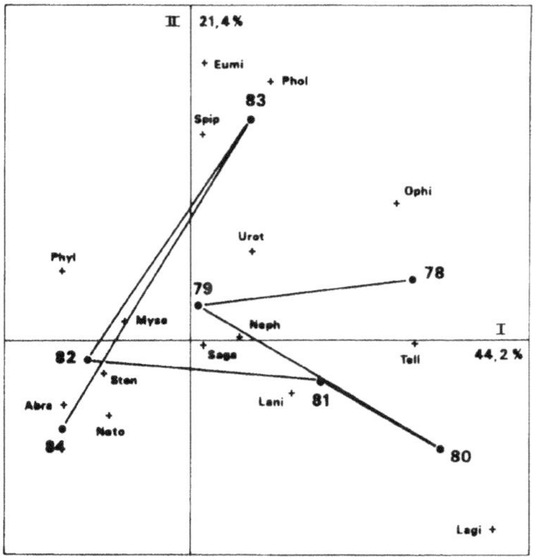

Fig. 4. Evolution du peuplement à *Abra alba* entre 1978 et 1984 (Analyse factorielle des correspondances) (Légendes Tableau 1).

Le plan des axes I et II de l'analyse factorielle extrait 65,6% de l'inertie totale (Fig. 4). Ce plan ne fait pas apparaître, en ce qui concerne l'évolution des cortèges spécifiques, de phénomène cyclique mais plutôt aléatoire. Les différents points-observations (années) semblent 'tirés' par des espèces qui ont bénéficié l'année précédente de recrutements efficaces. C'est le cas d'*Ophiura texturata* en 1977, de *Tellina fabula* en 1977 et 1979, de *Lagis koreni* en 1979, de *Phyllodoce mucosa* en 1981, *Eumida sanguinea, Pholoë minuta* et *Spiophanes bombyx* en 1982 et *Abra alba* en 1981 et 1983. Deux espèces ne semblent pas intervenir dans l'évolution du peuplement, il s'agit de *Sagartia troglodytes* et de *Nephtys hombergii.* Ces espèces sont caractérisées par une grande stabilité de leurs effectifs. Il faut noter que l'on observe depuis 1980 une dérive des saturations selon l'axe I vers des valeurs plus faibles. L'année 1984 se trouve à l'opposé de l'année 1978.

Les diagrammes rang-fréquence des prélèvements

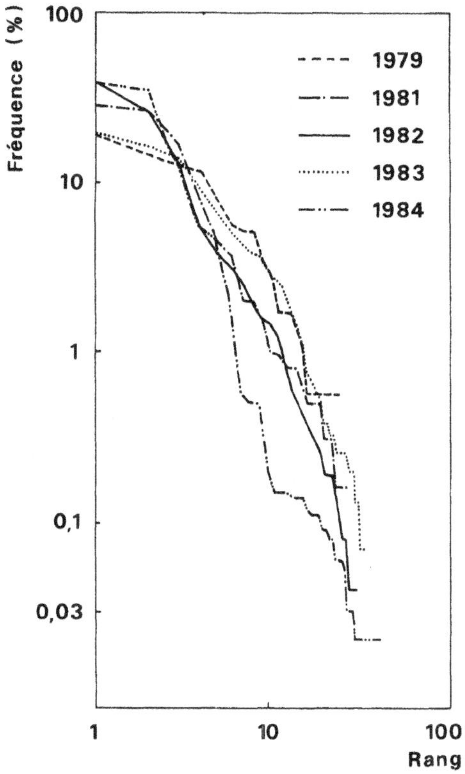

Fig. 5. Evolution des diagrammes rang-fréquence depuis 1979.

considérés font apparaître une certaine maturité du peuplement (Fig. 5) sauf en ce qui concerne l'année 1984 où le diagramme d'allure sigmoïde est caractéristique d'un peuplement juvénile. Notons toutefois que le point d'inflexion est situé au niveau de la dixième espèce (*Phyllodoce mucosa*: 0,2% de l'effectif total). Tout se passe donc comme si le peuplement avait une faculté de réponse très rapide à toute 'perturbation' matérialisée par des changements de l'importance relative des espèces principales.

Evolution de la granulométrie entre 1978–79 et 1983–84

L'analyse factorielle porte sur des analyses granulométriques réalisées sur deux cycles annuels (1978–79 et 1983–84). Le plan factoriel I–II (84% de l'inertie) fait apparaître une dérive entre les deux périodes d'étude vers des valeurs plus élevées selon l'axe I (Fig. 6). Les valeurs les plus faibles

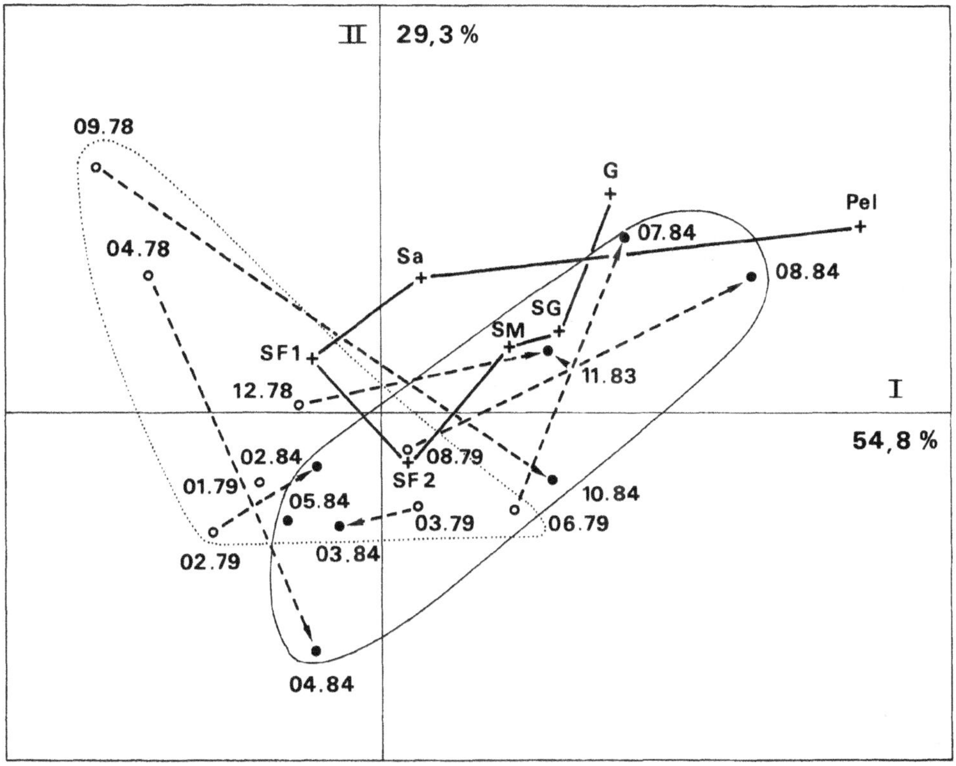

Fig. 6. Evolution de la granulométrie du sédiment entre deux cycles annuels: 1978–79 et 1983–84 (analyse factorielle des correspondances). (Légendes Tableau 1).

en 1984 correspondent aux valeurs les plus fortes de 1978. Les points-observations correspondants à des prélèvements effectués au même moment de l'année montrent un déplacement analogue sauf en ce qui concerne les prélèvements de mars 1979 et 1984. Les faibles saturations (respectivement +34 et −43) de ces points-observations rendent l'écart peu significatif. L'axe I semble étroitement lié à la teneur en particules fines. L'analyse factorielle met donc en évidence un enrichissement en pélites entre les deux périodes étudiées.

Etude biosédimentaire d'une période annuelle

L'étude porte sur l'évolution de la granulométrie du sédiment et des cortèges spécifiques pendant la période novembre 1983 – octobre 1984. Le plan factoriel I – II de l'analyse de ces résultats (66,8% de l'inertie) peut être décrit par la granulométrie. L'axe I est déterminé par les particules fines (jusqu'à 0,5 mm) tandis que l'axe II est lié à la fraction grossière (Fig. 7). Il est possible de mettre en évidence un cycle annuel centré sur certaines espèces connues pour être les plus contributives à l'organisation bionomique du peuplement. Il s'agit par ordre de poids décroissant d'*Abra alba, Tellina fabula, Mysella bidentata, Nephtys hombergii, Spisula subtruncata, Pholoë minuta* et *Sagartia troglodytes.* Autour de la surface déterminée par le cycle annuel on retrouve les espèces qui deviennent prépondérantes lors de certaines périodes du cycle, au moment de leur recrutement. Ce recrutement n'est

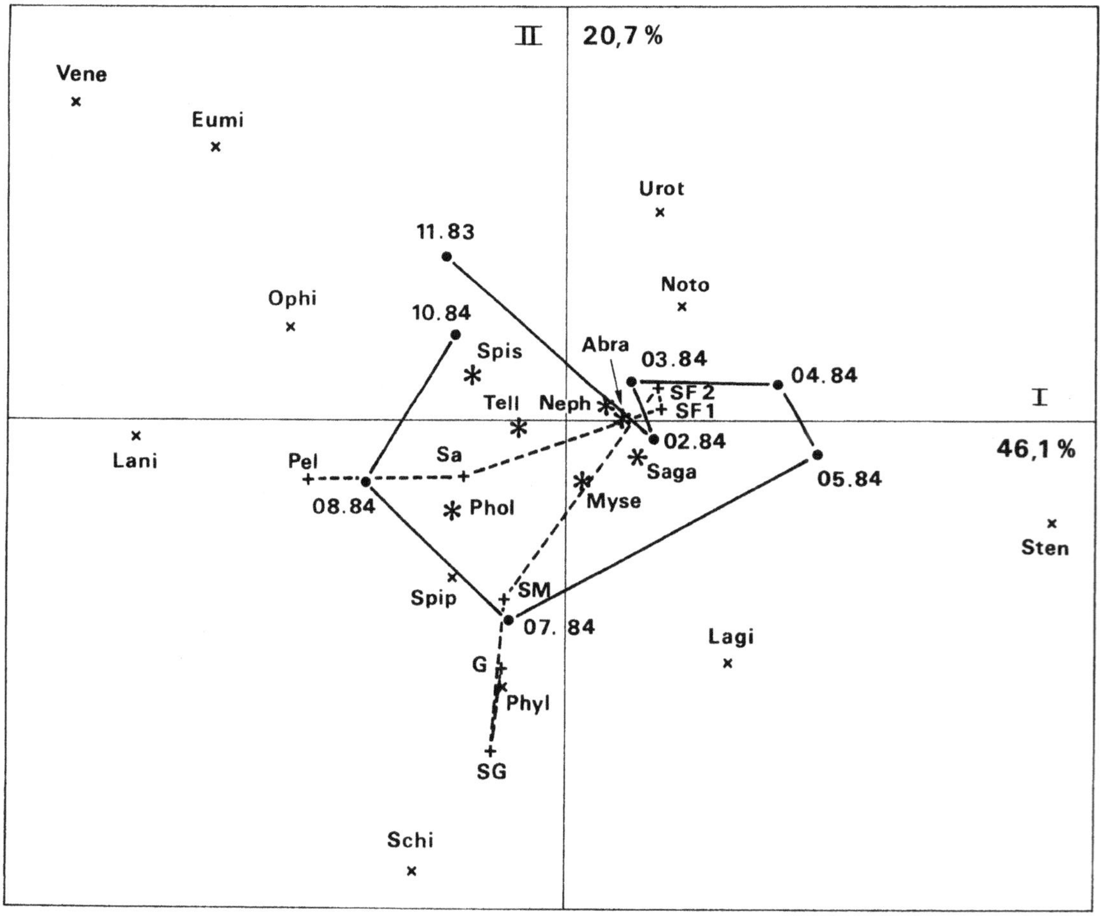

Fig. 7. Etude d'une période annuelle. Analyse factorielle des correspondances portant sur les espèces principales et la granulométrie. (Légendes Tableau 1).

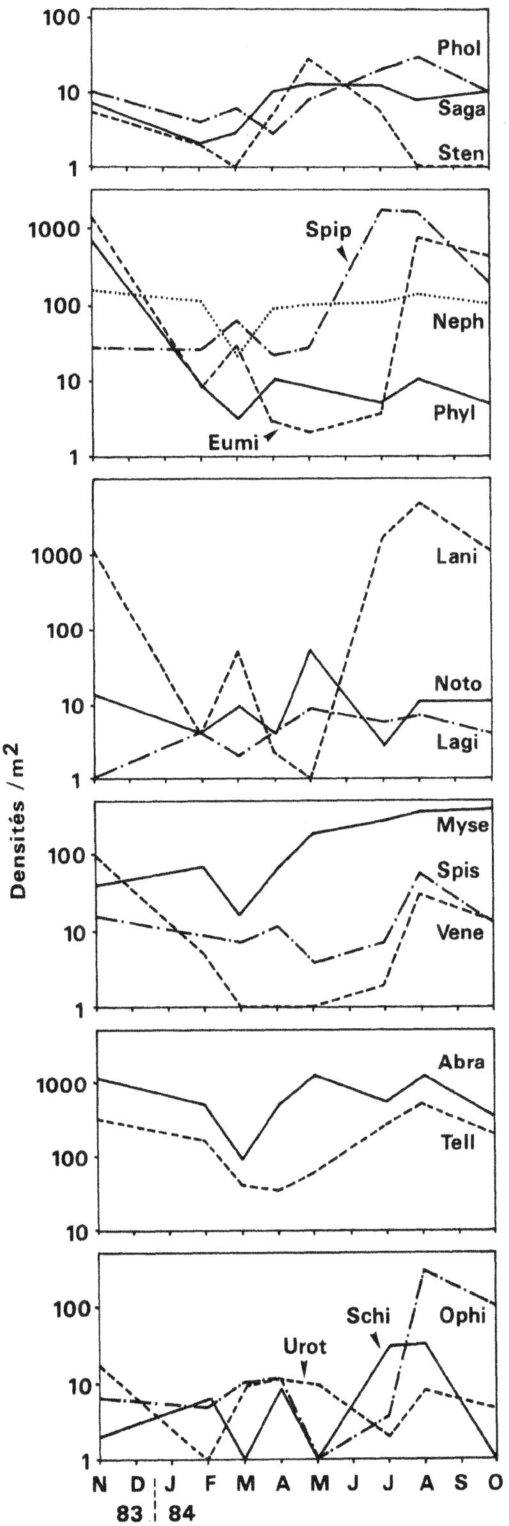

Fig. 8. Evolution des densités des 18 espèces principales pendant un cycle annuel. (Légendes Tableau 1).

pas toujours suivi d'un maintien à un haut niveau de la population, on note en général une diminution considérable de la densité de ces espèces peu de temps après (Fig. 8). Ainsi la densité de population d'*Eumida sanguinea* chute de 41% en 2 mois; *Spiophanes bombyx:* −89% en 3 mois; *Lanice conchilega:* −74% en 2 mois, *Ophiura texturata:* −63% en 2 mois.

Etude des populations d'Abra alba et de Tellina fabula

Abra alba est connue pour être une espèce opportuniste et pour présenter des variations de densités très importantes. Nos résultats confirment ces caractéristiques (Fig. 9). Les densités évoluent entre 15 et 4957 individus/m². Le recrutement a lieu au début de l'été (Dewarumez, 1979) et il est possible de mettre en évidence des augmentations brutales de la densité en dehors des périodes de fixations des larves (Décembre 1978, Novembre 1980, Janvier-Février 1982, Janvier 1984). La population s'est maintenue à un niveau élevé de densité en 1982 et au 1984: 900 individus/m². Les résultats partiels en 1985 montrent une chute considérable de la population après l'hiver 1984–85. Les densités printanières sont de l'ordre de 50 individus/m² en 1985.

Il est possible de mettre en évidence trois périodes dans le suivi démographique de *Tellina fabula* (Fig. 9):

– en 1976 et 1978–1979 la population se maintient à un niveau de 300 individus/m² en moyenne.

– en été 1979, après une mortalité printanière très importante, le recrutement estival fut excellent (623 individus/m² en juillet). La population est

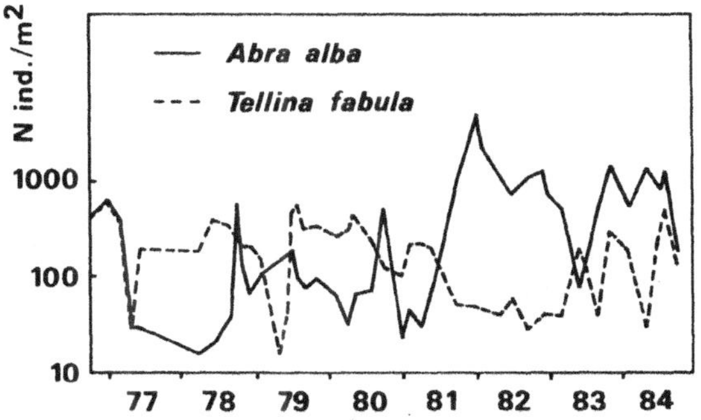

Fig. 9. Variation des densités des populations d'*Abra alba* et de *Tellina fabula.*

ensuite entrée dans une phase de déclin jusqu'au printemps 1983 (39 individus/m^2).

– depuis juillet 1983, les variations de densité inexpliquées font penser que *Tellina fabula* adopte une stratégie démographique identique à celle d'*Abra alba.*

Discussion et conclusion

Il apparaît de prime abord que le peuplement à *Abra alba* de la partie méridionale de la Mer du Nord est semblable à ce qui a pu être décrit par ailleurs en Manche ou dans les mers européennes. On peut sans conteste le rattacher par sa composition faunistique à la '*Syndosmya (Abra) alba* community' de Petersen (1914) et Thorson (1957). Ce peuplement est bien connu pour les modifications de sa structure bionomique (Gentil, 1976; Dewarumez, 1979; Arntz, 1981...). Dans le cas de la région étudiée, des espèces telles que *Lanice conchilega, Lagis koreni, Abra alba, Eumida sanguinea, Tellina fabula* et *Phyllodoce mucosa* peuvent prendre tour à tour des importances considérables. Ce sont surtout des espèces à stratégie démographique 'stratégie-r' au sens de Barbault (1976), (durée de vie courte, forte croissance, vie larvaire longue et à une seule période de reproduction), soit, en fait, des espèces colonisatrices.

En dépit de ces évolutions imprévisibles, la structure du peuplement est toujours celle d'un peuplement mature, les espèces semblant 'interchangeables' malgré un rôle fonctionnel différent. La domination d'une espèce se maintient pendant la durée de son cycle, bien que sa densité puisse chuter parfois d'une manière importante. La biomasse en poids sec décalcifié reste toujours très forte: 20 à 50 g/m^2.

L'étude du cycle annuel montre que la richesse du sédiment en pélites n'est pas constante. Les teneurs sont beaucoup plus importantes en période estivale (Fig. 10). Ceci correspond à un moment où les conditions hydrodynamiques sont les plus faibles et les conditions météorologiques les plus calmes. Il en résulte un brassage moins intense des eaux qui permet la sédimentation des MES. Les masses d'eaux locales sont caractérisées par une turbidité importante comme le montrent les valeurs du COP. La sédimentation de matière organique procure une ressource alimentaire capitale pour les peuplements

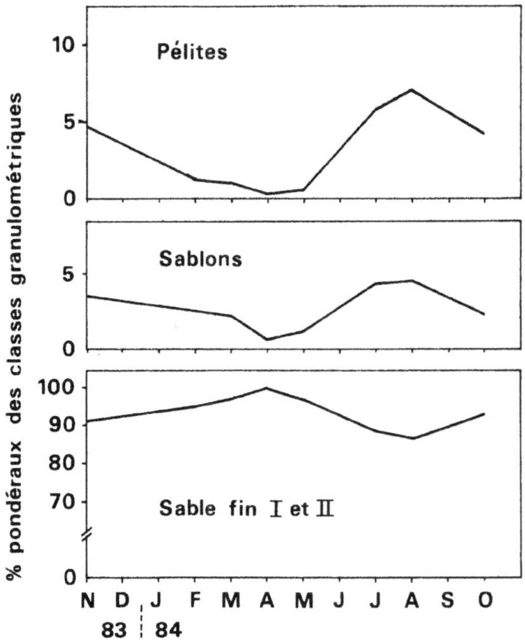

Fig. 10. Evolution de la fraction fine du sédiment.

établis sur des fonds côtiers au moment où le recrutement des espèces les plus contributives a lieu. La matière organique provient en période estivale des populations phytoplanctoniques (voir taux de la chla), zooplanctoniques et bactériennes comme le montre l'évolution du rapport C/N. Le moment de reprise de l'activité planctonique automnale correspond d'ailleurs avec la diminution de la teneur en fines du sédiment, malgré l'accroissement du débit des fleuves (Fig. 3). Le brassage des eaux plus intense favorise donc, dans un premier temps, les populations phyto et zooplanctoniques. La forte charge hivernale en MES est surtout due à de la matière minérale; la matière organique est représentée par des détritus de grande taille et le maillon bactérien représente à ce moment-là la source la plus importante de matière organique vivante pour le benthos. Cette production, pour importante qu'elle soit, n'alimente que faiblement les fonds en raison de l'hydrodynamisme. Cette diminution des apports nutritifs correspond d'ailleurs à un moment où les densités d'espèces benthiques sont les plus faibles et où manquent par exemple la plupart des Amphipodes et des Mysidacés à tendance benthique.

S'il est possible de mettre en évidence des varia-

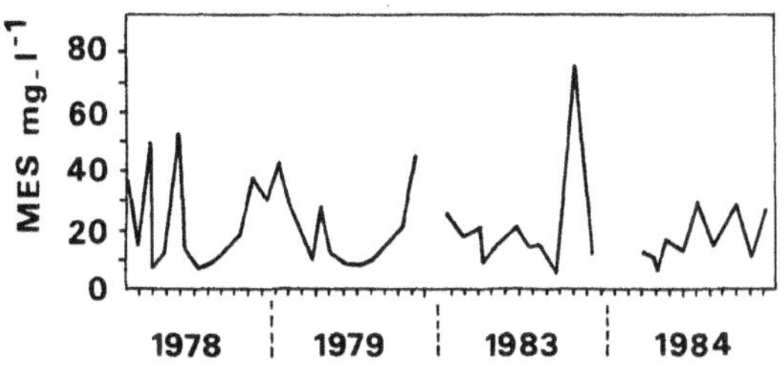

Fig. 11. Evolution des matières en suspension.

tions des teneurs en fines du sédiment, des phénomènes d'envasement à long terme sont également possibles. De telles tendances ont déjà été décrites dans la région (Souplet & Dewarumez, 1980). L'envasement mis en évidence par l'étude de l'évolution des granulométries entre 1978 et 1984 ne peut être imputé à une évolution des taux de matières en suspension dans la même période et à une sédimentation plus intense. En effet les teneurs estivales en MES en 1978 et 1979 sont en moyenne deux fois moindres qu'en 1983 et 1984 (Fig. 11). Cet envasement correspond certainement à une évolution générale à long terme des conditions sédimentaires.

Le comportement erratique du peuplement à *Abra alba* tel qu'il est démontré par l'analyse factorielle sur les espèces entre 1978 et 1984 ne semble pas être dû à des processus évolutifs du même type que ceux que décrit Hily (1983) en rade de Brest. Cet auteur met en évidence plusieurs phases correspondant à des réarrangements structuraux du peuplement par association d'espèces de groupes trophiques différents après l'établissement d'un peuplement pionnier constitué d'un faible nombre d'espèces. En accord avec les observations de Arntz et Rumohr (1982), on ne peut pas mettre en évidence de changements clairs au niveau de la succession d'espèces, envisagée selon leur mode alimentaire (Tableau 2). Dans le cas du peuplement étudié, la dynamique évolutive semble davantage être liée à la concordance entre certains facteurs édaphiques (teneur en fines des fonds par exemple) et la quantité de post-larves, prêtes à se fixer, disponibles sur place. Ce type de phénomène s'est d'ailleurs produit en 1985 dans le même secteur au sein du peuplement à *Ophelia borealis* (Dewarumez, inédit). Une forte population (1 000 ind/m^2) de *Magelona papillicornis* (espèce vasicole) s'est installée transitoirement à la faveur de rejets de vase (550000 m^3) au moment où les larves de cette espèce sont aptes à se fixer.

La population d'*Abra alba* intervient pour beaucoup dans les modifications bionomiques du peuplement. L'opportunisme de cette espèce a souvent

Tableau 1. Abréviations utilisées.

Abra	=	*Abra alba*	Eumi	=	*Eumida sanguinea*	Lagi	=	*Lagis koreni*
Lani	=	*Lanice conchilega*	Myse	=	*Mysella bidentata*	Neph	=	*Nephtys hombergii*
Noto	=	*Notomastus latericeus*	Ophi	=	*Ophiura texturata*	Phol	=	*Pholoë minuta*
Phyl	=	*Phyllodoce mucosa*	Saga	=	*Sagartia troglodytes*	Schi	=	*Schistomysis spiritus*
Spip	=	*Spiophanes bombyx*	Spis	=	*Spisula subtruncata*	Sten	=	*Stenelais boa*
Tell	=	*Tellina fabula*	Urot	=	*Urothoe poseïdonis*	Vene	=	*Venerupis pullastra*
Pel	=	Pélites	Sa	=	Sablons	SF	=	Sable fin
SM	=	Sable moyen	SG	=	Sable grossier	G	=	Gravier

Tableau 2. Répartition des espèces principales du peuplement à Abra selon leur mode alimentaire.
[(Etablie selon les travaux de Blegvad (1914); Fauchald & Jumars (1979); Arntz (1981); Hily (1984) et nos observations)].

Carnivores vagiles et sessiles	Suspensivores	Déposivores de surface	Déposivores de subsurface
Sagartia troglodytes	*Lanice conchilega*	*Spiophanes bombyx*	*Notomastus latericeus*
Pholoe minuta		*Mysella bidentata*	*Lagis koreni*
Stenelais boa		*Abra alba*	
Phyllodoce mucosa		*Tellina fabula*	
Eumida sanguinea		*Urothoë poseïdonis*	
Nephtys hombergii			
Ophiura texturata			

été signalée (Eagle, 1975; Rachor, 1980 et 1983...). Les variations de densités de cette espèce peuvent être causées par des facteurs édaphiques comme l'instabilité du sédiment (Rumohr *et al.*, 1982), par la pollution, par la prédation par des poissons (Arntz & Brunswick, 1975; Dewarumez *et al.*, 1976; Arntz, 1981; Delval & Desmarchelier, 1982). Les variations de densité de la population d'*Abra alba* sont dues, en fait, à une action de tous ces facteurs combinés. Il est aussi possible d'évoquer la répartition 'en taches' de cette espèce, et les problèmes d'échantillonnage, voire, peut être, des problèmes de migrations passives, les fortes densités notées en dehors des périodes de recrutement étant à cet égard tout à fait significatives.

Il se pose dans ce peuplement des problèmes de compétition au niveau de l'exploitation de la surface disponible. L'implantation de *Lanice conchilega* (jusqu'à 60000 postlarves/m^2 à la mi-mai – estimation par biovolumes) et le maintien temporaire de la population peut empêcher la fixation d'autres espèces dont le recrutement est plus tardif comme *Abra alba* et *Tellina fabula* (juin–juillet). Hily et Le Bris (1984) signalent un phénomène analogue en rade de Brest avec *Melinna palmata.*

La compétition interspécifique peut aussi être envisagée, notamment entre *Abra alba* et *Tellina fabula.* En effet certaines périodes de maximum de densité de l'une correspondant à un minimum de l'autre (Fig. 9). Ce problème de compétition ne doit pas se régler entre deux espèces proches mais au sein de la totalité du cortège spécifique en tenant compte du régime alimentaire de toutes les espèces. Les deux espèces les plus constantes du peuplement *Sagartia troglodytes* et *Nephtys hombergii* sont carnivores et sont donc, par nature, moins dépendantes des particules fines, déposées ou non. Pour ce qui est de *Nephtys hombergii,* le mode alimentaire est sujet à controverse mais en accord avec Retière (1979), nous estimons que le régime carnivore de cette espèce, même s'il n'est pas exclusif ne peut pas être mis en doute. *Nephtys hombergii* participe également aux processus de bioturbation du sédiment. Cet annélide chasseur creuse dans le sédiment des galeries très profondes et ramifiées (Retière, 1979) et sa population se trouve souvent sous échantillonnée, notamment en ce qui concerne les individus âgés.

Il paraît nécessaire de compléter ces données de bionomie benthique par une étude fine du devenir des particules en suspension et de la dissémination des populations de larves, corrélee à une étude courantologique adéquate. Le peuplement à *Abra alba,* très productif, qui caractérise l'ensemble des littoraux des côtes européennes, s'avère être un modèle d'écosystème très intéressant à étudier pour préciser le déterminisme du recrutement.

Remerciements

Ce travail a bénéficié du concours financier de l'Institut Français de Recherche pour l'Exploitation de la Mer.

Il a été réalisé grâce à l'aide technique de N. Degros, M. A. Janquin, D. Menu, et M. Pannier.

Références

Arntz, W. E., 1981. Zonation and dynamics of macrobenthos biomass in an area stressed by oxygen efficiency. In: G. W. Barret and R. Rosenberg (eds), Stress effects on natural ecosystems. John Wiley & Sons Ltd.: 215–225.

Arntz, W. E. & D. Brunswick, 1975. Studies on structure and dynamics of macrobenthos in the Western Baltic. Proceedings 10th European Marine Biology Symposium. Population dynamics 2: 17–42.

Arntz, W. E. & H. Ruhmor, 1982. An experimental study of macrobenthic colonization and succession, and the importance of seasonal variation in temperate latitudes. J. exp. Mar. Biol. Ecol. 64: 17–45.

Barbault, R., 1976. La notion de stratégie démographique en écologie. Bull. Ecol. 7: 373–390.

Blegvad, H., 1914. Food and conditions of nourishment among the communities of invertebrate animals found on or in the sea bottom in Danish waters. Rep. danish. biol. Stat. 24: 17–22.

Delval, C. & M. Desmarchelier, 1982. Biologie du flet (*Platichtys flesus* L.). Etude des chalutages effectués sur le littoral Nord/Pas-de-Calais. D.E.A. Université de Lille: 121 p.

Dewarumez, J. M., 1979. Etude biologique d'*Abra alba* (WOOD). Mollusque lamellibranche du littoral de la Mer du Nord. Thèse. Université de Lille: 139 p.

Dewarumez, J. M., F. Smigielski & A. Richard, (1976) 1978. *Abra alba.* (Mollusque lamellibranche) sa localisation en zone littorale de la mer du Nord. Haliotis 7: 13–19.

Eagle, R. A., 1975. Natural fluctuation in a soft bottom benthic community. J. mar. biol. Ass. U.K. 55: 867–878.

Fauchald, K. & P. Jumars, 1979. The diet of worms: a study of polychaete feeding guilds. Oceanogr. Mar. Biol. Ann. Rev. 17: 193–284.

Frontier, S., 1976. Utilisation des diagrammes rang-fréquence dans l'analyse des écosystèmes. J. Rech. oceanogr. 1: 35–48.

Gentil, F., 1976. Distribution des peuplements benthiques en Baie de Seine. Thèse. Université de Paris: 70 p.

Hily, C., 1983. Modifications de la structure écologique d'un peuplement à *Melinna palmata* (Annélide – Polychète) soumis aux effluents urbains et industriels en rade de Brest. Ann. Inst. océanogr. 59: 37–56.

Hily, C., 1984. Variabilité de la macrofaune benthique dans les milieux hypertrophiques de la rade de Brest. Thèse. Université de Brest: 696 p.

Hily, C. & H. Le Bris, 1984. Dynamics of an *Abra alba* population (Bivalve – Scrobiculariidae) in the Bay of Brest. Estuar. Coast. Shelf Sci. 19: 463–475.

Petersen, J., 1913. Valuation of the sea. II. The animal communities of the sea-bottom and their importance for marine zoogeography. Rep. danish. biol. Stat. 21: 44 p.

Quisthoudt, C., 1983. Hydrologie du détroit du Pas-de-Calais: Carbone – Azote – Chlorophylle. D.E.A. Université de Lille: 38 p.

Rachor, E., 1980. The inner German Bight. An ecological sensitive area as indicated by the bottom fauna. Helgoländer wiss. Meeresunters. 33: 522–530.

Rachor, E., 1983. Long term variability of *Abra alba* populations in several areas of the European coast. COST 47 Newslett 3: 6–7.

Retière, C., 1979. Contribution à la connaissance des peuplements benthiques du Golfe Normanno-Breton. Thèse. Université de Rennes: 421 p.

Richard, A., A. Souplet, J. M. Dewarumez & B. Vandorpe, 1978. Le littoral de la Mer du Nord: les problèmes d'environment liés à l'industrialisation et aux aménagements portuaires. In: CNRS (ed.), Villes et Ports, 587: 489–499.

Richards, F. A. & T. G. Thompson, 1952. The estimation and characterization of plancton populations by pigment analysis, II. A spectrophotometric method for the estimation of plancton pigments. J. mar. Res. 11: 156–171.

Souplet, A. & J. M. Dewarumez, 1980. Les peuplements benthiques du littoral de la région de Dunkerque. Cah. Biol. Mar. 21: 23–39.

Souplet, A., R. Glaçon, J. M. Dewarumez & F. Smigielski, 1980. Distribution des peuplement benthiques littoraux en Mer du Nord du Cap Blanc-Nez à la frontière de Belgique. C. r. Acad. Sci., Paris 290: 627–630.

Sverdrup, H. U., M. W. Johnson & R. H. Fleming, 1970. The Oceans. Prentice Hall Englewood Cliffs, N.J. 1087 p.

Thorson, G., 1957. Bottom communities (sublittoral and shallow shelf). In: J. W. Hedgpeth (ed.), Treatise on marine ecology and paleoecology. Geol. Soc. Amer. Mem. 67: 461–534.

Hydrobiologia 142: 199–207 (1986)

Common patterns in the fluctuations of macrozoobenthic species living at different places on tidal flats in the Wadden Sea

J. J. Beukema & K. Essink
Netherlands Institute for Sea Research, P.O. Box 59, 1790 AB Den Burg, Texel, The Netherlands & Rijkswaterstaat, Tidal Waters Division, Hereweg 99a, 9721 AA Groningen, The Netherlands

Keywords: macrobenthos, dynamics, parallel fluctuations, temperature, eutrophication

Abstract

During 17 successive years (1969–1985) the macrozoobenthos has been sampled quantitatively in the late-winter/early-spring period at 15 stations scattered over Balgzand (a 50 km^2 tidal flat area in the westernmost part of the Wadden Sea) and at 5 stations located in a small (3 km^2) area 150 km away from Balgzand in the eastern part of the Dutch Wadden Sea. In 25 species, numbers per m^2 were, in most years, sufficiently high at 2 or more of the Balgzand stations to allow between-station comparisons of fluctuation patterns. Comparisons were made by rank correlation. Out of a total of 1003 of such comparisons that could be made with the Balgzand data, 47% yielded significantly positive correlations and less than 1% significantly negative ones. Thus, nearly half of the fluctuation patterns of the populations living at the 15 stations within the Balgzand area showed a high similarity. Synchronization of population fluctuations was augmented particularly by the incidence of severe winters (causing low spring numbers in about a quarter of the species that were sensitive to low temperatures and high reproductive success in several species during the subsequent summer) and further by the increasing trends in numbers in about half of the species, probably as a consequence of eutrophication.

Similar results as on Balgzand were obtained within the restricted area Groninger Wad in the eastern part of the Dutch Wadden Sea (Essink & Beukema, this issue). Comparison of the fluctuation patterns between the 2 distant areas also yielded high numbers of significantly positive correlations, though the proportion of the patterns that were similar was lower than these proportions were within the 2 areas.

It is concluded that common patterns of fluctuation in numerical densities of macrobenthic species can be assessed over vast areas. Such common patterns will represent the 'normal' or 'base-line' fluctuations that may be used to distinguish (as departures from such patterns) the effects of local disturbing influences.

Introduction

When a student of population dynamics observes a striking change in numbers in one or more of the species living in his area of observation, he will be inclined to inquire of his colleagues working in adjacent areas whether or not a similar and simultaneous change took place there. Such an inquiry makes sense. If populations are found to fluctuate with similar (synchronized, parallel) patterns over vast areas, their numerical changes will be governed largely by factors that are effective at a large geographic scale, like broad climatic changes. When such parallel fluctuations do exist over large geographic areas, the observation of any local departure from the common pattern will, on the other hand, point to the influence of a strictly locally operating factor, like a local source of pollution. Thus, it is of some importance to demonstrate the existence of similar fluctuation patterns in numbers of animal populations over large areas. If such patterns did not generally exist, a local departure could not be identified and the attribution of a local change in numbers to a specific local cause would be more difficult.

Data are available from a 17-year sampling

programme in a 50 km^2 tidal flat area with 15 sampling stations (with distances between the stations varying from 1/2 to 10 km) to study in detail in a restricted area the degree of similarity in population fluctuation patterns. A similar data set is available from a small tidal flat area located 150 km away from the above area (Essink & Beukema, this issue). Intercomparison of the fluctuation patterns in the 2 areas shows the degree of their similarity over a longer distance.

Material and methods

The 15 sampling stations at Balgzand are scattered over a 50 km^2 tidal flat area in the westernmost part of the Wadden Sea (see Fig. 1 in Beukema, 1974; Beukema *et al.*, 1983). Among these 15 sampling stations, 3 are square plots of 900 m^2 each and 12 are transects of 1 km each. All stations are marked by iron poles.

These 15 stations have been sampled in a uniform way at least annually since 1969. Details of the sampling procedure can be found in earlier papers (Beukema, 1974, 1979). In short: 9 to 16 cores of 0.1 m^2 (randomly scattered within the squares) or 50 cores of 0.018 m^2 each (at regular distances along the transects) were taken and sieved in the field (1 mm mesh). All samples were sorted in the laboratory while the animals were still alive. Numbers are expressed per m^2 and these are close to the actual numbers found. Only data from samples taken in the late-winter/early-spring periods will be used in the following.

The 15 sampling stations cover nearly the full tidal range. This range is relatively small in the westernmost part of the Wadden Sea, with MWH at +58 and MLW at −78 cm from MTL. Tidal flats at MTL are drained for an average of 5½ h per cycle of 12½ h, but emersion is shortened at westerly winds and prolonged at easterly winds. The 15 sampling stations also cover nearly the whole range of sediments from almost pure fine sands (silt content below 1% and median grain size about 190 μm) at the most exposed stations to silty sands and muds (about 30% silt and a median grain size of about 90 μm) at the most sheltered stations.

Long-term monthly means of water temperatures in the westermost tidal inlet range from 3 °C for the coldest to 18 °C for the warmest month (Van der Hoeven, 1982). For possibly temperature-governed changes in numbers, the deviations from the mean are more relevant. Winters in the Wadden Sea area differ widely in character. In most winters, mild westerly winds prevail, air temperature never drops below −10 °C and only during less than 10 days is the air temperature below 0 °C the whole day in the westernmost part of the Wadden Sea (from monthly weather reports issued by KNMI, De Bilt, The Netherlands). During the 17-year period of observation, only 2 winters (viz. 1978/1979 and 1984/1985) were really cold with records below −10 °C on more than 5 days and with more than 20 days showing freezing temperatures all day (see also Fig. 1).

In the eastern part of the Dutch Wadden Sea (Groninger Wad), 5 square sampling plots were sampled in nearly the same way during the same period. These plots are located at short distances apart (1 to 2 km) in a restricted area that covers only part of the tidal range and sediments. In this area the outfall of a pipe-line for industrial wastewater is situated (Essink, 1978; Essink & Beukema, this issue). Though the climate in the eastern part of the Dutch Wadden Sea is slightly more continental (with more extreme temperatures), the same winters were classified as either mild or severe.

To compare the fluctuation patterns of the populations in the various species at different sampling stations, the actual numbers observed were transformed to rank numbers (1 to 17 for each of the years of observation). If, in more than half of the years, a species was absent at any station, the data for this station were not used in this species. The thus obtained rank numbers were compared for all possible pairs of stations (minimally 1 pair when the species was sufficiently numerous at 2 stations only, and maximally 105 when the species was numerous at all 15 stations at Balgzand) by calculating Spearman's *r*. This statistic was evaluated by Spearman's rank correlation test (with n = 17, i.e. the number of years of observation).

Results and discussion

a. Similarity of fluctuation patterns

In 25 species numbers were, in most years, sufficiently numerous at 2 or more Balgzand stations to allow at least 1 comparison of fluctuation patterns between stations. In most species many more comparisons were possible, up to 105 in 1 species (*Macoma balthica*), which was numerous in nearly all years at all 15 stations, yielding $1/2 \cdot 15 \cdot (15-1) = 105$ comparisons (Table 1). In this way a total of 1003 comparisons was available for comparison. A positive correlation was found in 897 comparisons, 473 of which were statistically significant at the 0.05 level and 273 at the 0.01 level (Table 1). Thus, nearly half (47%) of the comparisons showed a highly similar pattern of fluctuation at a station pair within the Balgzand area, with a value of Spearman's *r* exceeding +0.49.

Negative correlations were far less common (106 out of the 1003 comparisons) and were rarely significant (less than 1%). Moreover, 8 out of the 9 significantly negative correlations were confined to 1 species only (viz. *Arenicola marina*, see below). Though positive and negative correlations were distributed unevenly over the species, positive correlations prevailed in all species (Table 1). In half of the species all correlations observed were positive ones.

Particularly high proportions of significantly positive correlations (i.e. more than 50% better than $p < 0.05$ or more than 25% better than $p < 0.01$) were found in 10 species, viz. in the bivalves *Macoma balthica, Angulus tenuis* and *Abra tenuis*, in the polychaetes *Nephtys hombergii, Lanice conchilega, Scoloplos armiger, Nereis diversicolor, Heteromastus filiformis* and *Eteone longa*

Table 1. Numbers and proportions (in %) of similar (positive correlations) and opposite (negative correlations) fluctuation patterns in populations of macrozoobenthic species living at 2 to 15 (n_1) stations at Balgzand. Total number of comparisons for each species $n_2 = ½n_1 \cdot (n_1 - 1)$. The type of fluctuation pattern is indicated by either i (increasing), d (decreasing), w (winter sensitive) or n (no such trends or dependence).

species	n_1	n_2	number of correlations				proportion of n_2 statist. signif.				fluct. type
			positive		negative		positive		negative		
			tot.	sign.	tot.	sign.	$p<0.05$	<0.01	$p<0.05$	<0.01	
Macoma balthica (L.)	15	105	91	48	14	0	46	30	0	0	i
Cerastoderma edule (L.)	14	91	81	30	10	0	33	12	0	0	w
Arenicola marina (L.)	14	91	60	24	31	8	26	10	9	3	i
Eteone longa (Fabr.)	14	91	91	64	0	0	70	37	0	0	i
Heteromastus filiformis (Clap.)	13	78	78	57	0	0	73	47	0	0	i
Nephtys hombergii Sav.	13	78	78	68	0	0	87	71	0	0	w
Scoloplos armiger (O.F.M.)	12	66	66	52	0	0	79	50	0	0	i
Mya arenaria L.	12	66	59	12	7	0	18	12	0	0	n
Nereis diversicolor (O.F.M.)	12	66	65	49	1	0	74	47	0	0	i
Hydrobia ulvae (Penn.)	11	55	36	5	19	0	9	2	0	0	n
Anaitides spec. div.	10	45	38	8	7	1	18	4	2	0	i
Corophium volutator (Pall.)	10	45	34	9	11	0	20	4	0	0	n
Carcinus maenas (L.)	8	28	26	4	2	0	14	4	0	0	i
Antinoella sarsi (Kinb.)	8	28	28	13	0	0	46	14	0	0	w
Crangon crangon (L.)	7	21	20	8	1	0	38	19	0	0	w
Scrobicularia plana (Da Costa)	7	21	20	9	1	0	43	5	0	0	i
Mytilus edulis L.	5	10	10	3	0	0	30	10	0	0	i
Lanice conchilega (Pall.)	4	6	6	6	0	0	100	67	0	0	w
Scolelepis foliosa (A. & M−E.)	4	6	4	1	2	0	17	17	0	0	i
Angulus tenuis (Da Costa)	2	1	1	1	0	0	100	100	0	0	w
Abra tenuis (Montagu)	2	1	1	1	0	0	100	100	0	0	w
Littorina littorea (L.)	2	1	1	0	0	0	0	0	0	0	i
Bathyporeia spec.	2	1	1	1	0	0	100	0	0	0	d
Magelona papillicornis F.M.	2	1	1	0	0	0	0	0	0	0	n
Nemertini spec. div.	2	1	1	0	0	0	0	0	0	0	n
total (25 species)		1003	897	473	106	9					
proportion of total (%)			89	47	11	1	47	27	0.9	0.3	

and in the crustacean *Bathyporeia* spec. Among these species with highly similar fluctuation patterns, the following are known to be sensitive to low winter temperatures and, therefore, show low numbers after cold winters: *Angulus tenuis, Abra tenuis, Lanice conchilega* and *Nephtys hombergii* (Beukema, 1979; Beukema, 1984). Fig. 1 shows an example of highly synchronized fluctuation patterns in 3 different *N. hombergii* populations. These fluctuations are clearly related to the temperatures prevailing during the immediately preceding winter (Fig. 1b).

Another frequently occurring pattern, common to about half of the species, was an increasing trend during the period of observation, particularly during the last decade, probably as an effect of eutrophication (Beukema & Cadée, 1986). An example is shown in Fig. 2 for 3 populations of *H. filiformis*, showing parallel patterns of fluctuation with an overwhelming influence of the recent upward trend. The above list of species with highly similar fluctuation patterns contains more species with dramatically increasing trends during recent years, viz. *M. balthica, N. diversicolor, S. armiger* and *E. longa*. Only 1 species (*Bathyporeia* spec.) showed a (non-significantly) decreasing trend.

Thus, patterns of fluctuation are highly similar and synchronized especially in species in which numbers are related to an environmental factor as a cold winter (which is effective over wide areas) or in which numbers show a consistent trend (either upward or downward). Also in the cases of the (upwards) trends, a common environmental factor effective over wide areas may be responsible, e.g. better feeding conditions due to eutrophication. Out of the 25 species studied, no less than 19 were either sensitive to low winter temperatures (viz. 7 species: Beukema, 1979; Beukema, 1984) or showed a significantly increasing trend (viz. 12 species: Beukema & Cadée, 1986). Among these 19 species, 10 were listed above as showing a high proportion of similar fluctuation patterns. Among the remaining 6 species only one such species occurred (viz. *Bathyporeia* spec., the one species with a decreasing

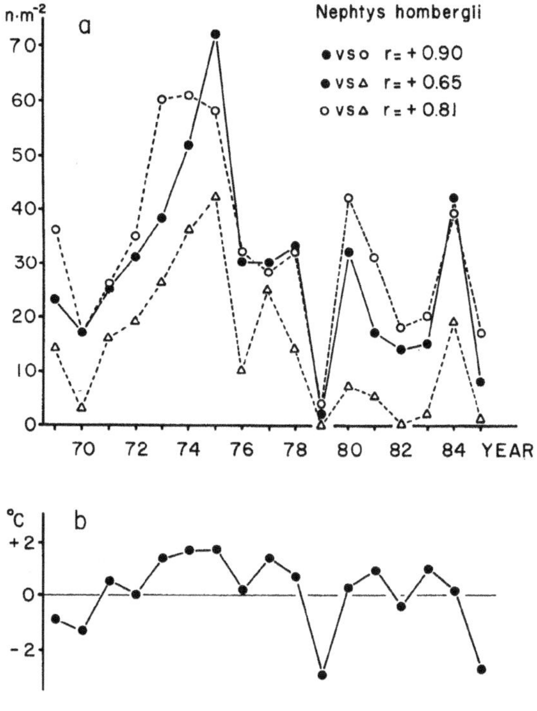

Fig. 1. a. Annual variation in the abundance of the polychaetous worm *Nephtys hombergi* (in $n \cdot m^{-2}$) as observed in spring during the 1969 to 1985 period at 3 tidal flat stations at Balgzand. The fluctuation patterns at the 3 stations are highly similar (values for Spearman's r between +0.65 and +0.90). b. Annual variation in mean winter temperature during the 1969 to 1985 period (as departures from the long-term average, in °C) at a nearby weather station. For all 3 stations the numbers of *N. hombergii* are significantly correlated with the mean temperature during the preceding winter ($p < 0.01$).

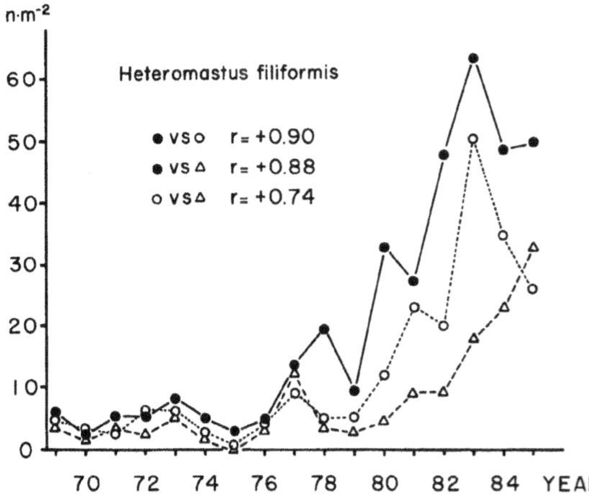

Fig. 2. Annual variation in abundance of the polychaetous worm *Heteromastus filiformis* (in $n \cdot m^{-2}$) as observed in spring during the 1969 to 1985 period at 3 tidal flat stations at Balgzand. At all 3 stations the numerical densities increased significantly ($p < 0.01$) during the period of observation. The fluctuation patterns at the 3 stations are highly similar (values for Spearman's r between +0.74 and +0.90).

trend). In all 5 remaining species without consistent trends in numbers and not showing sensitivity to low winter temperatures, similar fluctuation patterns were less common, none fulfilling the above criterium of 50% significantly positive correlations (none even exceeding 20% of such correlations, Table 1).

b. Negative correlations

Three species contributed in particular to the total of 106 negative correlations, viz. *Arenicola marina, Hydrobia ulvae* and *Macoma balthica* (Table 1). In these species several pairs of populations showed more or less opposite (mirrored) fluctuations in numbers. The characteristic these species have in common is the phenomenon of an annually recurrent migration within the tidal flat area studied (*M. balthica*: Beukema, 1973; Beukema *et al.*, 1978; De Vlas, 1973; *H. ulvae*: Dekker, 1979; *A. marina*: Beukema & De Vlas, 1979; Farke *et al.*, 1979). The mass movement from an area containing one group of sampling stations to an area containing another group of stations differed in intensity from year to year. Especially during severe winters a relatively high proportion of the juveniles of *M. balthica* and *A. marina* migrate from the high and sheltered tidal flats near the coast (i.e. the southwestern part of Balgzand) to the lower and more exposed offshore tidal flats (including the northern half of Balgzand). After winters with relatively high rates of migration, numbers will be relatively low (as compared to other years) in the coastal area and high in the offshore areas and vice versa. So on the one hand such migrations of variable intensity promote parallel changes in numbers (causing positive correlations) within each of the two subareas, but on the other hand cause opposite changes in numbers (and thus negative correlations) between the two subareas of Balgzand. Fig. 3 corroborates this view. In *M. balthica* significantly positive correlations were confined almost exclusively to the pairs of sampling stations within the same subarea (north and south, respectively, of the southern tidal stream), whereas negative correlations all referred to pairs of populations living on different sides of this stream (Fig. 3a, b). For *A. marina* a similar picture was obtained (Fig. 3c, d).

c. Similar patterns in distant areas

So far, only the similarity of fluctuation patterns within the 50 km^2 Balgzand area has been dealt with. The distances between the sampling stations at Balgzand ranged only from 1/2 to about 10 km (Fig. 3). The proportion of significantly similar patterns of 47% (or 52% when only the same species are taken into consideration) is close to the proportion (viz. 40%) observed in another small tidal flat area, viz. the Groninger Wad with distances of 1/2 to 3 km between the sampling stations (Essink, 1978; Essink & Beukema, this issue). Similarly, the proportion of negative correlations is small (viz. 8% at Groninger Wad and 7% for the same species at Balgzand, see Table 2).

In 7 species the fluctuation patterns could be compared between the two areas Balgzand and Groningen, which are about 150 km apart. Out of the 426 comparisons (7 species, up to 5 and 15 sampling stations), 335 yielded positive correlations, 64 of these being statistically significant ($p < 0.05$, Spearman's rank correlation test). So the proportion of similar fluctuation patterns in the two distant areas was 15%. This is significantly less than the percentages of 40 and 52 observed for comparisons within the two areas for the same 7 species (Table 2). The proportion of negative correlations (viz. 21%) is significantly higher ($p < 0.05$, χ^2 test) for the distant comparisons than for the within-area comparisons (viz. 7 and 8%). Note that the proportion of negative correlations of 21% is yet far below the percentage of 50 to be expected if the populations in the two areas fluctuated independently. Accordingly, the proportions of significantly negative correlations was negligible (less than 1%), whereas the value for the proportion of significantly positive correlations (15% for the distant comparisons) was much higher than could be expected from independent fluctuations (2½% at a 0.05-level of significance). In conclusion: similarity in fluctuation patterns of distant (150 km) tidal flat populations of benthic animals is an existing phenomenon.

The proportions of negative correlations were particularly high for the 2 sampling stations at Groninger Wad that were affected by the waste-water outfall, viz. 31 and 40%. For the other 3 stations in this area these proportions were 9, 10 and

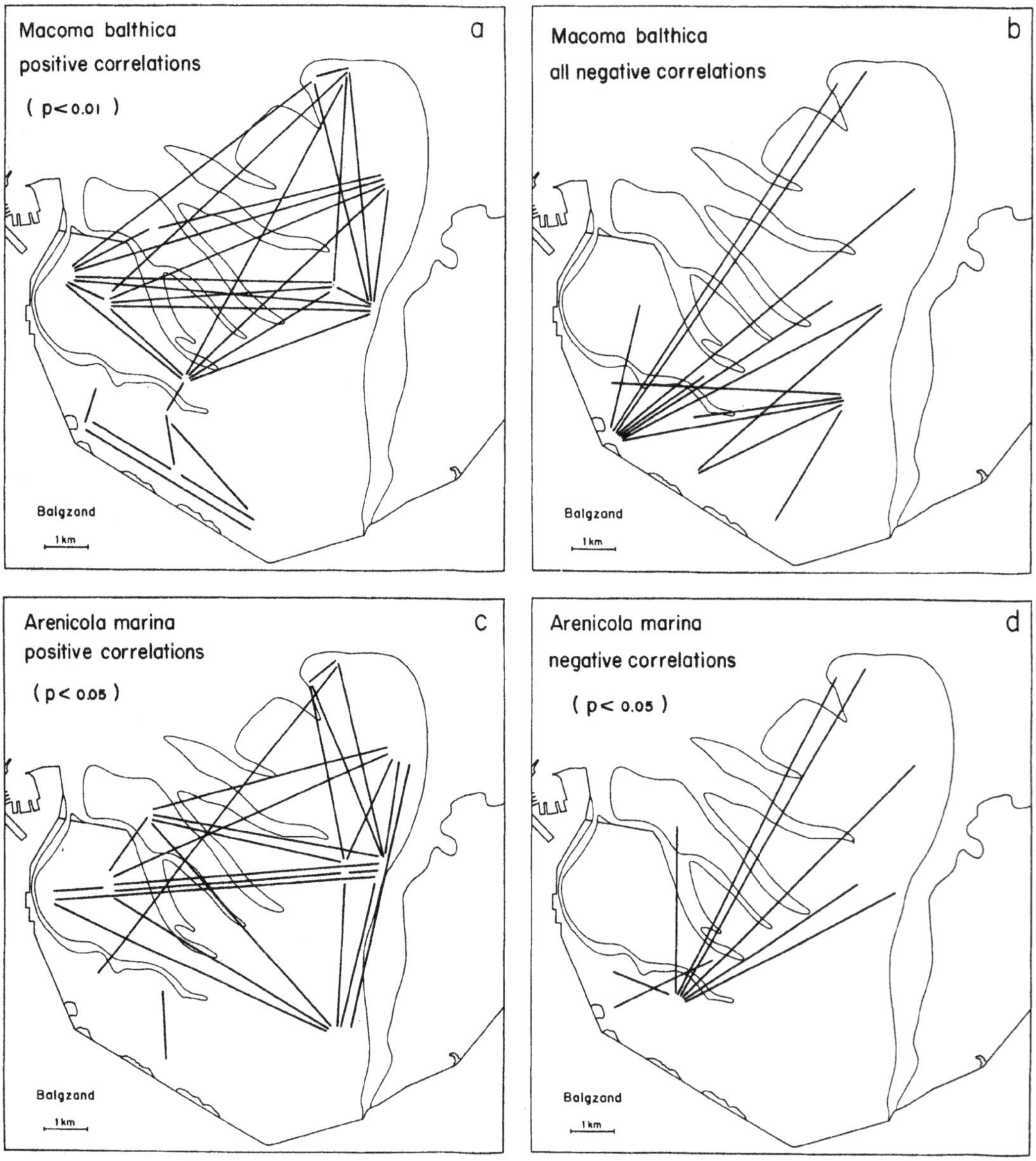

Fig. 3. Maps of the tidal flat area Balgzand with straight lines connecting sampling stations that show in 2 species either highly similar fluctuation patterns (a: *Macoma balthica*; c: *Arenicola marina*) or more or less opposite fluctuation patterns (b: *M. balthica*; d: *A. marina*). Number of sampling stations compared: 15 in *M. balthica* and 14 in *A. marina*. Pairs of sampling stations with similar patterns connected only when $p < 0.01$ in *M. balthica* (a) or < 0.05 in *A. marina* (c). Pairs with opposite patterns connected either whenever the correlation was negative in *M. balthica* (b) or when $p < 0.05$ in *A. marina* (d).

17%, i.e. mostly close to 7 and 8% observed for the within-area comparisons (Table 2). For these 3 less-affected stations, fluctuation patterns in most species (e.g. *Nephtys hombergii*, Fig. 4) were well-synchronized in the 2 distant areas. However, even for these stations the proportions of highly similar patterns (yielding significantly positive correlations) were lower (9, 19 and 23%) than those observed for the within-area comparisons (40 and 52%). Apparently, the similarities are generally weaker for distant (150 km) than for nearby (up to 10 km) comparisons.

Table 2. The degree of similarity of fluctuation patterns in 7 species of macrozoobenthos within two areas of observation (Balgzand and Groninger Wad) and between these areas. Similarity expressed as the proportions (%) of all station pairs (n) that showed a significantly positive correlation ($p<0.05$). A second column shows the proportions of negative correlations, with the significant ones between ( ).

comparison	percentages of correlations		n
	sign. positive	negative (sign.)	
within Balgzand	52	7 (0.2)	542
within Gron. Wad	40	8 (0.0)	62
Balgz. vs Gron. Wad	15	21 (0.7)	426

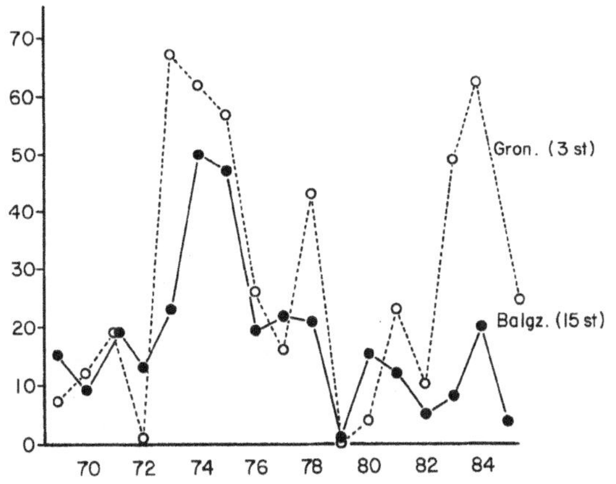

Fig. 4. Annual variation in abundance of the polychaetous worm *Nephtys hombergii* ($n \cdot m^{-2}$, spring data, 1969 to 1985 period) in two tidal flat areas about 150 km apart, viz. Balgzand (means of all 15 sampling station) and Groninger Wad (means of 3 sampling stations that were least affected by the waste water discharge, see Essink & Beukema, this issue).

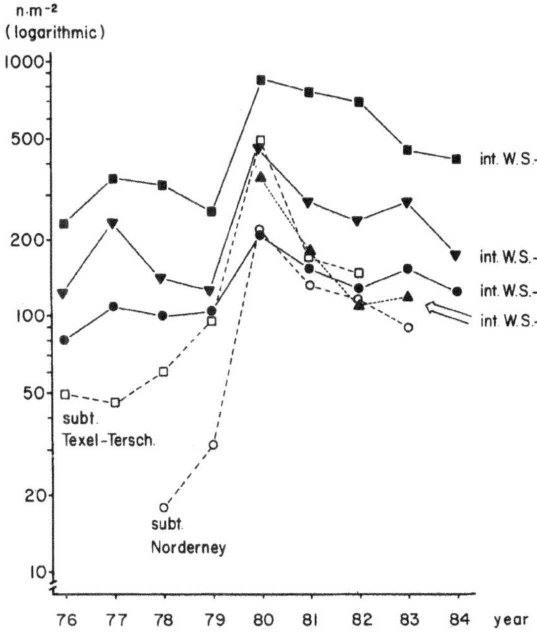

Fig. 5. Annual variations in abundance ($n \cdot m^{-2}$, spring data) of the bivalve *Macoma balthica* sampled in widely different parts of the Wadden Sea (solid points and lines) or nearby North Sea areas (open points, broken lines). The sampling areas from north to south: ▲ Danish Wadden Sea (int. W.S.-N); ▼ German Wadden Sea (int. W.S.-E); o North Sea off Norderney (Subt. Norderney); ■ Dutch Wadden Sea, Groninger Wad (int. W.S.-M); □ North Sea off Terschelling and Texel (subt. Texel-Tersch.); ● Dutch Wadden Sea, Balgzand (int. W.S.-W).

For one species (*Macoma balthica*) sufficient data are available to compare fluctuation patterns over an even wider area, including the entire Wadden Sea (Fig. 5). Data series on *Macoma* abundance were made available by courtesy of several members of the intertidal soft-sediments section of COST-647, including P. B. Madsen (Danish data), J. Dörjes, H. Michaelis and B. Rhode (German data) and the authors (Dutch data). Common trends in the numerical densities of *Macoma* living on tidal flats of the Wadden Sea (full lines in Fig. 5) include: more or less stable numbers during the second half of the seventies, a steep increase in 1979 and a slow decline during the first half of the eighties. The populations living in subtidal areas near the Wadden Sea (broken lines in Fig. 5) showed a similar pattern, with one major difference: the increase in 1979 started already before the annual spring sampling of 1979 and will have originated from winter-migration (Beukema, 1973). The further steep increase during 1979 that was observed in all sampling areas, will have originated from a highly successful recruitment during the summer of 1979, which followed the severe 1978/79 winter (Beukema, 1982).

Conclusions

Both within the restricted area of Balgzand and for the comparisons over a distance covering nearly the whole Dutch Wadden Sea, the number of positive correlations by far exceeded the number of negative correlations, pointing to the preponderance of environmental factors that govern the fluctuations of populations of benthic species over wide areas. One of these factors could easily be identified, viz. the severity of the winter preceding the annual sampling period in early spring. Severe winters are known to be co-incident over wide geographic areas, e.g. the one of 1962/63 caused high mortality in many sensitive species both along the British and mainland coasts of the Channel and the North Sea (as summarized by Cushing, 1982). The effect of a severe winter appears not to be restricted to heavy mortalities but may include an enhancement of recruitment in several (partly additional) species during the subsequent summer (Beukema, 1982). Another synchronizing factor will have to do with the increasing trend observed in about half of the species considered and may be related to the general eutrophication in Dutch coastal areas during the last few decades (Beukema & Cadée, 1986). Long-term oceanographic cycles also will govern parallel changes in numbers of marine species over vast areas (Gray & Christie, 1983).

Parallel changes during decades in populations of marine species over wide geographic areas are well known for fish (Cushing, 1982) and zooplankton species (Colebrook, 1978; this issue). In marine benthos fewer long-term series of population size records over extensive areas are available. However, the few cases published so far clearly point to similar fluctuation patterns in distant areas. In subtidal rocky shore species Svane (1984) observed synchronized fluctuation patterns along a 100-km stretch of the Swedish west coast and Lundälv & Christie (this issue) found the fluctuations observed along this coast to extend along the south coast of Norway. The synchronized success of recruitment in barnacles at sites of over 100 km on the Yorkshire coast provides an example from the intertidal rocky shore environment (Kendall *et al.*, 1985). Also in soft bottom benthic species, a few examples of synchronized population events over wide geographic areas have been published. Inflows of high-saline and oxygen-rich water into the Baltic and the subsequent decline of oxygen content in the stagnant water masses were found to go with large-scale increases and subsequent decreases of several benthic species in the deeper parts of the Baltic (Andersin *et al.*, 1978; Jensen, this issue). In the North Sea, successful recruitment of *Echinocardium cordatum* appears to be synchronized over vast areas, probably in relation to the incidence of severe winters (Beukema, 1985), as in the case of several species in the Wadden Sea (Beukema, 1982).

Further internationally co-ordinated studies of population fluctuations will surely reveal more examples of similar fluctuation patterns. Not only the results of the present study, indicating similar patterns in all species studied both in nearby and distant (150 km) populations, are promising. The first results from the comparison of fluctuation patterns over even longer distances within the Wadden Sea also point to parallel fluctuations. Examples are the cases of *Macoma balthica* (Fig. 5) and the dramatic increase of *Heteromastus filiformis* all over the Wadden Sea (Fig. 2 and Dörjes *et al.*, this issue). The COST-647 programme offers an excellent opportunity to extend this long-term co-ordinated study of population fluctuations to assess the size of the areas of synchronized fluctuations for several species and to unravel the processes by which environmental factors govern the parallel changes in numbers in marine benthic species (Lewis, this issue).

Gaining basic insight in the causes and extent of population fluctuations in the coastal marine environment is not the only reward obtained from such an internationally co-ordinated study. The assessment of common fluctuation patterns in a species over vast geographic areas offers a base-line of its changes. Any departure from such a common base-line pattern may indicate a local effect of e.g. a pollutant as shown for a waste water discharge in the Dutch Wadden Sea (Essink & Beukema, this issue). With such a sampling strategy, local changes in population numbers may be attributed to local disturbing factors with more confidence than would ever be possible by single local studies only.

References

Andersin, A.-B., J. Lassig, L. Parkkonen & H. Sandler, 1978. The decline of macrofauna in the deeper parts of the Baltic proper and the Gulf of Finland. Kieler Meeresforsch. 4: 23–52.

Beukema, J. J., 1973. Migration and secondary spatfall of *Macoma balthica* (L.) in the western part of the Wadden Sea. Neth. J. Zool. 23: 356–357.

Beukema, J. J., 1974. Seasonal changes in the biomass of the macrobenthos of a tidal flat in the Dutch Wadden Sea. Neth. J. Sea Res. 8: 94–107.

Beukema, J. J., 1979. Biomass and species richness of the macrobenthic animals living on a tidal flat area in the Dutch Wadden Sea: effects of a severe winter. Neth. J. Sea Res. 13: 203–223.

Beukema, J. J., 1982. Annual variation in reproductive success and biomass of the major macrozoobenthic species living in a tidal flat area of the Wadden Sea. Neth. J. Sea Res. 16: 37–45.

Beukema, J. J., 1984. Zoobenthos survival during severe winters at high and low tidal flats in the Dutch Wadden Sea. In J. S. Gray & M. E. Christiansen (eds), Marine biology of polar regions and effects of stress on marine organisms. John Wiley, Chichester: 351–361.

Beukema, J. J., 1985. Growth and dynamics in populations of *Echinocardium cordatum* living in the North Sea off the Dutch north coast. Neth. J. Sea Res. 19: 129–134.

Beukema, J. J. & G. C. Cadée, 1986. Zoobenthos responses to eutrophication of the Dutch Wadden Sea. Ophelia 26: in press.

Beukema, J. J. & J. de Vlas, 1979. Population parameters of the lugworm, *Arenicola marina*, living on tidal flats in the Dutch Wadden Sea. Neth. J. Sea Res. 13: 331–353.

Beukema, J. J., G. C. Cadée & H. Hummel, 1983. Differential variability in time and space of numbers in suspension and deposit feeding benthic species in a tidal flat area. Oceanol. Acta Vol. Sp.: 21–26.

Beukema, J. J., W. de Bruin & J. J. M. Jansen, 1978. Biomass and species richness of the macrobenthic animals living on the tidal flats of the Dutch Wadden Sea: Long-term changes during a period with mild winters. Neth. J. Sea Res. 12: 58–77.

Colebrook, J. M., 1978. Continuous plankton records: zooplankton and environment, North-East Atlantic and North Sea: 1948–1975. Oceanol. Acta 1: 9–23.

Colebrook, J. M., 1987. Environmental influences on long-term variability in marine plankton. Hydrobiologia 142: 309–325.

Cushing, D. H., 1982. Climate and fisheries. Academic Press, London, 373 pp.

Dekker, R., 1979. Numbers, growth, biomass and production of organic and calcareous matter of *Hydrobia ulvae* (Gastropoda: Prosobranchia) in the western Dutch Wadden Sea. Interne Verslagen, Ned. Inst. Onderz. Zee, Texel, 1979-15, 27 pp.

De Vlas, J., 1973. Migratie via getijstromen van jonge nonnetjes, *Macoma balthica* L., op en rond het Balgzand. Interne Verslagen, Ned. Inst. Onderz. Zee, Texel, 1973-14, 30 pp.

Essink, K., 1978. The effects of pollution by organic waste on macrofauna in the eastern Dutch Wadden Sea. Publication Series, Neth. Inst. Sea Res., Texel 1: 1–135.

Essink, K. & J. J. Beukema, 1987. Long-term changes in tidal flat macrozoobenthos as an indicator of stress by organic pollution. Hydrobiologia 142: 209–215.

Farke, H., P. A. W. J. de Wilde & E. M. Berghuis, 1979. Distribution of juvenile and adult *Arenicola marina* on a tidal mud flat and the importance of nearshore areas for recruitment. Neth. J. Sea Res. 13: 354–361.

Gray, J. S. & H. Christie, 1983. Predicting long-term changes in marine benthic communities. Mar. Ecol. Prog. Ser. 13: 87–94.

Jensen, K., 1987. Long-term changes of the macrozoobenthos in the Baltic Sea and the Kattegat area. Hydrobiologia 142: 129–135.

Kendall, M. A., R. S. Bowman, P. Williamson & J. R. Lewis, 1985. Annual variation in the recruitment of *Semibalanus balanoides* on the north Yorkshire coast 1969–1981. J. mar. biol. Ass. U.K. 65: 1009–1030.

Lewis, J. R., 1987. Latitudinal patterns and problems in the rocky littoral. Hydrobiologia 142: 1–13.

Lundälv, T. & H. Christie, 1987. Comparative trends and ecological patterns of rocky subtidal communities in the Swedish and Norwegian Skagerrak area. Hydrobiologia 142: 71–80.

Svane, I., 1984. Observations on the long-term population dynamics of the perennial ascidian, *Ascidia mentula* O. F. Müller, on the Swedish west coast. Biol. Bull. 167: 630–646.

Van der Hoeven, P. C. T., 1982. Observations on surface water temperature and salinity, State Office of Fishery Research (RIVO): 1860–1981. Scientific Report W.R. 82-2, KNMI, De Bilt: 118 pp.

Hydrobiologia 142: 209–215 (1986)

Long-term changes in intertidal flat macrozoobenthos as an indicator of stress by organic pollution

K. Essink[1] & J. J. Beukema[2]
[1]*Rijkswaterstaat, Tidal Waters Division, Hereweg 99a, 9721 AA Groningen, The Netherlands*
[2]*Netherlands Institute for Sea Research, P.O. Box 59, 1790 AB Den Burg (Texel), The Netherlands*

Keywords: macrobenthos, dynamics, Wadden Sea, pollution, indicator

Abstract

Organic waste-water from beetsugar factories is being discharged by pipeline into the eastern part of the Dutch Wadden Sea since 1969 mainly in September–December. The organic waste load was high in the first four years and diminished considerably from 1974 onwards.

From 1968 macrobenthos has been monitored at five permanent stations. Two of these stations were situated relatively close to the outfall of the pipe-line.

Data on numerical density of seven macrobenthic species were analysed for synchrony of yearly fluctuations in order to discriminate between pollution-induced changes in population densities and changes due to variation of natural environmental factors.

A great deal of concordance in density fluctuations was found between the five stations. Comparison with data from the western part of the Dutch Wadden Sea (150 km apart) revealed a relative good concordance of the density fluctuations in eastern (only 3 stations) and western Wadden Sea. The two stations close to the outfall showed an aberrant pattern of density variations due to the waste discharge.

Introduction

Organic wastewater from two beetsugar and some other factories is being discharged by pipeline into the eastern Dutch Wadden Sea since 1969 (see Fig. 1). The majority of the biodegradable waste is being discharged from September through December due to the production period of the beetsugar factories. The organic waste load was high in the years 1969–1973; it decreased from 1974 onwards due to sanitation at the factories. The effects of this waste discharge on water quality and macrozoobenthos have been studied from 1968 to 1975 (Essink, 1978). From 1975 onwards sampling of macrozoobenthos was continued at five stations.

The numerical densities of intertidal macrozoobenthos species vary from year to year. These variations are caused by variation in recruitment success, predation and winter mortality. Winter temperatures do have a synchronizing effect on the population fluctuations of certain intertidal species (Beukema, 1982, 1985; Beukema & Essink, 1986).

In addition the abundance of species may change as a consequence of non-natural causes, *e.g.* manmade disturbances such as pollution. In studies on the effect of pollution it is important to be able to discriminate between pollution-induced changes in the populations studied and changes due to variation in the natural abiotic environment. In this paper the patterns of variation in abundance of intertidal macrozoobenthos in the vicinity of a wastewater outfall in the eastern Dutch Wadden Sea ('Groninger Wad') are analysed in order to detect possible pollution-induced deviations from natural patterns.

The outfall of the wastewater pipe-line is situated in the upper half of the intertidal zone, in one of the sedimentation fields that are maintained along the coast (Fig. 1). Station 3 is situated close to the outfall in an area where poor survival of macrozoo-

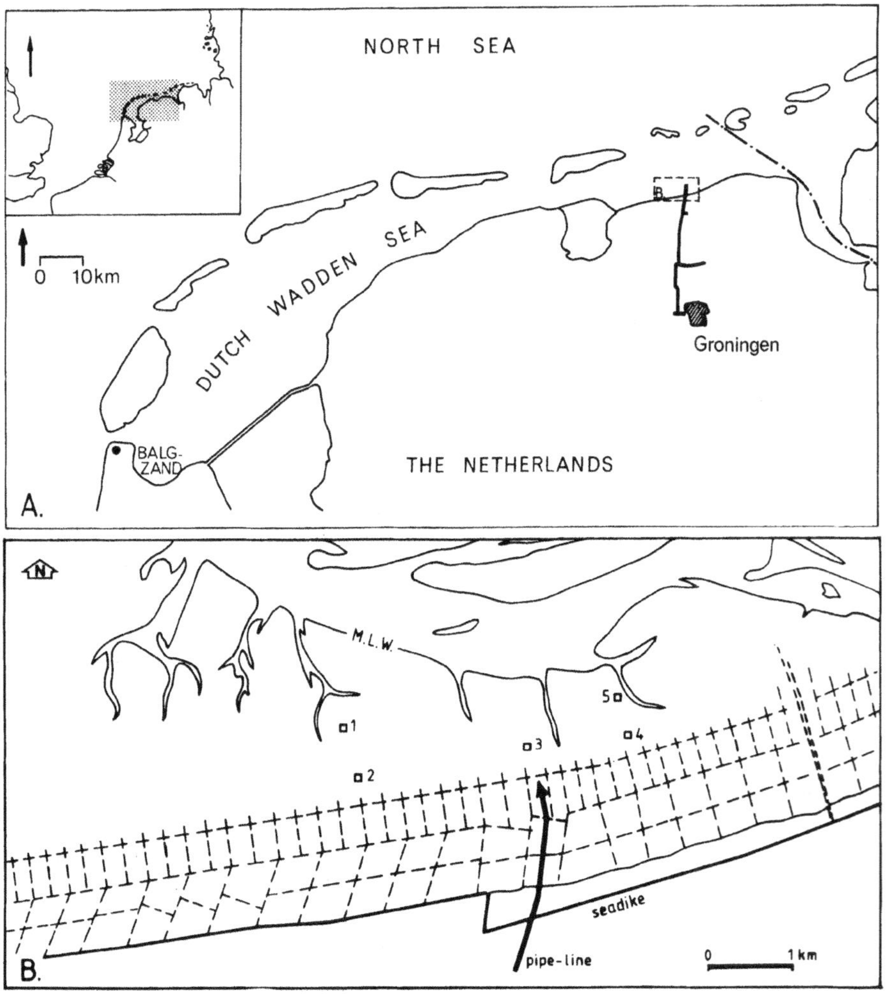

Fig. 1. Map of the Dutch Wadden Sea with the wastewater pipe-line from Groningen (A), and the position of the sampling stations (□) and the wastewater outfall (B).

benthos was observed during the period of maximal waste discharge. Sometimes the area of poor survival extended as far as station 4 (see below; Essink, 1978).

Methods

Samples were taken at 5 stations at different distances from the outfall (Fig. 1). Each station measures 30 × 30 m. Stations 1 and 5 have a silt content (fraction smaller than 16 μm) below 2%, whereas the other stations have a silt content of 5–25% (Essink, 1978).

Samples of macrozoobenthos were taken with a PVC corer of ca. 300 cm^2 to a depth of 20–25 cm. From 1980 onwards a corer of 230 cm^2 was used. On each collection 16 cores were taken per station. The samples were sieved in the field over a 1 mm sieve. For this study numerical densities from spring (ca. March–April) have been used. For station 3 no data are available for 1969. In 1981 no samples were taken in spring; therefore, spring densities have been estimated by interpolation.

Time-series of density data for 7 species will be analysed in this paper. These species comprise the bivalves *Macoma balthica, Cerastoderma edule* and *Mya arenaria*, and the polychaetes *Nephtys hombergii, Eteone longa, Anaitides maculata* and *Scoloplos armiger*. For any pair of stations the concordance of density variations was assessed using Spearman's rank correlation test.

Results

The macrobenthic species at the 5 stations showed large variations in numerical densities. A few examples will be given. In *Macoma balthica* (Fig. 2) density variations are characterised by an increase at all stations in 1969–1971 and particularly in 1979–1980, i.e. after relatively severe winters (see Beukema, 1985).

The densities of *Eteone longa* showed frequent variations (Fig. 3). At most of the stations these variations showed a highly similar pattern, except at station 3, where it lasted till 1974 before the density variations followed the pattern of the other stations. A similar pattern can be observed in *Nephtys hombergii*, where it lasted till about 1976 before densities at station 3 started to follow the pattern of the other stations (Fig. 4). The synchronising effect of the severe winter 1979 is clearly visible.

These examples show that the densities of the populations of a species fluctuated synchronously to a fairly large extent, and also that this synchrony was not present during the entire period at all stations. To obtain a measure for the synchrony of density variations Spearman's rank correlation test has been applied to the data sets of any pair of stations. The results are given in Table 1, showing some open spaces because hardly any *Scoloplos* has been found at station 3. The same holds for *Mya* at station 1. Out of the 62 correlations presented in Table 1 only 5 are negative. Out of the 57 positive correlations 25 are statistically significant ($p < 0.05$), i.e. 40% of all 62 correlations. None of the negative correlations is significant.

When for each of the stations the mean of all correlations to the other stations is calculated, station 3 and 4 show mean values for Spearman's r that are slightly lower than the mean r-values for

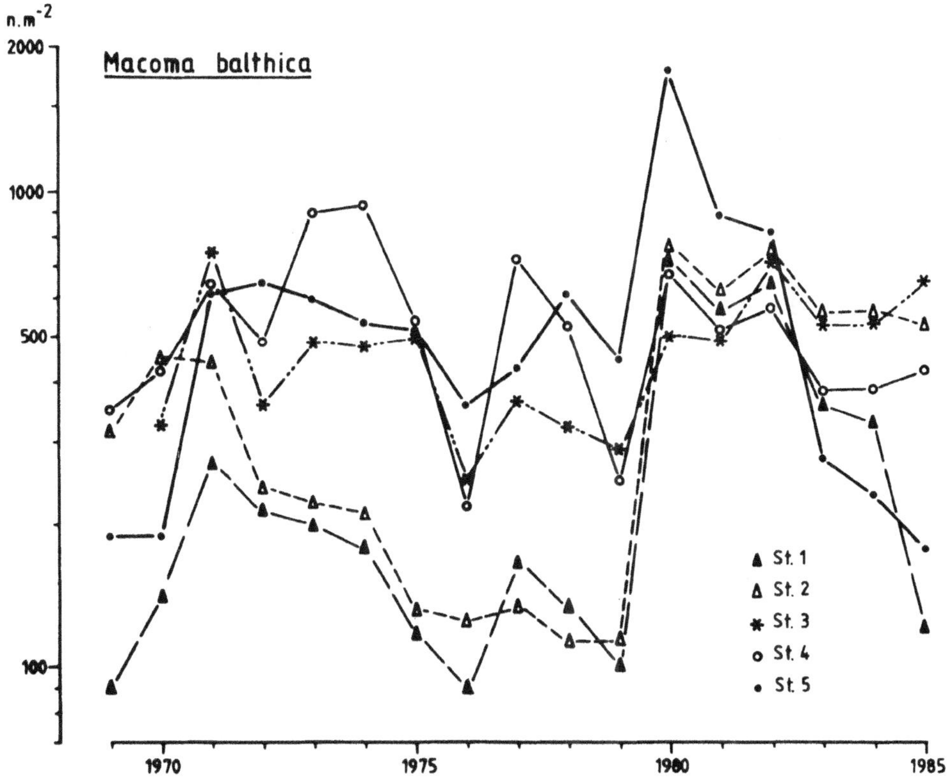

Fig. 2. Numerical density (n.m^{-2}) of *Macoma balthica* at 5 stations.

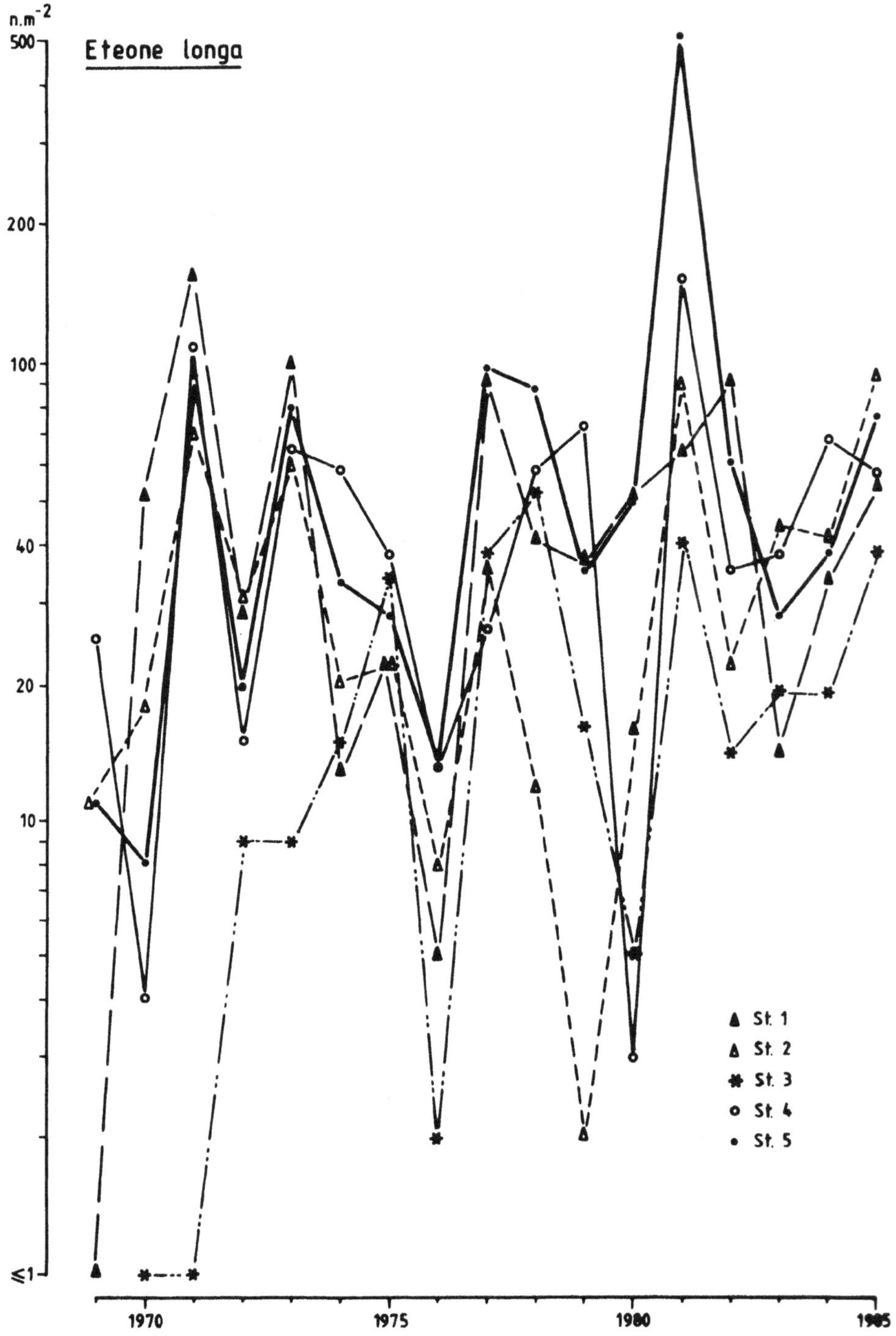

Fig. 3. Numerical density (n.m^{-2}) of *Eteone longa* at 5 stations.

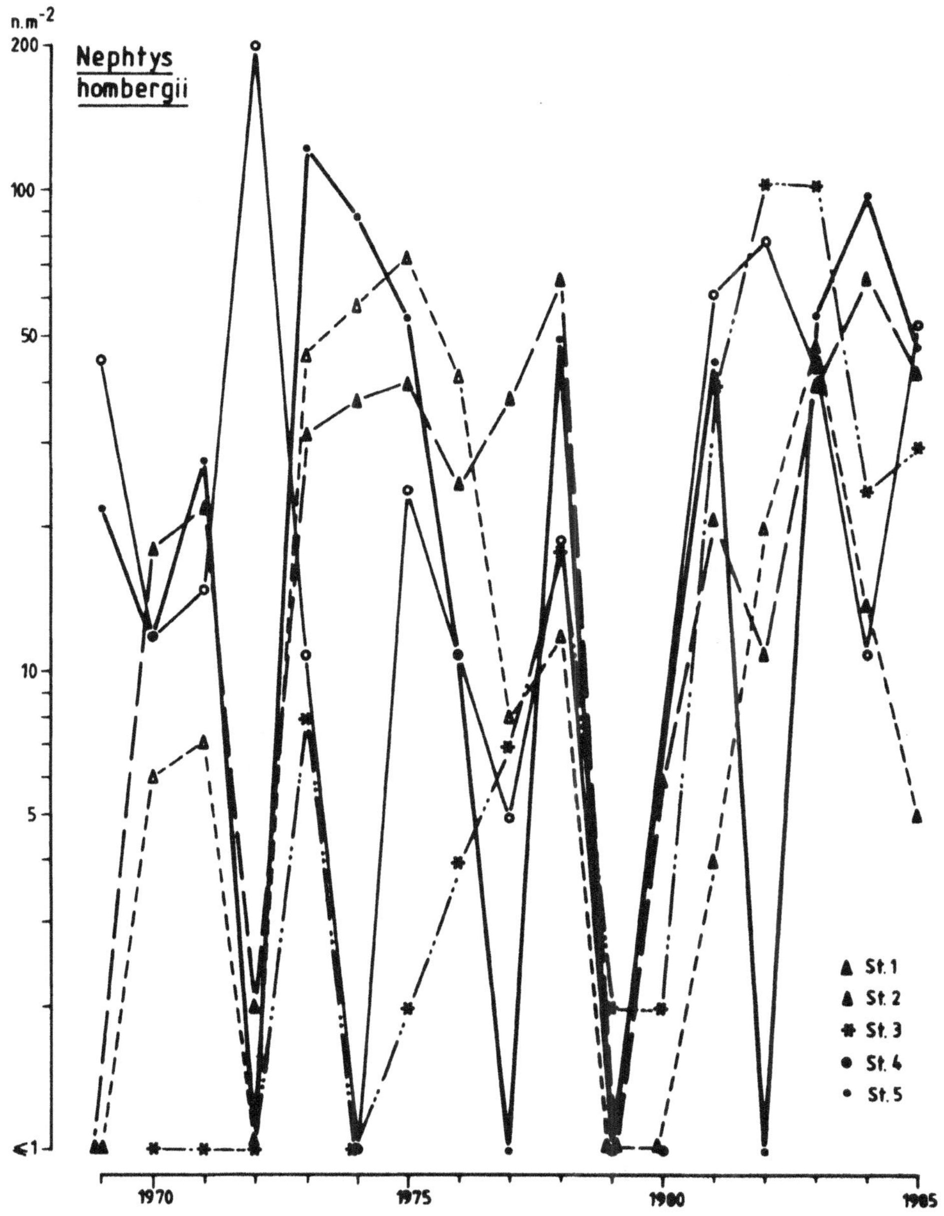

Fig. 4. Numerical density (n.m^{-2}) of *Nephtys hombergii* at 5 stations.

the other 3 stations (Table 2). The differences, however, are not statistically significant. When the same calculation is carried out for all correlations between any of the Groninger Wad stations and the 15 Balgzand stations in the same years (see Fig. 1 and Beukema & Essink, 1986) mean values for r are obtained for stations 3 and 4 that are significantly lower than each of the values for stations 1, 2 and 5 (Table 3).

So the fluctuation patterns observed at the stations 3 and 4 are significantly less similar to the Balgzand patterns than those found at the stations 1, 2 and 5. The similarity for the latter 3 stations to the distant (150 km) Balgzand area is only slightly lower than the similarity within the Groninger Wad area (compare the figures in Table 2 and 3).

Table 1. Spearman's rankcorrelation coefficient *r* (× 100) for correlations of numerical densities of 7 species between pairs of stations over a 16- or 17-year (n) period. MACO = *Macoma balthica,* ETEO = *Eteone longa,* CERA = *Cerastoderma edule,* NEPH = *Nephtys hombergii,* SCOL = *Scoloplos armiger,* MYA = *Mya arenaria,* ANAI = *Anaitides maculata.* + = significant.

stations	n	MACO	ETEO	CERA	NEPH	SCOL	MYA	ANAI	mean
1 vs. 2	17	76+	55+	-42	64+	49		64+	44
1 vs. 3	16	40	11	-38	48			34	19
1 vs. 4	17	44	32	13	-5	60+		15	26
1 vs. 5	17	60+	76+	11	71+	71+		58+	58
2 vs. 3	16	62+	33	23	30		49	87+	47
2 vs. 4	17	8	45	20	-11	8	50+	42	23
2 vs. 5	17	17	55+	41	61+	38	36	69+	45
3 vs. 4	16	19	43	44	31		63+	59+	43
3 vs. 5	16	6	53+	11	22		42	68+	34
4 vs. 5	17	56+	59+	44	-9	68+	83+	64+	52

Table 2. Mean values for Spearman's *r* for all correlations (n) between any of the 5 stations and the other 4 stations at Groninger Wad, and proportion of significant correlations (% signif.).

station :	1	2	3	4	5
n :	23	26	23	26	26
mean *r* :	0.37	0.40	0.36	0.36	0.47
% signif.:	43	38	26	35	58

Table 3. Mean values for Spearman's *r* for all correlations (n) between any of the Groninger Wad stations and the 15 Balgzand stations. The distance of the Groninger Wad stations to the wastewater outfall is indicated.

station :	1	2	3	4	5
distance (m):	2300	2100	500	1100	1300
n :	77	90	78	90	90
mean *r* :	0.25	0.30	0.15	0.07	0.28

Discussion and conclusions

The results presented in Table 1 show that within the Groninger Wad area positive correlations between numerical densities at different stations prevailed. Of all possible correlations 40% were significantly positive ($p < 0.05$). In this respect the data from Groninger Wad are in good agreement with data from Balgzand where 47% of all correlations (52% for the same 7 species) were found to be significantly positive (Beukema & Essink, 1986).

As already shown above for *Eteone longa* and *Nephtys hombergii* such common patterns were not present at all stations during the entire period of investigation. In particular the densities of some species at station 3, which is situated relatively close to the wastewater outfall, did not follow the patterns found at the other stations. This suggests a disturbance of the natural fluctuation pattern due to pollution at this station.

The degree of similarity of the population fluctuations in the 7 species considered was hardly different at the Groninger Wad stations (see Table 2). This may be caused by the fact that the 5 stations are situated within a relatively small area (less than 10 km^2) with only a restricted range of intertidal levels and sediments. The fluctuations at Groninger Wad showed a high degree of similarity to those found at Balgzand (Beukema & Essink, 1986). This was particularly so for the stations 1, 2 and 5. Therefore, the common fluctuation patterns at Balgzand and at these 3 stations may be considered 'natural' ones. At stations 3 and 4 (close to the outfall) the fluctuations concorded badly with the variations at Balgzand, as compared to the stations 1, 2 and 5 which have a greater distance to the outfall (see Table 3). In this respect it is of importance to note that this concordance applies to muddy (station 2) as well as sandy stations (1 and 5).

About the aberrant pattern of density variations at stations 3 and 4 the following can be said on the basis of observations by Essink (1978). From field observations on the distribution of the wastewater it could be concluded that stations 3 and 4 were directly affected by the wastewater. This effect was most pronounced in the autumn months of the

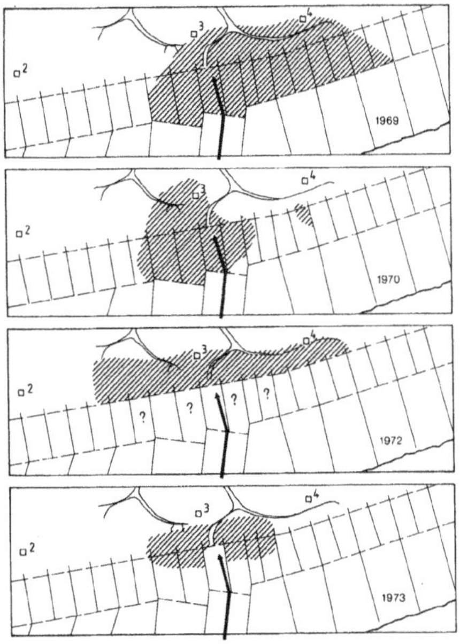

Fig. 5. Occurrence of crawling and moribund *Cerastoderma edule* at the tidal flats during autumn (after Essink, 1978).

years 1969–1973, when the waste load was high. In the vicinity of station 3 during part of the submergence period strongly reduced dissolved oxygen concentrations were measured. As a result of this pollution stress species richness in the vicinity of station 3 was found to decrease during the autumn months (Essink, 1978; Fig. 5). In the vicinity of station 3 a poor survival of O-group *Cerastoderma, Macoma, Mya* and *Nereis* was found in autumn 1969–1973. Mortality among *Cerastoderma* of different age classes due to the waste discharge was observed as far as station 4 (Fig. 5). The aberrant density pattern of *Eteone* and *Nephtys* at station 3 has already been introduced (Figs. 2 and 3).

All these observations indicate that during the first five years of the waste discharge at Groninger Wad disturbances of the natural variations in numerical densities occurred at stations 3 and 4 due to a high waste load. These local disturbances at Groninger Wad could be detected by comparing fluctuation patterns of numerical densities at Groninger Wad with those found at Balgzand.

References

Beukema, J. J., 1982. Annual variation in reproduction success and biomass of the major macrozoobenthic species living in a tidal flat area of the Wadden Sea. Neth. J. Sea Res. 16: 37–45.

Beukema, J. J., 1985. Zoobenthos survival during severe winters on high and low tidal flats in the Dutch Wadden Sea. In: J. S. Gray & M. E. Christiansen (eds), Marine biology of polar regions and effects of stress on marine organisms. J. Wiley & Sons Ltd., Chichester etc.: 351–361.

Beukema, J. J. & K. Essink, 1987. Common patterns in the fluctuations of macrozoobenthic species living at different places on tidal flats in the Wadden Sea. (this volume).

Essink, K., 1978. The effects of pollution by organic waste on macrofauna in the eastern Dutch Wadden Sea. Netherlands Institute for Sea Research, Publications Series No. 1: 1–135.

Hydrobiologia 142: 217–232 (1986)

Long-term studies of macrozoobenthos in intertidal and shallow subtidal habitats near the island of Norderney (East Frisian coast, Germany)

J. Dörjes*, H. Michaelis** & B. Rhode**
**Forschungsinstitut Senckenberg, Abteilung für Meeresgeologie und Meeresbiologie, Schleusenstraße 39A, D-2940 Wilhelmshaven, FRG*
***Niedersächsisches Landesamt für Wasserwirtschaft – Forschungsstelle Küste –, An der Mühle 5 D-2982 Norderney, FRG*

Keywords: macrozoobenthos, long-term development, eulittoral, sublittoral, North Sea

Abstract

Near the East Frisian island of Norderney two sites are investigated permanently in order to study long-term fluctuations of macrozoobenthos: one transect (since 1977) at the northern side in shallow subtidal waters and another one (since 1976) at the sheltered southern side in the intertidal area of the Wadden Sea. Since 1980 the investigations have been continued in the frame of COST-47, sedimentary intertidal programme (including the shallow subtidal habitats colonized by the *Macoma balthica* community).

The results up to 1984, respectively to 1985, are presented and the changes of abundance of the dominant species are discussed. Discussed are also the influences of water temperatures and sediment disturbances caused by wave action. There is evidence that the intertidal variety of the *Macoma balthica* community shows a greater stability than the subtidal variety.

Introduction

Although mankind has always experienced in nature the alternation of 'fat' and 'lean' years there were quite static ideas of the marine littoral benthos for a long time. Many former authors found that structures and densities of animal communities have changed over the years (e.g. Thamdrup, 1935; König, 1943; Jepsen, 1965). Also the population dynamics of selected animal species were the object of investigations at an early stage (e.g. Dales, 1951; Kristensen, 1958; Gibbs, 1968). However, the increasing pollution of the sea since the 1960's has channeled our stronger interest in the changes within a given time, which the littoral communities are subjected to. The first to undertake an approach to such a complex of questions in the area of the Wadden Sea were Hauser (1973), Beukema (1974), Beukema *et al.* (1978) and Essink (1978).

Long-term investigations were started on the East Frisian coast in 1976 (eulittoral) and 1977 (sublittoral). Since 1980 they have been continued within the scope of the COST programme.

The habitats which were investigated belong to the distribution area of the *Macoma balthica* community, which is characteristic for the North Sea littoral. In the intertidal area it forms the 'intertidal variety of the *Macoma* community', which is poor in species, while the sublittoral is inhabited by the 'subtidal variety of the *Macoma* community', the species richness of which is considerably higher.

The following report contains a first analysis and evaluation of the data and above all shows the development of the dominant species.

For the part of the work dealing with the intertidal bottom fauna we received financial support from the Federal Ministry of Research and Technology.

Area of investigation and methods

Intertidal and subtidal habitats near the East Frisian island of Norderney are the area of investigation. Norderney is a 'barrier' island between the Wadden Sea and the North Sea. As can be seen in

Fig. 1 the stations 1 to 4 lie south of the island on a transect in the intertidal area.

Since 1976 samples were taken in intervals of one month (till 1977), of two months (till 1980), and three months (from 1981 on). For each year only the data for January, April, July, and October will be presented in this paper. Station 2 is identical with a permanent plot which was already inves-

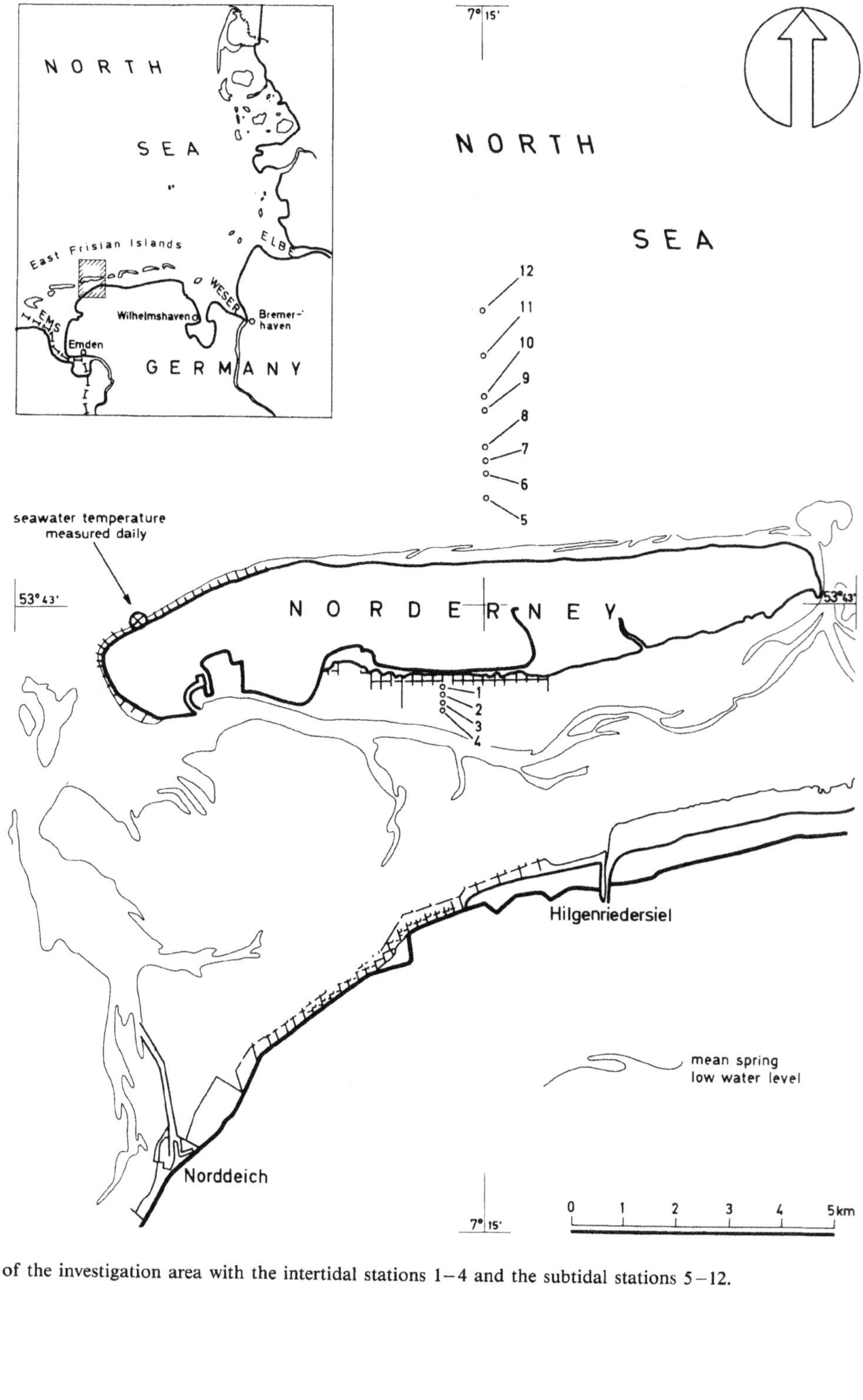

Fig. 1. Map of the investigation area with the intertidal stations 1–4 and the subtidal stations 5–12.

tigated by Hauser (1973) in 1970 till 1972.

At each station one sample of 0.125 m^2 was taken till 1977; from 1978 on 7 samples of 177 cm^2 were taken at each station. The bottom fauna was sieved with a 1 mm mesh aperture and then sorted out and counted in a live state. The biomass was determined as ash-free dry weight (ADW), at first according to Beukema (1974), from 1984 on with a reduced combustion temperature according to Lappalainen & Kangas (1975). Mollusc shells were separated from the flesh, apart from small gastropods (*Hydrobia, Littorina* juv.) and the spat of bivalves.

The following abiotic parameters were investigated: sea-level for calculating the inundation time, grain size distribution of the sediment, organic matter as loss on ignition, water content, salinity of the interstital water, and depth of the oxidized layer. The weather station on Norderney put data on water temperature at our disposal, which were measured once per day at the northern edge of the island (see Fig. 1). Data on ice formation were also received from the weather station and completed by our own observations. At the sublittoral stations 5 to 12 a first survey was carried out in 1970, the results of which have been published by Dörjes (1976). Since 1977 samples were taken in monthly intervals. Due to rough seas and ice-drift there have been some gaps. The Van Veen grab with a sampling area of 0.2 m^2 was used. At each station only 1 sample was taken. The bottom fauna was sieved with 0.63 mm mesh aperture and fixed in formaldehyde. The biomass was not determined. Sedimentological data (grain size distribution) have been placed at our disposal by an investigation of Reineck (1976) which is running parallel to ours.

Results

Environmental conditions

The characteristic water temperatures for the area vary (according to monthly averages of long-term measurements on the seaward side) between a maximum of 18.5 °C in July/August and a minimum of 2.0 °C in January/February. The deviations from the average which have appeared between 1976 and 1985 can be seen in Fig. 2. Above average temperatures prevailed in the hot summers of 1976, 1982, and 1983. The cold winters of 1978/79, 1981/82 and 1984/85 are distinguished by ice-drift and a long-lasting ice-cover in the intertidal area.

According to long-term measurements monthly averages of salinity at the south side of Norderney fluctuate between 31‰ (in July) and 28‰ (in January). Altogether the deviations extend from 34‰ to 21‰. The lower values show the influence of the Ems estuary which lies to the west. In the seaward area salinities are higher and more constant and they generally don't drop below 30‰.

The mean tidal amplitude at Norderney is 2.4 m. As the intertidal stations lie on a transect with decreasing level the inundation time increases from ca. 35% at Station 1 to ca. 55% at station 4.

The sediments of the intertidal stations can be classified as sand between fine and medium sand with an average medium value of about 200 μm, a small portion (up to 4%) of silt (< 63 μm), an average loss on ignition of 1–2% and an average water content of 20–30%. The sediment composition is very constant. Since the investigations of 1970 to 1972 (Hauser, 1973) the grain size distribution shows practically no changes (Fig. 3).

The oxidized layer is rather thin, only 0.5 to 3.0 cm deep in summer and autumn, and 0.5 to 8.0 cm deep in winter and spring. This fact can be explained by the sheltered position of the habitat.

The sublittoral stations are situated in water of 5 m (station 5) to 14 m depth (station 12). The sediments consist of fine sand and medium sand (Fig. 3).

Composition of the bottom fauna

In the intertidal stations 35 species occurred (18 annelids, 9 molluscs, 7 crustaceans, and 1 insect), not including several unidentified nemertines (Table 1). The species which were already found in 1970–1972 by Hauser (1973) at station 2 are marked with an asterisk. The most constant members of the community are also dominant ones with respect to density and/or biomass. *Nereis diversicolor, Scoloplos armiger, Heteromastus filiformis, Cerastoderma edule, Macoma balthica, Tubificoides benedeni*, and in a less frequent occurrence, *Pygospio elegans* and *Hydrobia ulvae* belong to the most dominant species in numbers (Fig. 4b). Dominant species according to weight are

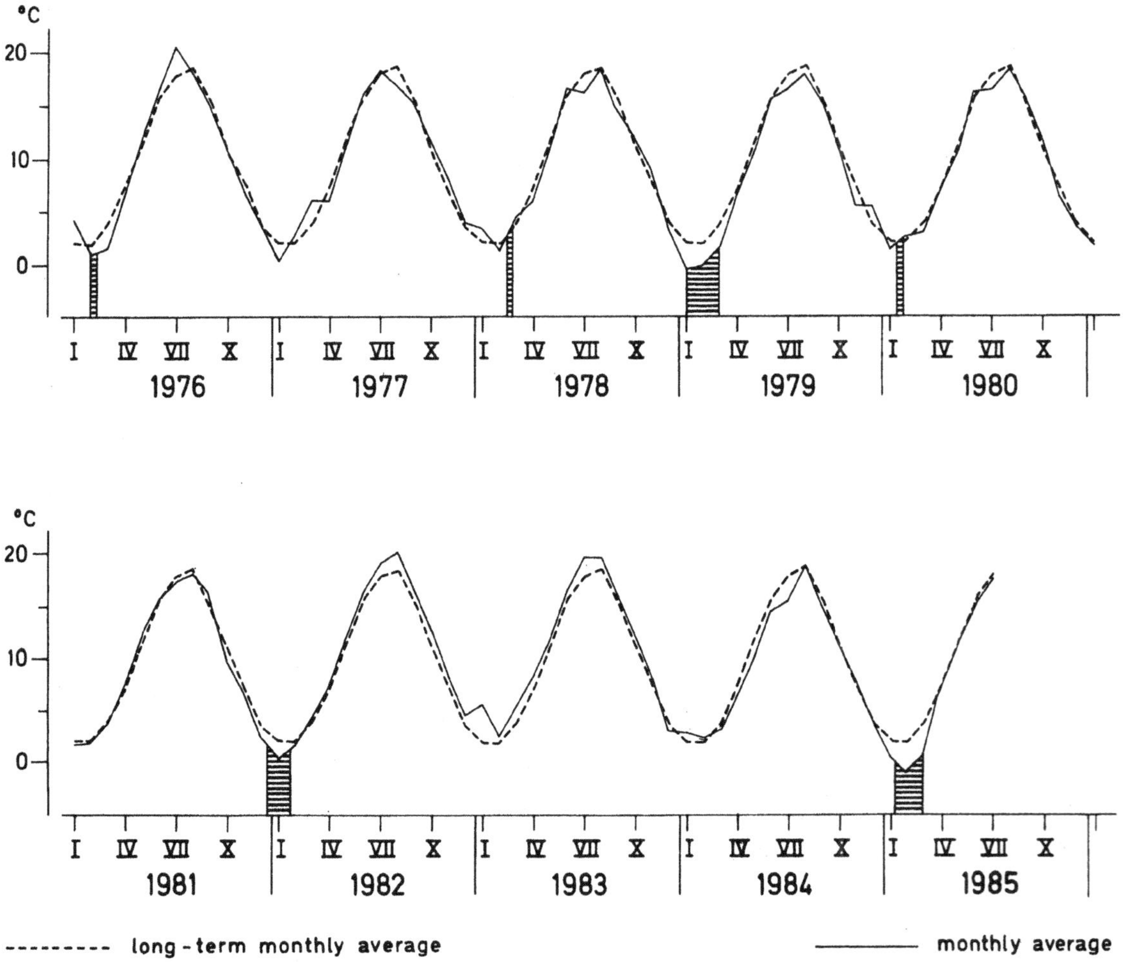

Fig. 2. Mean water temperature measured at the seaward side of the island (monthly means of 1971–1984) and deviations in the period 1976–1984. Hatched: ice-cover on tidal flats.

Arenicola marina, Nereis diversicolor, Scoloplos armiger, Heteromastus filiformis, Cerastoderma edule, and *Macoma balthica* (Fig. 4c).

The sublittoral bottom fauna at stations 5 to 12 is almost three times as rich in species (Table 2). The nemertines excluded, 103 species were found, which are distributed among the various taxonomic groups as follows: 36 polychaetes, 32 crustaceans, 24 molluscs, 5 echinoderms, 2 coelenterates, 2 pisces, 1 phoronid, and 1 pantopod. In the sublittoral there is also a restricted number of species which appear regularly and create dense populations. These are *Magelona papillicornis, Macoma balthica, Scoloplos armiger, Nephtys hombergi* and – less frequent and abundant – *Lanice conchilega, Bathyporeia pelagica, Spio filicornis, Urothoe grimaldii var. poseidonis,* and *Pectinaria koreni.* There are 23 species which occur commonly in the intertidal and in the subtidal habitats (Tab. 1, 2).

Changes in abundance and biomass of dominant intertidal species

Figures 4, 5 and 6 demonstrate the periodic seasonal changes for a number of species in abundance and biomass. This seasonal variation generally shows a maximum in summer or autumn and a minimum in late winter or early spring, often in

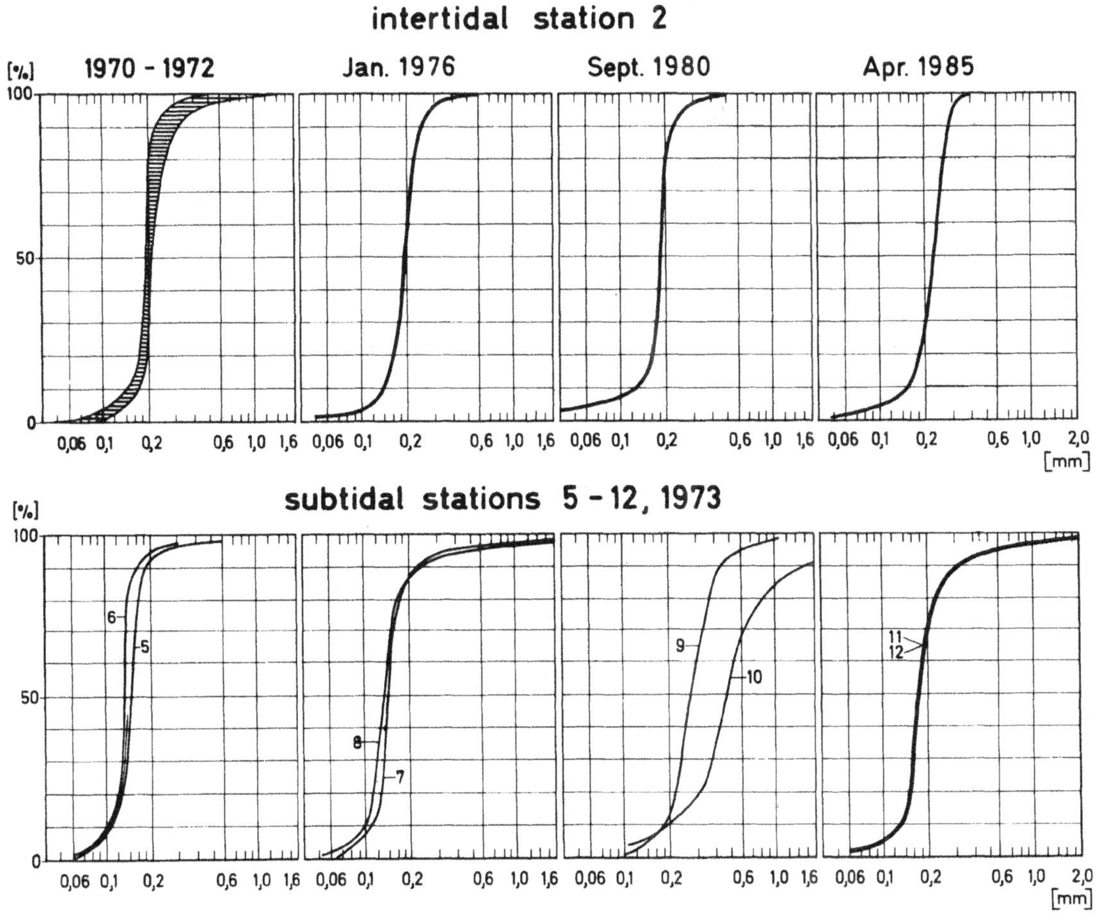

Fig. 3. Above: grain size distribution at station 2. Conditions found by Hauser (1973) compared with later investigations. Below: grain size distribution at station 5–12 in 1973.

April. The total maximum densities amount to 15000–40000 individuals · m^{-2}. At the time of the minimum densities, there are still about 4000–10000 specimens · m^{-2} present.

The maximum values of the total biomass are about 50 to 80 g ADW · m^{-2}, in some years even 150 g and more. The minima lies in general at 20 and 30 g, measured against the maximum weight one fifth to one third.

Figure 4b shows the share of 8 dominant species in the total population density. Figure 4c shows how the total biomass is distributed among the 6 species which are dominant in weight. With respect to the different species the following can be said.

The lugworm *Arenicola marina,* with maximal densities of 40 to 60, is not dominant in numbers (Fig. 5). However, it is significant with respect to weight (10–20% of the total biomass). It has survived the severe winters 1978/79 and 1984/85 with surprisingly high densities (25–35 animals · m^{-2}). Relatively low numbers and biomass values were found in the years 1982–1984.

The ragworm *Nereis diversicolor* has a weak dominance by numbers but a great constancy. Its contribution to the total biomass remains mostly under 10% (Fig. 5).

Scoloplos armiger is a very typical inhabitant of the biotope that was investigated, with densities of mostly more than 600 animals · m^{-2}. The biomass contributes to the total weight with a maximum of 10% (Fig. 5). The mass production of egg capsules in every spring (March/April), which cover the bottom surface, is spectacular. In April 1976 a weight of 0.7 g ADW was determined for an aver-

Table 1. Species found at the intertidal stations 1 – 4 from 1976 to 1985. With asterisk: Species already found by Hauser (1973) at station 2. Underlined: Species common to the intertidal and subtidal transect.

Station No.	1	2	3	4
Nemertini				
Several unidentified species				
Annelida				
*Anaitides mucosa	+	+	+	+
*Arenicola marina	+	+	+	+
*Capitella capitata	+	+	+	+
*Eteone longa	+	+	+	+
*Harmothoe sarsi	+	+	+	+
*Heteromastus filiformis	+	+	+	+
*Lanice conchilega	+	+	+	+
Lumbricullus lineatus	+	+	+	+
*Magelona papillicornis		+		
*Nephtys hombergii	+	+	+	+
*Nereis diversicolor	+	+	+	+
*Paranais litoralis	+			
Polydora ligni	+	+	+	+
*Pygospio elegans	+	+	+	+
Scolelepsis foliosa			+	+
*Scoloplos armiger	+	+	+	+
*Tharyx marioni	+	+	+	+
*Tubificoides benedeni	+	+	+	+
Mollusca				
*Cerastoderma edule	+	+	+	+
Ensis directus	+	+		+
*Hydrobia ulvae	+	+	+	+
*Littorina littorea	+	+	+	+
*Macoma balthica	+	+	+	+
*Mya arenaria	+	+	+	+
*Mytilus edulis		+	+	+
*Retusa obtusa	+	+		
Scrobicularia plana		+	+	+
Crustacea				
*Bathyporeia sarsi	+	+	+	+
*Carcinus maenas	+	+	+	+
*Corophium cf. volutator	+	+	+	+
*Crangon crangon	+	+	+	+
*Gammarus sp.	+	+	+	+
Jaera albifrons		+	+	+
*Urothoe grimaldii var. poseidonis	+	+	+	
Insecta				
Chironomidae (larvae)			+	
Number of species	29	32	31	30

age of 152 capsules · m^{-2}; compared with the coincident weight of the population of 1.3 g ADW · m^{-2} it is a remarkable amount. The high loss of substance due to spawning results to minimal values of the biomass after the spawning time in only a few years (1979, 1981, 1983).

Heteromastus filiformis (Fig. 5) is generally considered to be a species which preferably colonizes muddy, anaerobic sediments. It kept to this rule at the time of Hauser's (1973) investigation and did not exceed densities of 100 individuals · m^{-2} at station 2. In our period of investigation the densities exceed 1000 · m^{-2} for the first time in 1976, and till 1980 rise above 7000 specimens · m^{-2}. After occurrence of relatively low densities in 1981 and 1982 very high densities occur again in 1983. From 1984 on there seems to be an indication of a decline.

In respect of the seasonal fluctuations it is remarkable that maximal and submaximal densities often coincide with the minimum temperature in January, as in 1977, 1978, 1980, 1983 and 1984. The biomass increased to values of 9 g ADW and more after 1980 and contributed 10 to 20% to the total biomass.

Cerastoderma edule is in principle a constant species of the habitat with densities which mostly amount to several hundreds, often 1000 individuals · m^{-2} (Fig. 6). It was temporarily wiped out due to ice-drift in the severe winters of 1978/79 and 1984/85, and was strongly decimated in winters with only short-term ice formation (1975/76, 1979/80, and 1981/82). A new population was established at the deserted spots only due to spatfall. Strong spatfalls led to densities of 8000 to 20000 animals · m^{-2} in the summers of 1976, 1981, and 1984.

The significance of *Cerastoderma* for the total biomass of the habitat is outstanding. 20 to 40 g ADW · m^{-2} are predominant values, the maxima can exceed 100 g. The share of the total biomass amounts to 40–60% over several years. After severe winters the biomass was temporarily reduced to zero, as in 1978/79. After that the biomass improved till 1981 to a level of 10 g · m^{-2}.

The population of *Macoma balthica* shows an extraordinary constant minimum density of 100 to 300 individuals · m^{-2} (Fig. 6). Spatfalls of significant size resulted in the summer 1979 after 2 hard winters and also in 1976, 1981 and 1984 after moderate or mild winters. The maximum of 1979 amounted to almost 8000 individuals · m^{-2}. The biomass shows minima of 1 to 3 g ADW · m^{-2} in late winter or early spring. The maxima, which mostly result in July, reach values of 4 to 14 g

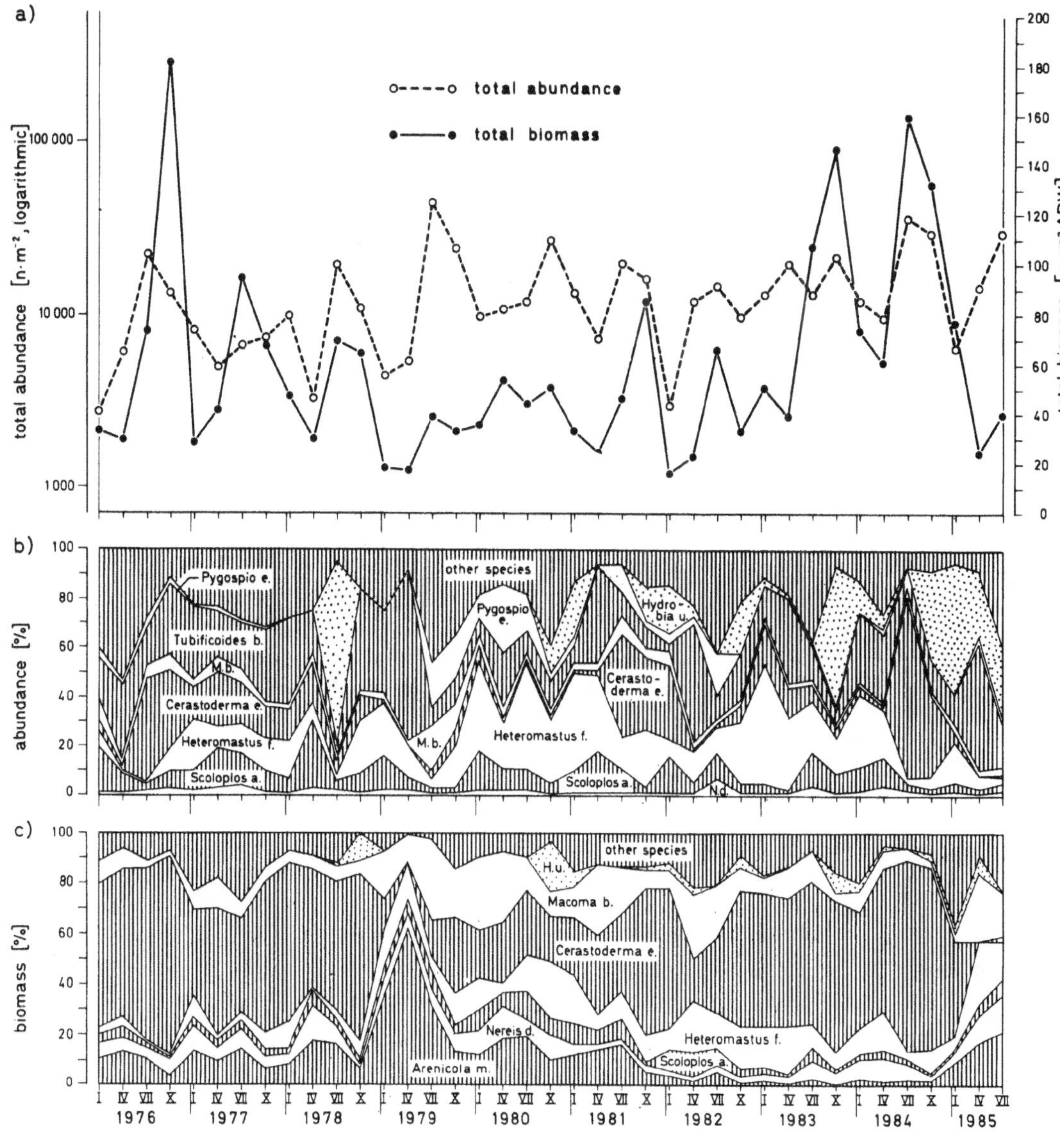

Fig. 4. Intertidal macrobenthos. a) Variations in total abundance (number of individuals) and total biomass (ash-free dry weight). b) Share of dominant species in total abundance. c) Share of dominant species in total biomass. a, b, c: means of the stations 1–4 are demonstrated. N.d.: *Nereis diversicolor*; M.b.: *Macoma balthica*; H.u.: *Hydrobia ulvae*.

ADW · m^{-2}. Correspondingly the share of the total weight fluctuates between 5 and 25%.

Of the other dominant species *Tubificoides benedeni* is the only one which is continuously present in densities of 600–5000 individuals · m^{-2} (20 to 40% of the total abundance). Its contribution to the biomass is insignificant (<3%).

Members of the community with a strongly irregular occurrence are *Pygospio elegans* and *Hydrobia ulvae*. In the short term *Hydrobia* can become a predominant species as happened in July 1978, January 1982, and October 1983 to April 1984 (Fig. 4b) and then it can be of great significance for the biomass (Fig. 4c).

Changes in the abundance of dominant subtidal species

Only very few species are really abundant in the foreshore and the upper sublittoral. Some of them

Table 2. Species found at the subtidal stations 5 – 12 from 1977 to 1984. + rare; + + medium frequent; + + + frequent. Underlined: Species common to the intertidal and subtidal transect.

Station No.	5	6	7	8	9	10	11	12
Water depth (m)	5	6	7	8	9	10	12	14
Nemertini								
Several unidentified species	+	+	+	+	+	+	+	+
Annelida								
Anaitides groenlandica	+	+	+	+	+	+	+	+
Anaitides mucosa	+	+	+	+	+	+	+	(+)+
Capitella capitata	+	+	+	+	+	+	+	+
Eteone lactea			+		+	+	+	+
Eteone longa	+	+	+	+	+	+	+	+
Eulalia viridis	+	+	+	(+)+	+ +	+	+	+ +
Eumida sanguinea			+					
Goniada maculata	+	+	+	+	+	+	+	+
Harmothoe longisetis		+						
Harmothoe lunulata	+	+	+	+	+	+	+	+
Harmothoe sp.	+	+	+	+	+		+	+
Heteromastus filiformis	+	+	+		+			+
Lanice conchilega	+	+	+ +	+ +	+ + +	+ +	+ +	+ +
Magelona papillicornis	+ + +	+ + +	+ + +	+ + +	+ + +	+ + +	+ + +	+ + +
Microphthalmus similis						+		
Nephtys hombergi	+ + +	+ + +	+ + +	+ + +	+ + +	+ + +	+ + +	+ + +
Nereis succinea	+	+					+	+
Nereis virens				+	+			
Ophelia limacina	+	+	+	+	+	+ +	+	+
Owenia fusiformis		+	+					
Pectinaria koreni	+ +	+ +	+ +	+ +	+	+	+	+
Pholoe minuta	+	+	+	+	+		+	+
Pisione remota						+		
Polydora caeca							+	
Polydora ligni	+		+					
Polydora pulchra				+	+			
Pygospio elegans			+					
Scalibregma inflatum								+
Scolelepis bonnieri	+	+	+	+	+	+	+	+
Scolelepis foliosa						+		
Scoloplos armiger	+ +	+ +	+ +	+ +	+ +	+ +	+ + +	+ + +
Sphaerodorum flavum						+	+	
Sigalion mathildae				+				
Spio filicornis	+ +	(+)+	+ +	+ +	+	+	+	+
Spiophanes bombyx	+ +	+ +	+ +	+ +	+ +	+ +	+ +	+ +
Tharyx marioni		+	+	+	+	+	+	+
Mollusca								
Abra alba	+	+	+	+	+		+	+
Abra nitida								+
Abra prismatica					+			
Angulus fabula	+	+	+ +	+ +	(+)+	+	+ + +	+ +
Angulus tenuis		+				+		+
Astarte triangularis	+	+		+	+	+	+	+
Cerastoderma edule	+	+	+					+
Donax vittatus	+	+	+	+	+	+	+	+
Ensis directus	+	+	+	+	+	+	+	+
Hydrobia ulvae	+			+				+
Lunatia nitida		+			+			+
Lunatia sp.			+	+		+	+	
Macoma balthica	+ + +	+ + +	+ + +	+ + +	+ + +	+ + +	+ +	+ +

Table 2. Continued.

Station No. / Water depth (m)	5 / 5	6 / 6	7 / 7	8 / 8	9 / 9	10 / 10	11 / 12	12 / 14
Mactra corallina	+			+	+		+	+
Montacuta ferruginosa	+	(+)+	+	+	+	+	+	+
Mya arenaria	+		+		+		+	+
Mya truncata								+
Mysella bidentata	+	+	+	+	+	+	+	+
Mytilus edulis	+	+	+	+	+	+	+	+
Retusa obtusa	+							
Spisula elliptica				+	+		+	+
Spisula solida	+	+	+	+	+	+	+	+
Spisula subtruncata					+		+	+
Venus gellina					+			
Crustacea								
Ampelisca sp.							+	+
Bathyporeia gracilis							+	
Bathyporeia pelagica	++	++	++	++	++	+	++	++
Bodotria scorpioides							+	+
Caprella linearis				+	+	+	+	+
Carcinus maenas					+	+		
Corophium cf. volutator	+		+		+			+
Corystes cassivelaunus				+				
Crangon crangon	+	+	+	+	+	+	+	+
Cumopsis goodsiri	+	+	+	(+)+	++	+	+	++
Diastylis bradyi	+	+	+	+	+	+	+	++
Eurydice pulchra	+	+	+	+	+	+	+	+
Gammaridea sp. I	+	+	+	+	+	+	+	+
Gammaridea sp. II	++	+	+	(+)+	++	++	(+)+	+
Gammaridea sp. III	+	+	+	+	+	+	+	+
Gammaridea sp. IV	+	+	+	+	+	+	+	+
Gammaridea sp. V		+	+	+	+		+	+
Gammaridea sp. VI			+		+		+	
Gastrosaccus sanctus						+		
Gastrosaccus spinifer		+	+	+	+	+	+	+
Hyas araneus					+			
Idothea linearis	+					+		
Iphinoe trispinosa		+	+	+	+	+	+	+
Isopoda sp.						+		
Lamprops fasciata	+	+	+	+	+	+	+	+
Liocarcinus holsatus	+	+	+	+	+	+	+	+
Mesopodopsis slabberi	+	+	+	+		+	+	+
Pagurus bernhardus	+		+	+	+	+	+	+
Paramysis kervillei	+	+	+	+	+	+	+	+
Pinnotheres pisum	+				+			
Processa canaliculata		+						+
Urothoe grimaldii var. poseidonis	+	+	+	(+)+	++	+	++	++
Echinodermata								
Amphiura filiformis						+		
Asterias rubens	+	+	+	+	+	+	+	+
Echinocardium cordatum	+	+	+	+	+	+	+	+
Ophiura albida	+	+	+	+	+	+	+	+
Ophiura texturata							+	
Coelenterata								
Edwardsia cf. danica								+
Sagartia troglodytes					+			

Table 2. Continued.

Station No.	5	6	7	8	9	10	11	12
Water depth (m)	5	6	7	8	9	10	12	14
Pisces								
Hyperoplus lanceolatus		+		+		+	+	+
Pomatoschistes minutus							+	
Phoronida								
Phoronis mülleri					+	+		
Pantopoda								
Nymphon sp.								+
Number of species	58	59	61	61	69	62	68	73

are *Macoma balthica, Scoloplos armiger, Nephtys hombergi* and *Magelona papillicornis,* the annual cycles of which (mean values of stations 5–12) are shown in Fig. 7. The first two species do also occur abundantly in the tidal flat habitat. Therefore comparisons of fluctuations and dynamic processes of these species are possible.

Magelona papillicornis is the most abundant species in the upper sublittoral. In certain years densities of more than 2000 specimens · m^{-2} are reached. In contrast to the winter minima the overall low population densities in 1977 to 1979 as well as 1981 are difficult to explain. Maxima occur only in 1980, 1982, and 1983.

Nephtys hombergi is represented in the upper sublittoral in much lower numbers than *Magelona papillicornis.* The highest numbers with more than 100 specimens·m^{-2} are reached in summer or early autumn. In 1977, 1978 as well as in 1980 and 1981 no distinct maxima are developed.

After *Magelona papillicornis* the polychaete *Scoloplos armiger* is the second common species in the foreshore and the upper sublittoral with distinct maxima in summer and distinct minima during winter times. An exception of this cycle occurred in 1983 probably due to a low spatfall. Generally for *Scoloplos armiger* a decreasing trend in the number of individuals can be determined since 1978.

In the early years of investigation *Macoma balthica, Nephtys hombergi,* and *Magelona papillicornis* show low densities, and recruitment is failing or rather low. Since 1979 (*Nephtys*) respectively 1980 (*Macoma, Magelona*) recruitment in these species is more successful and their abundance is raised to higher levels.

Discussion

The investigation of permanent sampling stations in the East Frisian intertidal and shallow subtidal area resulted in two habitats which are rather different in species richness and community structure. The intertidal bottom fauna is distinguished by great constancy, in spite of the seemingly unstable environment. It is on the whole poor in species (35), and only a small group of species is significant for the formation of the stock (total abundance, total biomass). This is a general feature of the intertidal macrofauna of the Wadden Sea as described by Hauser (1973), Beukema (1974, 1981), Meyer & Michaelis (1980), Obert (1982), and others. In the habitat investigated here, the basic pattern of composition is formed by 8 species and has maintained itself unchanged from 1976 to 1985. Of course, there have been certain alternations in the order of precedence.

In contrast to the eulittoral, constancy of species composition in the sublittoral is considerably less pronounced. Many of the great number of species (103) only have a sporadic appearance. Others, like *Pectinaria koreni* and *Donax vittatus,* develop large populations in certain years but in other years they are unimportant in quantitative respect, or disappear completely. Possibly instability of the substrate is a controlling factor. Also in the sublittoral the number of species which occur regularly is restricted. However, with exception of *Magelona papillicornis,* they cannot be called abundant or dominant in the same sence as the dominating species of the eulittoral. The abundance of *Macoma balthica* and *Scoloplos armiger,* for instance, is one

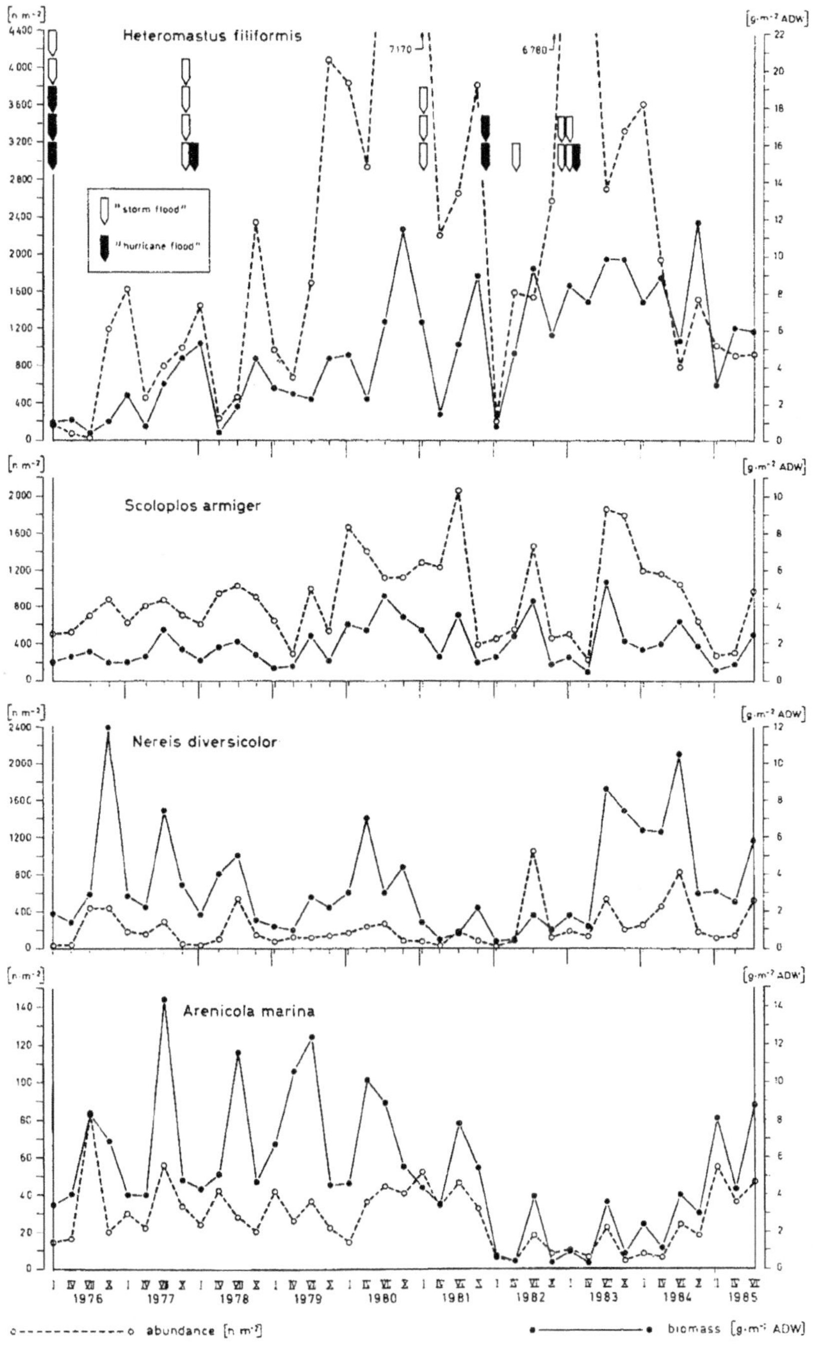

Fig. 5. Intertidal macrobenthos. Variations in abundance and biomass (means of stations 1–4) of dominant species (polychaetes). Each arrow indicates one 'storm flood' respectively 'hurricane flood'. Several storm surges occurring within a short time are represented by arrows standing above each other.

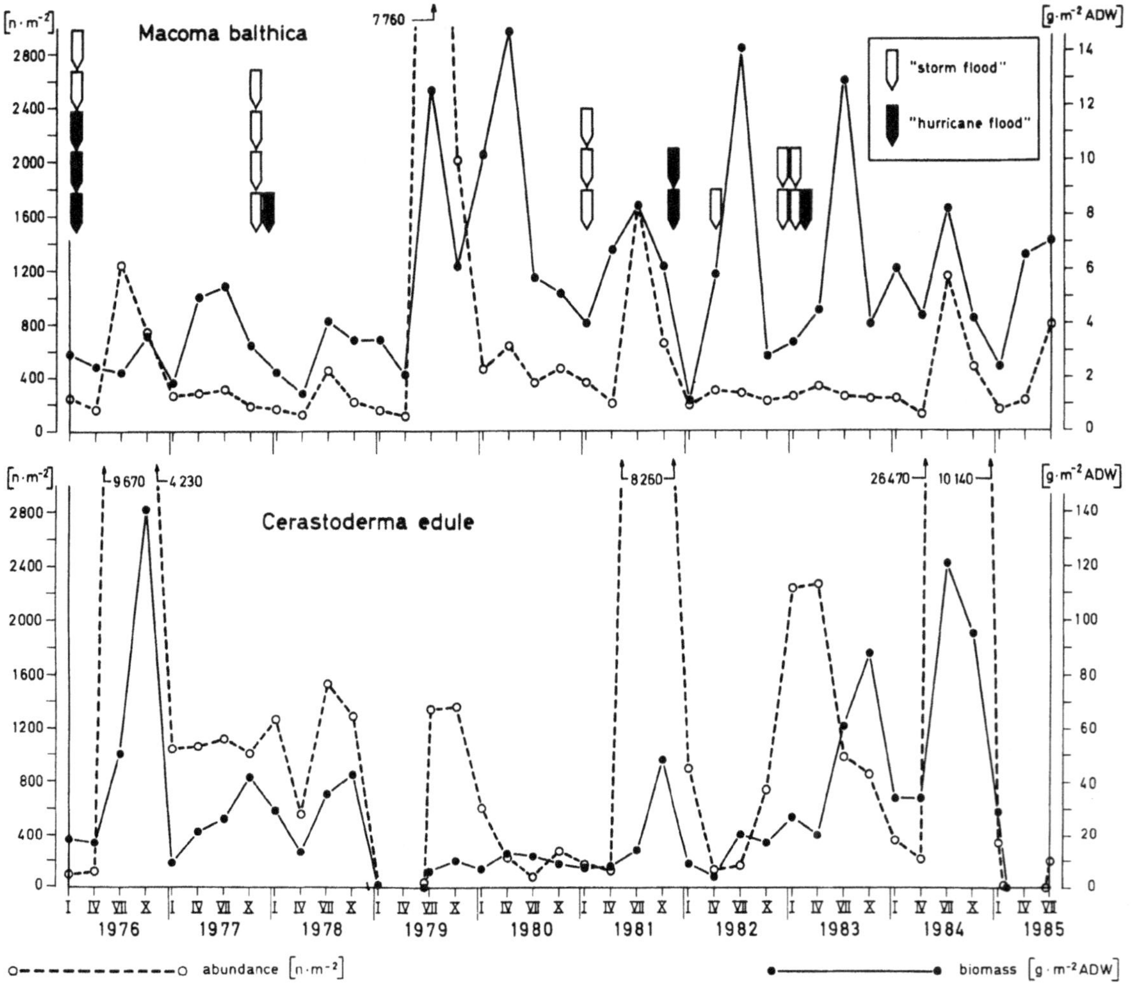

Fig. 6. Intertidal macrobenthos. Variations in abundance and biomass (means of stations 1–4) of dominant species (bivalves). Each arrow indicates one 'storm flood' respectively 'hurricane flood'. Several storm surges occurring within a short time are represented by arrows standing above each other.

order of magnitude lower in the sublittoral than in the intertidal area.

In the variation of abundance and biomass the seasonal fluctuations are the most obvious ones, as well in the individual species as in the whole communities. The underlying processes are (Beukema 1974): from spring to summer and autumn a net increase due to settlement of larvae, to immigration and to growth of the individuals; from autumn to winter and spring a net loss due to the predominance of mortality, migration and the losses of weight of the individuals. In bivalves it is considered to be the rule, that particularly successful spatfalls follow directly after severe winters (literature reviewed by Beukema, 1982). In the present investigation that applies only to one of the cases observed: *Macoma* had a rich spatfall in 1979 in the intertidal, but not in the subtidal stations. High spatfalls of *Cerastoderma* in the intertidal area were preceded either by moderate (1976) or by mild, ice-free winters (1981, 1984).

Referring to the total biomass of the intertidal habitat, the highest summer values (50–180 g ADW $\cdot$ m^{-2}) and the lowest winter values (15–60 g ADW $\cdot$ m^{-2}) are above the average, certainly due to the dense cockle population. By comparison, the

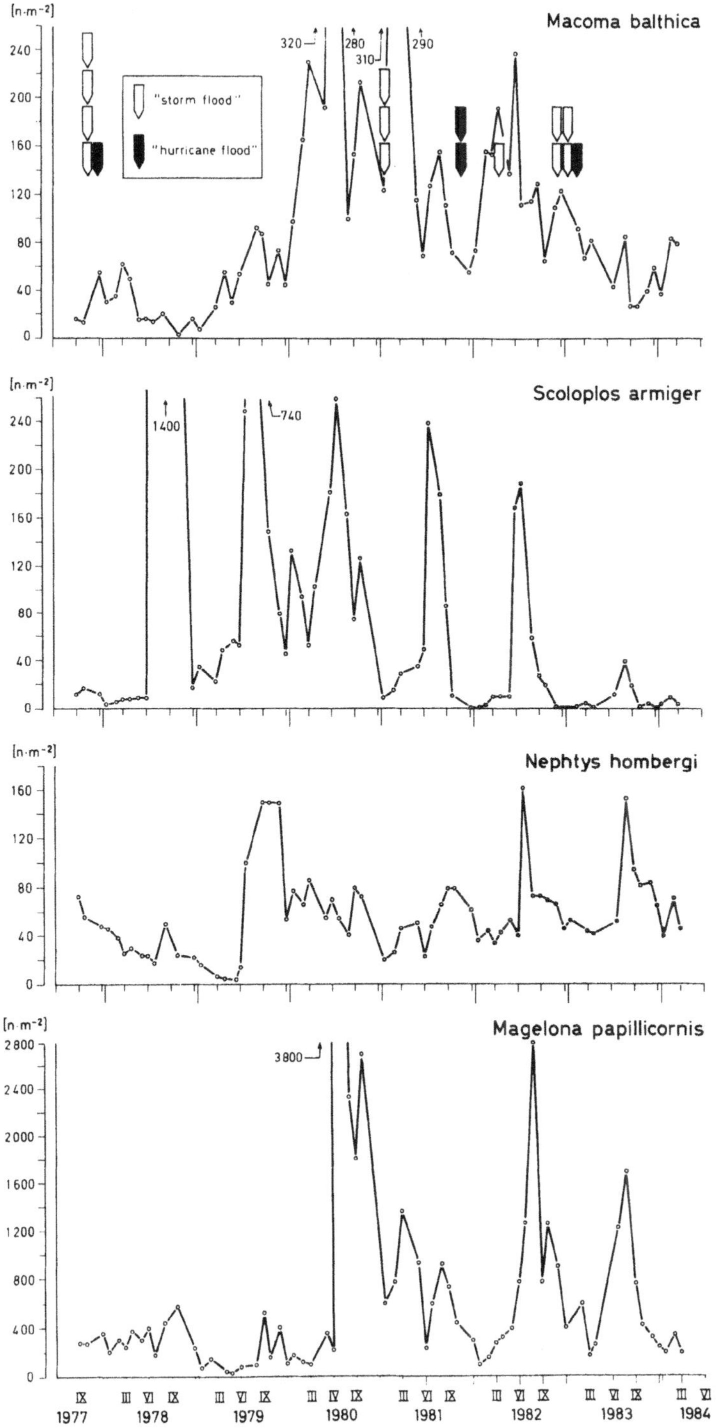

Fig. 7. Subtidal macrobenthos. Development of abundance (means of stations 5–12) of dominant species. Each arrow indicates one 'storm flood' respectively 'hurricane flood'. Several storm surges occurring within a short time are represented by arrows standing above each other.

flats near the island Borkum, west of Norderney, revealed a mean summer biomass of 36 g ADW · m^{-2} (Obert, 1982) and from the western Dutch Wadden Sea winter values varying from 14 to 24 g ADW · m^{-2} were reported (Beukema, 1981). In the shallow subtidal area biomass has not been measured. Informations from similar habitats in the southern North Sea are not available.

Except the more or less regular seasonal variations there are single extreme situations which cause the most drastic changes. In the case of temperature the benthic community experiences great shocks due to severe winters, as has often been described for the eulittoral as well as the sublittoral (Smidt, 1944; Ziegelmeier, 1964; Hauser, 1973; Beukema, 1979; Dörjes, 1980; and others). In the time of investigation there are two of such winters with an iceing-up of the intertidal flats over a longer period, namely in 1978/79 and in 1984/85. In the intertidal *Scoloplos, Nereis,* and *Macoma* had to sustain slight losses, but hardly affected was the predominant species in the habitat, the cockle. The same about this species has been reported by numerous other authors (e.g. Kristensen,1958; Hauser, 1973; Essink, 1978; Beukema, 1979; Dörjes, 1980; Madsen, 1984). The cockle, which was previously the second in numbers and the first according to biomass, disappeared in 1979 and 1985 until the new spatfall in the summer.

Concerning the sublittoral, it would seem that severe winters cause greater damage to the fauna than in the eulittoral. This was the case in the winter 1978/79. The abundance of *Nephtys* and *Magelona* dropped close to zero, while *Macoma* and *Scoloplos* were less clearly affected. From the species represented in Table 2 a great part disappeared and the regeneration of the whole community required a number of years. According to Ziegelmeier (1964, 1978), Rachor & Gerlach (1978), and Dörjes (1980) large parts of the benthos were destroyed in the winters of 1962/63 and 1978/79 in depths of up to 30 m. A new colonization of completely different structure follows and former conditions are only reestablished if the development is undisturbed over several years (Rachor & Gerlach 1978).

The other factor leading to irregular and drastic changes in the benthic communities is wave action caused by strong gales. In the period of our investigations some series of 'storm floods' and 'hurricane floods' (according to the classification used in Germany) have occurred at times which can be seen in Figs 5, 6, and 7. In the intertidal bottom fauna they have caused noticeable declines in the development of *Heteromastus* and *Macoma* during the winters 1975/76, 1977/78 and especially after the 'hurricane floods' of November 1981 (Fig. 5, 6). *Arenicola* only reacted upon these last events (November 1981) with a serious decrease. There followed a period of almost three years with very low densities, but that may also be attributed to other reasons. Slight effects can be supposed in *Cerastoderma* (Fig. 6) while *Scoloplos* and *Nereis* seem to be able to withstand the short-term effects of heavy wave action without suffering greater losses (Fig. 5).

Total abundance and biomass are reduced to values belonging to the lowest of the whole investigation period (Fig. 4). Therefore, it may be concluded, that exceptionally strong gales, like in the winters 1975/76 and 1981/82, have similar effects as low temperatures in severe winters, like 1978/79 and 1984/85.

There is evidence, that great damage has also been done to the sublittoral bottom fauna by gales, probably because sediments are strongly disturbed and reworked by wave action. *Macoma, Scoloplos,* and *Magelona* show deep depressions in their abundance especially after the two 'hurricane floods' in November 1981. An exception is given by *Nephtys,* probably due to its ability of quick movement in the sediment.

As in the case of the severe winter 1978/79, also after the storms of November 1981 the total number of species was reduced considerably and did not recover completely until 1984, the present state of the evaluation.

The significance of the wave action factor for the spatial distribution of benthic animals has already been pointed out earlier by Muus (1967) and Wolff (1973). Therefore, equally strong influences by this factor must be accepted in the temporal demension. As far as the Wadden Sea fauna is concerned, questions have rarely been asked in this direction. In the case of the cockle the observation was made that it is susceptible to erosion, is swept out of the bottom by wave action, and is transported down to the channels by currents (Schäfer, 1962; Ohde, 1981).

A study focussing directly on this subject was

carried out by Madsen (1984) in the Danish Wadden Sea, using the opportunity of the severe storm flood in November 1981. He found a reduction of total abundance and biomass of 20–25%. Compared to Norderney, there were partly similar, partly different reactions of the individual species. Madsen concludes, that on the whole the losses of macrofauna caused by gales may be relatively small in the Danish area. The sublittoral communities of the southern North Sea are known to be rather sensitive to the influence of storms. Rachor & Gerlach (1978) assume that the seasonal decline of the bottom fauna is not only caused by low temperature and a lack of food, but also by increased intensity of wave action. In addition, they consider strong storms to be a frequent cause of destruction of the sublittoral bottom fauna in depths of up to 30 m. For example, Rachor (1980) described the devastating effects of heavy gales in January 1976 on a mud bottom community in the German Bight.

At the attempt of interpreting the fluctuations of individual species and the whole community, biotic interactions remained largely unconsidered, as this investigation was restricted to the endobenthic sector of the ecosystem. Neither the changing food supply can be referred to as explanation, nor the predation pressure, due to shrimps, crabs, fishes and birds, the outstanding significance of which for the structure, density and biomass of the intertidal fauna has been shown by Reise (1977a, b, 1985), de Vlas (1979), and Smit (1980). Some observations could be made on parasitic infestation of intertidal bivalves. Infection by trematode larvae resulted in mortality of *Macoma* and *Cerastoderma* in certain years (Michaelis 1981). During the summer months animals were found which had left the bottom and lay weakened, dying or dead on the surface. However, the losses did not cause obvious declines of the populations.

As far as long-term changes are concerned, only in one species a clear trend of a recently deviating development can be recognized: the intertidal population of *Heteromastus filiformis* started in 1977 to spread out from its traditional muddy habitat to sandy sediments and establish itself there as a dominant species (Fig. 5). It would appear to be a widespread phenomenon, since 1978 a multiplication of density was determined near Texel in the Dutch Wadden Sea (Cadée, 1979) and this species was distributed over a large area and in great densities in 1981 on sandy flats of the Ems estuary (Obert, 1982). With regard to the other species and to the whole community the range of seasonal and annual fluctuations is too wide as to distinguish long-term changes after the period of ten, respectively seven years of investigation. However, it may be finally mentioned, that Beukema and Cadée (in press) recently could prove benthos responses to eutrophication in the western Dutch Wadden Sea. During 1970–1984 they found in half of the common species an increase in numbers and biomass and a doubling of the total biomass.

Acknowledgement

We are indebted to Mr H. H. Kramer for the ten years of patient sorting, counting and weighing, to Mr B. Loveridge for the translation from German into English, and to Mr G. Meyer for the drawings.

References

Beukema, J. J., 1974. Seasonal changes in the biomass of the macrobenthos of a tidal flat area in the Dutch Wadden Sea. Neth. J. Sea Res. 8: 94–107.

Beukema, J. J., 1979. Biomass and species richness of the macrobenthic animals living on a tidal flat area in the Dutch Wadden Sea: effects of a severe winter. Neth. J. Sea Res. 13: 203–223.

Beukema, J. J., 1981. Quantitative data on the benthos of the Wadden Sea proper. In N. Dankers, H. Kühl & W. J. Wolff (eds), Invertebrates of the Wadden Sea. Balkema, Rotterdam: 134–142.

Beukema, J. J., 1982. Annual variation in reproductive success and biomass of the major macrozoobenthic species living in a tidal flat area of the Wadden Sea. Neth. J. Sea Res. 16: 37–45.

Beukema, J. J., W. de Bruin & J. J. M. Jansen, 1978. Biomass and species richness of the macrobenthic animals living on the tidal flats of the Dutch Wadden Sea: long-term changes during a period with mild winters. Neth. J. Sea Res. 12: 58–77.

Beukema, J. J. & G. C. Cadée, in press. Benthos responses to eutrophication of the Dutch Wadden Sea.

Cadée, G. C., 1979. Sediment reworking by the polychaete *Heteromastus filiformis* on a tidal flat in the Dutch Wadden Sea. Neth. J. Sea Res. 13: 441–456.

Dales, R. P., 1951. An annual history of a population of *Nereis diversicolor*. Biol. Bull. Woods Hole 101: 131–138.

Dörjes, J., 1976. Primärgefüge, Bioturbation und Makrofauna als Indikatoren des Sandversatzes vor Norderney (Nordsee). II. Zonierung und Verteilung der Makrofauna. Senckenbergiana marit. 8: 171–188.

Dörjes, J., 1980. Auswirkungen des kalten Winters 1978/79 auf das marine Makrobenthos. Natur u. Museum 110: 109–115.

Essink, K., 1978. The effects of pollution by organic waste on macrofauna in the eastern Dutch Wadden Sea. Netherlands Institute for Sea Research, Publication Series No. 1, 144 pp.

Gibbs, P. E., 1968. Observations on the population of *Scoloplos armiger* at Whitstable. J. mar. biol. Ass. U.K. 48: 225–254.

Hauser, B., 1973. Bestandsänderungen der Makrofauna an einer Station im ostfriesischen Watt. Jber. Forschungsstelle f. Insel- u. Küstenschutz Norderney 24: 171–204.

Jepsen, U., 1965. Die Struktur der Wattbiozönosen im Vormündungsgebiet der Elbe. Arch. Hydrobiol., Suppl. 29: 252–370.

König, D., 1943. Vergleichende Bestandsaufnahme an bodenbewohnenden Wattieren im Bereich des Sicherungsdammes vor Friedrichskoog 1935–39. Westküste 1943: 120–172.

Kristensen, I., 1958. Differences in density and growth in a cockle population in the Dutch Wadden Sea. Arch. Neerl. Zool. 12: 351–453.

Lappalainen, A. & P. Kangas, 1975. Littoral benthos of the northern Baltic Sea. II. Interrelation of the wet, dry and ashfree dry weights of macrofauna in the Tvärminne area. Int. Revue ges. Hydrobiol. 60: 297–312.

Madsen, P. Brinch, 1984. The dynamics of the dominating macrozoobenthos in the Danish Wadden Sea 1980–1983. Rep. Mar. Pollut. Lab. 7: 1–35.

Meyer, M. & H. Michaelis, 1980. Das Makrobenthos des westlichen 'Hohen Weges'. Jber. Forschungsstelle f. Insel- u. Küstenschutz Norderney 31: 91–155.

Michaelis, H., 1981. Recent Changes in the Communities of the Wadden Sea – natural Phenomena or Effects of Pollution? In S. Tougaard & C. Helweg Ovesen (eds), Environmental Problems of the Waddensea-Region. Fiskeri-og Søfartsmuseets, Saltvandsakvariets Biologiske Meddelelser 5: 87–95.

Muus, B. J., 1967. The fauna of Danish estuaries and lagoons. Distribution and ecology of dominating species in the shallow reaches of the mesohaline zone. Medd. Danmark Fisk. Havundersøg. (n. ser.) 5: 1–316.

Obert, B., 1982. Bodenfauna der Watten und Strände um Borkum – Emsmündung. Jber. Forschungsstelle f. Insel- und Küstenschutz Norderney 33: 139–162.

Ohde, J., 1981. Entstehung von Besiedlungsmustern der Makro-Endofauna im Wattenmeer der Elbe-Mündung. Thesis, University of Hamburg, 445 pp.

Rachor, E., 1980. The inner German Bight – an ecologically sensitive area as indicated by the bottom fauna. Helgol. Meeresunters. 33: 522–530.

Rachor, E. & S. A. Gerlach, 1978. Changes of macrobenthos in a sublittoral sand area of the German Bight 1967–1975. Rapp. P.- v. Réun Cons. perm. int. Explor. Mer 172: 418–431.

Reineck, H.-E., 1976. Primärgefüge, Bioturbation und Makrofauna als Indikatoren des Sandversatzes im Seegebiet vor Norderney (Nordsee). I. Zonierung von Primärgefügen und Bioturbation. Senckenbergiana marit. 8: 155–169.

Reise, K., 1977a. Predation pressure and community structure of an intertidal soft-bottom fauna. In B. F. Keegan, P. O'Céidigh & P. J. S. Boaden (eds), Biology of Benthic Organisms. Pergamon Press, New York: 513–519.

Reise, K., 1977b. Predation exclusion experiments in an intertidal mud flat. Helgol. wiss. Meeresunters. 30: 263–271.

Reise, K., 1985. Tidal Flat Ecology. Springer-Verlag, Berlin. 191 pp.

Schäfer, W., 1962. Aktuo-Paläontologie nach Studien in der Nordsee. Verlag Waldemar Kramer, Frankfurt a. M., 666 pp.

Smidt, E. L. B., 1944. Das Wattenmeer bei Skallingen, 3. The effects of icewinters on marine littoral fauna. Folia geogr. dan. 2: 1–36.

Smit, C. J., 1980. Production of biomass by invertebrates and consumption by birds in the Dutch Wadden Sea area. In C. J. Smit & W. J. Wolff (eds), Birds of the Wadden Sea. Balkema, Rotterdam: 290–301.

Thamdrup, H. M., 1935. Beiträge zur Ökologie der Wattenfauna auf experimenteller Grundlage. Medd. Danmarks Fisk. Havundersøg., Ser. Fisk. 10(2): 1–125.

Vlas, J. de, 1979. Annual food intake by plaice and flounder in a tidal flat area in the Dutch Wadden Sea, with special references to consumption of regenerating parts of macrobenthic prey. Neth. J. Sea Res. 13: 117–153.

Wolff, W. J., 1973. The estuary as a habitat. An analysis of data on the soft bottom macrofauna of the estuarine area of the rivers Rhine, Meuse, and Scheldt. Zoologische Verhandelingen 126: 1–242.

Ziegelmeier, E., 1964. Einwirkungen des kalten Winters 1962/63 auf das Makrobenthos im Ostteil der Deutschen Bucht. Helgol. wiss. Meeresunters. 10: 276–282.

Ziegelmeier, E., 1978. Macrobenthos investigations in the eastern part of the German Bight from 1950 to 1974. Rapp. P.-v. Réun. Cons. perm. int. Explor. Mer 172: 432–444.

Hydrobiologia 142: 233–248 (1986)

Recruitment and year-to-year variability in a population of *Macoma balthica* (L.)

Guy Bachelet
Institut de Biologie Marine, Université de Bordeaux I, 2, rue du Professeur Jolyet, 33120 Arcachon (France)

Keywords: *Macoma balthica*, population dynamics, recruitment, juvenile growth, long term changes

Abstract

Because of methodological problems, macrobenthic studies usually neglect the juvenile stages of invertebrate communities, due to the fact that appearance of recruits in samples is only detected some weeks or even months after their true recruitment. During this period, the temporary meiobenthos undergoes high rates of mortality. From year to year, juvenile survival rate is thus responsible for temporal patterns observed in adult population densities.

The results presented here relate to the population dynamics of the tellinid bivalve *Macoma balthica* (L.). A study of temporary meiobenthos was conducted over two consecutive years in an intertidal *Macoma*-community located at the mouth of the Gironde Estuary in southwest France. Sampling of juvenile stages required short intervals (2 weeks) between successive samplings and a fine sieving mesh size (63 μm). Other population parameters, such as temporal patterns in density, reproductive cycle, and individual growth, were recorded.

Recruitment processes showed a year-to-year variability, with regard to settlement density, settlement period, and survival rate. In 1983, recruitment was moderate and protracted over several months. Only one main recruitment period was detected in 1984, resulting in a high juvenile density. In a previous study (1977), by contrast, recruitment was almost non-existent.

This variability is discussed as a function of climatic and sedimentological conditions which prevailed in the estuary throughout the study period. However, none of these physical factors appeared to underlie the recruitment fluctuation in *Macoma balthica*. It is suggested that biological interactions are of prime importance in regulating population densities in this community.

Introduction

It is a well known fact that the abundance of marine and estuarine benthic invertebrate populations fluctuate seasonally. It follows that in a natural environment, without any source of pollution or man-induced changes, benthic communities may present a different specific composition from year to year, particularly when the species have a life span of about one year (Eagle, 1975). Most of the variability in the seasonal and long-term dynamics of benthic communities can be attributed to the success or failure of a larval recruitment and the subsequent survival of juvenile stages. Long-term monitoring of temporal fluctuations has often shown different levels of recruitment during consecutive years (Glémarec, 1978; Beukema, 1980, 1982; Vahl, 1982; Persson, 1983; Diaz, 1984; Holland, 1985; Nichols, 1985). It appears then that the size of the settlement as well as the survival of recruits and individual growth rate reflect good and bad conditions.

Adequate knowledge of the causes of natural variability in marine communities requires detailed information on all the repopulation processes (including gonad development, survival of larvae, settlement density and post-settlement survival and growth) which are potentially the most sensitive to environmental constraints. Quantitative studies on recruitment are still scarce and long-term data are lacking. This lack of information on recruitment events proceeds in a great part from inaccurate sampling methods (Bachelet, 1984). In this respect, two main features in the biology of benthic invertebrates must be kept in mind. First, larval settlement is a transient phenomenon and survivorship of newly settled juveniles is usually low (Cattaneo & Massé, 1983; Diaz, 1984; Luckenbach, 1984; Powell *et al.*, 1984). Secondly, recruits are of meiofaunal-size and thus should be extracted from soft sediments with a fine sieving mesh (Williams, 1980; Bachelet, 1985). With usual methods employed by macrobenthologists, the first appearance of recruits in samples may sometimes be detected some weeks or even months after their true recruitment. During this period, the temporary meiobenthos (i.e. the post-larvae) undergoes high rates of mortality and the level of recruitment can thus be considerably under-estimated.

The tellinid bivalve *Macoma balthica* (L.) is one of the species for which the largest amount of data on biology and ecology has been published. Occurring both intertidally and in shallow subtidal marine and estuarine waters, this species is widely distributed in North-West Europe and on the two northern coasts of North America (Bachelet, 1980; Beukema & Meehan, 1985), covering a large scale in latitude. Though long-term surveys on *M. balthica* are available, particularly in the Dutch Wadden Sea (Beukema & De Bruin, 1977; Beukema *et al.*, 1977, 1985; Beukema, 1980, 1982), precise data relating to recruitment have only been reported by Caddy (1969) and Ankar (1980).

This paper reports on a two-year study of recruitment in an intertidal population of *Macoma balthica* located in the Gironde estuary, SW France, which represents the southern limit of the geographical distribution of the species in European waters (Bachelet, 1980). Yearly variations in reproductive cycle, recruitment and growth were examined for 1983 and 1984, and are compared with a previous survey in 1977.

Material and methods

Macoma balthica were sampled from an intertidal mud flat at Le Verdon (Station 1 in Bachelet, 1980) near the mouth and on the left shore of the Gironde Estuary (45°33′N, 1°03′W). This station was located approximately 2 m above MLW (percentage of time drained: about 50%) in a sheltered bay protected from the strongest waves by a rocky dam.

For a study of recruitment processes, a survey strategy must contain frequent samplings to monitor the large, rapid changes that occur in spatfall densities. Weekly collections are probably an ideal strategy, however, sorting of so many samples is too time-consuming. Moreover the distance between the sampling site and the laboratory (150 km) did not allow for weekly sampling. Samples were thus taken approximately every fortnight from January 1983 to January 1984. During the second year survey (1984), the sampling frequency was reduced to monthly intervals.

Faunal samples were collected on each occasion using cores of two sizes. To sample large *Macoma balthica*, three to five replicate cores were taken with an aluminium box corer (20 cm × 20 cm × 30 cm deep) from random locations within a fixed 20 × 20 m plot. *M. balthica* were never found deeper than 10 cm. The depth of excavation was thus limited to 15 cm. Juveniles were usually too numerous to be counted in large core samples, so smaller cores were used to collect them. Three to eight (mean = 5) plexiglass cores (inner diameter 3.2 cm, sampling area 8 cm^2, depth 2 cm) per sampling occasion were analysed for the meiofaunal-sized spat.

All cores of both sizes were fixed with 10% buffered formalin before sieving. Rose Bengal was added to small cores in order to facilitate the sorting of juveniles. Sieving was performed in the laboratory. The 400 cm^2 macrofaunal samples were sieved through a 400 μm screen that retained all individuals with lengths $\geq$0.8 mm. The 8 cm^2 cores were first sieved through a 400 μm mesh and then a 63 μm screen. Trials with a 100 μm screen showed that this size was sufficient enough to collect the smaller spat, so the 63 μm screen was omitted. A similar mesh size was used by Ankar (1980) with *M. balthica*, and by Williams (1980) and Luckenbach (1984) for other bivalves.

Sorting, identification of juveniles and measurements were made under a dissecting microscope with an eyepiece micrometer. The maximum shell length of all specimens was recorded with a 0.01 mm accuracy (spat <1.5 mm length) or to the nearest 0.1 mm (other specimens). Growth rings on the shell were measured in the same way. Clams of less then 10 mm length were separated into cohorts based on plots of shell lengths vs frequency through time; in this size range, the number of rings were used to separate the generations. Unlike a previous survey (Bachelet, 1980, 1982), growth cessation rings could not be used to separate year classes when shells were marked by more than one ring, because there were no distinct rings or they were too numerous to indicate a clear biological pattern.

To determine gonad condition, 12 *Macoma* over 10 mm in length were selected from each sample and fixed in Bouin's fluid. Two approaches were used (Caddy, 1967). First, the visceral mass was examined under a dissecting microscope after removal of the shell valves and the macroscopic appearance of the gonad was used to assess grossly the state of development. When the gonad tissue was developing and imbricated in the dark brown digestive gland, the animal was considered mature and when gonad regression was observed it was ascribed to spawning. After macroscopic examination, the visceral mass was embedded in paraffin (melting point 56 °C). Transverse 7 μm sections were cut along the dorso-ventral axis at the regions of the foot, crystalline style and the gut. Sections were mounted on slides and stained with hematoxylin and eosin. The state of gonad development was determined according to Caddy (1967) and Lammens (1967). The arbitrary scheme of gonad classification (spent, developing, ripe, spawning) and the mean gonad index of Brown (1982) were used to describe the seasonal gonad changes in *M. balthica*.

Analysis of grain size and organic carbon in the sediment was performed on the top 2 cm. At each sampling date, four cores (3.5 cm in diameter) were taken at the same site as the biological samples. Silt and clay content (particles <63 μm) was estimated by wet sieving the sediment from two cores. Samples for organic carbon content were treated with N hydrochloric acid to remove carbonates and bicarbonates. As a rough estimate of organic content, the loss on ignition (L.O.I.) was determined as the difference between the dry weight (100 °C, 48 h) and ashfree dry weight (600 °C, 2 h) of decalcified sediment. Silt and clay and organic matter contents are given as percentage of the dry sediment.

The only physical measurements taken on the field were sediment temperature near the surface ($\approx$5 mm) and at 15 cm depth. Meteorological data were gathered from the monthly weather reports issued by the Météorologie Nationale (Bulletin Climatologique du Sud-ouest); data were used from the weather station of Le Verdon. Hydrological data (water flow, temperature, salinity, nutrient contents) were obtained from stations 1 and 2 of the Réseau National d'Observation in the Gironde estuary.

Results

Environmental variables

Monthly average values (mean of the years 1976–1983) for hydrological data at the mouth of the Gironde estuary are presented in Fig. 1, together with mean monthly water flow for the period 1961–1970.

Mean monthly salinity of surface waters vary between 19.4 and 29.5‰. Highest salinities are recorded in August–October, during a period with little rainfall and low fluvial discharge; lowest surface salinities occur in January–May, due to swellings of the Gironde estuary. At Le Verdon, the extremes of the salinity range throughout the year are 33.7–18.0‰.

Mean monthly surface water temperatures vary between 7.2 (January) and 20.1 °C (August) but shallow water moving onto or off the flat with the tide may reach more extreme temperatures. Measurements of sediment temperatures at low tide showed a thermic inertia at 15 cm depth while surface sediment temperatures were always higher than air or water temperatures, due to solar irradiation during daytime. Linear relationships with highly significant correlations were found between air (T_A) and sediment temperatures (T_S):

$$T_A = 1.682\ (T_S \text{ deep}) - 13.309 \quad (r = 0.962,\ P > 0.99)$$

and

$$T_S \text{ surface} = 1.418\ (T_S \text{ deep}) - 3.778 \quad (r = 0.901,\ P > 0.99).$$

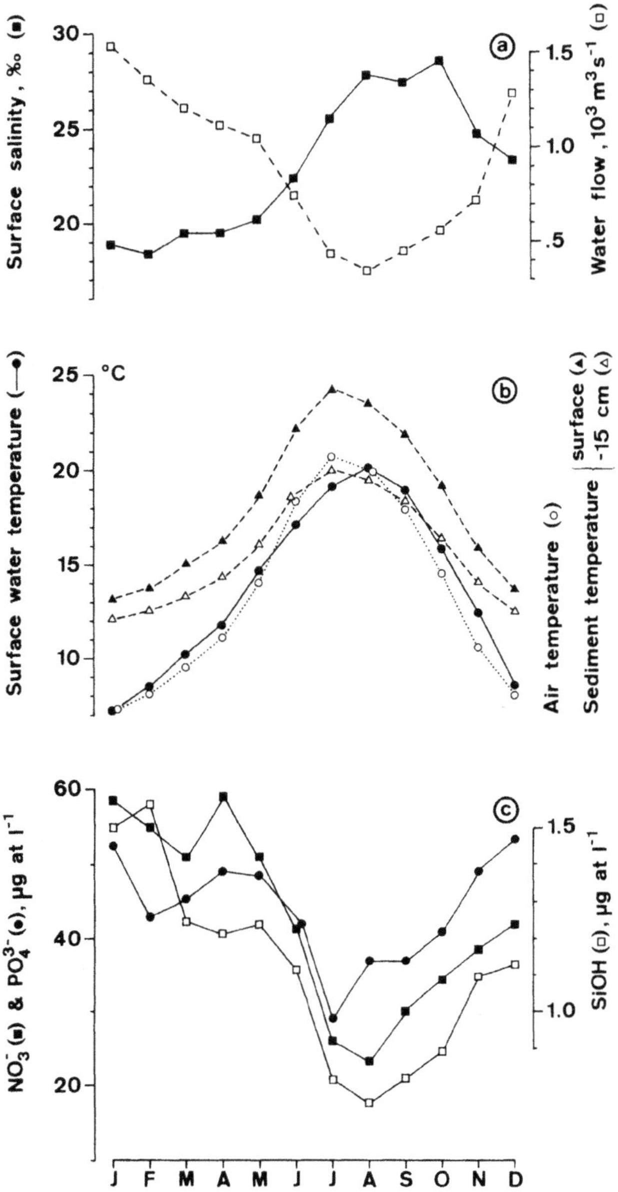

Fig. 1. Summary of data on (a) surface water salinity and water flow, (b) surface water, air, and sediment (surface and 15 cm depth) temperatures, (c) phosphate, nitrate, and silicate contents in surface waters, at Le Verdon, Gironde estuary. Data are average monthly values calculated for the period 1976–1983, except for estuarine water flows averaged for the period 1961–1970.

The range of values used for air temperature was large (between −7.5 and 28.0 °C), but only pertinent to day-time and to emerged sediment.

Nutrient concentrations in surface estuarine waters show a seasonal evolution. Highest levels are measured in winter: 1.5 µg at P l^{-1} in December–January, 58.4 µg at N l^{-1} in January and 58.0 µg at Si l^{-1} in Febuary and remain stable until April–May, when a second peak is obvious. Lowest values are recorded in July (phosphates) and August (silicates and nitrates), then increase again until winter.

During 1983 and 1984, fine particles and organic matter contents showed a great temporal variability in the sediments: 28–90% for the former, 4.2–10.9% for the latter (Fig. 2). A high silt and clay content during the summer and autumn months was the mark of a calm weather and a low fluvial discharge. In contrast, high river runoff generated strong tidal currents which accounted for the erosion of mud flats from January to May–June. Silt plus clay and organic contents were highly correlated. During the 1983–84 survey, three peaks of organic matter could be distinguished: two around 10% in December–January and August–October, and a third, less obvious, in May.

Seasonal gonad changes and spawning

In spite of the low number of individuals examined at each sampling occasion, a fairly clear pattern in reproductive condition of *Macoma balthica* occurred during the two consecutive years of detailed investigation (Table 1). However, analysis of data for individual years reveals some variation from one year to another.

Gross examination of the visceral mass showed

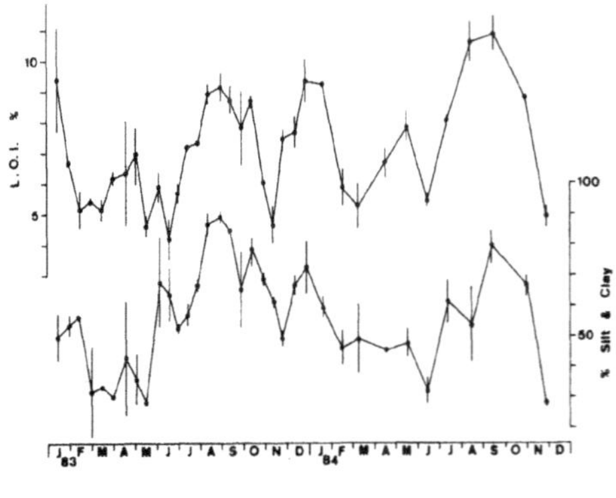

Fig, 2. Organic matter (L.O.I.) and silt plus clay contents (mean ± standard error) in the top 2 cm of sediment at station 1, Le Verdon, during the years 1983–1984.

that almost all individuals had well developed gonads, imbricated in the digestive gland, from November to March. Two drops in the percentage of mature animals were observed in early March and early June in 1983, in February and April 1984. No gonads were obvious in summer and early autumn. These observations are, however, insufficient to precisely determine the exact spawning time, because gonads are well developed when they are either ripe or in the stages 3 and 4 of both developing and spawning states (Brown, 1982). Thus the scheme of gonad classification and the calculation of a gonad index as described by Brown, *loc. cit.*, was used to describe with more details the seasonal gonad changes (Table 1).

In January 1983, nearly all the adult population of *M. balthica* was in the ripe condition, with individuals having a swollen foot filled with ripe ova and sperm. After that, the proportion of ripe individuals declined rapidly, as the spawning state became dominant. Around mid-April, there were no more clams with ripe gametes, but a small increase

Table 1. Reproductive condition of *Macoma balthica* in the Gironde estuary during 1983 – 1984. Gonad classification and gonad index according to Brown (1982). The mean gonad index varies from 0 (all spent) to 5 (all ripe).

Date	Gonads visible by macroscopic examination %	Developing %	Ripe %	Spawning %	Spent %	Gonad index
1983						
14 Jan.	100	13	87	0	0	4.8
31 Jan.	100	0	92	8	0	4.7
15 Feb.	100	0	64	36	0	4.5
01 Mar.	100	0	42	58	0	4.0
15 Mar.	67	0	18	82	0	3.2
30 Mar.	58	0	8	84	8	2.5
18 Apr.	42	0	0	92	8	2.2
02 May	33	0	8	75	17	2.0
16 May	8	0	9	55	36	1.4
03 June	25	0	25	42	33	1.8
17 June	0	20	0	30	50	0.6
29 June	0	18	0	6	66	0.5
13 July	0	17	0	8	75	0.3
27 July	0	17	0	16	67	0.3
11 Aug.	0	21	0	0	79	0.5
29 Aug.	0	21	0	4	75	0.8
13 Sep.	0	33	0	0	67	1.0
27 Sep.	0	92	0	0	8	2.3
12 Oct.	0	100	0	0	0	2.5
28 Oct.	17	83	17	0	0	2.9
12 Nov.	67	67	25	8	0	3.3
25 Nov.	92	50	50	0	0	4.0
12 Dec.	100	17	83	0	0	4.8
27 Dec.	100	8	75	17	0	4.7
1984						
19 Jan.	100	0	75	25	0	4.5
17 Feb.	92	0	84	8	8	4.3
09 Mar.	100	0	67	33	0	4.0
16 Apr.	92	0	21	71	8	2.3
17 May	75	0	10	65	25	1.5
15 June	0	13	0	29	58	1.0
11 July	0	17	0	0	83	0.3
14 Aug.	0	25	0	0	75	0.5
17 Sep.	0	33	0	0	67	1.1
29 Oct.	30	75	25	0	0	3.0
27 Nov.	100	58	42	0	0	4.3

occurred again in May and early June. No ripe specimens were found from mid-June to mid-October. Large numbers of spawning individuals were dominant from February to April, but were still found until July in a few females which exhibited gonads with residual eggs. Since the first spawning specimens were noticed on January 31 and the mean gonad index markedly declined up to May 16, the February–May period could be characterized by a first intense spawning. A rising proportion of ripe individuals, together with a new increase in the gonad index on June 3, followed by a quick drop, suggests that a second, minor spawning period took place in June–July.

Animals with developing gonads had already appeared in June. Their proportion remained stable (around 20%) until late August, then abruptly increased to 100% in October. The first specimens with ripe spermatozoa and eggs were noticed in late October. Their maximum numbers were reached from December 1983 to February 1984, so that most animals had ripe gametes from November to March. During this period, no clear decline in the mean gonad index was ascertained, but spawning individuals were present in a noticeable proportion in late December and early January, suggesting a spawning early in the year. *M. balthica* in a ripe state declined in numbers from February to June 1984. As seen by a steady decrease in the gonad index, the main spawning period during this year can be considered to be the months March–June or July.

As in the preceding year, the development started in June, as soon as spawning was completed. Recovery of condition continued throughout the summer months (increase of gonad index) so that mature gametes were again present in late October.

Age distribution and abundance

Size-frequency histograms of *Macoma balthica* were obtained by pooling 8 cm^2 and 400 cm^2 samples from January 1983 to November 1984 as shown in Fig. 3 and 4. Because of their three-dimensional shape (length > height > thickness), the same size juvenile clams were not equally retained at a given sieving mesh size, so that the smallest clam found on the 400 μm sieve was 0.54 mm in length, whereas the largest retained on the 100 μm sieve was 0.71 mm. In this overlapping size region (0.54 to 0.71 mm), a mean density was thus determined from the data of both cores. Frequency data for each 0.5 mm length intervals were finally normalized to a standard area of 1 m^2.

From January 1983 to January 1984, the density of animals older than about one year remained relatively high (around 500 ind. m^{-2}) and formed a single size group between 10 and 15 mm shell length. It has been shown that these individuals belonged to three or four age groups (Bachelet, 1980). A few clams were larger (the maximum recorded length was 20.9 mm) but lengths greater than 16 mm were rarely found. During their growth, new recruits formed broad frequency histograms with several peaks that overlapped with the older individuals into the 10–15 mm grouping in the summer months of the next year. This size group, though well represented in January 1984, showed reducing densities thereafter, and from May onwards nearly completely disappeared. Thus, some factor led to the total extinction of the oldest generations, in such a manner that by November, the population was only composed of individuals less than 10 mm in length.

Recruitment features

In 1983, the first sign of recruitment occurred by mid April. From this date and onwards, decomposition of the size frequency histograms shown in Fig. 3 and analysis of more detailed histograms built with size classes of 0.1 mm for juvenile shell lengths led to the partition of the newly recruited generation into four successive waves named A, B, C and D (Fig. 5). A single cohort (A) was present from mid April to early June. The other three cohorts (B, C and D) were first observed at 17 June, 13 July and 27 July, respectively. Densities were almost identical at the time of settlement and relatively low (about 500 ind. m^{-2}) in these four cohorts; their maximal cumulated numbers, found in late July, were 3250 ind. m^{-2}. A steady decrease in the abundance of the 0+ year class resulted in about 1000 ind. m^{-2} in December, when the four cohorts became undistinguishable because of the junction of their modal size. Autumnal densities remained stable, then a new mortality occurred in spring months. Approximately one year after their settlement, the density of juveniles was found to be only 100/300 ind. m^{-2} at the most (see generations '82 and '83 respectively in June 1983 and 1984, Fig. 5).

During the year 1984 spat fall began sooner than in the preceding year. Densities of juveniles of up to 15000 ind. m^{-2} were found in February and a high mortality occurred after that. Only one cohort could be distinguished throughout the course of the year, with possibly a second, poorly individualized cohort in June when densities rose again (Fig. 5). However the fact that only one or two cohorts were found in 1984 could be a consequence of the monthly interval between two successive samplings, since it was fortnightly in 1983.

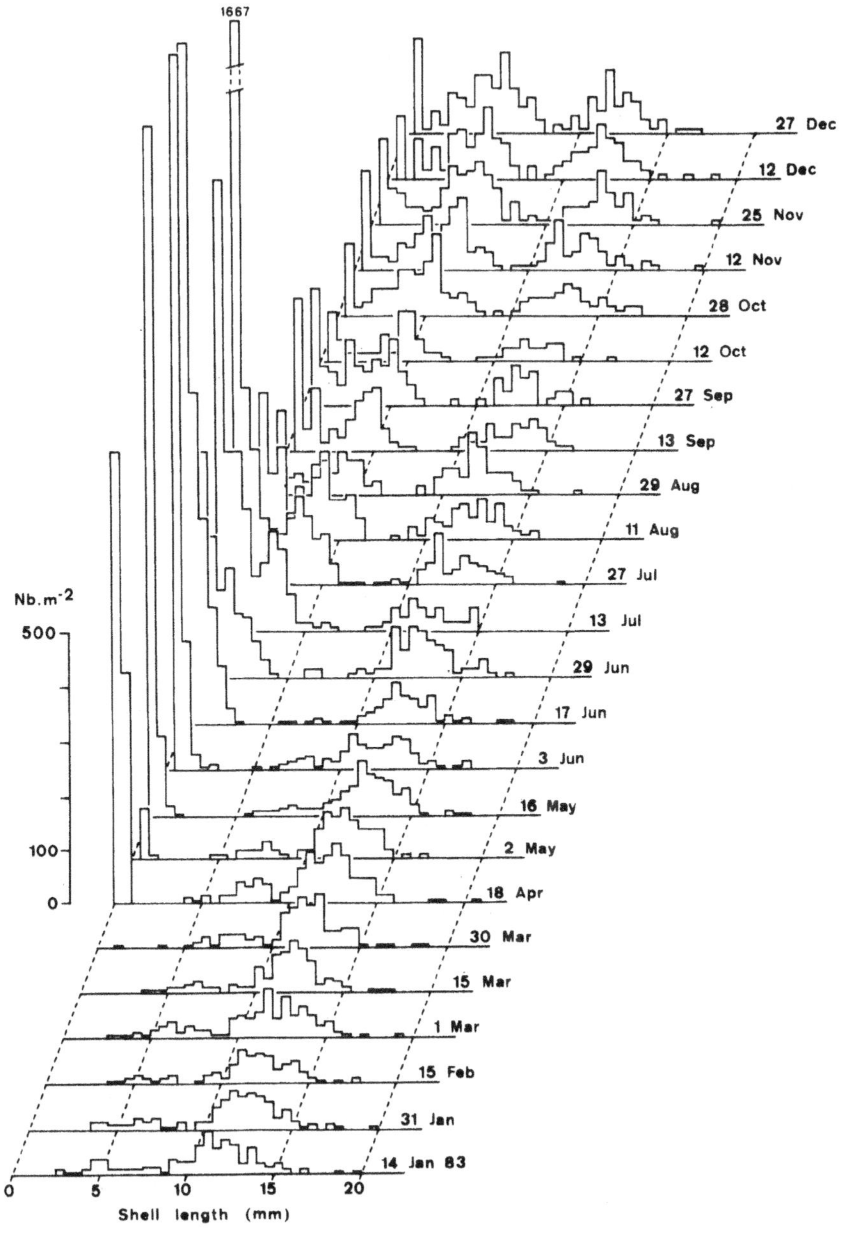

Fig. 3. Length-frequency distributions of *Macoma balthica* from Le Verdon, Gironde estuary, in 1983 (size classes 0.5 mm). 8 cm^2 cores and 400 cm^2 quadrats are pooled and the densities of *M. balthica* converted for 1 m^2.

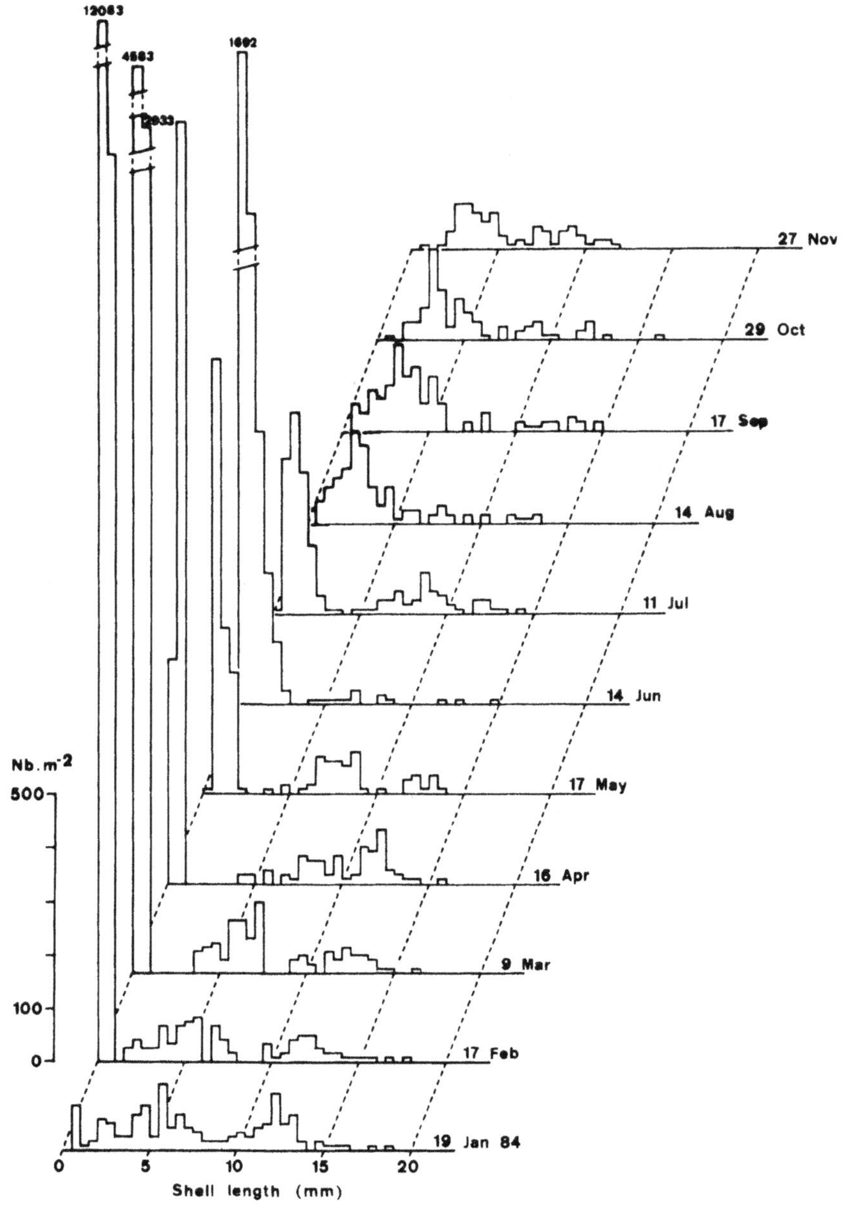

Fig. 4. Length-frequency distributions of *Macoma balthica* from Le Verdon in 1984.

Postlarval and juvenile growth

At the time of settlement, the smallest benthic stages of *M. balthica* were usually 270 to 310 μm long, but the minimal size recorded was 200 μm. Measurements of prodissoconch II shells (Pd) in postlarval clams of about 1 mm shell length showed a range of 240–300 μm for Pd length (mean = 272 μm). Metamorphosis therefore occurred at around 250–300 μm shell length.

The seasonal increase in shell length of the postlarval and juvenile *M. balthica* (age classes 0+ and 1+) can be seen in Fig. 6. For all the newly recruited cohorts, mean length at first sampling was near

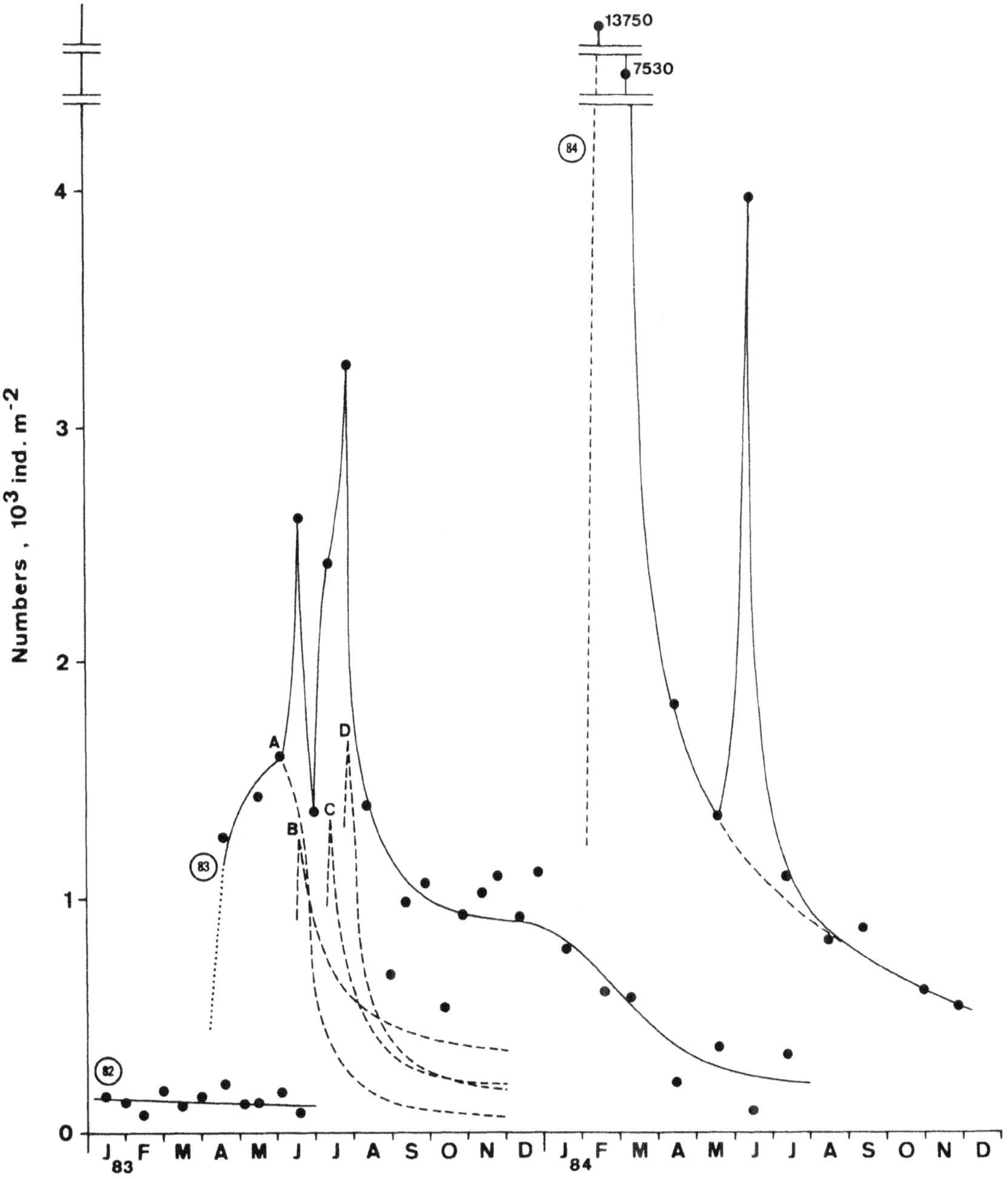

Fig. 5. Survivorship curves (ind. m^{-2}) for the generations of *Macoma balthica* born in 1982, 1983 and 1984, followed throughout 1983 and 1984. Curves are fitted by eye and are shown only for the first year of life. Points are mean densities at each sampling date. The four cohorts identified for the '83 year class are marked by a letter (A to D) and their survivorship curves are delineated with dashed lines (observed densities not shown).

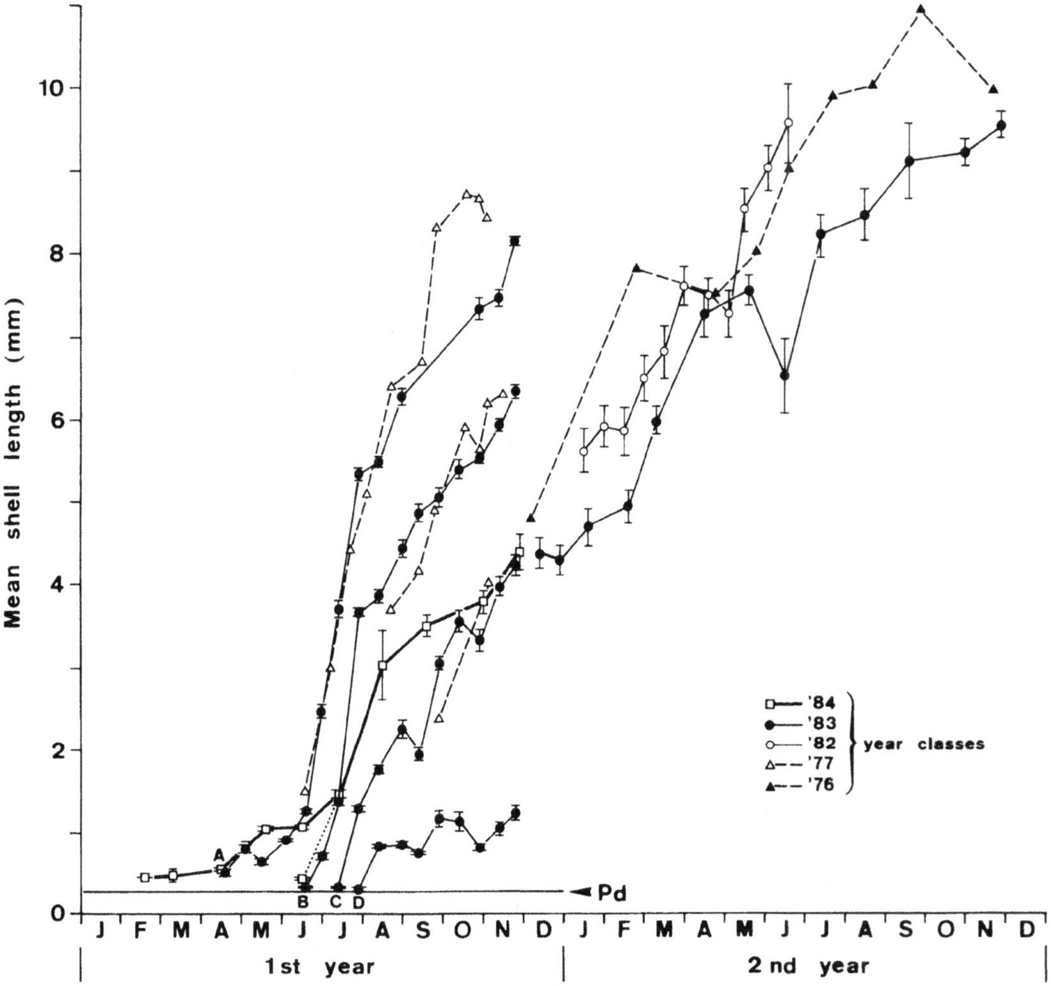

Fig. 6. Seasonal growth in shell length (mm) of juvenile *Macoma balthica* during their first and second growing seasons. Symbols are means ± standard errors. Cohorts of the 0+ year class are marked by letter. Pd = mean size of prodissoconch shell.

the Pd size; 0.51 mm (S.D. = 0.05), 0.33 mm (0.01), 0.32 mm (0.01), and 0.30 mm (0.02) respectively for the four '83 cohorts, 0.44 mm (0.01) and 0.42 mm (0.02) for the two '84 cohorts, which confirms the validity of sampling strategy.

Separate analysis of each cohort growth curve of the age class 0+ gives interesting information on growth rates as a function of settlement dates. 1) Cohort '83$_A$ showed three distinct growth periods: slow growth (0.0125 mm d^{-1}) from 18 April to 17 June, rapid growth (0.1014 mm d^{-1}) from 17 June to 27 July, and moderate growth (0.0269 mm d^{-1}) until 25 November. 2) Length increments for the cohort '83$_B$ were similar (0.0808 mm d^{-1} from 17 June to 27 July, with 0.1629 mm d^{-1} during the period 13 July–27 July), but as mean length by late July was lower than that of cohort '83$_A$ (3.7 against 5.3 mm) the final length by late November was also lower (6.3 against 8.1 mm); shell length increase for cohort '83$_B$ (0.0220 mm d^{-1}) was the same as for '83$_A$ from 27 July to 25 November. 3) During a fortnight following its settlement, cohort '83$_C$ exhibited good growth in shell length (0.0693 mm d^{-1}), but from 27 July forward its daily increase was of the same magnitude as cohorts '83$_A$ and '83$_B$ (0.0246 mm d^{-1}). 4) The main growth of cohort '83_D took place from 27 July to 11 August (0.0343 mm d^{-1}). After this date, it was almost null (0.0047 mm d^{-1}). 5) Generation '84, that had appeared already in February, showed small growth until July (0.0071 mm d^{-1}); it increased in July–August (0.0456 mm d^{-1}), then decelerated (0.0136 mm d^{-1}).

As cohorts were undistinguishable after the end of the first calendar year, they could be pooled at that time into one age group whose individuals were about 4.5 mm (generations 1983 and 1984) or 5.5 mm length (generation 1982). Growth more or less ceased in the colder months. The second growing season had already begun by the end of February and the first ring on the shells became obvious in the samples from 1983, 1st March and 1984, 9 March, when mean shell length was about 6 mm (mean length of 1st growth ring: 6.6 mm, Bachelet, 1980). For generation '83, a rapid increase in shell length (0.0429 mm d^{-1}) took place from 17 February to 16 April; growth did not end at this date but continued at a slower rate (0.0102 mm d^{-1}), at least until late November 1984. Growth of the generation '82 began by 15 February and did not show any slowing until 17 June (average increased: 0.0343 mm d^{-1}); from this date and onwards, the shells could not be aged with accuracy.

Discussion

During the course of species development, several stages are decisive for the maintenance of benthic communities. In macrofaunal organisms with benthoplanktonic life-cycle, there are at least five such crucial stages: 1) maturation of sexual products, 2) spawning, 3) planktonic larval life, 4) larval settlement, and 5) post-larval growth, as observed from time of settlement to a size large enough to be retained by sieves used for macrobenthos studies (i.e. a 0.5- or 1 mm mesh-size). Strictly defined, the term 'recruitment' only refers to the last two stages, but its size closely depends on the first ones. In order to ascertain the reasons for a success or a failure in recruitment therefore requires the study of potential contributory factors, not simply restricted to the time of settlement, but also encompassing preceding events and early benthic life.

The timing of reproductive cycle and recruitment

During the first two years of a long-term study on recruitment in a population of *Macoma balthica*, a certain year-to-year variability in the reproductive cycle as well as in the production and survival of newly settled spat was observed. Results concerning seasonal gonad changes and settlement may be recapitulated for each year as follows:

Year 1983: ripe individuals: January/March, May/early June;
spawning: February/mid May, June/mid July;
recruitment: April, June, July (4 cohorts).

Year 1984: ripe individuals: November (1983)/May;
spawning: January, March/mid July;
recruitment: February, June (2 cohorts).

With regard to gonadal regeneration, no significant difference was found between the two years; gonads started to develop by mid June, and proliferation of the sex cells was accelerated from September to November and ceased by the end of December.

In a previous survey (1977), spawning was detected in May/July, then in October (Bachelet, 1980). At that time, only macroscopic examination of the visceral mass was performed and an arbitrary state of gonad development was assigned according to the extent to which the gonad tissue had migrated downwards or upwards from the line of attachment of palps and gills (Caddy, 1967). As has already been mentioned, this method does not allow us to distinguish between ripe specimens, developing specimens approaching sexual maturity, and specimens that have already started to spawn. Thus, the actual onset of spawning periods in 1977 probably occurred sooner than recorded.

A correction must also be made for the dates of recruitment in 1977, because a sieve with 0.5 mm mesh size was employed. Three cohorts of newly recruited spat were found from June to September. When the subsequent growth in shell length of these cohorts (Bachelet, 1980: 109, Fig. 4) is superimposed on to the growth of the recruits in 1983 (Fig. 6), a clear parallel is found between growth of the cohorts A, B, and C of both years. Hence, the recruitment periods in 1977 may be assumed to be the same as in 1983. A further settlement was observed in autumn at site 1 (Bachelet, *loc. cit*).

In summary, the most likely sequence of events in 1977 was as follows:
spawning: March or April/July, October;
recruitment: April, June, July, November (4 cohorts).

Comparison of reproductive cycle in 1977, 1983 and 1984 shows that *M. balthica* spawns twice a year in the Gironde estuary. The animals always undergo a prolonged spawning period that lasts from March through July, but a time-lag may occur in its onset. For instance, spawning started in February during 1983 and in January 1984, which involved the spliting of the spawning period in two. In other European waters, intertidal populations of *M. balthica* shows a spring spawning that more or less coincides with our observations: late February/April in the Ythan estuary (Chambers & Milne, 1975); March/April (Lammens, 1967) or April/May (De Wilde & Berghuis, 1978) in the Dutch Wadden Sea; April/June in the Thames estuary (Caddy, 1967). The Gironde population has a more protracted spawning time (4–5 months) than the others (2–3 months).

As a result of the generally extended periods over which spawning took place, settlement occurred several times a year, at different periods and with a variable intensity from year to year. Recently settled spat can usually be found from February to July, but occasionally in autumn as well, such as in 1977. The present data on spawning and recruitment corroborate the estimate made by Caddy (1969) who determined a planktonic life of 2–5 weeks in *M. balthica.*

It has been established that spawning normally occurs in the period March/July in the Gironde estuary, however two major deviations must be emphasized: the early spawning in January 1984 and the late one in October 1977. Such a variation in the timing of the spawning season from one year to another has been found for subtidal bivalves (Brown, 1982). Intertidal invertebrates are directly submitted to climatic factors and though these organisms are ecologically adapted to fluctuating parameters, they are subjected to environmental stimuli which are likely to start up physiological processes when such stimuli exceed a certain threshold. Temperature is probably the most fluctuating parameter, both annually and daily, in intertidal areas. Thus it is not surprising that a relationship was found between spawning activity and temperature in *M. balthica* (De Wilde, 1975; De Wilde & Berghuis, 1978). Expulsion of the gametes in field populations is restricted to seawater temperatures in the range 7–14 °C (Caddy, 1967). Such a range normally occurs between November to May in the Gironde estuary (Fig. 1) where the lowest monthly average temperature is recorded in January (7.2 °C), and may explain why *M. balthica* is able to spawn there in January/February. However spawning occurs until mid July when water temperatures are about 19 °C and air temperatures may be up to 30 °C on tidal flats, which is clearly above optimal temperature for spawning. De Wilde & Berghuis (*loc. cit.*) showed that in the laboratory, spawning in *M. balthica* was induced by administration of a temperature shock, consisting of a rise in temperature from 5 to 12.5 °C in 5 min. Thermic stress could be the factor inducing spawning, rather than a slow increase in temperature. Thus, when specimens still posess ripe gametes, as is the case in June/July, sudden changes of temperature (daily or nycthemeral) possibly could induce some egg release. The complete exhaustion of the sexual products in July coincides with the maximal sediment temperature (Fig. 1).

An autumn spawning took place in October 1977. Similar observations have been previously described in some populations where a spring spawning also occurred (Shaw, 1965; Caddy, 1967; Rasmussen, 1973; Nichols & Thompson, 1982). Developing gonads were present in this period in 1983 and 1984 as well, but no ripe specimens were found. In fact, climatic conditions in 1977 were quite different from the other years in that maximum air temperatures were unusually low from April to September (Fig. 7). Furthermore, water nutrient concentrations were relatively high during the entire year 1977, whereas they generally show a marked drop in the summer months. Reduced thermic variations and a high level of metabolites could have favoured a second gonad development in the second half of this year.

Juvenile growth

At the time of settlement, the larvae have a size of about 300 μm in shell length. The mean size of Pd II (270 μm) is close to that reported by various

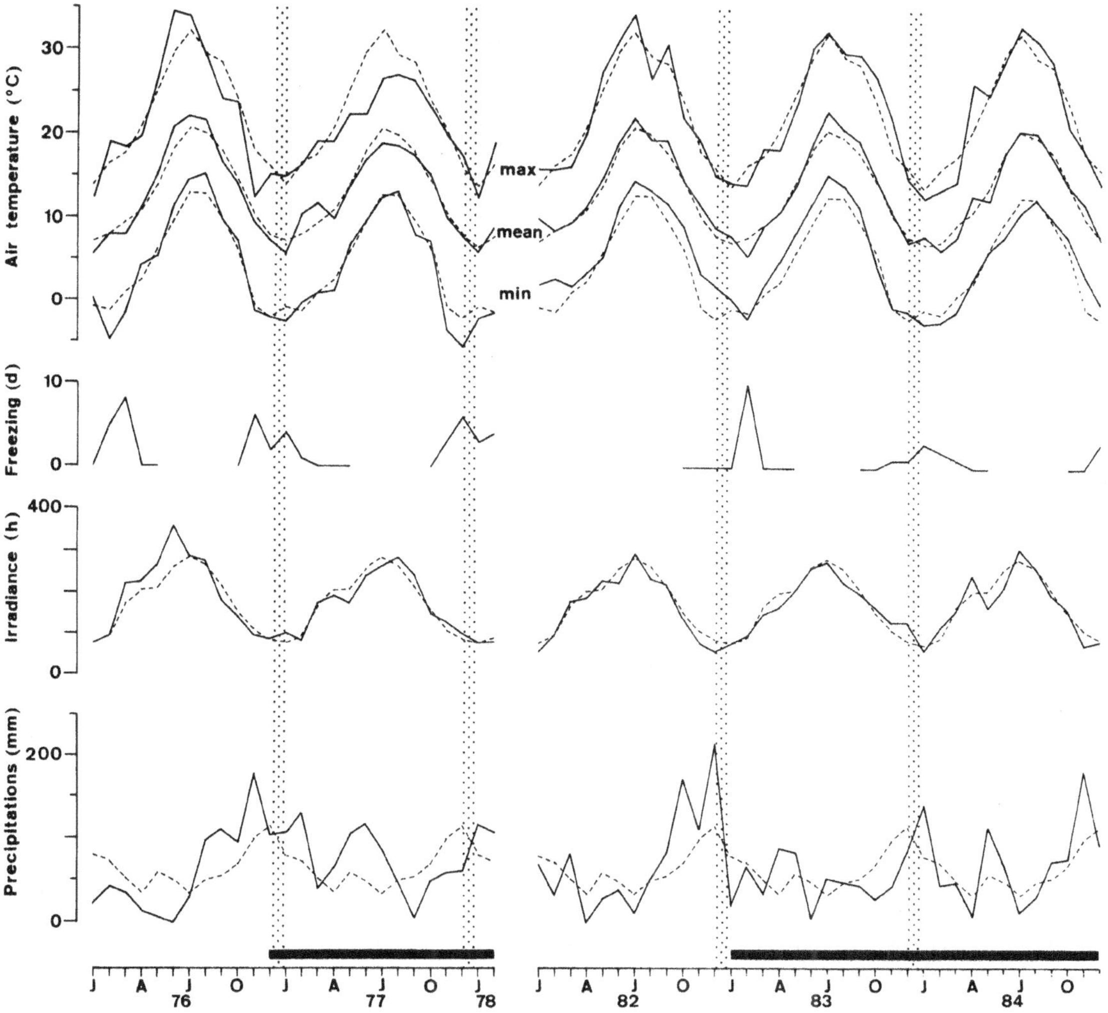

Fig. 7. Air temperature, irradiance, and rainfall fluctuations recorded near the sampling station during the years 1976–1978 and 1982–1984. Monthly numbers of freezing days are also shown. Mean values for ten years (1974–1984) are represented by a dotted line and thick horizontal bars indicate periods of sampling.

authors: 294 μm in the Øresund (Jørgensen, 1946), 255 μm at Prince Edward Island (Sullivan, 1948), 285–301 μm in the Thames estuary (Caddy, 1969). Depending on the period of recruitment, juvenile growth and size attained by the first winter were largely variable (Fig. 6). For the cohorts '83, the earlier they settled, the better they grew; individuals of each cohort were about 8, 6, 4 and 1 mm long respectively in December 1983. In this year-class, the highest growth rate was found from mid-June to the end of July, but the growing season continued at least until the end of November. After this date, cohorts were not separable one from the other. The last cohort ('83_D) showed a very low growth rate. The three cohorts '77 showed the same increase in shell length as the first three cohorts '83.

The 1984 recruitment seems rather peculiar in that it occurred very early in the year (February) and had a slow growth rate until June which increased from July forward. There is an uncertainty however about growth rate during this year for the following reasons: 1) the spat had reached a length of around 4 mm in November which was the mean length of the whole generation '83 in the preceding year and 2) because a second settlement probably occurred in June.

Growth was almost null during the first winter after settlement. The second growing season began by mid-February (year-classes '82 and '83) or in March/April (year-class '76). The timing of maximal increase in shell length varied from year to year: April/June (1977), February/June (1983), February/April (1984). As was suggested for the spawning period, temperature may have an influence on growth rate as well. From laboratory experiments, De Wilde (1975) showed that *M. balthica* thrives best in the temperature range 0–15 °C, with an optimum at about 10 °C. When maximum growth occurs in the Gironde estuary, the range of seawater temperatures is 8–18 °C, and some shell growth is even observed when water temperature exceeds 20 °C. This species shows a cyclic burying at Le Verdon that might be an adaptation to high summer temperatures (unpubl. observ.): localized in the top 2 cm of sediment during the colder months, *M. balthica* buries deeper when air temperature increases above 15 °C (i.e. from April/May to September). This vertical migration was also found by De Wilde (*loc. cit.*) and could allow the animals to find lower temperatures in order to prolong their growing season.

A range of water temperatures between 8 and 15 °C also occurs from October to December, but there is no indication of a second annual growing period (Beukema & De Bruin, 1977). From a long-term survey in the Dutch Wadden Sea, Beukema *et al.* (1977) did not find a statistically significant correlation between temperatures and growth rates in *M. balthica*. Hence, temperature is not the only growth regulating factor. The presence of sufficient living food is probably another important factor (De Wilde, 1975; Ankar, 1980; Beukema *et al.*, 1985; Hummel, 1985a, b). Indeed, high food availability coincides with growing period in the Gironde estuary, where dissolved nutrients are at a high level in January/May and large quantities of benthic diatoms have developed in the summer months.

Growth of the cohorts '83_D and '84_A (until June) was surprisingly low, in comparison to the other cohorts (Fig. 6). This is probably a consequence of a switch in feeding habits, as Caddy (1969) demonstrated that metamorphosis to the adult form and function was only completed after a shell length of 2 mm was attained. According to the anatomical changes that occur during juvenile development, Ankar (1980) offered an explanation for a slow-growing Baltic population of *M. balthica* that may be applied in our case study. When it is less than about 1.5 mm long, spat has rudimentary siphons and gills and behaves as a filter-feeding organism. For cohorts '83_D and '84_A, this size corresponds to periods (respectively August/November and January/March) with low phytoplankton concentration in the Gironde estuary (chlorophyll *a* measurements, J. Castel, pers. commun.), unfavourable to their growth. When the development of nutrition organs is completed, i.e. at 2 mm long, spat can apply both suspension- and deposit feeding. Inversely, plankton concentration was optimal when cohorts '83_B and '83_C settled and they immediately exhibited good growth.

Variability in the magnitude of recruitment

Data on the importance of recruitment in populations of *M. balthica* is scarce. Caddy (1969) and Chambers & Milne (1975) found spat densities of about 5000 ind. m^{-2}, whereas Ankar (1980) mentioned 80000 ind. m^{-2} using a 100 μm sieving mesh size. The greatest abundance of recruits found in the Gironde estuary was about 15000 ind. m^{-2} in February 1984, but in 1983 four settlements occurred with an average density of 1500 ind. m^{-2}. Survival, however, was lower at the end of 1983 than for the same period in 1984: one juvenile for 4 recruits and one for 23, respectively. In 1977, recruitment was almost non-existent, although specimens were detected with a 500 μm sieve (Bachelet, 1982). Hence, recruitment patterns appear to be highly variable in different years, not only with regard to the magnitude of settlement, but also the survival of recruits. Moreover, the example of 1984 shows that there is no relation between spawning intensity and the strength of settlement.

Settlement and survival of benthic invertebrates may be affected by several factors: mortality in the plankton, environmental factors, competition for resources, predation on post-larvae, and interactions such as amensalism. Mortality of planktonic larvae is difficult to ascertain, so this factor will not be considered. Inspection of climatic events (Fig. 7) and sediment characteristics (Fig. 2) at the possible recruitment periods does not show any outstanding features. Thus, recruitment is apparently not related to obvious environmental factors. It can be pointed out that an unusually large number of freezing days in February 1983 could explain why no recruits were found before mid-April while spawning started in early February. These low temperatures might have killed planktonic larvae or early juvenile stages.

It may be suggested that most of the variability in the magnitude and survival of recruitment must be examined in relation to biological interactions. Though still barely understood, competition (for space and food) and predation have been demonstrated to influence population sizes of marine bivalves (Peterson & Andre, 1980; Williams, 1980; Blundon & Kennedy, 1982; Weinberg, 1985; and others). Particularly interesting in this respect is the co-occurrence of relatively dense populations of *M. balthica* and dense assemblages of spionids on the tidal flats of the Gironde estuary. Some species, e.g. *Polydora ligni* (Breese & Phibbs, 1972), were found to ingest bivalve molluscan larvae, but a probably intense competition does exist between spat and spionids (Woodin, 1976). For example, when *M. balthica* settled by mid-July 1983, *Streblospio shrubsolii* were up to 200000 ind. m^{-2}; in February 1984, densities of *S. shrubsolii* were only 16000 ind. m^{-2} when the settlement of *M. balthica* was 10 times greater than for the same period in 1983. Future field and laboratory manipulations with these species are promising.

Acknowledgements

This work is a contribution to the 'Programme National sur le Déterminisme du Recrutement' (A.T.P. 95-93-62 to Dr. M. Bhaud). Financial support was in part provided through an 'Aide à la Recherche Universitaire' from the 'Ministère de l'Education Nationale' to Professor J. Boisseau. I would like to thank Dr. J. Castel for critically reading the manuscript and to Prof. J. Boisseau for the discussions which have contributed to this work. I gratefully acknowledge G. Imbert for laboratory assistance and C. Kirk for improving on the English text.

References

Ankar, S., 1980. Growth and production of *Macoma balthica* (L.) in a northern Baltic soft bottom. Ophelia Suppl. 1: 31–48.

Bachelet, G., 1980. Growth and recruitment of the tellinid bivalve *Macoma balthica* at the southern limit of its geographical distribution, the Gironde Estuary (SW France). Mar. Biol. 59: 105–117.

Bachelet, G., 1982. Quelques problèmes liés à l'estimation de la production secondaire. Cas des bivalves *Macoma balthica* et *Scrobicularia plana*. Oceanol. Acta 5: 421–431.

Bachelet, G., 1984. Le recrutement des populations annélidiennes sur substrat meuble: aspects méthodologiques. Océanis 10: 735–746.

Bachelet, G., 1985. Influence de la maille de tamisage sur les estimation d'abondance du macrobenthos marin. C. R. Acad. Sc. Paris 301, Ser. 3: 795–798.

Beukema, J. J., 1980. Calcimass and carbonate production by molluscs on the tidal flats in the Dutch Wadden Sea: I. The tellinid bivalve *Macoma balthica*. Neth. J. Sea Res. 14: 323–338.

Beukema, J. J., 1982. Annual variation in reproductive success and biomass of the major macrozoobenthic species living in a tidal flat area of the Wadden Sea. Neth. J. Sea Res. 16: 37–45.

Beukema, J. J., G. C. Cadée & J. J. M. Jansen, 1977. Variability of growth rate of *Macoma balthica* (L.) in the Wadden Sea in relation to availability of food. In B. F. Keegan, P. O'Ceidigh & P. J. S. Boaden (eds), Biology of Benthic Organisms. Pergamon Press, Oxford: 69–77.

Beukema, J. J. & W. De Bruin, 1977. Seasonal changes in dry weight and chemical composition of the soft parts of the tellinid bivalve *Macoma balthica* in the Dutch Wadden Sea. Neth. J. Sea Res. 11: 42–55.

Beukema, J. J., E. Knol & G. C. Cadée, 1985. Effects of temperature on the length of the annual growing season in the tellinid bivalve *Macoma balthica* (L.) living on tidal flats in the Dutch Wadden Sea. J. exp. mar. Biol. Ecol. 90: 129–144.

Beuekema, J. J. & B. W. Meehan, 1985. Latitudinal variation in linear growth and other shell characteristics of *Macoma balthica*. Mar. Biol. 90: 27–33.

Blundon, J. A. & V. S. Kennedy, 1982. Refuges for infaunal bivalves from blue crab, *Callinectes sapidus* (Rathbun), predation in Chesapeake Bay. J. exp. mar. Biol. Ecol. 65: 67–81.

Breese, W. P. & F. D. Phibbs, 1972. Ingestion of bivalve molluscan larvae by the polychaete annelid *Polydora ligni*. Veliger 14: 274–275.

Brown, R. A., 1982. Reproduction of *Abra nitida* (Müller) (Bivalvia) in the southern Skagerrak. Sarsia 67: 55–60.

Caddy, J. F., 1967. Maturation of gametes and spawning in *Macoma balthica* (L.). Can. J. Zool. 45: 955–965.

Caddy, J. F., 1969. Development of mantle organs, feeding, and locomotion in postlarval *Macoma balthica* (L.) (Lamellibranchiata). Can. J. Zool. 47: 609–617.

Cattaneo, M. & H. Massé, 1983. Importance du recrutement de *Spisula subtruncata* (da Costa) sur la structure et les fluctuations d'un peuplement benthique. Oceanol. Acta N° sp.: 63–67.

Chambers, M. R. & H. Milne, 1975. The production of *Macoma balthica* (L.) in the Ythan Estuary. Estuar. coast. mar. Sci. 3: 443–455.

De Wilde, P. A. W. J., 1975. Influence of temperature on behaviour, energy metabolism, and growth of *Macoma balthica* (L.). In H. Barnes (ed.), Proc. 9th Europ. mar. biol. Symp. Aberdeen University Press: 239–256.

De Wilde, P. A. W. J. & E. M. Berghuis, 1978. Laboratory experiments on the spawning of *Macoma balthica*; its implication for production research. In D. S. McLusky & A. J. Berry (eds), proc. 12th Europ. mar. biol. Symp. Pergamon Press, Oxford: 375–384.

Diaz, R. J., 1984. Short term dynamics of the dominant annelids in a polyhaline temperate estuary. Hydrobiologia 115: 153–158.

Eagle, R. A., 1975. Natural fluctuations in a soft bottom benthic community. J. mar. biol. Ass. U.K. 55: 865–878.

Glémarec, M., 1978. Problèmes d'écologie dynamique et de succession en baie de Concarneau. Vie Milieu 28–29: 1–20.

Holland, A. F., 1985. Long-term variation of macrobenthos in a mesohaline region of Chesapeake Bay. Estuaries 8: 93–113.

Hummel, H., 1985a. Food intake of *Macoma balthica* (Mollusca) in relation to seasonal changes in its potential food on a tidal flat in the Dutch Wadden Sea. Neth. J. Sea Res. 19: 52–76.

Hummel, H., 1985b. An energy budget for a *Macoma balthica* (Mollusca) population living on a tidal flat in the Dutch Wadden Sea. Neth. J. Sea Res. 19: 84–92.

Jørgensen, C. B., 1946. Reproduction and larval development of Danish marine bottom invertebrates. 9. Lamellibranchia. Medd. Komm. Havundersøg., Kbh Ser. d, Plankton 4: 277–311.

Lammens, J. J., 1967. Growth and reproduction in a tidal flat population of *Macoma balthica* (L.). Neth. J. Sea Res. 3: 315–382.

Luckenbach, M. W., 1984. Settlement and early post-settlement survival in the recruitment of *Mulinia lateralis* (Bivalvia). Mar. Ecol. Prog. Ser. 17: 245–250.

Nichols, F. H., 1985. Abundance fluctuations among benthic invertebrates in two Pacific estuaries. Estuaries 8: 136–144.

Nichols, F. H. & J. K. Thompson, 1982. Seasonal growth in the bivalve *Macoma balthica* near the southern limit of its range. Estuaries 5: 110–120.

Persson, L. E., 1983. Temporal and spatial variation in coastal macrobenthic community structure, Hanö Bay (Southern Baltic). J. exp. mar. Biol. Ecol. 68: 277–293.

Peterson, C. H. & S. V. Andre, 1980. An experimental analysis of interspecific competition among marine filter feeders in a soft-sediment environment. Ecology 61: 129–139.

Powell, E. N., H. Cummins, R. J. Stanton Jr. & G. Staff, 1984. Estimation of the size of molluscan larval settlement using the death assemblage. Estuar. coast. shelf. Sci. 18: 367–384.

Rasmussen, E., 1973. Systematics and ecology of the Isefjord marine fauna (Denmark). Ophelia 11: 1–507.

Shaw, W. N., 1965. Seasonal setting patterns of five species of bivalves in the Tred Avon River, Maryland. Chesapeake Sci. 6: 33–37.

Sullivan, C. M., 1948. Bivalve larvae of Malpeque Bay, P.E.I. Bull. Fish. Res. Bd Can. 77: 1–36.

Vahl, O., 1982. Long-term variations in recruitment of the Iceland scallop, *Chlamys islandica*, from northern Norway. Neth. J. Sea Res. 16: 80–87.

Weinberg, J. R., 1985. Factors regulating population dynamics of the marine bivalve *Gemma gemma*: intraspecific competition and salinity. Mar. Biol. 86: 173–182.

Williams, J. G., 1980. The influence of adults on the settlement of spat of the clam, *Tapes japonica*. J. mar. Res. 38: 729–741.

Woodin, S. A., 1976. Adult-larval interactions in dense infaunal assemblages: patterns of abundance. J. mar. Res. 34: 25–41.

Hydrobiologia 142: 249–270 (1986)

Fluctuations naturelles et évolution artificielle des biocénoses macrozoobenthiques intertidales de trois estuaires des côtes françaises de la Manche

M. Desprez*, J-P. Ducrotoy* & B. Sylvand**, Groupe d'Etude des Milieux Estuariens et Littoraux
* *Antenne IFREMER, 115 quai Jeanne d'Arc, 80230 Saint Valery/Somme, France*
** *CNRS, Station Marine, rue Dr Charcot, 14530 Luc/Mer, France*

Keywords: estuary, tidal flats, benthic macrofauna, biosedimentology, population dynamics, environmental conditions

Abstract

The study of the intertidal benthic population dynamics in three estuaries of the English Channel (Baie des Veys, Seine estuary, Baie de Somme:France) brings to light two types of species:

- key-species which directly respond to the local disturbance of the environmental conditions in their densities (Spionidae, Capitellidae) and in their growth rates (*Cerastoderma edule*);
- target-species such as *Macoma balthica* which can endure brief changes in the environmental factors and shows no sign of long-lasting consequences on its population dynamics; yet, it fully integrates long-term changes through its numbers and productivity.

The parallel between such a regular study of the seasonal variations on selected sites and various base line surveys allows the authors to discuss the COST 647 sampling programm in order to select natural fluctuations (storms, range of temperature) from human disturbances (embankments, chemical pollution, eutrophication).

Diverse hypothesis are suggested which bring about several research topics to be developed within a european cooperation.

Introduction

Les études portant sur la cinétique et la dynamique des populations du macrozoobenthos intertidal des principaux estuaires de la Manche centrale et orientale n'ont démarré que dans les années 1970–1972 pour la Baie des Veys (Sylvand, 1986), 1978 pour l'estuaire de la Seine (Desprez, 1981) et la Baie de Somme (Simon *et al.*, 1981). La constitution du Groupe d'Etude des Milieux Estuariens et Littoraux remonte à 1981, année de lancement du programme européen COST 647. En compilant les données acquises de façon discontinue sur l'ensemble de ces trois écosystèmes estuariens jusqu'en 1985 et sur quelques stations-pilotes retenues dans le cadre du COST, il est maintenant possible de s'adonner à la comparaison de l'évolution de ces embouchures macrotidales mais aussi de comprendre leur évolution en la confrontant à celle d'autres estuaires européens ou d'un vaste ensemble intertidal comme la Mer des Wadden où le suivi du benthos couvre près d'un siècle: Van Der Baan *et al.* (1958) aux Pays-Bas, Mobius (1893) et Wolhenberg (1937) en Allemagne, Thamdrup (1935) au Danemark. Nos travaux s'inspirent d'ailleurs largement des recherches menées plus récemment par Beukema (1974, 1979, 1982, 1985), Michaelis (1976) et Madsen (1984), et ont bénéficié des conseils précieux de ces auteurs.

Présentation des sites

Les trois sites estuariens étudiés, les plus importants des côtes françaises de la Manche centrale et orientale, sont régulièrement distribués d'ouest en est (Fig. 1):

- la *Baie des Veys* est le seul site à présenter une

ouverture vers le nord, les deux autres s'ouvrant à l'ouest. La marée, d'une amplitude de 8 mètres, découvre 37 km^2 de zone intertidale entaillés par les chenaux de la Vire et de la Douve. Les modifications des communautés benthiques ont justifié le remplacement de la station VE 0 (1977–1978) par VE 1 (1982–1985) puis VE 2 (1983–1985). Les cotes bathymétriques de ces trois stations sont respectivement de 6 m, 5.50 m et 5 m au-dessus du 0 des cartes marines françaises.

– *l'estuaire de la Seine* se singularise par l'importance de l'arrivée d'eau douce. Malgré la taille de son embouchure et un marnage de 8 m, cet estuaire ne compte plus aujourd'hui que 30 km^2 de zone intertidale par suite des importants aménagements dont il fait l'objet. Ces derniers ont entraîné la création d'une seconde station COST en 1982 (HON 4) en complément de la station HON 5 qui est suivie depuis 1979, mais a été profondément perturbée en 1981 par suite de la réalisation d'un déversoir dans la digue submersible proche. Ces deux stations, dont les cotes bathymétriques sont respectivement de 5 m et de 4 m, se situent à l'extrémité de la fosse de flot sud de l'estuaire. Des détails sur l'évolution bathymétrique de ce secteur apparaissent dans l'article de Desprez & Dupont (1985).

– la *Baie de Somme* présente les marées de plus forte amplitude (10 m) et la plus vaste zone intertidale (70 km^2) caractérisée par une progression très rapide du schorre. La première reconnaissance du benthos effectuée par Simon *et al.* (1978) a permis de poursuivre l'étude du comblement artificiel de la zone sud (radiale HH). Deux stations ont été ajoutées dans le cadre du COST en 1982 (HHS) et en 1983 (LCS). Toutes ces stations se situent à un niveau marégraphique élevé (HHA: 8 m; HHC: 6.2 m; HHS: 6 m; LCS: 6.50 m), alors qu'elles se trouvent dans les deux autres sites au-dessous du niveau des hautes mers de morte eau.

Materiel & méthodes

Jusqu'en 1981, les prélèvements sont effectués sur une surface de 0.25 m^2 et une profondeur de 25 cm; à partir de 1982, suivant la nature du peuple-

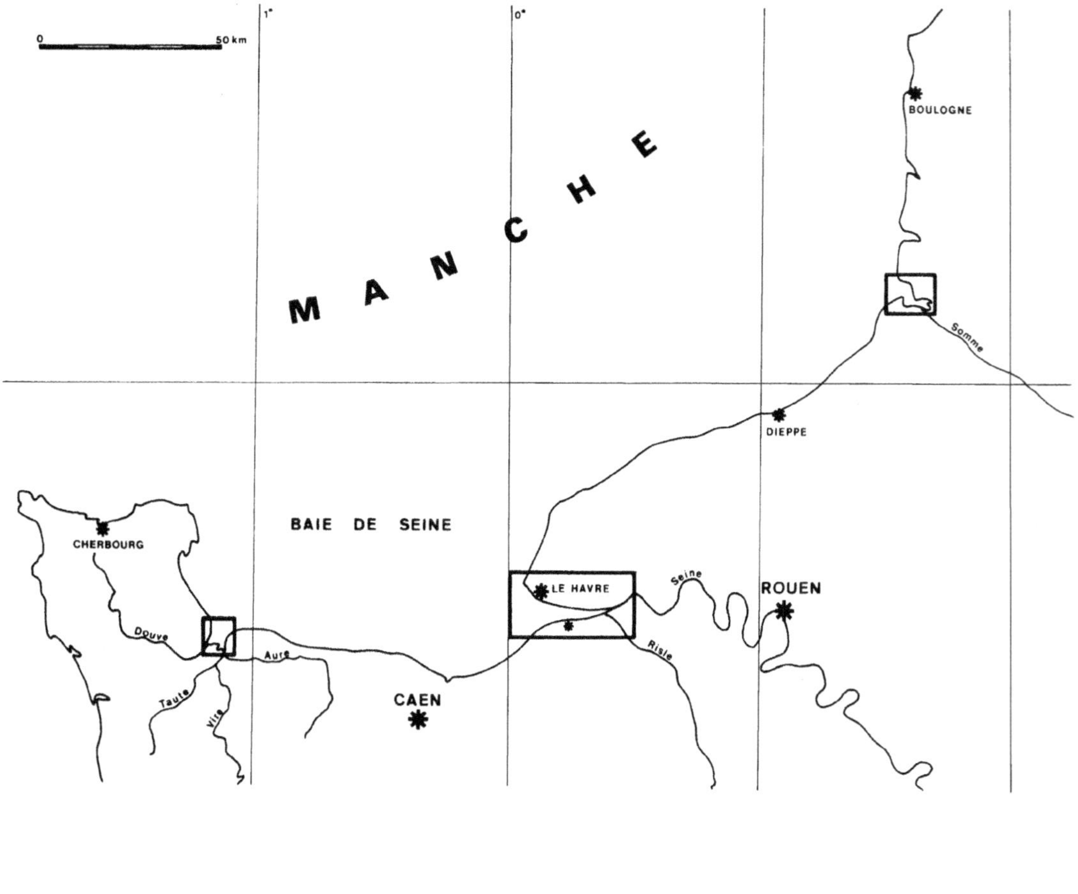

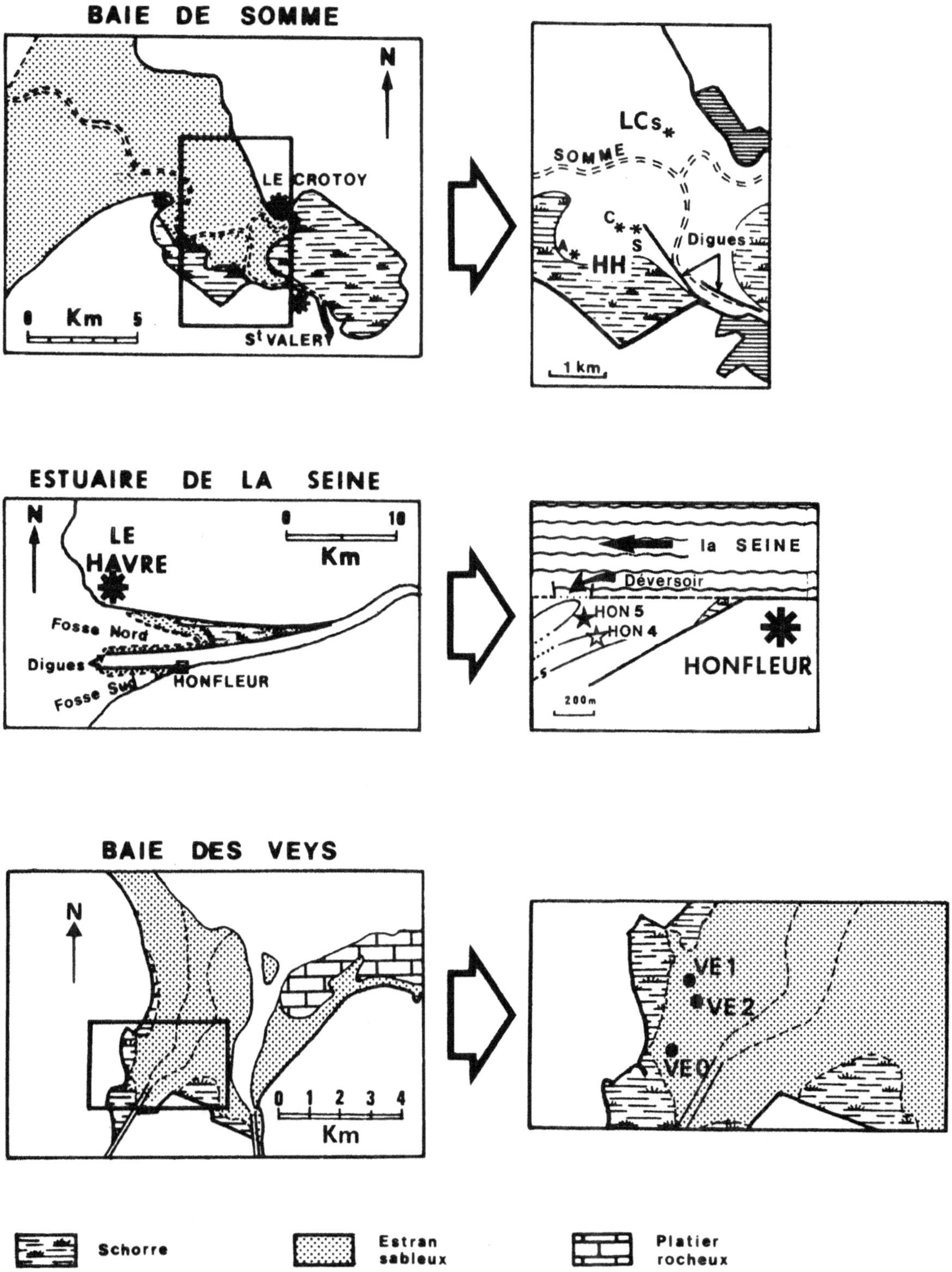

Fig. 1. Localisation des trois sites estuariens en Manche centrale et orientale. Emplacement des stations de prélèvement du macrozoobenthos intertidal.

Tableau 1. Espèces rencontrées dans les stations étudiées et classées selon l'ordre taxinomique.
Les croix indiquent la présence de l'espèce, les nombres précisent le rang de chacune d'elles, classées selon leur biomasse annuelle moyenne (exprimée en grammes de matière sèche sans cendre par mètre carré).

	VE 0	VE 1	VE 2	HON 4	HON 5	HH S	LC S
Pholoe synophtalmica				+	+		
Phyllodoce mucosa	+ 12	+	+				
Eteone longa	+ 6	+	+	+	+	+ 6	+
Nereis diversicolor	+ 3	+	+	+	+	+ 3	+
Nephtys cirrosa	+ 16				+		
Nephtys hombergii		+	+	+	+	+ 9	
Scoloplos armiger		+				+ 11	
Nerine cirratulus					+		+
Spiophanes bombyx	+ 21						
Spio filicornis					+		
Pygospio elegans	+ 5	+	+	+	+	+ 10	+
Polydora ciliata				+	+		
Tharyx marioni				+	+		
Capitomastus minimus				+	+		
Heteromastus filiformis	+ 11	+	+			+	+
Capitella capitata	+ 16	+	+	+	+	+ 5	+
Arenicola marina	+ 2	+	+	+	+	+ 11	+
Tubifex sp.				+	+		
Tubificoides benedeni	+ 16	+	+	+	+	+ 11	
Hydrobia ulvae	+ 16	+				+ 11	+
Mytilus edulis		+					
Cerastoderma edule	+ 1	+	+	+	+	+ 1	+
Abra tenuis	+ 10	+	+				
Scrobicularia plana	+ 4	+	+	+	+		
Macoma balthica	+ 9	+	+	+	+	+ 2	+
Mya arenaria				+	+		
Mysella bidentata					+		
Eurydice pulchra						+ 11	+
Haustorius arenarius							+
Urothoe grimaldii	+ 14	+	+				
Bathyporeia sarsi	+ 16	+	+		+		+
Bathyporeia pilosa	+ 13	+		+	+	+ 7	+
Corophium volutator	+ 6	+	+	+	+	+ 11	+
Corophium arenarium							+
Cyathura carinata	+ 14	+	+	+	+		
Crangon crangon		+	+		+	+ 8	+
Leander vulgaris							+
Carcinus maenas	+ 8	+	+	+	+	+ 4	+
Biomasse annuelle moyenne ($g \cdot m^{-2}$)	16.4 (1977)					37.1 (1983)	

ment étudié, cinq à trente prélèvements cylindriques de 1/30 ou 1/50 m^2 ont été réalisés à chaque station avec un carotteur spécial (T.A.S.M.: voir Ducrotoy & Desprez, 1986). Après tamisage sur maille carrée de 1 mm, le refus est fixé au formol salé additionné de phloxine B.

Les travaux de biométrie (mesure de l'axe antéro-postérieur au 1/10 mm) des coquilles de bivalves ont permis, en liaison avec un comptage des stries annuelles d'arrêt de croissance hivernal, de suivre l'évolution de taille des diverses classes d'âge.

Les mesures de biomasses pour la Baie de Somme ont été réalisées au Nederlands Instituut voor Onderzoek der Zee par le Dr J. J. Beukema. Le sédiment, prélevé parallèlement à la faune, a fait l'objet d'une analyse granulométrique classique (tamisage à sec sur colonne AFNOR) pour la fraction sableuse et d'une analyse microgranulométrique (compteur COULTER) pour la fraction fine (inférieure à 60 μm).

En Baie des Veys, quatre stations supplémentaires, correspondant à quatre faciès caractéristiques, ont été suivies mensuellement durant un an (de mars 1977 à mars 1978) dans le but d'appréhender le comportement global des biocénoses. Une étude de corrélation a été faite avec les facteurs climatiques.

Résultats

1. Structure des peuplements

Le tableau 1 récapitule les espèces rencontrées dans les stations COST des trois estuaires et, lorsque les biomasses ont été calculées, précise le rang de ces espèces. Sur les 38 recensées, 15 espèces sont communes aux trois sites alors que certaines caractérisent un site particulier; c'est le cas d'*Abra tenuis, Urothoe grimaldii* et *Phyllodoce mucosa* pour la Baie des Veys, de *Tharyx marioni, Mya arenaria, Tubifex sp, Pholoe synophtalmica, Capitomastus minimus* et *Spio filicornis* pour l'estuaire de la Seine, et de *Eurydice pulchra, Haustorius arenarius, Corophium arenarium* et *Leander vulgaris* pour la Baie de Somme. Il est important de préciser ici les limites d'une telle comparaison inter-sites effectuée à partir de quelques stations, si judicieusement choisies fussent-elles.

En Baie des Veys et Baie de Somme, le calcul des biomasses sur un cycle annuel met en évidence l'importance de *Cerastoderma edule* et *Nereis diversicolor,* voire *Eteone longa*, dans les deux sites, le rôle prédominant de *Macoma balthica*, des Capitellidés, de *Bathyporeia sp* et *Crangon crangon* dans l'estuaire picard, enfin la place importante de *Arenicola marina, Scrobicularia plana, Pygospio elegans* et *Corophium volutator* dans l'estuaire bas-normand.

Enfin, ce tableau fait ressortir la faible biomasse annuelle moyenne de la Baie des Veys par rapport à la Baie de Somme qui se situe au niveau d'autres sites estuariens européens (Ducrotoy, 1984).

2. Cinétique des espèces et données édaphiques

2.1. En Baie des Veys, malgré la brièveté de la période échantillonnée, on observe nettement à la station VE 1 une diminution des effectifs de la plupart des espèces (Fig. 2):
– chute brutale en 1983 de *Cerastoderma edule, Corophium volutator, Tubificoides benedeni* et *Heteromastus filiformis*;
– chute brutale en 1984 de *Nereis diversicolor*;
– tendance à la régression de *Abra tenuis* et *Arenicola marina* de 1983 à 1985.

Seules *Pygospio elegans* et *Urothoe grimaldii* présentent une évolution contraire, tandis que *Macoma balthica* et *Scrobicularia plana* témoignent d'une certaine stabilité, avec des densités relativement faibles. Par contre, en VE 2, aucune tendance évolutive générale ne peut apparaître sur une aussi courte période. On doit cependant noter les faibles effectifs en 1984–1985 de *Corophium volutator, Nereis diversicolor* et *Heteromastus filiformis.*

Du point de vue sédimentologique (Tableau 2), on observe un désenvasement synchrone des deux stations, plus net toutefois en VE 2 dont la teneur en particules fines est 3 à 4 fois supérieure à celle de VE 1. Cette tendance est confirmée par l'intrusion, sur le flanc ouest de la baie, de sédiments d'origine marine qui contrarient une progression côtière de l'envasement. En effet, les deux stations suivies se trouvent, ajourd'hui, dans une zone de friction entre ces deux phénomènes contradictoires.

2.2 En estuaire de Seine, nous observons également une chute des densités de nombreuses espèces (Fig. 3), à partir de 1981, aux deux stations prospectées (*Cerastoderma edule, Scrobicularia plana, Mya arenaria, Pygospio elegans, Polydora ciliata, Corophium volutator* et oligochètes). A la station HON 5, *Macoma balthica, Nereis diversicolor* et *Eteone longa* sont également en régression alors que leurs effectifs sont relativement stables à la station HON 4. Les évolutions de *Nephtys hombergii* et *Arenicola marina* s'avèrent difficiles à interpréter. Quant aux crustacés, s'ils tendent à disparaître des prélèvements dès 1981 à HON 5, nous observons à HON 4 une évolution synchrone des pics d'abondance en 1982–1983 qui conduit, là encore, à la disparition de ces espèces, exception faite de *Cyathura carinata.*

Au cours de la même période, les paramètres sédimentologiques du substrat ont profondément évolué, notamment la teneur en particules fines. A la station HON 5, cette teneur chute de 32% à 4% dans les douze mois qui suivent la réalisation du

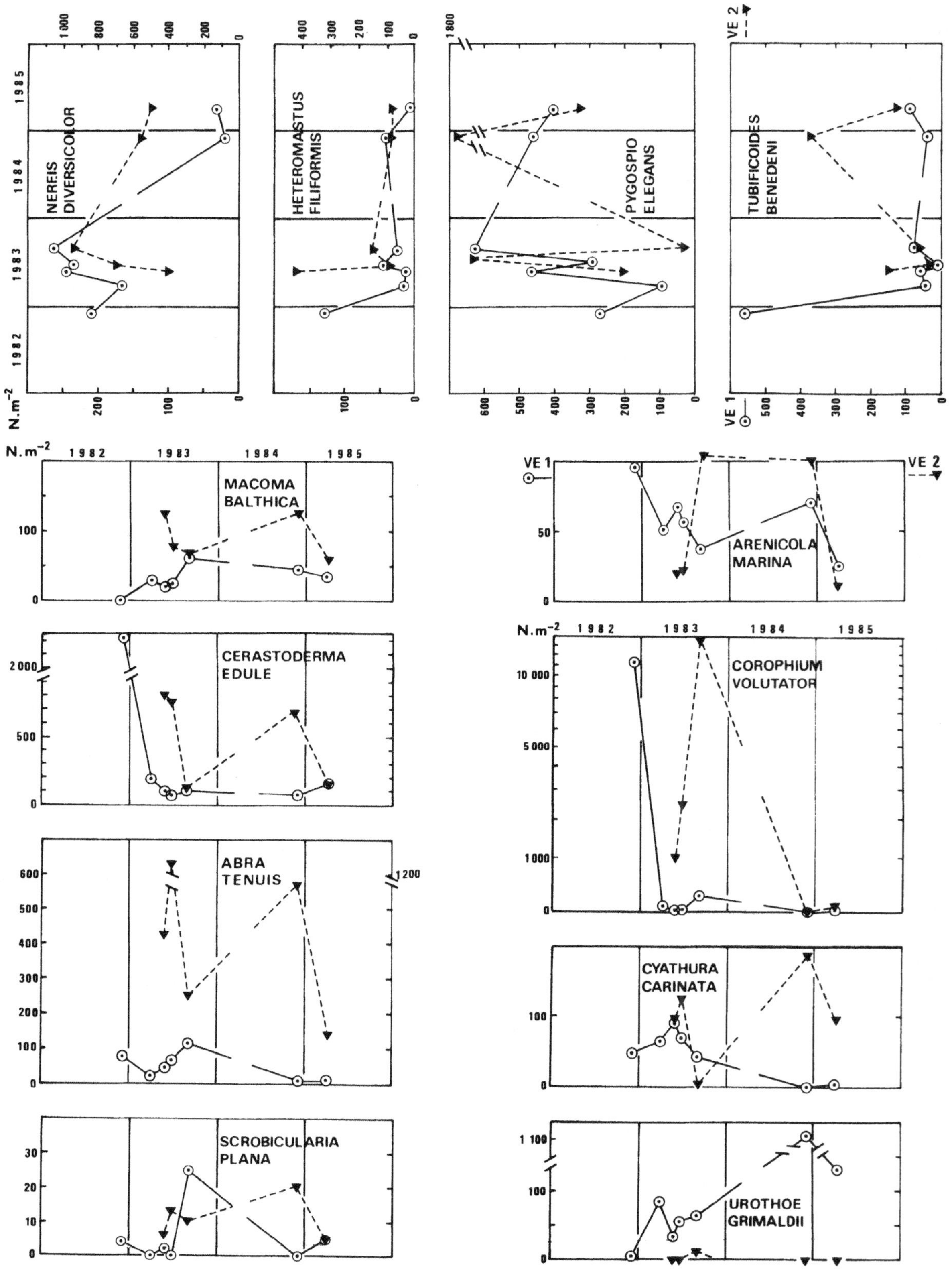

Fig. 2. Evolution temporelle des densités au mètre carré ($N \cdot m^{-2}$) de principales espèces du macrozoobenthos intertidal prélevées en Baie des Veys aux deux station (VE 1 et VE 2) suivies de décembre 1982 à mars 1985.

Tableau 2. Principales caractéristiques sédimentologiques des stations étudiées en Baie des Veys.

		11/82	03/83	05/83	06/83	08/83	11/84	03/85
	Médiane (μm)	150	180	180	-	160	170	185
VE 1	Mode (μm)	130	165	165	-	150	170	160
	Fines (%)	7,3	4,7	2,9	3,1	5,2	0,9	1,5
	Médiane (μm)			125	-	125	130	140
VE 2	Mode (μm)			150	-	145	140	140
	Fines (%)			14,6	13,5	14,8	10,7	6,6

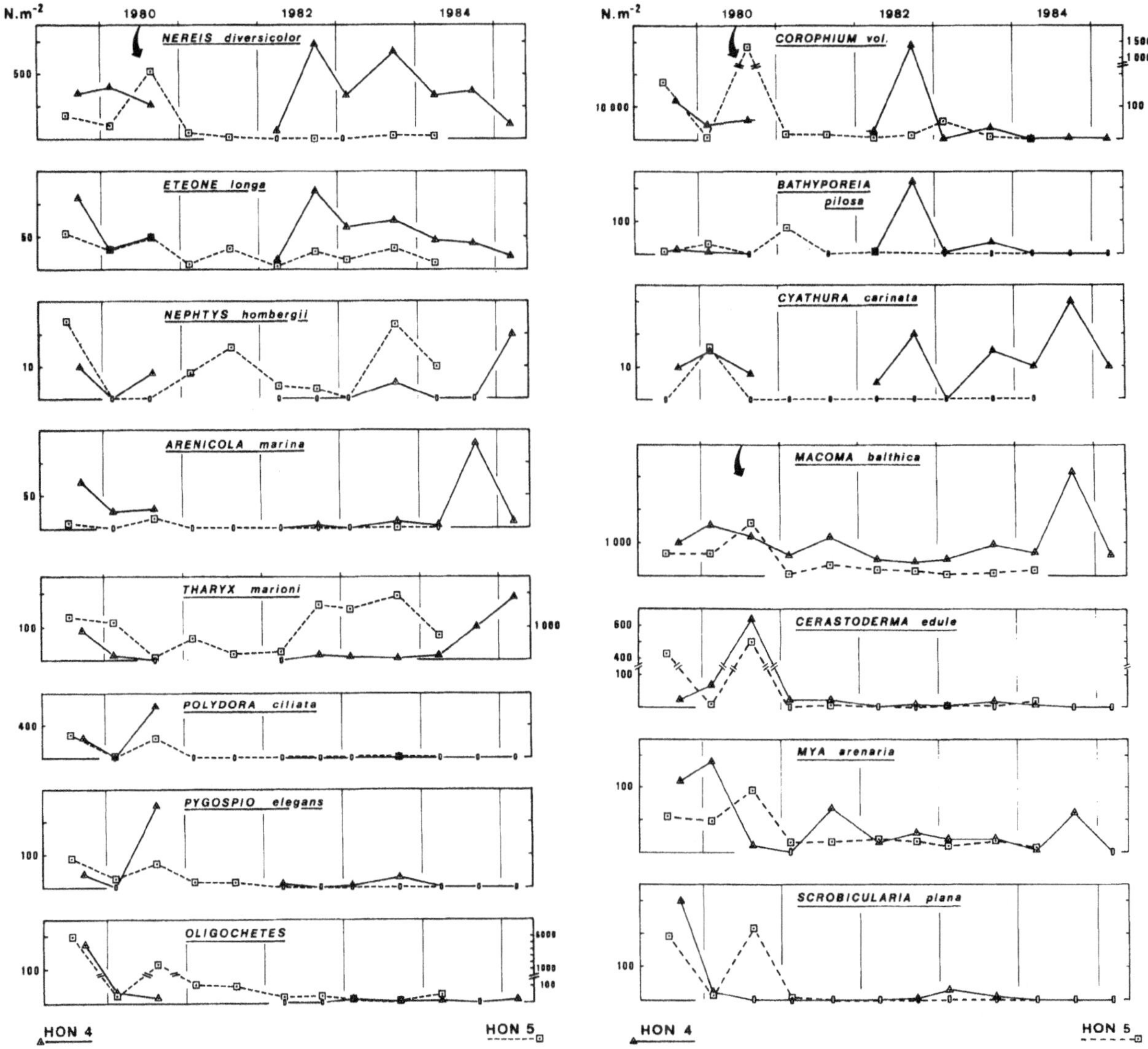

Fig. 3. Evolution temporelle des densités au mètre carré ($N \cdot m^{-2}$) des principales espèces du macrozoobenthos intertidal prélevées dans l'estuaire de la Seine aux deux stations (HON 4 et HON 5) suivies de juillet 1979 à mars 1985. Pour les années où un suivi bimestriel a été effectué, seules ont été reportées les valeurs minimales et maximales des abondances. Le chiffre 0 indique les effectifs nuls; la réalisation du déversoir (juillet 1980) est figurée par la flèche.

déversoir. Par contre, à la station HON 4, à une période de relative stabilité de ce paramètre (plus de 20% de fines) succède à partir de 1984 une phase de désenvasement du sédiment superficiel, résultat des importants apports sableux qui caractérisent l'emsemble de la fosse de flot (Desprez & Dupont, 1985).

2.3. En Baie de Somme (Fig. 4), le schéma évolu-

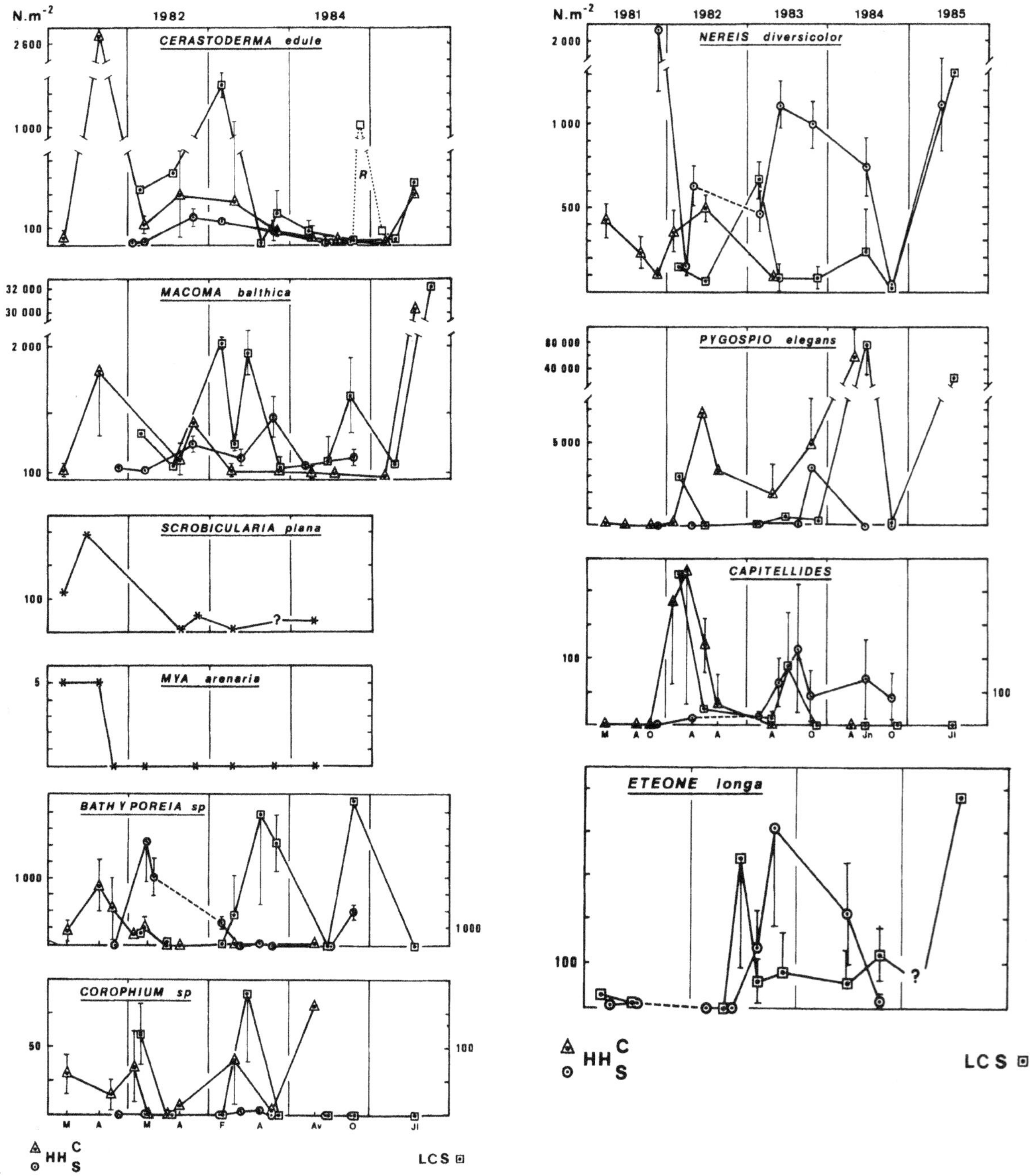

Fig. 4. Evolution temporelle des densités au mètre carré (N · m^{-2}) des principales espèces du macrozoobenthos intertidal prélevées en Baie de Somme aux stations HH C, HH S et LC S, suivies de février 1981 à juillet 1985. Pour *Scrobicularia plana* et *Mya arenaria*, les courbes sont le résultat du suivi d'une station annexe (HH A: astérisques).

Pour *Cerastoderma edule*, le réensemencement en naissain effectué en novembre 1984 est matérialisé par la courbe en pointillés et la lettre R.

tif des densités des espèces benthiques est différent des précédents. Trois espèces de bivalves (*Cerastoderma edule, Scrobicularia plana* et *Mya arenaria*) montrent une baisse très nette de leurs effectifs à toutes les stations où elles étaient représentées en 1982, tandis que d'autres espèces semblent globalement en augmentation (*Pygospio elegans, Eteone longa, Macoma balthica*).

En HHC, les effectifs de *Macoma balthica* diminuent régulièrement de 1981 à 1984, période suivie d'un spectaculaire redressement à l'issue de l'hiver 1984–1985. En HHS se produit le phénomène inverse avec une bonne installation en 1982–1983, mais qui demeure sans suite. En LCS, les variations saisonnières montrent, malgré des densités parfois élevées, l'insuccès des recrutements successifs; en été 1985 on observe des densités exceptionnelles de 32000 individus · m^{-2}.

Alors que, en 1981, *Pygospio elegans* apparaissait à de très faibles densités, depuis 1982 se succèdent des pullulations de cette espèce qui atteignent de très fortes valeurs (près de 80000 individus · m^{-2} en 1984).

Le schéma évolutif des effectifs de *Eteone longa* montre un parallélisme évident avec celui observé chez *Nereis diversicolor*; les années 1983 et 1985, particulièrement favorables, succèdent à de mauvaises années 1982 et 1984.

Nous remarquons qu'en 1982–1984 la chute des effectifs de bivalves, spécialement *Cerastoderma edule*, s'accompagne d'un développement momentané des Capitellidés de 1983 à la fin de l'hiver 1984–1985. Quant aux crustacés, un relai des cycles saisonniers s'observe entre les trois stations pour *Bathyporeia pilosa* et *B. sarsi*, selon une succession bathymétrique depuis les plus hauts niveaux (HHC, HHS et LCS). *Corophium volutator* est toujours peu abondant en HHS et soumis à des poussées saisonnières dans les autres stations où il est accompagné par *C. arenarium* (LCS). A partir de 1984, ces espèces sont en nette régression dans les stations étudiées.

L'évolution granulométrique du secteur sud de la baie (radiale HH) est inverse en 1981–1982 de celle qui caractérise les années 1975–1980 où un envasement important avait sanctionné les travaux de chenalisation (Dupont, 1983); la teneur en particules fines des sédiments diminue lentement (ainsi qu'en LC), tandis que d'importants apports sableux viennent combler les bas niveaux (Dupont & Lafite, 1984), accélérant le colmatage de cet estuaire.

Les années 1983 et 1984 ont connu des chaleurs estivales importantes, la température des sédiments atteignant 40 °C en surface. Dans le même temps, les eaux réchauffées ont été le siège de développements particulièrement importants de phytoplancton parfois toxique (*Dinophysis acuminata*).

2.3. *Comparaison inter-sites de quelques espèces significatives*

Cette comparaison permet de mettre en évidence des tendances évolutives communes ou au contraire contradictoires:

– chez *Nereis diversicolor*, un effondrement quasi-général des effectifs se produit après l'hiver 1981–1982 (Baie de Somme et estuaire de Seine), ainsi qu'en été 1984 dans les trois sites. Dans l'estuaire de la Seine, la régression de cette espèce est cependant plus lente, sur fond de cycle saisonnier, à la station HON 4, alors qu'à la station HON 5 la raréfaction de ce polychète est synchrone de la réalisation du déversoir.

– *Cerastoderma edule* est en mauvaise position dès 1981 dans l'estuaire de la Seine; en Baie de Somme, le mauvais recrutement du printemps 1982 semble compensé, à l'automne suivant, mais on observe un effondrement dans le courant de l'année 1983, phénomène qui s'observe également en Baie des Veys. La tentative de repeuplement artificiel, visible sur LCS (Fig. 4), n'est suivie d'aucun effet; par contre, le recrutement 1985 est satisfaisant, phénomène confirmé en Baie des Veys où une pullulation de très jeunes coques de 5 mm à 1 cm a été observée au cours de l'été 1985 (malheureusement non quantifiée).

– Deux espèces de Capitellidés, *Capitella capitata* et *Heteromastus filiformis*, absentes de l'estuaire de la Seine, montrent en Baie de Somme une évolution opposée à celles de *Nereis diversicolor* et *Cerastoderma edule*. On observe au contraire un parallélisme d'évolution entre *Cerastoderma edule* et *Heteromastus filiformis* en Baie des Veys.

– Alors que les fluctuations de *Macoma balthica* observées dans les deux baies se révèlent difficilement interprétables, cette espèce se maintient dans l'estuaire de la Seine à des niveaux d'abondance relativement stables: seul le mauvais recrutement de 1982 et le très bon recrutement de 1984 rompent la régularité du cycle saisonnier.

– Les fluctuations considérables des densités de

Corophium volutator observées tant en Baie des Veys qu'en estuaire de Seine, montrent, avec l'absence de synchronisme des pics d'abondance dans les trois sites, le comportement imprévisible de cette espèce.

– La cinétique de *Pygospio elegans* est elle aussi difficilement interprétable. Cependant, la simple observation des niveaux de densité montre les caractéristiques propres de chaque écosystème. En estuaire de Seine, les maxima observés ne dépassent pas quelques centaines d'individus · m^{-2} en 1979 et 1980; en Baie des Veys, les effectifs oscillent considérablement de quelques dizaines d'individus à 1800 individus · m^{-2} en 1984; mais c'est en Baie de Somme que ce polychète atteint les plus fortes densités (plusieurs dizaines de milliers d'individus · m^{-2}), constituant des banquettes qui peuvent coloniser de grandes surfaces (Dupont, 1975).

3. Evolution des abondances globales et du nombre d'espèces

3.1. Abondances (Fig. 5)

Les fluctuations importantes des densités de *Corophium volutator* (estuaire de Seine, Baie des Veys) et de *Pygospio elegans* (Baie de Somme) se

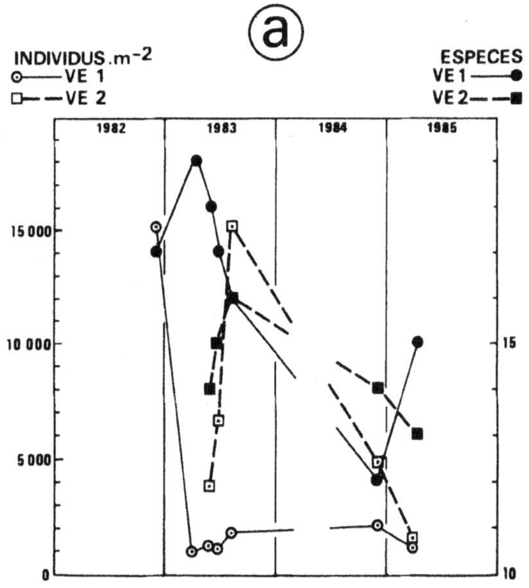

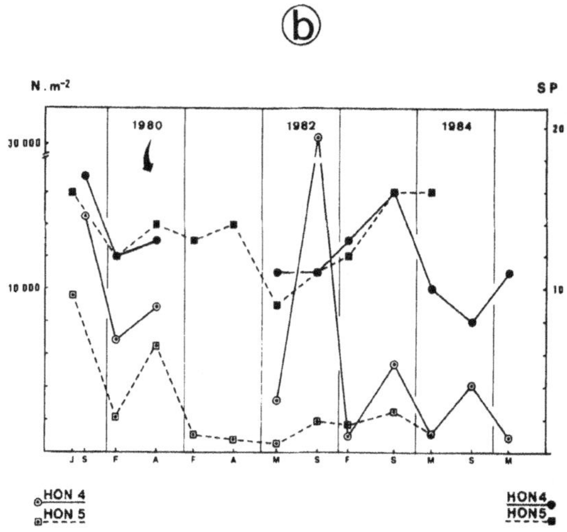

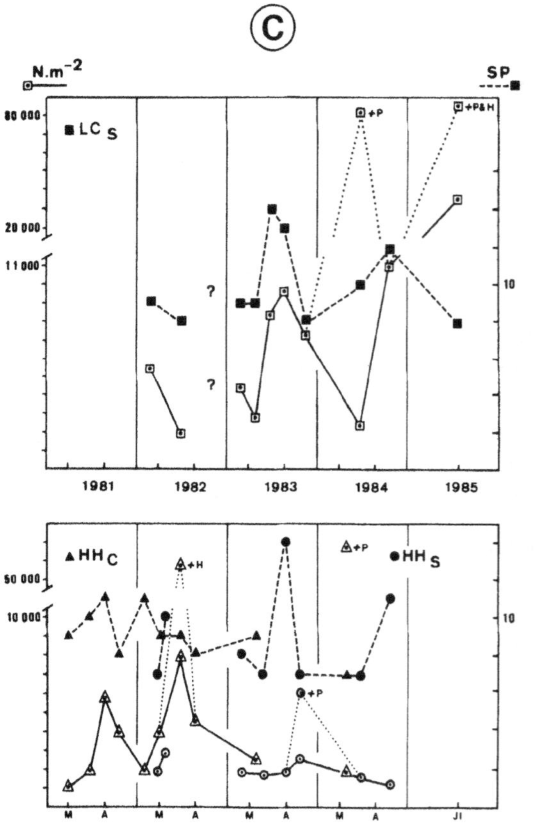

Fig. 5. Evolution temporelle des abondances globales (N · m^{-2}) et de la richesse spécifique (SP) du macrozoobenthos intertidal en Baie des Veys (a), en estuaire de Seine (b) et en Baie de Somme (c).

Dans ce dernier site, lors de pullulations d'*Hydrobia ulvae* ou de *Pygospio elegans*, la valeur inférieure de la densité globale ne tient pas compte de leurs effectifs, au contraire de la courbe supérieure en pointillés (+H et +P).

En estuarie de Seine, le flèche indique la réalisation du déversoir (juillet 1980).

répercutent d'une façon magistrale sur les fluctuation des abondances globales. Les pullutations de *Corophium sp* se produisent en 1982 en HON 4 et VE 1, et en 1983 en VE 2, alors que *Pygospio elegans* dépasse 50000 individus au mètre carré en 1983 en HHS et en 1984 en HHC et en LCS.

Ces observations montrent le rôle essentiel de ces espèces dans la dynamique des peuplements. Alors que *Pygospio elegans* se comporte en véritable espèce opportuniste, *Corophium volutator* est parfois victime, en Baie des Veys, du transport passif par les houles levées par les tempêtes qui occasionnent des concentrations de l'espèce.

Dans l'estuaire de la Seine, on observe, dès la réalisation du déversoir, une chute rapide des densités à la station HON 5 en 1981 et une régression régulière de ces densités à la station HON 4, modulée par le rythme saisonnier. Dès 1982, on assiste à une restructuration du peuplement de la station HON 5 qui conduit au doublement des densités minimales observées en 1981 (Desprez & Dupont, 1985).

3.2. Richesse spécifique (Fig. 5)

En Baie des Veys, la chute du nombre d'espèces répertoriées en VE 1 et VE 2 rend compte de l'évolution des communautés estuariennes du flanc ouest. Cette régression régulière, qui s'observe à partir de 1983, traduit une déstabilisation du système. Aucune espèce nouvelle n'apparait mais il se produit un changement de dominance à VE 1, *Urothoe grimaldii* prenant le relais de *Nereis diversicolor, Corophium volutator* et *Cerastoderma edule*.

Dans l'estuaire de la Seine, à la disparition d'espèces (1981–1982) succède (1983) un enrichissement spécifique par arrivée d'espèces nouvelles mieux adaptées aux nouvelles conditions hydrodynamiques et sédimentologiques (Desprez & Dupont, 1985).

De nouveau, une forte régression s'observe en 1984 en HON 4.

En Baie de Somme, un accroissement comparable de la richesse spécifique se produit en 1983, année où de nombreuses espèces ont vu chuter leurs effectifs.

Dans ces deux derniers sites, l'augmentation du nombre d'espèces correspond à une intrusion marine qui se traduit par l'apparition de *Bathyporeia sarsi* et *Nerine cirratulus*, espèces auxquelles s'ajoute *Haustorius arenarius* en Baie de Somme. Alors que dans l'estuaire picard, on assite à un renouvellement de ce phénomène, apparemment saisonnier, avec retard cependant, au cours de l'été suivant (1984), dans les estuaires normands, nous venons de voir que la diversité a stagné à un faible niveau pour amorcer, semble-t-il, une récupération en 1985, sauf en VE 2.

En été 1983, une perturbation (qui pourrait être dûe aux conditions climatiques) semble se manifester dans les trois sites par un enrichissement spécifique suivi d'une déstabilisation des systèmes qui ne récupéreraient pas en synchronisme, que ce soit dans un même estuaire suivant la station considérée ou d'un estuaire à l'autre.

4. Croissance des bivalves

Alors que, comme nous venons de le voir, les fluctuations d'abondance des espèces ne permettent pas toujours de comprendre le rôle joué par les facteurs de l'environnement – soit à cause de la dynamique propre de ces espèces, soit à cause de leur sensibilité respective aux agressions naturelles ou artificielles du milieu –, un paramètre comme la croissance apparaît de première importance pour compléter l'information fournie par l'étude de l'évolution des densités.

Ainsi, dans l'estuaire de la Seine, l'évolution des effectifs de *Macoma balthica* et de plusieurs autres espèces ne permet pas de juger de l'impact de la réalisation du déversoir sur le peuplement de la station HON 4. Par contre, le suivi de la croissance de ce bivalve (Fig. 6) met clairement en évidence l'influence négative des nouvelles conditions hydrodynamiques locales: la croissance des individus de la classe 1 se trouve interrompue dès juillet 1980 alors qu'elle se poursuit d'ordinaire jusqu'à l'entrée de l'hiver; elle ne retrouve un rythme normal qu'au printemps suivant. Cette perturbation affecte, selon le même schéma, l'ensemble de la population du bivalve (Ducrotoy *et al.*, 1985).

En Baie de Somme, les perturbations mises en évidence en 1982–1983 avec l'étude de l'évolution des effectifs de bivalves, se trouvent confortées par l'étude de la croissance de *Cerastoderma edule* (Fig. 7). Les années de crise 1982 et 1983 se traduisent par un rythme de croissance de la classe 1 inférieur à celui de 1981, et surtout par l'absence de recrutement (classe 0). Ces problèmes de recrutement se sont traduits, à partir de 1984, par une crise de la production commerciale de ce bivalve qui constituait un stock exploitable. En 1985, à la suite de

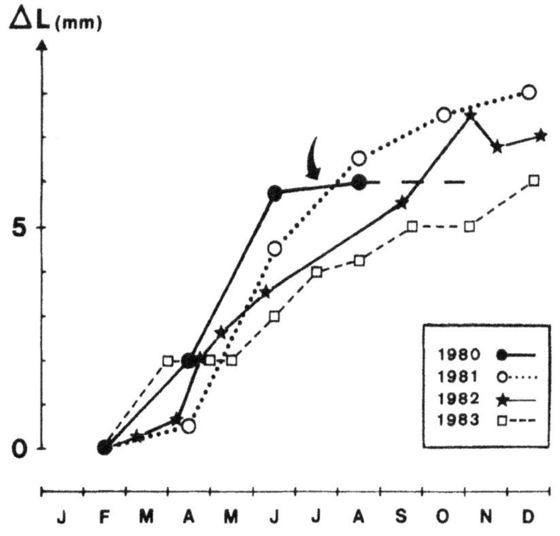

Fig. 6. Accroissement annuel en longueur (ΔL) des individus de la classe 1 (animaux âgés d'un an) de *Macoma balthica* dans l'estuaire de la Seine (station HON 4) entre 1980 et 1983.

La flèche indique le mois de réalisation du déversoir (juillet 1980).

l'hiver exceptionnellement rigoureux, un bon recrutement a été observé et le suivi de la classe 0 montre un rythme de croissance nettement supérieur à celui observé en 1981, avant la crise.

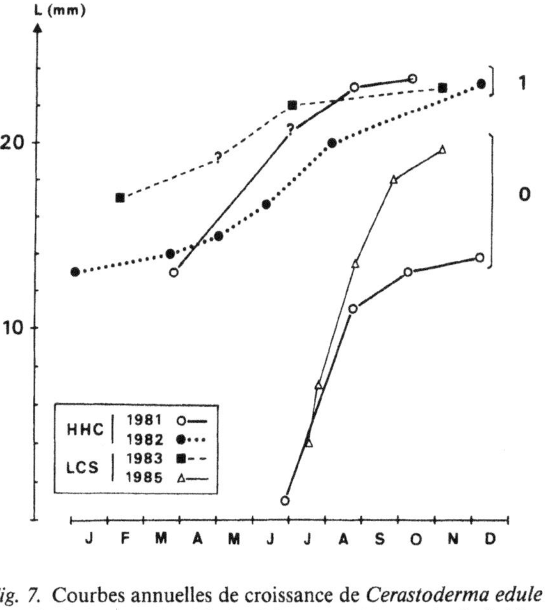

Fig. 7. Courbes annuelles de croissance de *Cerastoderma edule* en Baie de Somme de 1981 à 1983 et en 1985 pour les individus des classes d'âge 0 et 1 (animaux de l'année et âgés d'un an).

Le point d'interrogation correspond à des valeurs obtenues à une station voisine de même niveau marégraphique.

5. Evolution globale des peuplements

L'ensemble des résultats précédents s'inscrit dans un contexte évolutif global des biocénoses que nous avons cartographiées.

5.1. En Baie des Veys (Sylvand, 1986), les peuplements de type estuarien se cantonnaient en 1974 (Fig. 8) principalements dans le fond sud-ouest de la baie qui était aussi le secteur préférentiel d'implantation de *Macoma balthica*, tandis que le flanc est affichait des affinités marines homogènes avec un peuplement caractérisé par *Scoloplos armiger.* Le secteur central de la baie, bien développé, abritait un peuplement de transition est-ouest entre les deux précédents.

On observe, jusqu'en 1985, une réduction importante des surfaces occupées par le peuplement estuarien dans le secteur sud-ouest; cette évolution est liée d'une part à l'extension du schorre de la côte vers le centre, d'autre part à la progression, en moyenne et basse-plage, vers le sud d'un peuplement de type marin diversifié (caractérisé par *Urothoe grimaldii*) tandis que *Abra tenuis* remplace *Macoma balthica.* En 1974, deux espèces caractéristiques, l'une estuarienne (*Macoma balthica*), l'autre marine abritée (*Scoloplos armiger*) dominaient largement les peuplements. L'apparition d'espèces intrusives *Urothoe grimaldii* et *Abra tenuis* ou en cours d'extension importante (*Pygospio elegans* et *Macoma balthica* sur le flanc est) s'est faite progressivement et il est important de suivre, plus que les effectifs de ces espèces, l'évolution de l'importance relative des espèces nouvelles par rapport aux espèces initiales.

Plusieurs types de comportements cénotiques ont pu être définis grâce à l'outil statistique multivarié appliqué au suivi réalisé en 1977–78 (Fig. 9):

– le flanc est (Géfosse) est fondamentalement stable et abrité; par conséquent, les transformations survenant dans cette zone auront un caractère durable;

– le fond sud-ouest (Grand Vey), bien qu'abrité, présente une fragilité face aux houles du nord-est: le point 13 (janvier 1978) se singularise par rapport aux deux ensembles saisonniers et montre l'influence de type catastrophique des houles levées par les coups de vent de nord-est;

– sur le flanc ouest, cette influence néfaste apparait encore à la Petite Dune, de type perturbable ouvert, au même point 13, isolé par rapport à un ensemble groupé;

– plus au nord, en milieu marin ouvert (La Madeleine), les fortes houles sont mieux tolérées par des peuplements adaptés à ces conditions, et les deux sous-ensembles saisonniers ne sont pas perturbés.

5.2. Dans l'estuaire de la Seine (Desprez & Dupont, 1985), l'évolution des peuplements entre 1980 et 1984 est masquée par les conséquences brutales des aménagements constants dont il fait l'objet. Sur la rive nord, l'importante régression des surfaces occupées par le peuplement estuarien supérieur (caractérisé par *Tubifex costatus* et *Manayunkia estuarina*) résulte de la progression spectaculaire du schorre vers l'aval; cette dernière intervient elle-même après le comblement naturel de l'estuaire, phénomène amplifié par les endiguements (Fig. 10). Sur la rive sud, le peuplement estuarien inférieur a été perturbé par la modification des conditions hydrodynamiques locales liée à la réalisation du déversoir. Le désenvasement consécutif de la fosse de flot a entraîné la disparition d'espèces typiquement estuariennes comme *Scrobicularia plana, Corophium volutator* et *Tubificoides benedeni*, tandis qu'apparaissaient des espèces d'affinité plus marine comme *Bathyporeia sarsi, Spio filicornis, Nephtys cirrosa* ou *Nerine cirratulus* (Desprez & Dupont, 1985).

5.3. En Baie de Somme, l'évolution des peuplements entre 1978 et 1985 est essentiellement marquée par la régression spectaculaire du peuplement estuarien vrai alors que, après 1969, la chenalisation du fleuve avait artificiellement permis le développement d'une slikke importante dans le secteur sud de la baie, par suite du comblement de l'ancien chenal (Ducrotoy *et al.*, 1985). A partir de 1978 se produit une remarquable progression des phanérogames maritimes dans tout le secteur interne de

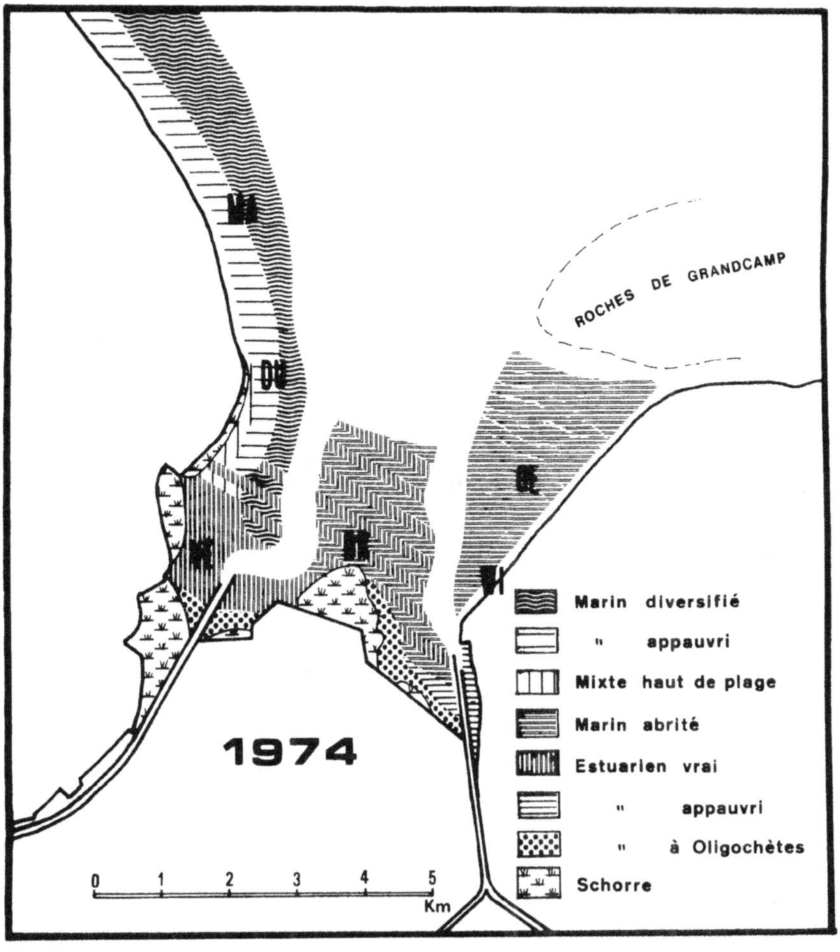

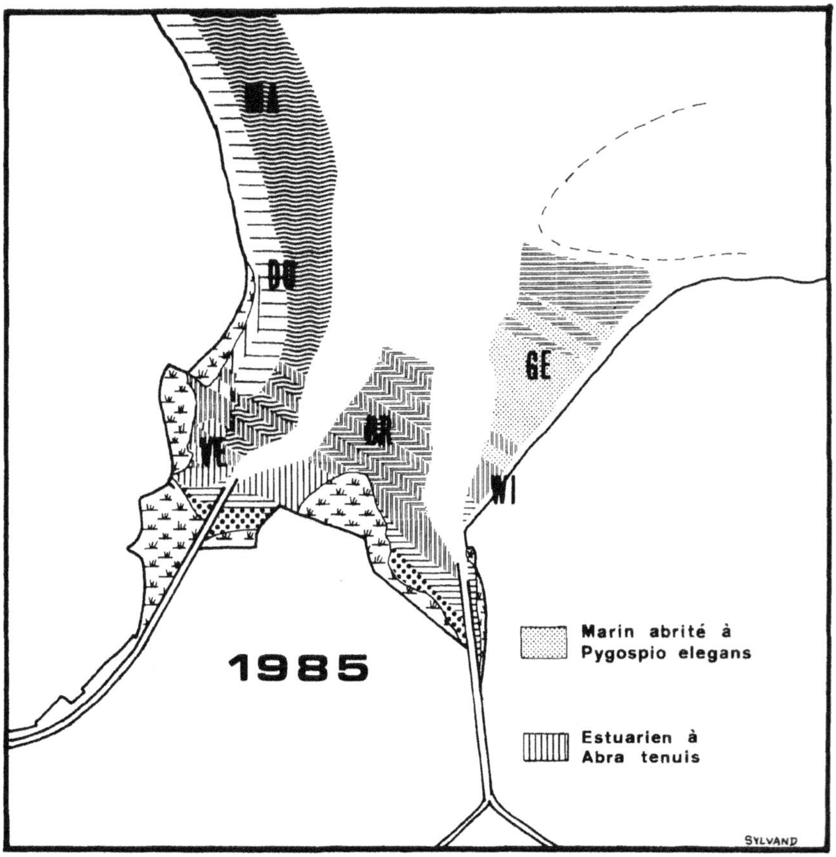

Fig. 8. Evolution spatiale des peuplements du macrozoobenthos intertidal de la Baie des Veys entre 1974 et 1985.
Sites étudiés: MA = la Madeleine; DU = la petite Dune; VE = le grand Vey; BR = pointe de Brévand; WI = le Wigwam; GE = Gefosse.

la baie ainsi que l'extension d'un faciès appauvri en son centre. Cette évolution naturelle s'est traduite notamment par une crise de la pêche à pied des coques, les sables propres à *Bathyporeia pilosa, B. sarsi* et *Haustorius arenarius* ayant remplacé les sables envasés à *Cerastoderma edule* qui occupaient les abords des chenaux du milieu de la baie. Parallèlement, s'est produite une migration des champs de *Pygospio elegans* des zones périphériques vers des zones plus centrales de la baie, phénomène traduisant peut-être l'exhaussement important des fonds de l'estuaire.

Discussion

L'évolution numérique des effectifs des espèces du macrozoobenthos intertidal de la Baie des Veys, de l'estuaire de la Seine et de la Baie de Somme, intègre les variations des facteurs de l'environnement des trois écosystèmes. Comme le soulignent Elliott & Mc Lusky (1985), les variations de densité des animaux estuariens reflètent, à long-terme, celles liées aux conditions du milieu, qu'elles soient naturelles ou artificielles. Alors que de nombreux travaux ont porté sur la recherche d'espèces indicatrices de pollution organique (Leppakowski, 1975; Pearson & Rosenberg, 1978), il y en a peu qui aient réellement cherché à expliquer l'impact des variations climatiques saisonnières sur l'abondance du macrozoobenthos comme l'ont fait Beukema (1974, 1985) et Dankers & Beukema (1981). La difficulté rencontrée dans ce genre d'étude vient essentiellement du besoin de suivre de nombreux cycles annuels sans discontinuité.

Replacés dans le contexte d'évolution générale de

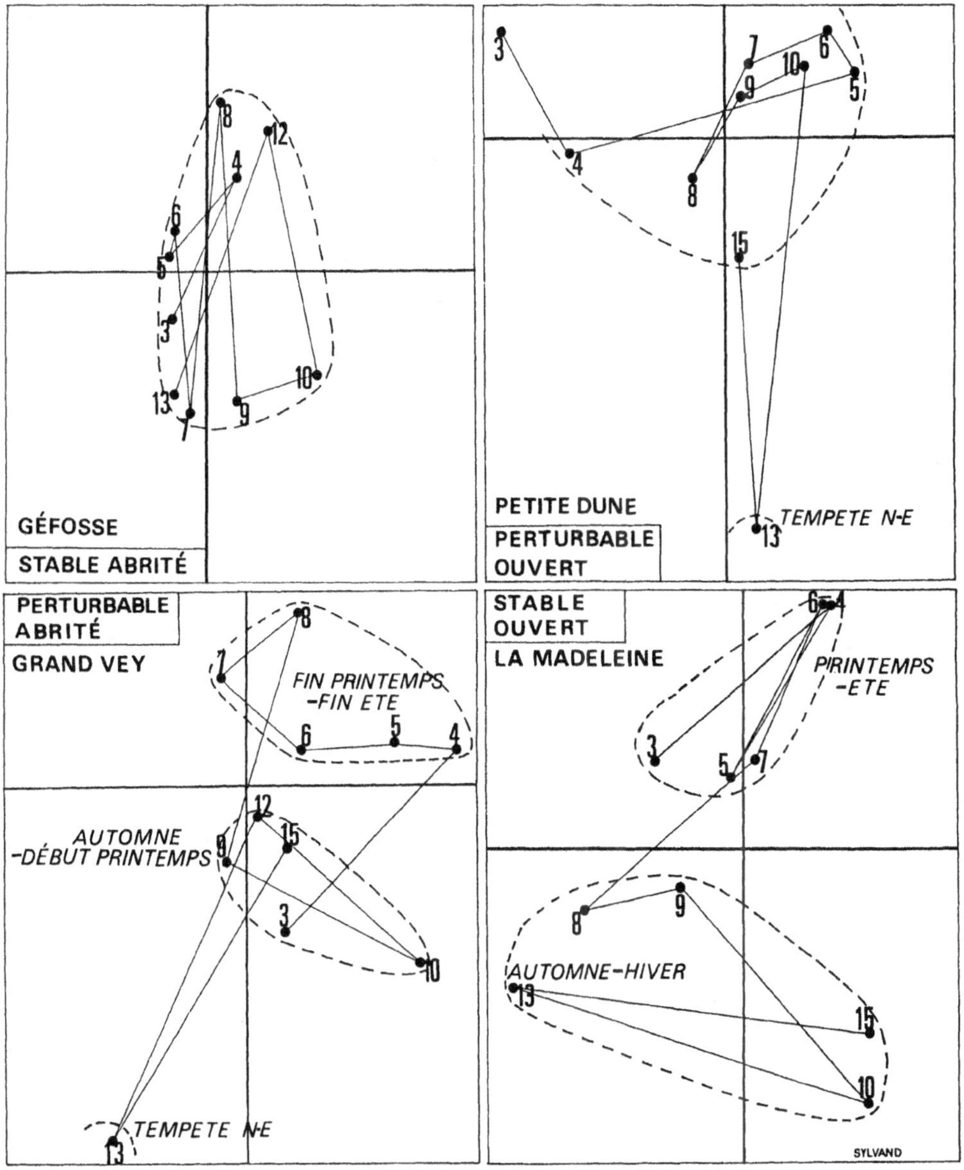

Fig. 9. Analyse factorielle des correspondances des données portant sur les effectifs globaux du macrozoobenthos intertidal de la Baie des Veys. 3 à 12: mois de mars à décembre 1977, 13 à 15: mois de janvier à mars 1978.

chaque estuaire, les résultats concernant les stations pilotes choisies dans le cadre du programme COST 647 montrent l'existence d'espèces dont les fluctuations numériques traduisent les tendances générales du milieu et que l'on peut reconnaître comme 'espèces-clés':

1. Dans l'estuaire de la Seine, les espèces-clés sont *Scrobicularia plana, Cerastoderma edule, Polydora ciliata, Pygospio elegans* et *Tubificoides benedeni*, espèces qui reflètent la tendance évolutive générale de l'estuaire au désenvasement (Desprez & Dupont, 1985).
2. En Baie de Somme, les espèces-clés sont *Heteromastus filiformis* et *Capitella capitata* (Capitellidés), *Scrobicularia plana, Cerastoderma edule, Bathyporeia pilosa* et *B. sarsi*, espèces qui reflè-

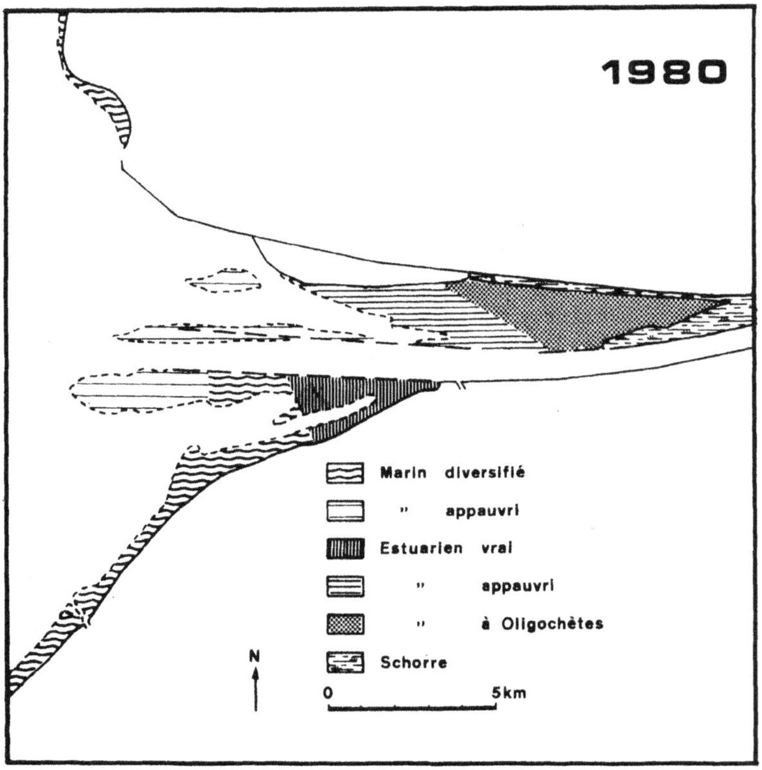

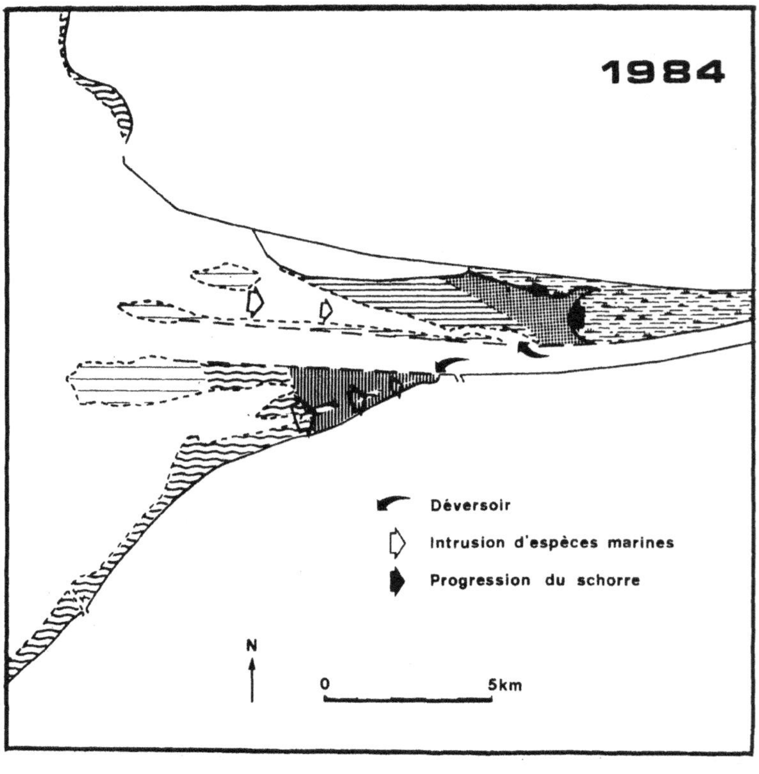

Fig. 10. Evolution spatiale des peuplements du macrozoobenthos intertidal de l'estuaire de la Seine entre 1980 et 1984.

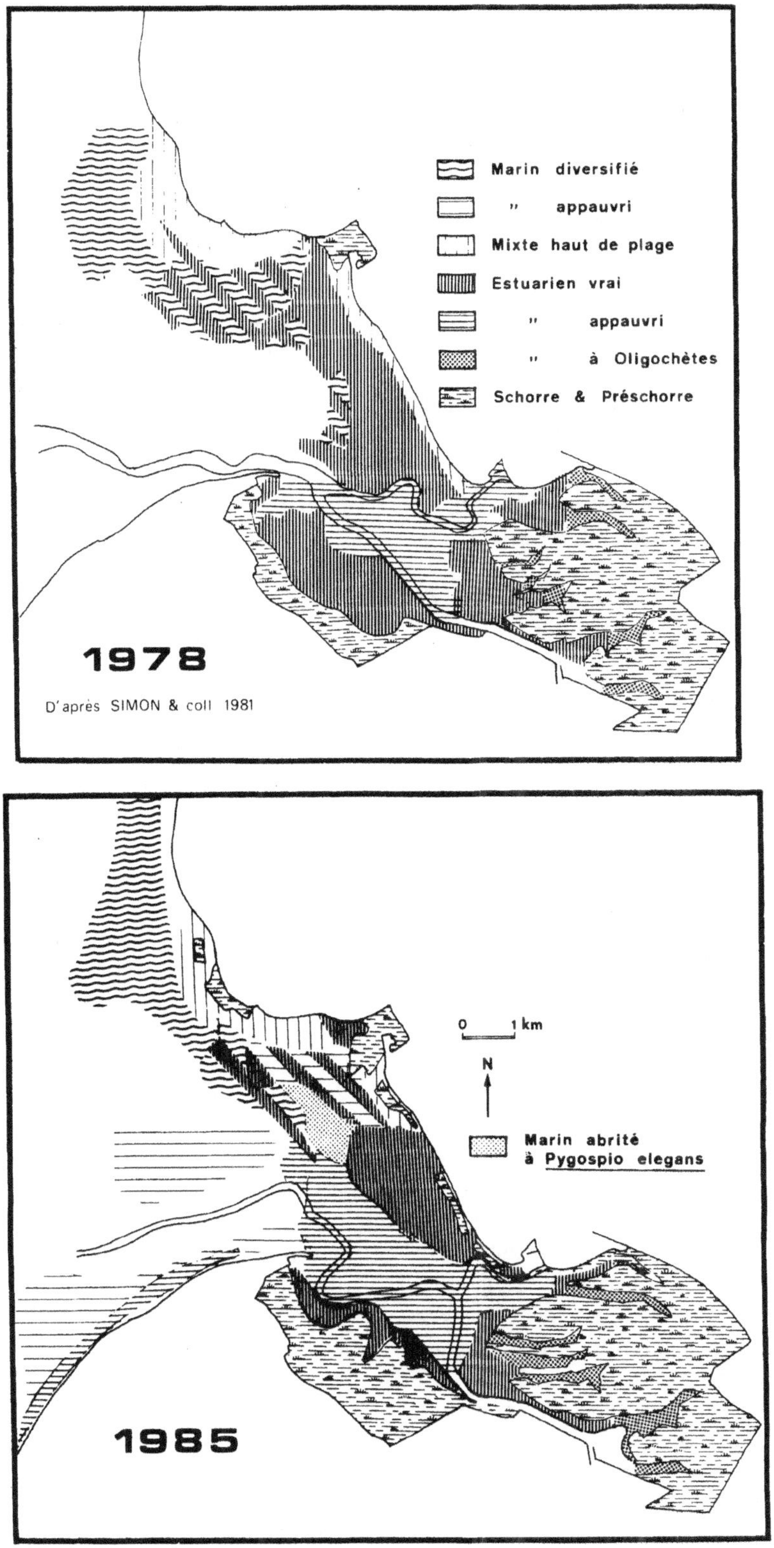

Fig. 11. Evolution spatiale des peuplements du macrozoobenthos intertidal de la Baie de Somme de 1978 à 1985.

tent, comme pour le site précédent, l'évolution de l'écosystème (évolution sédimentaire et de la qualité des eaux).

3. En Baie des Veys, six espèces reflètent l'évolution du peuplement pour la période étudiée: *Nereis diversicolor, Heteromastus filiformis, Tubificoides benedeni, Cerastoderma edule, Corophium volutator* et *Urothoe grimaldii*. Par contre, les espèces qui montrent le mieux les tendances évolutives de l'écosystème depuis 1972 (Sylvand, 1986) sont *Pygospio elegans, Scoloplos armiger, Macoma balthica* et encore *Cerastoderma edule*. Cette différence dans la nature des espèces-clés tient à deux raisons essentielles:
   - la brièveté du suivi COST et le manque de prélèvements en 1984 ne permettent pas de dégager des tendances évolutives significatives;
   - l'échantillonnage d'un nombre restreint de stations rend difficile l'extrapolation des résultats à l'ensemble du site étudié.

Parmi les espèces-clés de l'évolution des écosystèmes, certaines jouent véritablement un rôle de 'signal' mettant en évidence une perturbation particulière des conditions de l'environnement. Ainsi, en Baie de Somme, l'apparition brutale des Capitellidés en 1982 témoigne d'un apport massif de matière organique en suspension qui a indirectement contribué à la régression catastrophique des gisements de coques. Ce signal se surimpose au bruit de fond des variations saisonnières des abondances et marque des périodes-charnières de l'évolution des écosystèmes. Cependant, comme nous l'avons vu dans le cas de la Baie des Veys, plusieurs écueils sont à éviter dans le cadre du suivi COST:

* En matière de périodicité de prélèvements, le plan d'échantillonnage doit obligatoirement reposer sur la connaissance d'un cycle saisonnier complet qui, seul, permet de préciser les périodes les plus significatives du cycle annuel du peuplement étudié. En Baie des Veys, le plan d'échantillonnage mensuel réalisé en 1977–1978 met en évidence trois périodes principales de prélèvements impératives: mars (minima des effectifs et biomasses), juin et octobre (maxima des effectifs et biomasses) alors que les recommandations du COST, s'inspirant de l'observation du rythme saisonnier des peuplements de la Mer des Wadden, préconisent des prélèvements en février-mars et août-septembre. Malgré tout, seuls des prélèvements mensuels peuvent mettre en évidence l'influence de facteurs catastrophiques comme les tempêtes de nord-est qui s'avèrent d'importance primordiale en Baie des Veys (Sylvand en préparation).
* La représentativité des stations choisies suppose un nombre minimum de stations judicieusement localisées dans les principaux faciès reconnus pour chaque site (Hancock & Franklin, 1972). En Somme, la simple exploitation des résultats d'une station unique située dans le faciès à *Cerastoderma edule* traduit parfaitement la crise contemporaine de la pêche à pied de ce coquillage dans ce site. Les tendances évolutives comparables de plusieurs espèces de la Baie de Somme ne sont donc pas de simples analogies visuelles entre les différents graphiques comme en atteste le calcul des intervalles de confiance des moyennes des densités.

A la notion d'espèces-clés, caractéristiques de chacun des sites, s'oppose celle d'espèces-cibles. Les fluctuations d'abondance de ces dernières ne répondent que peu ou pas aux perturbations propres à chacun des écosystèmes (Gray, 1979) et leur large distribution permet de suivre à grande échelle l'effet de grands accidents climatiques comme les hivers rigoureux de 1979–1980 ou 1984–1985 (Beukema, 1985). A cet égard, le bivalve *Macoma balthica* peut être considéré comme une espèce d'intérêt exceptionnel ayant justifié le choix de nombreux auteurs (Elliott & Mc Lusky, 1985; Bachelet même recueil) qui ont étudié tant les fluctuations annuelles de l'abondance de cette espèce que celles de son rythme de croissance.

Le suivi régulier des peuplements à l'échelle de l'estuaire s'avère indissociable de celui des quelques stations COST pour permettre l'analyse globale de l'écosystème en ajoutant une dimension spatiale aux résultats.

Afin de déterminer les potentialités biologiques et économiques de milieux lagunaires (parmi lesquels prennent place les estuaires), Guelorget *et al.* (1983) on fait appel à la notion de confinement qui leur paraît constituer le paramètre fondamental de l'organisation du domaine paralique. Appliqué à nos estuaires, ce concept rend compte effectivement d'une zonation biologique quantitative des écosystèmes étudiés en faisant apparaître clairement les discontinuités qui s'installent le long du gradient qui s'établit depuis le milieu marin jusqu'au milieu terrestre.

Parallèlement, Wilson *et al.* (1986) ont appliqué

deux indices de qualité des estuaires (l'un chimique, l'autre biologique: Jeffrey *et al.*, 1985) à la Seine et à la Somme. La zonation repose ici sur la structure des peuplements et l'importance relative des surfaces occupées par les espèces opportunistes ou, au contraire, indiquant un certain degré de stabilité du système.

Dans les deux types de recherches évoqués ci-dessus, il s'agit de proposer une image globale et instantanée de l'état de l'estuaire à un moment donné en vue de l'aménager au mieux de ses potentialités (première méthode) ou de mettre à jour des perturbations graves de l'écosystème (deuxième méthode). Dans les deux cas, les investigations mènent à des résultats complémentaires qui font ressortir essentiellement les effets dûs aux activités humaines. Au contraire, le suivi de stations pilotes, à intervalle de temps régulier et à long-terme, comme dans le cadre du programme COST 647, permet, au travers des fluctuations d'abondance du benthos, de découvrir de quel ordre est la variabilité dûe aux facteurs naturels ou aux substances polluantes. A notre sens, l'application des ces résultats précis, mais ne portant que sur quelques stations pilotes, ne peut rendre compte de l'évolution du système dans son entier que s'il est possible de recaler les résultats du COST avec ceux provenant d'études plus synthétiques comme celles évoquées ci-dessus.

Afin de mieux cerner la causalité des phénomènes biologiques observés, les résultats concernant la cinétique des populations peuvent être complétés par l'évaluation de la biomasse. En tant que telle, comme nous l'avons mis en évidence en Baie des Veys, l'évaluation des variations mensuelles de la biomasse totale n'apporte pas d'information complémentaire utile puisque les cycles observés se superposent en quelque station que ce soit à ceux des variations numériques des espèces de grande taille comme *Arenicola marina* et *Cerastoderma edule*. Par contre, le décours temporel des productions d'une espèce comme *Macoma balthica*, tel que nous avons réalisé son suivi en estuaire de Seine (Ducrotoy & Desprez, 1986), apporte des renseignements précieux sur les performances du bivalve dans l'estuaire considéré et constitue un moyen d'évaluation de l'importance de cette espèce vis-à-vis de la structure et du fonctionnement de l'écosystème auquel elle appartient (Bachelet, 1982). D'autre part, Dauer & Conner (1980) ont montré sur les Polychètes qu'il est très utile de déterminer la production en plus de la structure de la communauté et du mode de reproduction de l'espèce dans le cas de pollutions légères comme celles que nous connaissons dans nos estuaires.

Dans le même ordre d'idées, la biométrie des bivalves débouche, au travers de l'étude de la croissance du test, sur une mesure des 'stress' subléthaux subis par ces organismes et permet de détecter des perturbations naturelles ou artificielles qui n'atteignent pas la structure des peuplements (Gray, 1979; Elliott & Mc Lusky, 1985; Ducrotoy *et al.*, 1985).

Ces études biométriques montrent que le rythme de croissance des populations de *Macoma balthica* des estuaires de Seine et de Somme est optimal parmi les sites étudiés tant en Europe qu'en Amérique du Nord (Bachelet,1980; Desprez, 1981; Elliot & Mc Lusky, 1985) et comparable à celui observé dans la Mer des Wadden (Lammens, 1967). Nous avons cependant observé que l'estuaire de la Seine semble plus favorable à la croissance des jeunes et la Baie de Somme à celle des adultes; cette situation s'explique sans doute par la différence de qualité granulométrique du substrat de ces deux sites plutôt que par la durée d'immersion plus importante en Seine (Hummel, 1985). En effet, les sédiments de la station HON 4 contiennent de 20 à 50% de particules fines alors que ceux de Baie de Somme présentent toujours une teneur inférieure à 10%. Cette hypothèse est confortée par les résultats obtenus en 1982 le long de la radiale HH qui présente un gradient d'affinement du bas vers le haut de l'estran (Ducrotoy *et al.*, 1985).

Le rythme de croissance de *Cerastoderma edule*, tel qu'il a été établi pour les années 1982 et 1983, traduit la crise qu'a connu le gisement de la Baie de Somme lorsqu'on le compare avec le rythme observé en 1985. Ce dernier est identique à celui observé par Orton (1926) à Plymouth, ce qui placerait la Baie de Somme au rang des meilleurs sites européens en matière de croissance de ce bivalve. Mais ce rythme est-il celui que l'on pouvait observer avant la crise? Ou bien s'agit-il d'une croissance de type exceptionnel liée à des conditions climatiques particulières (hiver extrêmement rigoureux, été sans fortes chaleurs)? Le rythme observé en 1981, semblable à ceux décrits dans de nombreux gisements des îles britanniques (Cole, 1956), appuie cette dernière hypothèse, à moins que la crise mise en évidence par l'évolution des effectifs en 1982–1983 n'ait en fait commencé dès 1981, la diminution de la vitesse de croissance annonçant la chute des effectifs. Ces questions, qui sont la conséquence d'un manque de données antérieures, plaident en faveur d'études à long terme du type de celles menées dans le cadre du programme COST 647.

Conclusion

A l'issue du suivi réalisé dans le cadre du programme COST 647 en Baie des Veys depuis 1983, en estuaire de Seine depuis 1979 et en Baie de Somme depuis 1981, diverses hypothèses sur l'évolution comparée des écosystèmes ont été générées par divers types d'exploitation des données numériques portant sur l'évolution temporelle des densités des espèces macrobenthiques en quelques stations pilotes mais aussi sur la production et la croissance des bivalves. Parmi les quelques techniques de calcul que nous avons appliquées, ce sont les techniques multivariées (Green, 1980) et en particulier l'analyse factorielle de correspondance qui permettent au mieux de les étayer (pour l'estuaire de la Seine, voir Desprez *et al.*, 1985; pour la Baie de Somme, Ducrotoy & Lafite en préparation).

Comme l'avaient montré Desprez *et al.* en 1983, l'incidence de l'hydrodynamisme local constitue un facteur prépondérant dans les estuaires côtiers tels que ceux du nord-ouest de la France. Confrontées aux résultats obtenus depuis 1972, les informations acquises dans le cadre du COST 647 mènent aux hypothèses complémentaires suivantes sur les mécanismes évolutifs du milieu:

* En Baie des Veys, on pressent une 'rotation' des faciès de l'ouest vers l'est de l'estuaire. Cela est reflété par le déplacement des dépôts fins vers le flanc est de la baie et s'accompagne de l'installation de caractères biocénotiques estuariens stables, la progression des espèces et des sédiments marins sur le flanc ouest s'effectuant, au contraire, aux dépens des faciès estuariens.

A ces facteurs hydrosédimentaires, s'ajouterait une perturbation d'origine anthropique comme la diffusion de polluants d'origine continentale (nitrates et détergents anioniques notamment) qui se manifeste à plus grande échelle sur le flanc est.

* Dans l'estuaire de la Seine, l'évolution des biocénoses traduit celle qui résulte essentiellement des travaux de chenalisation du fleuve. Ainsi, la progression vers l'aval du schorre se déroule aux dépens des biocénoses benthiques estuariennes, l'accélération du colmatage naturel des deux fosses de flot situées de part et d'autre du chenal de navigation s'accompagne de la pénétration d'espèces marines et le désenvasement des fosses de flot, amplifié par la réalisation de déversoirs visant à les maintenir en eau, entraîne la régression des espèces estuariennes. Dans les mécanismes décrits, il reste à démontrer le rôle de substances chimiques telles que les polychlorobiphényls (PCB) et de la concentration alarmante en métaux lourds (cadmium) mais aussi en sulfate de calcium (phosphogypse), en nitrates et phosphates (dans une moindre mesure).

* En Baie de Somme, à l'évolution rapide du milieu qui achève de subir le contre-coup de la construction de digues submersibles en 1969 et de polders en 1976, s'est ajoutée en 1982–1984 une perturbation d'origine anthropique qui s'est traduite par un effondrement de la pêche à pied.

Parmi les facteurs explicatifs, il faudra tester l'influence des phosphates (celle des nitrates étant moindre ici), celle des métaux lourds (nickel, fer, zinc) et des hydrocarbures. Quant à la matière organique, qu'elle soit dissoute ou particulaire, elle favorise une pollution bactériologique dont la dynamique en milieu estuarien commence seulement à être abordée (Dupont & Lafite, 1984).

Si l'on rassemble les observations effectuées dans les trois estuaires français, il apparaît que:

- les facteurs hydrodynamiques façonnent les bassins estuariens en entraînant une dissymétrie de la répartition des faciès par rapport à un axe amont-aval et modifient cette répartition par à-coups bien observés au niveau des stations COST;
- l'aire occupée par le schorre augmente rapidement et, en même temps, des sables envahissent les bas niveaux, entraînant une sévère diminution de l'emprise spatiale des faciès estuariens (slikke) en Seine et en Somme.

Deux facteurs primordiaux modulent alors la cinétique des populations benthiques:

- parmi les facteurs climatiques envisagés, la température joue un rôle régulateur vis-à-vis de nombreuses espèces mais le lien avec le gradient de salinité reste à préciser;

– la pollution organique (matières dissoutes ou particulaires) modifie la répartition et l'abondance d'espèces opportunistes qui traduisent ainsi la perturbation en cause.

Parmi d'autres mécanismes qu'il faudra tester dans un avenir proche, mentionnons:

– les facteurs gouvernant le recrutement des espèces benthiques (concentration de l'eau en oligo-éléments ...),
– la pathologie de ces animaux (parasitoses, viroses ...),
– l'effet des prédateurs et de la pêche,
– la toxicité de substances exogènes naturelles ou artificielles (eaux rouges, pollution bactérienne, polluants chimiques ...).

Pour ces raisons, notamment, il s'avère indispensable, d'une part de continuer le suivi de l'abondance de toutes les espèces (chacune pouvant se révéler être une *espèce-clé* a posteriori, suivant le type d'évolution du milieu), d'autre part d'affiner la connaissance de la production et de la croissance d'*espèces-cibles* comme *Macoma balthica* et enfin d'effectuer un suivi régulier des biofaciès de chaque écosystème estuarien s'appuyant sur des techniques de télédétection (comme cela a été fait dans la Somme, Demarcq *et al.*, 1979).

Les estuaires macrotidaux dont nous avons exposé l'étude du benthos, s'avèrent bien plus instables que les zones intertidales de la Mer des Wadden (Zijlstra, 1972; Beukema, 1982; Reise, 1982; Madsen, 1984) sachant que l'aire qu'ils occupent reste très faible par rapport à la vaste étendue de la zone de balancement des marées des Pays-Bas, Allemagne et Danemark. Le gradient de confinement y est nettement plus contracté et l'impact d'un phénomène comme l'eutrophisation, plus perturbant, risque de compromettre des atouts économiques importants. Sans savoir si nos estuaires étaient originellement oligotrophes, ce caractère serait, de toutes façons, largement contrarié par l'apport massif de nutrients d'origine agricole qui favorisent des 'blooms' phytoplanctoniques exceptionnels. A cela s'ajoute une pollution chimique (métaux lourds, hydrocarbures) qui rend difficile l'interprétation de facteurs naturels comme le climat agissant globalement et en synchronisme d'un site européen à l'autre.

Remerciements

Le suivi des stations-pilotes COST retenues pour cet article a été financé par le Groupe d'Etude des Milieux Estuariens et Littoraux. A partir d'août 1985, l'Institut Francais de Recherche pour l'Exploitation de la Mer a demandé à cette équipe de lui apporter, dans le cadre du IXème plan quinquennal, les données de base concernant l'évolution biosédimentologique de l'estuaire de la Somme (contrat n° 85 6 810381).

Le travail scientifique a été accompli à la Station d'Etudes en Baie de Somme et au Laboratoire Maritime de Luc-sur-Mer. Les auteurs remercient les Universités de Picardie et de Basse-Normandie pour les facilités procurées.

References

Bachelet, G., 1980. Growth and recruitment of the tellinid Bivalve *Macoma balthica* at the southern limit of its geographical distribution, the Gironde estuary (South-Western France). Mar. Biol. 59: 105–117.

Bachelet, G., 1982. Quelques problèmes liés à l'estimation de la production secondaire: cas des bivalves *Macoma balthica* et *Scrobicularia plana*. Oceanol. Acta 5, 4: 421–431.

Beukema, J. J., 1974. Seasonal changes in the biomass of the macro-benthos of a tidal flat area in the Dutch Wadden Sea. Neth. J. Sea Res. 8, 8: 94–107.

Beukema, J. J., 1979. Biomass and species richness of the macrobenthic animals living on a tidal flat area in the Dutch Wadden Sea: effects of a severe winter. Neth. J. Sea Res. 13, 2: 203–223.

Beukema, J. J., 1982. Annual variation in reproductive success and biomass of the major macrozoobenthic species living in a tidal flat area of the Wadden Sea. In J. W. De Blok (ed.), Dynamic Aspects of Marine Ecosystems. Neth. J. Sea Res. 16: 37–45.

Beukema, J. J., 1985. Zoobenthos survival during severe winters on high and low tidal flats in the Dutch Wadden Sea. In J. S. Gray (ed.), Marine Biology of Polar Regions and Effects of Stress on Marine Organisms. John Wiley and sons, N.Y.: 351–361.

Cole, H. A., 1956. A preliminary study of growth-rate in cockles (*Cardium edule* L.) in relation to commercial exploitation. J. Cons. perm. int. Explor. Mer., 22: 77–90.

Dankers, N. & J. J. Beukema, 1981. Distributional patterns of macrozoobenthic species in relation to some environmental factors. In N. Dankers, H. Kühl & W. J. Wolff (ed.), Ecology of the Wadden Sea. I. Balkema, Rotterdam: 69–103.

Dauer, D. M. & W. G. Conner, 1980. Effects of moderate sewage input on benthic Polychaete populations. Estuar. coast. mar. Sci. 10, 3: 335–346.

Demarcq, Y., R. Regrain & J. Rudelle, 1979. Deux méthodes d'analyse des données de télédétection: application à la géomorphologie et à l'aménagement du littoral picard. Mémoires du laboratoire de géomorphologie de l'Ecole Pratique des Hautes Etudes, Paris, 32: 105 pp.

Desprez, M., 1981. Etude du macrozoobenthos intertidal de l'estuaire de la Seine, Thèse de Troisième Cycle, Rouen: 186 pp.

Desprez, M., J.-F. Brulard, J.-P. Dupont, S. Simon, B. Sylvand & P. Duval, 1983. Etude des biofaciès intertidaux de l'estuaire de la Seine. C. r. Acad. Sci., Paris, 296, III: 521–526.

Desprez, M., J.-F. Brulard, J.-P. Dupont, S. Simon & B. Sylvand, 1985. Etude du macrozoobenthos intertidal de l'estuaire de la Seine: relations avec le substrat. Actes du Muséum de Rouen, 3: 37–79.

Desprez, M. & J.-P. Dupont, 1985. Impact biosédimentaire d'aménagements portuaires en estuaire de Seine. In L. Cabioch (ed.), La Baie de Seine: fonctionnement, conséquences en matière d'environnement et d'utilisation de la mer, Caen, 2: 273–280.

Ducrotoy, J.-P., 1984. Partage des ressources en milieux estuarien et littoral de Picardie. In J.-L. Mériaux & P. Tombal (ed.), l'Environnement en Picardie, Amiens: 145–150.

Ducrotoy, J.-P., M. Desprez & J.-P. Dupont, 1985. Short- and long-term biosedimentary evolution of two *Macoma balthica* communities. In J. G. Wilson & W. Halcrow (ed.), Estuarine Management and Quality Assessment, Plenum Press, N.Y.: 105–130.

Ducrotoy, J.-P. & M. Desprez, 1986. Evolution spatio-temporelle de populations estuariennes de bivalves, liée à des perturbations naturelles ou artificielles. Haliotis, 15. A paraître.

Ducrotoy, J.-P., R. Lafite, B. El Kaim & J.-P. Dupont, 1986. Discussion des notions de confinement appliquées à la Baie de Somme. J. Rech. Oceano. 11, 1. A paraître.

Dupont, J.-P., 1975. Présence de biolithosores à *Pygospio elegans* Claparède (Polychètes Sédentaires) dans la Baie de Somme (80). C. r. Acad. Sci., Paris, 280, D; 1773–1776.

Dupont, J.-P., 1983. Les séquences biosédimentaires de la Baie de Somme. Actes de Muséum de Rouen, 3: 62–102.

Dupont, J.-P. & R. Lafite, 1984. Principales caractéristiques hydrosédimentaires de la Baie de Somme. In J.-L. Mériaux & P. Tombal (ed.), l'Environnement en Picardie, Amiens: 141–144.

Elliott, M. & D. S. McLusky, 1985. Invertebrate production ecology in relation to estuarine quality management. In J. G. Wilson & W. Halcrow (ed.), Estuarine Management and Quality Assessment, Plenum Press, N.Y.: 85–103.

Green, R. H., 1980. The assessment of ecologic similarity. Annual Review of Ecological Systems, 11: 1–14.

Gray, J. S., 1979. Pollution-induced changes in populations. Phil. Trans. r. Soc., Lond., B, 286: 545–561.

Guelorget, O., G. Frisoni & J.-P. Perthuisot, 1983. La zonation biologique des milieux lagunaires: définition d'une échelle de confinement dans le domaine paralique méditerranéen. J. Rech. Océanograph., XIII, 1: 15–35.

Hancock, D. A. & A. Franklin, 1972. Seasonal changes in the condition of the edible cockle (*Cerastoderma edule*). J. appl. Ecol., 9: 567–579.

Hummel, H., 1985. Food intake of *Macoma balthica* (Mollusca) in relation to seasonal changes in its potential food on a tidal flat in the Dutch Wadden Sea. Neth. J. Sea res., 19, 1: 52–76.

Jeffrey, D. W., J. G. Wilson, C. R. Harris & D. L. Tomlinson, 1985. The application of two simple indices to Irish estuary pollution status. In J. G. Wilson & W. Halcrow (ed.), Estuarine Management and Quality Assessment, Plenum Press, N.Y.: 147–161.

Lammens, J. J., 1967. Growth and reproduction in a tidal flat population of *Macoma balthica* (L.). Neth. J. Sea Res., 3, 3: 315–382.

Leppakowski, E., 1975. Assessment of degree of pollution on the basis of macrozoobenthos in marine and brackish water environments. Acta Academia Aboensis, B, 35: 1–89.

Madsen, P. B., 1984. The dynamics of the dominating macrozoobenthos in the Danish Wadden Sea 1980–1983. Report of the Marine Pollution Laboratory (Miljostyrelsens Havforureningslaboratorium), 1: 35 pp.

Michaelis, H., 1976. Die Makrofauna des nördlichen Eversandes (Wesermündung). Jahresber. Forschungstelle Norderney 1975.

Mobius, K., 1877. Die Auster und die Austernwirtschaft. Berlin: 126 pp.

Orton, J. H., 1926. On the rate of growth of *Cardium edule*. Part I. Experimental observations. J. mar. biol. Ass. New Series, XIV, 2: 239–279.

Pearson, T. & R. Rosenberg, 1978. Macrobenthic succession in relation to organic enrichment and pollution of the marine environment. Oceanography and marine Biology annual Review, 16: 229–311.

Reise, K., 1982. Long-term changes in the macrobenthic invertebrate fauna of the Wadden Sea: are polychaetes about to take over? In J. W. De Blok (ed.), Dynamic Aspects of Marine Ecosystems. Neth. J. Sea Res., 16: 29–36.

Simon, S, M. Desprez & J.-P. Dupont, 1981. Distribution du macrozoobenthos intertidal de la Baie de Somme. C. r. Acad. Sci., Paris, 292, III: 1013–1016.

Sylvand, B., 1986. La Baie des Veys (Baie de la Seine, Manche centrale): évolution récente et incidence des aménagements. Colloque Bordomer 1985: 112–127.

Thamdrup, H. M., 1935. Beiträge zur Ökologie des Wattenfauna auf experimenteller Grunglage. Meddelel. Komm. Danmarks Fisk., 10: 1–125.

Van der Baan, G., A. Blok, P. Nijhoff & C. Swennen, 1958. Een inleidend onderzoek naar de betrekkingen tussen wadvogels en bodemfauna. Verslag werkkamp N.J.N. sectie vogelwerkgroep KNNV en NJN: 27 pp.

Wilson, J. G., J.-P. Ducrotoy & M. Desprez, 1986. Application d'indices de qualité écologique des estuaires en Manche Centrale et Orientale: comparaison de la Seine et de la Somme. Vie et Milieu, 37, 1: (A paraître).

Wohlenberg, E., 1937. Die Wattenmeer-Lebensgemeinschaften im Konogshafen von Sylt. Helgoländer wiss. Meeresunters, 1: 1–92.

Zijlstra, J. J., 1972. On the importance of the Wadden Sea as a nursery area in relation to the conservation of the southern North Sea fishery ressources. Symp. Zool. Soc., Lond., 29: 233–258.

Hydrobiologia 142: 271–279 (1986)

Long-term variability of meiobenthos: value, synopsis, hypothesis generation and predictive modelling*

Bruce C. Coull
Belle W. Baruch Institute for Marine Biology & Coastal Research, Marine Science Program and Department of Biology, University of South Carolina, Columbia, SC 29208, U.S.A.

Keywords: meiobenthos, long-term, synopsis, hypothesis, modelling

Abstract

Few long-term data sets exist for meiofauna. Such data sets are expensive to collect, sort and identify; continuous meiofauna data for a period of greater than two years are limited to one site in Belgium (7 yrs) and two sites (one mud, one sand) in South Carolina, USA (11 yrs). The Belgian study concentrates on benthic copepod abundances whereas data from South Carolina includes major taxa and benthic copepods as well as 4 years of concurrent macrofauna abundance and 3 years of nematode species abundances.

In South Carolina, the variance associated with meiofaunal abundance had 6 or 12 month recurrent cycles. Similar analyses on 4 years of macrofauna from the same 2 sites indicated the same cyclicity: one year. Seasonality of the South Carolina major taxa and the 6 most abundant mud copepod species was pronounced at the mud site, but absent or less pronounced at the sand site. Similar results were also found for the nematode species over three years. Variability in meiofaunal abundance was greater year-to-year than within a year.

Many such long-term data sets are analysed and abandoned. Herewith, I use our long-term results to hypothesize the causes of the high temporal variance in mud and the lower temporal variance in sand. Is it because the mud fauna is controlled by seasonal inputs of natant predators while at the hydrodynamically active sand site temporal variability is homogenized by constant physical activity?

By appropriate statistical modelling long-term data sets can also be used to assess the appropriateness of the sampling schedule (spatial and temporal) and as a predictor of future trends.

Introduction

Consistently collected, replicated data sets are rare in most fields of ecology and even rarer in soft bottom benthic ecology. Such data sets are, of course, important for understanding ecosystem functioning (Wiens, 1977; Callahan, 1984) and critical to determining if and how populations and communities fluctuate. With an increasing number of perturbations infringing on the world's coastral zone, long-term studies to illustrate how an 'ecosystem' (or a portion of it) should naturally function are requisite if such ecosystems are ever to be 'repaired' following perturbation. While long-term data sets are thus extremely useful in and of themselves, it is important to use them for more than just 'before' and 'after' studies or for simply describing cycles and their correlations with some biological of physical forcing function(s).

If long-term data analysis is to be more than descriptive ecology then the data must also be used to generate testable hypotheses. The need for accurate hypothesis testing through controlled experimentation has been reiterated throughout the literature (e.g. Platt, 1970; Connell, 1974, 1975; Dayton & Oliver, 1980) and long-term sets are ideal to generate hypotheses to be so tested.

In this paper I provide a synopsis of the known long-term meiofauna data sets published to date

*Contribution No. 618 from the Belle W. Baruch Institute for Marine Biology and Coastal Research

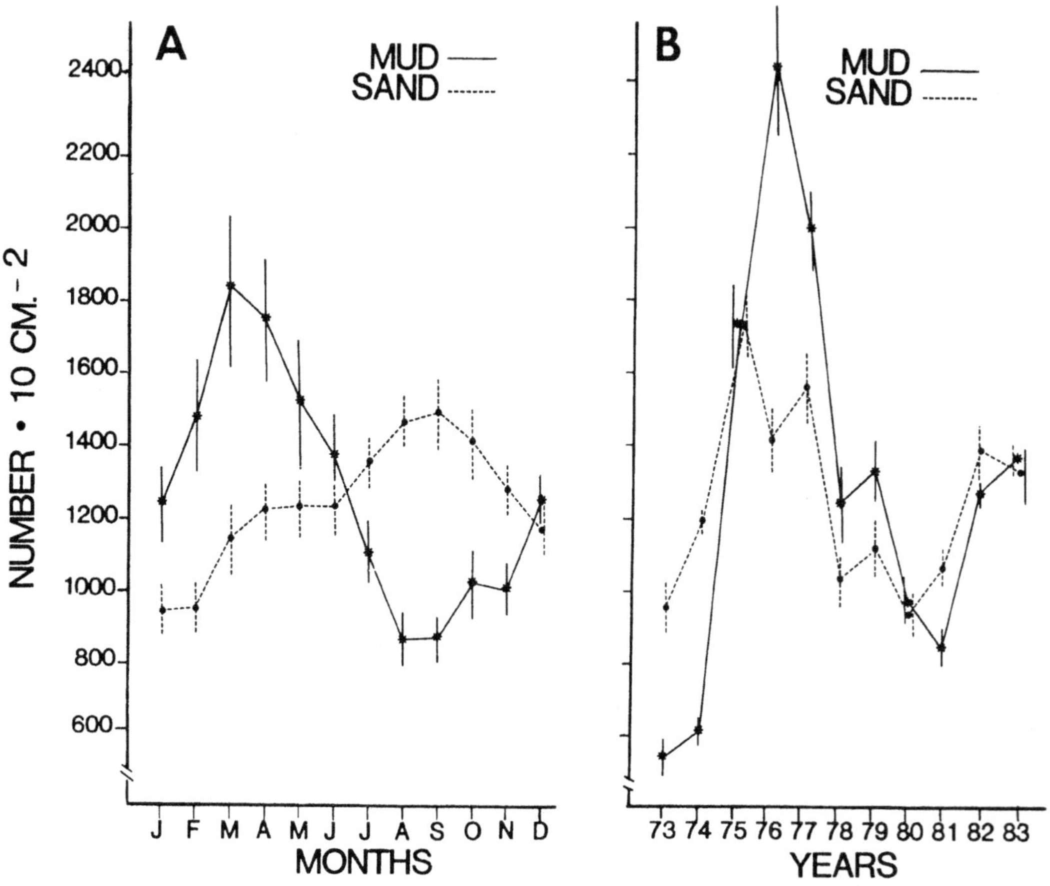

Fig. 1. A) Monthly mean abundance (±SE) of total meiobenthos at a mud and sand site in South Carolina, USA. The monthly values are the means of 11 years of January's, etc.
B) Yearly mean abundance (±SE) of total meiofauna from the mud and sand South Carolina sites (from Coull, 1985a).

and suggest a new hypothesis that can be tested based on long-term South Carolina data. An overview of a method to quantitatively alter a sampling regime without loosing much precision as well as an example of predictive forecasting of meiofaunal abundance are also provided.

Synopsis of long-term meiofauna studies

Typically, meiofaunal community/population studies have been temporally limited to a year or two in duration (e.g. Stripp, 1969; Tietjen, 1969; Coull, 1970; de Bovee & Soyer, 1974; Rudnick *et al*, 1985). Except for McIntyre & Murison's (1973) study where meiofauna were sampled monthly for 1 year then irregularly for 9 years and my recent study (Coull, 1985a) where meiofauna were sampled at two sites monthly for 8 years and fortnightly for 3 years, I am unaware of any other study that enumerates all meiofauna taxa for a period of greater than two years. There are, however, two data sets on the abundance of copepod species for longer periods i.e. a Belgian study for 7 years (Heip, 1980; Herman & Heip 1983; Heip & Herman 1985) and our South Carolina study over 11 years (Coull & Dudley, 1985). Complementary to our 11 years of meiofauna taxa and copepod species, Eskin (1985) has studied the population dynamics and abundance of nematode species for 3 years at two sites and 4 years of data are available on macrofauna abundance at the same two sites.

The 11 year South Carolina meiofauna taxon study was conducted at a mud and a sand site in the North Inlet estuary. Variability in abundance at the mud site was approximately twice that of the sand site (Fig. 1), and year-to-year variability was greater

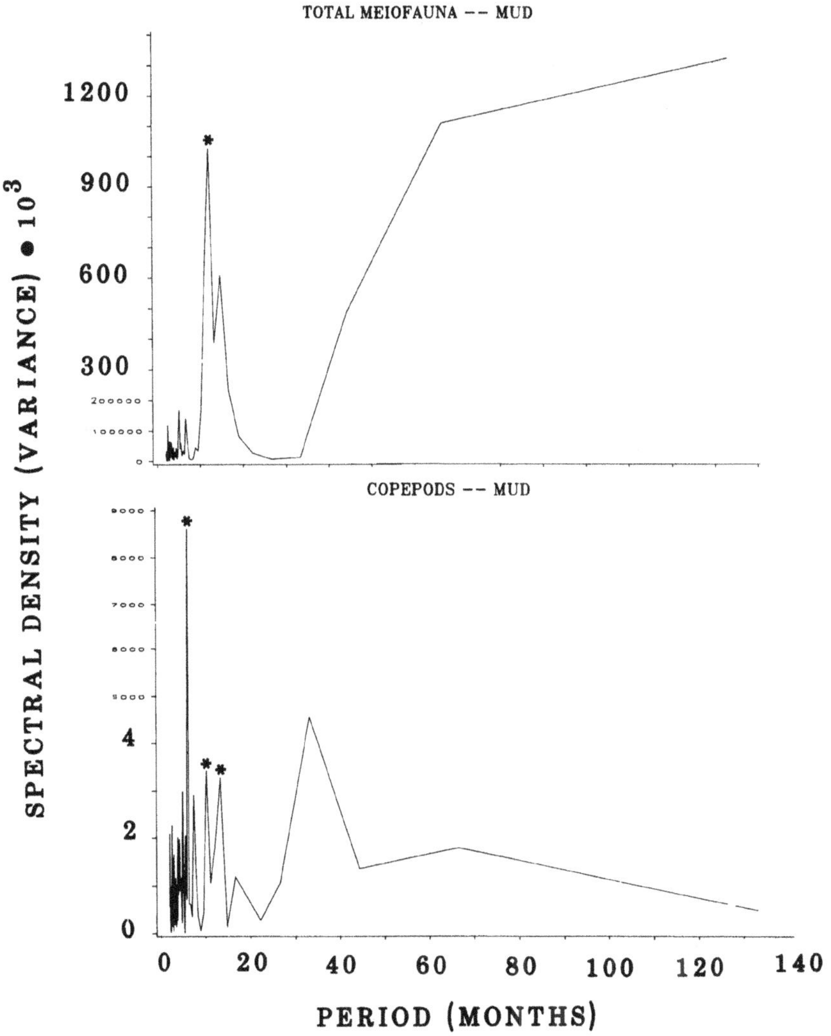

Fig. 2. Spectra of total meiofauna and total copepods at the long-term South Carolina mud site. *Peaks that are significant (by Fishers G_{crit}statistic) and are not harmonic anomalies.

at both sites than the inherent seasonality. There were no long-term (>1 yr) cycles (spectral time series analysis) in the fauna (Figs. 2, 3), temperature (Fig. 4), or salinity at either site; 12 months was the recurrent cycle throughout the data, with total copepods at the mud site having six month and one year (actually 13&15 month) cycles (Fig. 2). Total nematodes at the sand site had no recurrent periodicity (Fig. 3). The years 1975–1977 had the highest abundances at both sites (Fig. 1B), but these peaks were not correlated with any of the measured physical variables. This mid-1970's abundance maxima, which had the same seasonal cycle as other years, was not part of a longer term trend; but demonstrates the unexplainable variability in natural meiofauna abundance.

While macrofaunal cycling is not officially within the framework of this meiofauna paper; the macrofauna, too, had a statistically repeatable cycle of but one year at both South Carolina sites. Fig. 5 illustrates the spectral analysis plot of polychaete abundance at the sand site, a variance plot typical of the total macrofauna and polychaetes at the long-term meiofauna sand and mud sites.

When the data set length is long enough for cylces to be repeated (i.e. replicated), the copepod

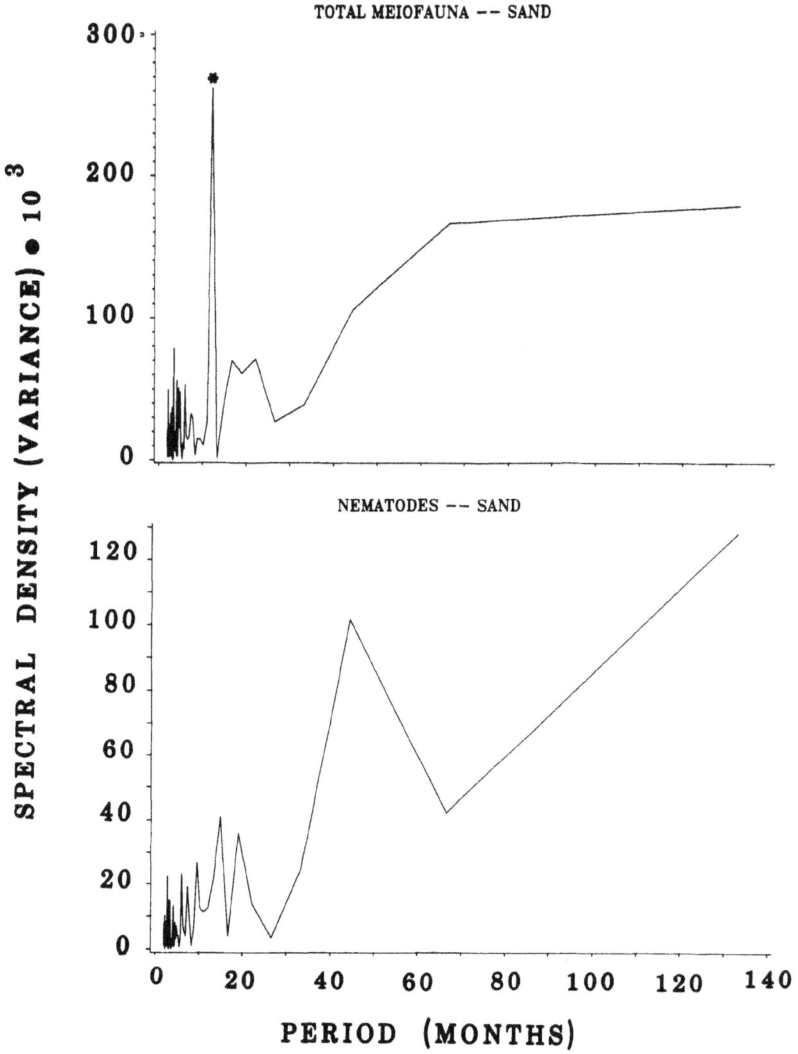

Fig. 3. Spectra of total meiofauna and total nematodes at the long-term South Carolina sand site. *Peaks that are significant (by Fishers G_{crit}statistic) and are not harmonic anomalies.

species data from the Belgian site and the South Carolina sites are quite similar. Of the sixteen species for which spectral analyses are available (4 from Belgium, Heip & Herman, 1985; 12 from South Carolina, Coull & Dudley, 1985), only one has a statistically repeatable cycle of greater than one year i.e. *Canuella perplexa* in Belgium appears to have a 3.5 yr cycle (Herman & Heip, 1983). Heip & Herman (1985) report cycles of 4.6 years for *Paronychocamptus nanus* and *Halicyclops magniceps*, but since the data set length is only seven years, these can only be predicted cycles (Kirk *et al.*, 1979, p. 72) not empirically replicated ones. Certainly there are oscillations of greater than one year in the Belgian and South Carolina copepods but no time series method can resolve cycle greather than ½ the length of the time series because the cycle must be repeated to determine if it is indeed a true cycle.

In South Carolina three of the dominant species at the sand site (*Leptastacus macronyx, Kliopsyllus constrictus* and *Arenosetella spinicauda*) declined significantly in abundance over the eleven years. All three are interstitial forms. However, two of the

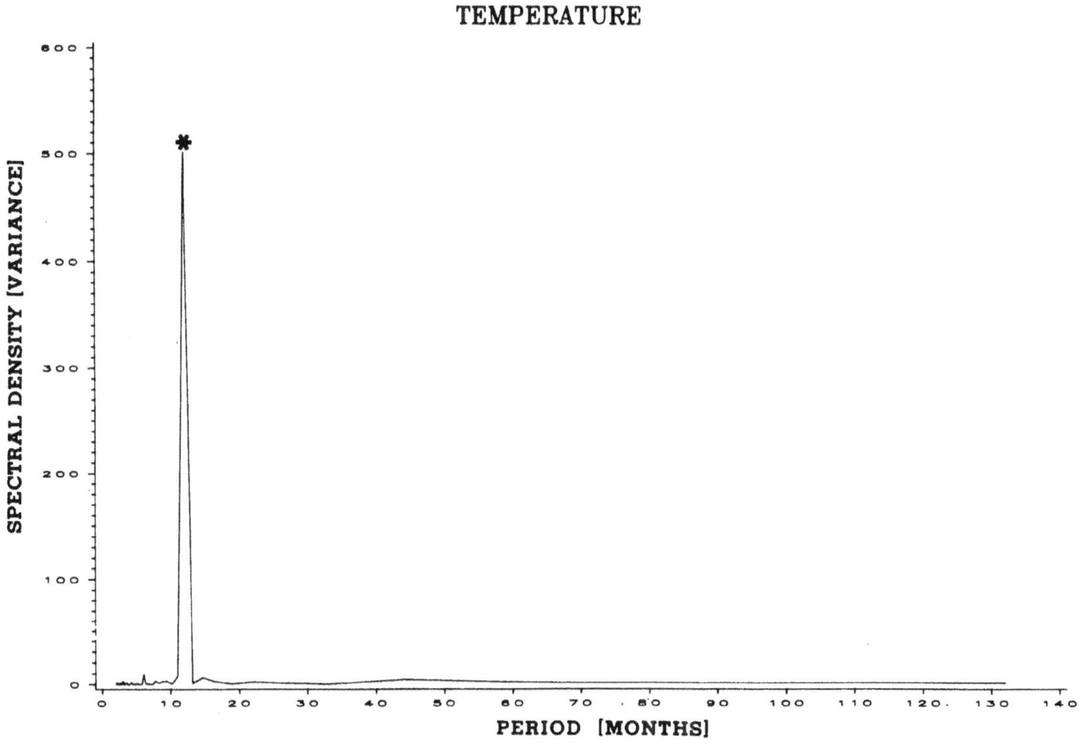

Fig. 4. Spectra of temperature from South Carolina showing the significant 12 month peak.

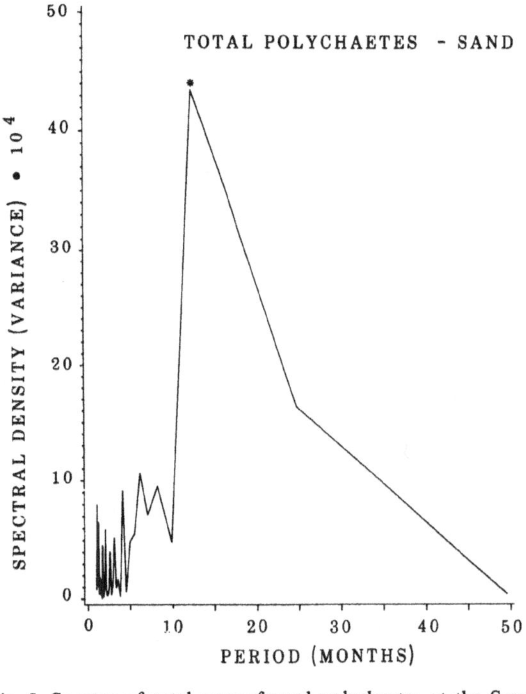

Fig. 5. Spectra of total macrofaunal polychaetes at the South Carolina long-term sand meiofaunal site. *Significant peak.

three species that did not decline significantly (*T. hyaenae* and *H. winonae*) are epibenthic forms which occupied the sediment surface. The number of species did not change seasonally or over the eleven years at the mud site, but at the sand site, the number of species significantly decreased over the study period, concomitant with the lower median grain size and decreased sediment sorting (Coull, 1985a; 1985b). The finer, sediments apparently eliminated the interstitial habitat necessary for *Leptastacus macronyx*, *Kliopsyllus constrictus*, *Arenosetella spinicauda* and other less abundant interstitial species.

Nematode species data from the two South Carolina sites for 1980 through 1983 demonstrated the same general patterns as those of the eleven years copepod species data set (see Coull & Dudley, 1985) in that the mud nematode species were much more seasonal than were the sand species (Table 1). Four of the six most common mud species were distinctly seasonal; the two non-seasonal species were present throughout the year. At the sand site, only *Paracomesoma hexasetosum* was clearly seasonal

Table 1. Seasonal rank abundance of the most abundant nematode species at the South Carolina mud and sand sites over 3 years. Seasons with common underlines are not significantly different in abundance (Tukey's studentized range test, α = 0.05). Seasons are Su = Summer; Sp = Spring; A = Autumn; W = Winter. (From Eskin 1985).

Species	Season			
MUD:				
Daptonema erectum	Sp	W	A	Su
Sabatieria americana	Sp	W	Su	A
Sabatieria kelleti	F	Sp	W	Su
Ptycholaimellus hibernus	W	Sp	A	Su
Dorylaimopsis metatypica	W	Sp	Su	A
Viscosia brachylaimoides	Su	Sp	A	W
SAND:				
Daptonema sp.	Su	A	W	Sp
Paracomesoma hexasetosum	Sp	Su	A	W
Sabatieria armata	Su	Sp	A	W
Parapomponema macrospiralis	Su	Sp	W	A
Oncholaimoides striatus	W	Sp	A	Su

of the most abundant species (Eskin, 1985). With but three years of nematode species data the maximum period discernable by spectral analysis would have been 1.5 years; thus time series analysis was not conducted on the nematode species.

Hypotheses suggested by the long-term data

I (Coull, 1985b) have previously suggested that the South Carolina meiofauna data set be used as the basis for two testable hypotheses i.e. 1) the seasonal abundance patterns with peak abundance in spring at the mud site and peak abundance in summer at the sand site (see Fig. 1A) were controlled by different factors. Specifically, I proposed that the mud faunal pattern was controlled by juvenile fish predation that was absent in the sand. We have tested this hypothesis experimentally and the results will be published shortly (Smith & Coull, in press); and 2) the decline in copepod abundance and species diversity over the eleven years at the sand site was a result of decreasing grain size and elimination of the interstitial space. This hypothesis has yet to be tested experimentally.

In my 'hypothesis generating' paper (Coull, 1985b), in addition to stating the proposed hypotheses I suggested ways to test them. Below, I provide another (third) hypothesis and ideas on how to test it experimentally; an hypothesis generated, again, only because of the long term South Carolina data set.

It is obvious that the mud meiofauna is much more variable temporally than the sand meiofauna (Fig. 1). Without repeating every such bit of information reported by Coull (1985a), Coull & Dudley (1985) and Eskin (1985), the mud meiofauna had a mean annual range of 979 animals $\cdot$ 10 cm^{-2}; the sand fauna = 536 $\cdot$ 10 cm^{-2}; over the eleven years the mud meiofauna ranged from 93–6462 $\cdot$ 10 cm^{-2}, the sand fauna from 313–3498; the coefficient of variation at the mud site was 69.7 ($\overline{X}$ of 11 yr) and 40.2 at the sand site; nematode and copepod species were distinctly seasonal at the mud site and either aseasonal or with just a few species seasonal at the sand site. Listing just these few (of many such) measures, and despite the fact that the sand site granulometry changed while the mud site granulometry remained constant, it is clear that even in the more hydrodynamically benign mud, some factor(s) is (are) causing greater temporal variability in mud than in sand.

A priori I would have predicted the mud habitat to be 'less disturbed-more stable' than the hydrodynamically active sandy habitat with its megaripples and significant amount of crossbedding, and therefore, mud populations would be expected to show less fluctuation in temporal variance. I propose as an hypothesis, however, that the frequent hydrodynamic disturbances in the sand habitat 'homogenizes' the fauna maintaining it relatively constant over time, while at the mud site biological, not hydrodynamic, factors influence the greater temporal variability. A general corollary to this hypothesis, then, is that assemblages that are biologically controlled are more variable in time than those under constant physical perturbation. All hypotheses, must of course, be falsifiable and below I outline two potential ways to falsify the above stated hypothesis.

One way to test the hypothesis is by a natural experiment, where one monitors the variability of the fauna (*sensu* Coull, 1985a; Coull & Dudley, 1985) as a site changes from sand to mud or vice-versa. Changing a site from mud to sand would probably require some catastrophic event (hurricane etc.) to deposit a significant amount of sandy sediment on top of an already existing mud site. Such a catastrophic change would, I fear, be so drastic that a natural experiment would not provide the gradual progression necessary for quantifying changes over the seasons. A change from sand to mud is more realistic and indeed is already occurring at our sand site. This then seems to be the ideal situation to *naturally* test the hypothesis. The sand flat which we sample has undergone several large physiographic changes in the last decade. *Spartina* islands have now invaded and the large megaripples reported by Grant (1981) have decreased. The sand flat *was* (1972–1977) essentially in the main inlet channel from the Atlantic Ocean (see Fig. 1 in Grant, 1981). Now, the barrier island that fronts the ocean has prograded south and the sand flat has become more protected. Concommitant with the prograding barrier island has been a 93 μm decrease in median grain size and increased sediment sorting (better sorted). It appears that the sand flat is one in transition from sand to sandy mud. If it continues on this trajectory it should, in time, become a muddy site. The hypothesis then is falsified if the muddy fauna (on the original sand site) maintains low variability; supported if it becomes highly variable as it is in the present mud site.

A second way to test the hypothesis is by sediment manipulation i.e. transplanting mud sediments to the sand site and vice-versa. Such transplant experiments are, of course, not easy to do but if logistically feasible and persistent, would allow one to monitor temporal variability to determine if, indeed, the hydrodynamic regime or the sediment type controls the observed temporal variability. For example, if the transplanted sand (in the mud site) persisted with low faunal variability or the transplanted mud (at the sand site) persisted with high variability, it would seem clear (with proper replication etc.) that sediment type *per se* was the primary factor. If the temporal faunal patterns did not persist, but reversed, then hydrodynamic factors would be implicated.

I have not conducted these experiments nor do I know if they will work and certainly alternative hypotheses could be proposed to account for my observations. My point, however, is not to provide fully designed, infallible experiments or the *only* hypothesis, but to demonstrate that long-term data sets can, and should, be used to generate testable hypotheses. The above hypothesis (see Coull, 1985b for 2 others), would not have been possible without my long-term data set.

Predictive modelling

Prediction of future events and efforts should be one of the goals of long-term ecological research. Such questions as how many samples need be collected at each sampling interval to derive a predetermined variance; how should these samples best be spaced temporally and can one predict the abundance (or some other parameter) with some level of confidence into the future need be asked if a long-term data set is to be useful for future endeavors. If for example, monitoring is to be continued can it be reduced both temporally and spatially to make it more cost efficient but still maintain population estimates within a reasonable range of error? Edwards & Coull (submitted MS) outline an autoregressive technique (ARIMA) to determine the number of, and spacing, of samples necessary to continue cost effective monitoring without loosing much statistical precision in estimating trends. Using the eleven year South Carolina data set they found that to determine the number of replicates necessary one has to calculate the ratio of replication variance to overall variance. For the eleven year data set two replicates provide ratios of 0.15–0.30; 3 replicates 0.10–0.20. Reducing this ratio provides greater statistical sensitivity. One needs to determine the acceptable ratio for each data set.

Suppose after sampling monthly or biweekly for some period of time it is decided to reduce the number of samples taken. One must determine the standard error (SE) of the trend estimator (estimated by an autoregressive procedure – Edwards & Coull, submitted). This calculation takes a fairly large (mainframe) computer, and is designed to consider rational choices in sample reduction (e.g. dropping every other sample date, etc.). Edwards & Coull (submitted) tested this methodology by comparing the SE of the trend estimator from the actual data set and the model. Table 2 lists the results for data collected in the long-term South Carolina meiofauna study. The *modelled* eleven year values check very well with the *observed* 11 year values (Column 1). The loss of precision is switching from a monthly sampling regime to, for example, a quarterly sampling regime, is not great; the standard error only changes from 0.0478 to 0.0510 for total meiofauna at the mud site. The larger the data set (e.g. total meiofauna, nematodes, etc.) the more ac-

Table 2. Standard errors (actual & projected) for the trend estimator (β) under different temporal sampling plans.

MEASUREMENT	Actual measured Value	Projected value at different sampling intervals		
		Monthly	Bimonthly	Quarterly
A. MUD SITE				
LN (TOTAL MEIOFAUNA +1)	0.0478	0.0478	0.0486	0.0510
LN (NEMATODE +1)	0.0622	0.0623	0.0632	0.0650
LN (COPEPOD +1)	0.0392	0.0392	0.0442	0.0509
TEMPERATURE	0.1301	0.1304	0.1441	0.1587
SALINITY	0.2058	0.2062	0.2192	0.2396
B. SAND SITE				
LN (TOTAL MEIOFAUNA +1)	0.0187	0.0187	0.0201	0.0223
LN (NEMATODE +1)	0.0241	0.0241	0.0264	0.0278
LN (COPEPOD +1)	0.0395	0.0396	0.0433	0.0459
LN (GASTROTRICH −1)	0.0216*	0.0220*	0.0267*	0.0320*
TEMPERATURE	0.1307	0.1309	0.1358	0.1743
SALINITY	0.0963	0.0966	0.1247	0.1329

* 9 years

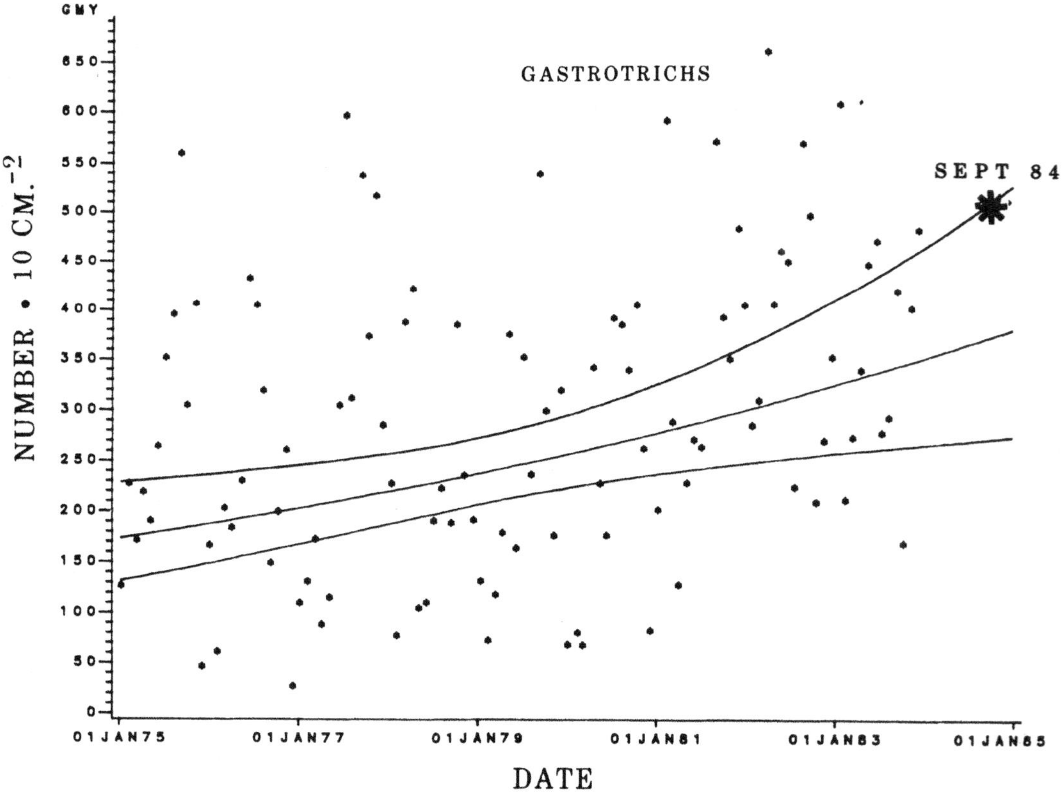

Fig. 6. Median abundance of gastrotrichs at the sand site (retransformed to No. 10 cm^{-2}) and 95% confidence bands, extrapolated beyond the original data set to January 1985. The September 1984 'real' data point is indicated.

curate the prediction. Programs are available for this calculation.

Besides determining sampling schemes for future reseach, long-term data sets can also be used, using ARIMA, for prediction of future trends. Using this technique Edwards & Coull (submitted), predicted ± simultaneous 95% confidence bands for the abundance trend of various taxa for a year beyond the last sampling data. Figure 6 illustrates the confidence bands and median number of gastrotrichs for the period Jan 1975 – Dec 1983 *and* the predicted median number (±95% CI) for 1984. Note the confidence bands become wider the further removed from the empirical end point, thus predictions far into the future are not very precise. To roughly estimate the precision of the model, we sorted samples from September 1984 (i.e. in the predicted region). Median gastrotrich abundance (2 replicates) at the sand site was 510·10 cm^{-2} just within the confidence band (see Fig. 6), fitting well within our predicted upward trend.

Summary

Long term meiofauna data sets are rare, but of those available it appears that one year is the primary recurrent cycle in meiofaunal assemblages. Long-term data sets should be used beyond their descriptive phase for generating testable hypotheses and predicting future sampling regimes and trends.

Acknowledgements

Bettye W.: Dudley sorted >90% of the samples for the South Carolina data set. This research was supported by the Biological Oceanography section of the National Science Foundation through grants DES 7201573, OCE 7617584, OCE 7809450, OCE8007968, OCE 838114 (B. C. Coull, principal investigator) and the LTER Program of the Ecosystems section of NSF, Grant DEB 8012175 (F. J. Vernberg, principal investigator).

References

de Bovée, F. & J. Soyer, 1974. Cycle annuel quantitive du méiobenthos des vases terrigénes côtières. Distribution vertical. Vie Milieu 24: 141–157.

Callahan, J. T., 1984. Long-term ecological research. BioScience 34: 363–367.

Connell, J. J., 1974. Field Experiments in Marine Ecology. In R. Mariscal (ed.), Experimental Marine Biology. Academic Press, New York: 21–54.

Connell, J. H., 1975. Some mechanisms producing structure in natural communities. In M. L. Cody & J. M. Diamond, (eds.), Ecology and Evolution of Communities. Belknap Press, Cambridge, Mass.: 460–488.

Coull, B. C., 1970. Shallow water meiobenthos of the Bermuda platform. Oecologia (Berl.) 4: 325–357.

Coull, B. C., 1985a. Long-term variability of estuarine meiobenthos: an 11 year study. Mar. Ecol. Prog. Ser. 24: 205–218.

Coull, B. C., 1985b. The use of long-term biological data to generate testable hypotheses. Estuaries 8: 84–92.

Coull, B. C. & B. W. Dudley, 1985. Dynamics of meiobenthic copepod populations: a long-term study (1973–1983). Mar. Ecol. Prog. Ser. 24: 219–229.

Dayton, P. K. & J. S. Oliver, 1980. An Evaluation of Experimental Analyses of Population and Community Patterns in Benthic Marine Environments. In K. R. Tenore & B. C. Coull, (eds.), Marine Benthic Dynamics. University of South Carolina Press, Columbia, S. C.: 93–120.

Edwards, D. & B. C. Coull, Submitted Ms. Autoregression in trend analysis: An example using long-term ecological data. Oikos.

Eskin, R. A., 1985. Population dynamics and ecology of the meiobenthic nematodes of North Inlet, South Carolina. Ph.D. Thesis, Univ. of South Carolina.

Grant, J., 1981. Factors affecting the occurrence of intertidal amphipods in reducing sediments. J. exp. mar. Biol. Ecol. 49: 203–216.

Heip, C., 1980. The influence of competition and predation on production of meiobenthic copepods. In K. R. Tenore & B. C. Coull (eds.), Marine Benthic Dynamics. University of South Carolina Press, Columbia, S.C.: 167–177.

Heip, C. & P. M. J. Herman, 1985. The stability of a benthic copepod community. In P. E. Gibbes (ed.), Proc. 19th European Marine Biology Symp. Cambridge Univ. Press, Cambridge, U.K.: 255–264.

Herman, P. M. J. & C. Heip, 1983. Long-term dynamics of meiobenthic populations. Oceanol. Acta 1983: 109–112.

Kirk, B. L., B. W. Rust & W. Van Winkle, 1979. Time series by maximum entropy method. ORNL-5332, Oak Ridge National Laboratory, Oak Ridge, Tennessee.

McIntyre, A. D. & D. J. Murison, 1973. The meiofauna of a flatfish nursery ground. J. mar. biol. Ass. U.K. 53: 93–118.

Platt, J. R., 1964. Strong Inference. Science 146: 347–353.

Rudnick, D. T., R. Elmgren, & J. B. Frithsen, 1985. Meiofaunal prominence and benthic seasonality in a coastal marine ecosystem. Oecologia (Berl.) 67: 157–168.

Smith, L. D. & B. C. Coull, in press. Juvenile spot (pisces) and grass shrimp predation on meiofauna in muddy and sandy substrates. J. Exp. Mar. Biol. Ecol.

Stripp, K., 1969. Jahreszeitliche Fluktuationen von Makrofauna und Meiofauna in der Helgoländer Bucht. Verröff. Inst. Meeresforsch. Bremerhaven 12: 65–94.

Tietjen, J. H., 1969. The ecology of shallow water meiobenthos in two New England estuaries. Oecologia (Berl.) 2: 251–291.

Wiens, J. A., 1977. On competition and variable environments. Am. Sci. 65: 590–597.

Hydrobiologia 142: 281–290 (1986)

The predictability of biological populations and communities: an example from the meiobenthos

Peter M. J. Herman & Carlo Heip
Marine Biology Section, Zoology Institute, Rijksuniversiteit Gent, Ledeganckstraat 35, B-9000 Gent, Belgium

Keywords: benthos, time series, monitoring, prediction, copepods

Abstract

The predictability of temporal changes was analysed in the population density of three meiobenthic copepod species, and for three parameters characterizing this community. For *Paronychocamptus nanus* and *Tachidius discipes* a large part of the variation in time scales longer than 1 year is explainable as cyclic factors. In *P. nanus* clear cyles with periods of 2 years and 1.5 years were found. These cyclic factors can be extrapolated for prediction. In *Canuella perplexa* almost all the variation was due to unpredictable long-term drift. Community parameters are generally better predicted than population densities. In our data set total density of the copepods was the best parameter to monitor. Monitoring does not necessitate a strict sampling scheme; irregular quarterly sampling from the *P. nanus* series showed that the essential features of the series were still discernible.

Introduction

Correct assessment of the biological consequences of pollution in coastal waters demands prior and adequate understanding of the scales and causes of natural ecological variation (COST 47 Activity Report 1983). Predictability of monitored variables chosen to describe the community is a central assumption of ecological monitoring strategies. Effects of pollution can only be inferred when the variates differ significantly from what is expected based on previous experience; that is, when the observation(s) fall(s) outside the predicted range.

Ideally, prediction should be based on a complete understanding of the intricate ecological interactions governing the dynamics of populations and communities, but this is hardly possible. However, even when the causes of variation are unknown or unquantified, prediction may still be possible on a purely statistical basis. For example, it is not necessary to understand the laws governing the movements in the planetary system to predict when the sun will rise tomorrow. When the variation in a variable is (partly) regular, this regularity can provide the basis for prediction.

Populations are variable in time and space. Both sources of variation must be studied for monitoring purposes. Here we will consider temporal variability of meiobenthic copepods from a shallow brackish-water pond. This environment has some disadvantages: large temperature fluctuations and a long-term trend in salinity. However, both can be quantified and considered as forcing functions. Sometimes catastrophic events occur: ice formation or drying. On the other hand there is no advection, a factor which has recently been shown to be of importance for meiofauna (Palmer, 1984), and temporal variability is much larger than spatial variability (Heip & Engels, 1977). As is typical for the meiofauna (Warwick, 1984), reproduction is much more predictable than for most macrofauna populations. The phenology of reproduction, measured as the date when the populations start to increase exponentially, is very constant from year to year (Heip, 1980a).

Harpacticoid copepods are small organisms,

around 1 mm or less long and weighing between 0.5 and 5 μg dry weight. Development times are typically around two–three weeks and they are short-lived. One therefore expects that the relevant time-scales will be in the order of less than to about one year. This argument has frequently been used in favour of using harpacticoid copepods in pollution studies (Heip, 1980b).

In previous studies of this community we have shown that long-term periodicity is important, at least in some of the populations. The time series were studied by Maximum Entropy Spectral Analysis (Herman & Heip, 1983a; Heip & Herman, 1985). The highest peaks in the spectra were often found for periodicities of two to five years. These results differ basically from those reported for estuarine copepods in South Carolina, USA by Coull (1985) and Coull & Dudley (1985), who found no periodicity longer than one year.

In this paper we will address the following questions:

a) How predictable (and how truly cyclical) is the temporal variation on time scales in excess of 2 months in our populations and communities?

b) What parameters can be used for prediction of future states of the system?

c) What is the sampling effort required?

Material and methods

The sampling site and techniques are described in many previous papers (e.g. Heip, 1980a). The time-series is based on fortnightly samples from August 1968 till December 1976. One sample was analysed on each date. Copepod species density and population composition were determined, from which biomass and respiration were calculated. Species richness is low and attention in this paper will be focused on the three dominants *Paronychocamptus nanus, Tachidius discipes* and *Canuella perplexa.* For more information on these species we refer to Heip (1980a), Herman & Heip (1983a, b; 1985) and Herman, Heip & Guillemijn (1984).

The analysis in this paper is independent from the spectral analysis performed earlier. Here we use a simple but very robust procedure, described by Kendall (1976). The data series were grouped per two months, so that, e.g. the 'Jan.–Feb.' value of 1973 is the mean from all the samples that were taken in these two months. The series is split into three elements: a) a long-term trend (which can further be split into a linear trend and more or less regular fluctuations around it); b) cyclic components (including seasonal effects); c) irregular or random effects. These elements are analysed after log-transformation of the series (the rationale being that multiplicative effects are more important than additive effects).

Kendall's (1976) method for the extraction of seasonal effects proceeds as follows:

a) a moving average with a length of one year is calculated.

b) this long-term trend is subtracted from the series

c) from the residual series seasonal components are calculated by taking the average values over the first two months of each year, over the second two months of each year and so on. These six 'raw' cyclic effects are also averaged, and their average is subtracted from the six values. These final values express the seasonal effects, and are subtracted from the residuals. These new residuals are then checked to see if any periodicity remains.

Examination of the residuals of several of our series showed that an important cyclicity with period 2 years remained. The analysis was, in these cases, adapted to extract 2-yr. cyclic effects (and, at the same time, all effects with periods of an integer fraction of 2 years, i.e. 1 year, 0.5 year etc.). In this analysis we calculated from the original series a 2-yr. moving average, instead of a 1-yr. moving average, and proceeded to extract 12 raw and adjusted cyclic components (analogous to the seasonal components) by averaging the residuals of the first, second,..., twelfth two-month interval. Thus an average was made of the residuals in Jan.–Feb. of every second year, Mar.–Apr. of every second year, etc. The seventh period is again Jan.–Feb., the eighth again Mar.–Apr., but these are other cyclic components than the first six values.

For the density of *Paronychocamptus nanus,* and the community parameters influenced by this density, a periodicity of 1.5 year was found in the residuals after extraction of 2 years period components. This periodicity was further subtracted in a similar way, whereafter the residuals were irregular (note that the subtraction of the 1.5 year periodicity was performed after the extraction of the 2 years period components, and did not replace it, as the 2 year component extraction replaced the seasonal component extraction).

To check whether a series is random or irregular one studies the correlogram of the series, constructed on the basis of autocorrelation (see Platt & Denman, 1976), the correlation of the series with itself at a certain time-lag.

When a series is decomposed in this way, it can be used in forecasting: by reassembling the trend, the seasonal and other cyclic effects and the oscillations around the trend we can project the series forward in time.

Results

1. Paronychocamptus nanus

Figure 1A shows the autocorrelation values of the density of *P. nanus* after extraction of a 2 yr. moving average. It is apparent that cyclical elements are present in the series. These were not removed by extraction of the average seasonal cycle (average within-year effect). We therefore proceeded to extract a cyclic component with period 2 yr. from the series. This component is highly asymmetric between one year and the next (Fig. 2C). The residuals after extraction of the 2 yr. moving average (Fig. 2B) and the 2 yr. period cyclical component are shown in Fig. 2D. It immediately appears from inspection of the data and the correlogram (Fig. 1B) that another cyclic phenomenon, with period 1.5 yr., dominates this residual series. The 1.5 yr. component was extracted, and added to the 2 yr. component (Fig. 2E). A kind of interference pattern is produced, that mimicks very well the detrended series. The final residuals are shown in Fig. 2F. This series is produced by subtracting from the original series (Fig. 2A) the 2 yr. moving average (Fig. 2B) and the summed cyclic components (Fig. 2E). Note that, because we do not force the 'cyclic components' to take the form of sine waves, cycles with frequencies that are integer multiples of the basic frequencies, are also filtered out. Thus the extraction of a cyclical component with period 2 yr., in principle also extracts components with periods 1 yr., 2/3 yr., 1/2 yr., 2/5 yr., 1/3 yr.

The correlogram of the final residuals (Fig. 1C) reveals no particular structure left. The division of

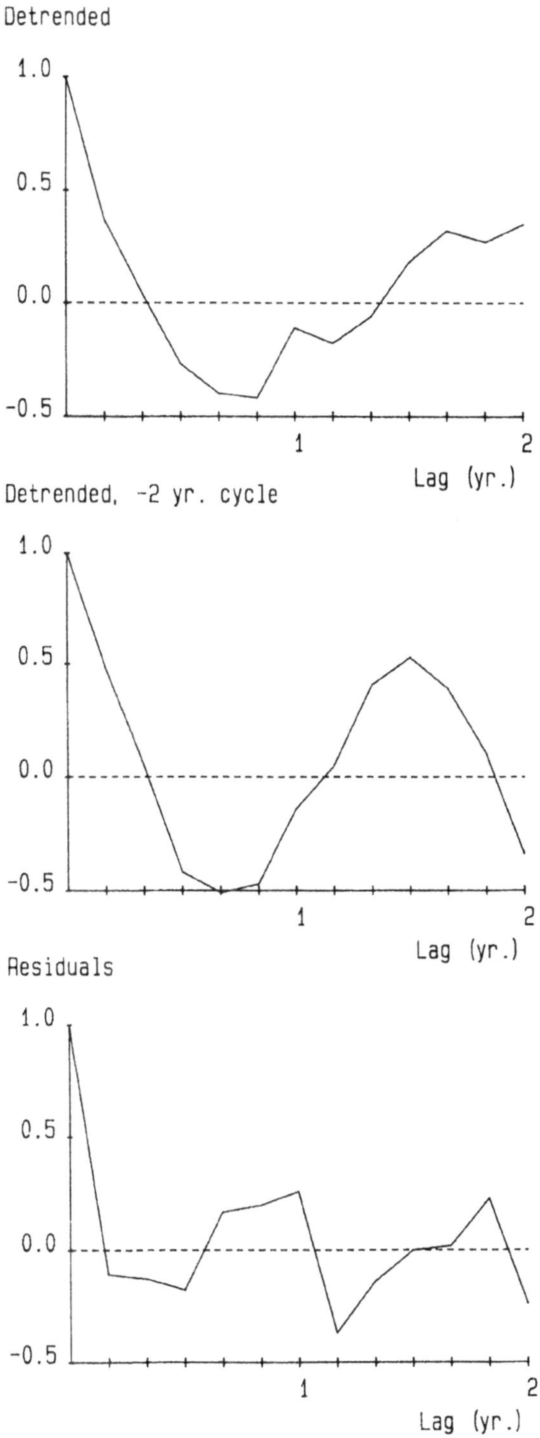

Fig. 1. Density of *Paronychocamptus nanus:* autocorrelogram of the original series (Fig. 1A), after extraction of the 2-years period components (Fig. 1B), and residuals after further extraction of the 1.5 year period.

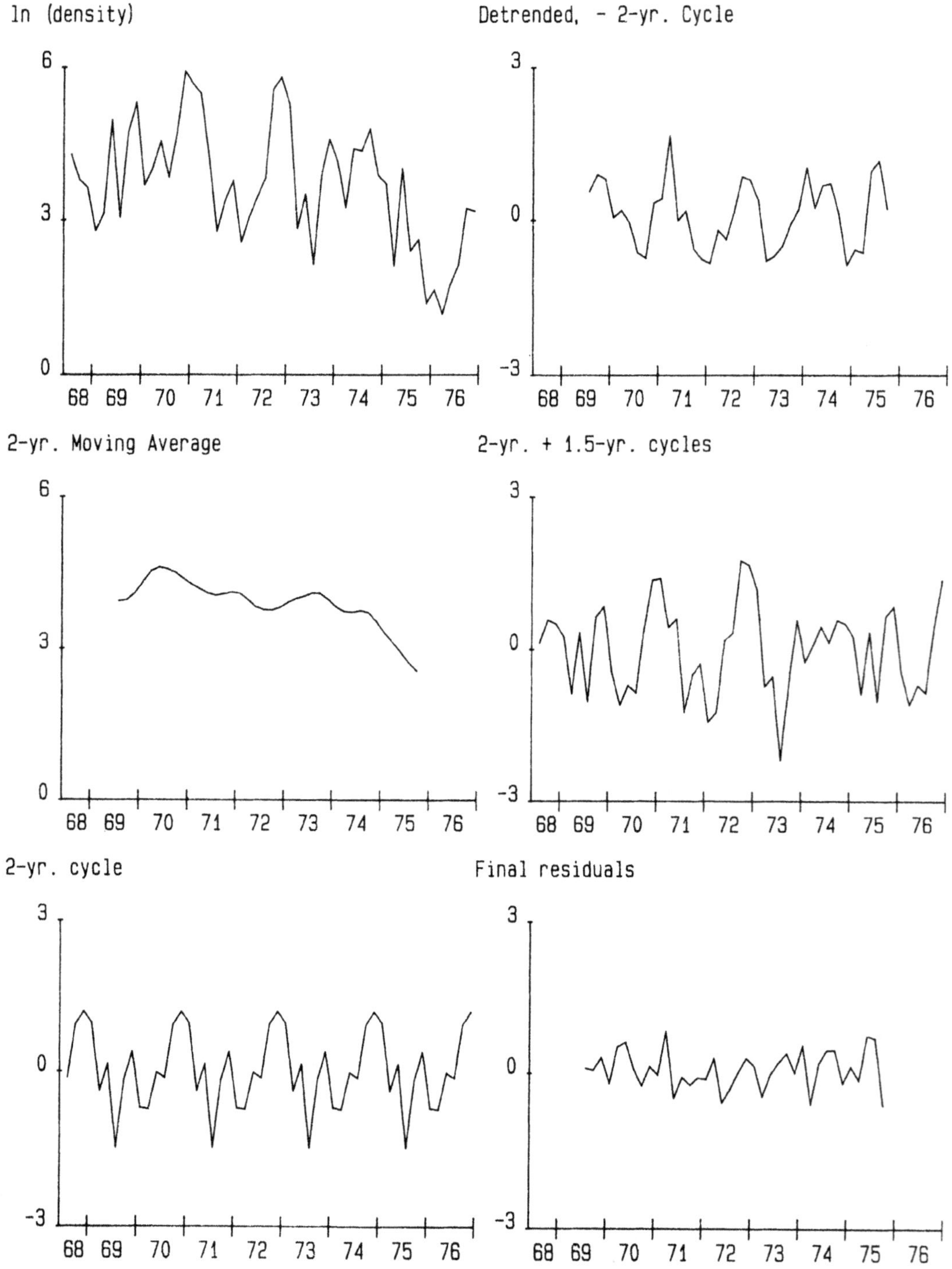

Fig. 2. Density of *Paronychocamptus nanus:* original series (Fig. 2A), 2-yr. moving average calculated from this series (Fig. 2B), extracted 2-yr. cycle (Fig. 2C), residuals after extraction of the 2-yr. moving average and the 2-yr. cycle (Fig. 2D), sum of the 2-yr. and 1.5-yr. cycles (Fig. 2E), and final residuals (after extraction of the 2-yr. moving average, and the 2-yr. and 1.5-yr. cycles).

the Sums of Squares of the original series between the different components is shown in Table 1.

The SS of the different components add to a higher amount than the SS of the original series, indicating that there is a (small) negative covariance between the components.

A relatively large proportion of the variance of the original series (appr. 75%) is explained by the summed cyclic components. For predictive purposes, these can in principle be extrapolated to the future. As a test we estimated the harmonic component from the series between 1968 and 1974, thus neglecting the last two years. It is seen from Fig. 3 that the pattern coincides reasonably well with that estimated for the whole series (Fig. 2E).

Table 1. 7-years series of the density of *Paronychocamptus nanus*: division of the Sums of Squares of the original series between the different components. The first and last year of the series have been left out, since we have no values for most components here.

Series	SS
Original series ('69 – '75)	40.33
Moving average	8.20
2 yr. cycle	22.53
1.5 yr cycle	11.00
Residuals	5.32

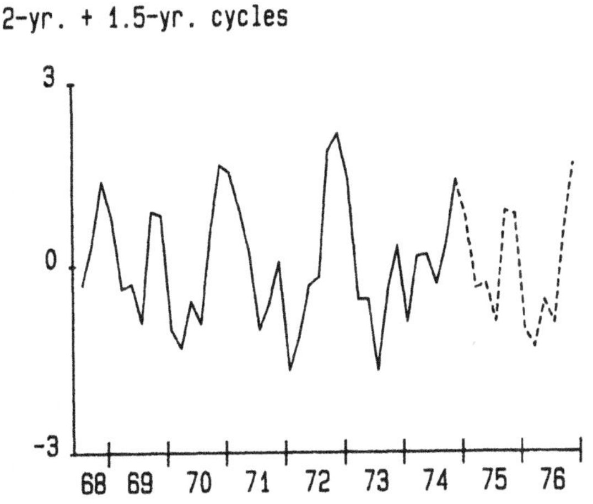

Fig. 3. Density of *Paronychocamptus nanus:* Cyclic components (sum of 2-yr. cycle and 1.5-yr. cycle) determined on part of the original series: 1975 and 1976 are left out of the analysis. Compare with Fig. 2E.

2. Other series

As in *P. nanus,* the only identifiable structure in the other series are the cycles that were already revealed by the spectral analysis (Herman & Heip, 1983a; Heip & Herman, 1985). In *Canuella perplexa* the spectrum was largely dominated by a peak in the very low frequencies (the peak value corresponded to a period of 3.8 yr.). Correspondingly, we find very little structure in the autocorrelogram after extraction of the trend (Fig. 4). A very small (yearly) seasonal factor was found (Fig. 5B). Inclusion of the years 1968 and 1969 has a profound influence on the interpretation of the long-term trend (dashed line in Fig. 5A). From 1970 onwards a kind of cycle, such as shown by the spectral analysis, can be seen. However, as the species was completely absent in the winter '68–'69, this pattern is abruptly changed. *C. perplexa* was also completely absent in 1980. The population seems to crash and be refounded at irregular intervals, which makes prediction of its numbers a very difficult task.

For Tachidius discipes the spectrum is largely dominated by a peak on 1 yr. Small peaks correspond to periods of 2 yr., 1/2 yr. and 1.5 yr. (note that MESA does not provide a test for the statistical significance of the peaks). Extraction of components with period 2 yr. or integer fractions thereof explains about 70% of the variance of the

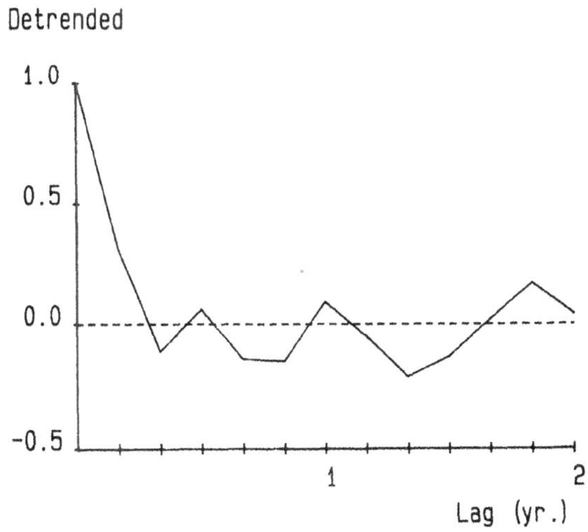

Fig. 4. Density of *Canuella perplexa:* autocorrelogram after extraction of a 1-yr. moving average from the series.

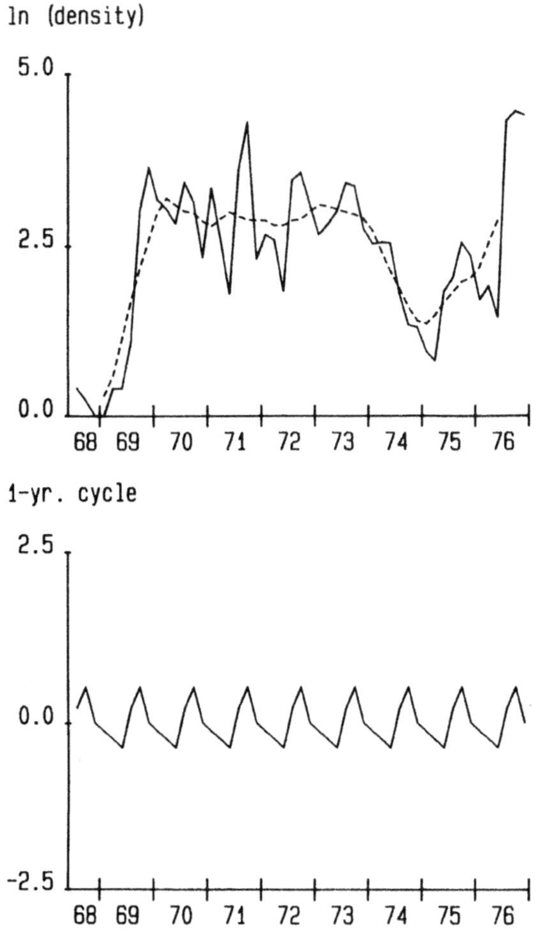

Fig. 5. Density of *Canuella perplexa:* original series (Fig. 5A, full line), with superimposed 1-yr. moving average (dashed line). Extracted 1-yr. cycle is shown in Fig. 5B.

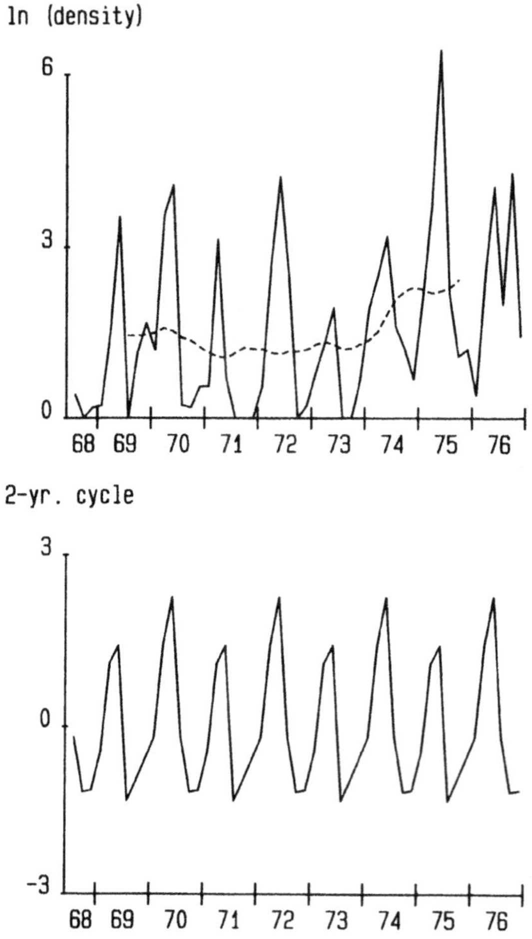

Fig. 6. Density of *Tachidius discipes:* original series (Fig. 6A, full line), with superimposed 2-yr. moving average (dashed line). Extracted 2-yr. cycle is shown in Fig. 6B.

detrended series. (Fig. 6). The residual variance is further reduced by one third after the extraction of a 1.5 yr. period component. Trend is relatively unimportant (Fig. 6A: dashed line). Trend and residual variance contain about 25% of the variance of the original series. In fact, this series is very predictable, except for the unusually high peak of 1975, and the second peak in 1976, which is, however, attributable to a special event (part of the pond dried out in this very warm summer).

3. Community parameters

As we stressed earlier (Heip & Herman, 1985) the community parameters 'total density', 'total biomass' and 'community respiration' show little or no trend (Figs 7, 8, 9: dashed lines). This is especially so for respiration, the most stable parameter of the three. The community parameters are also much more stable than the densities of the constituent species in another sense. The coefficients of variation ($CV=100\ s/\bar{x}$) are much smaller than for the population densities (Table 2).

About an equal fraction of the SS (2/3) in total density and in respiration is explained by cyclic factors (periods 2 yr. and 1.5 yr. summed). In biomass this is much less (<1/4) and this series is almost completely non-seasonal.

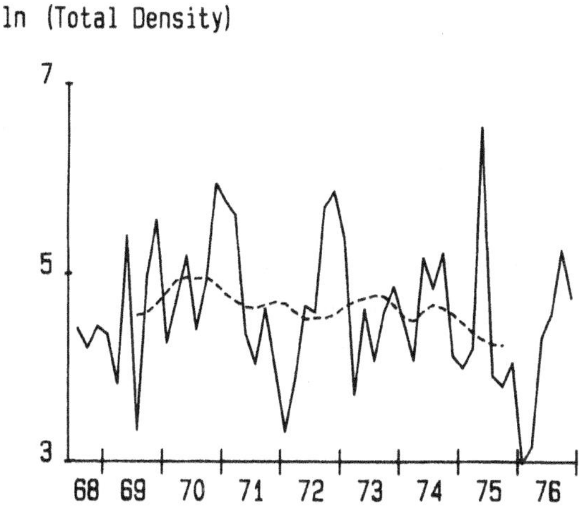

Fig. 7. Total copepod density in the community studied: original series (full line) and 2-yr. moving average (dashed line).

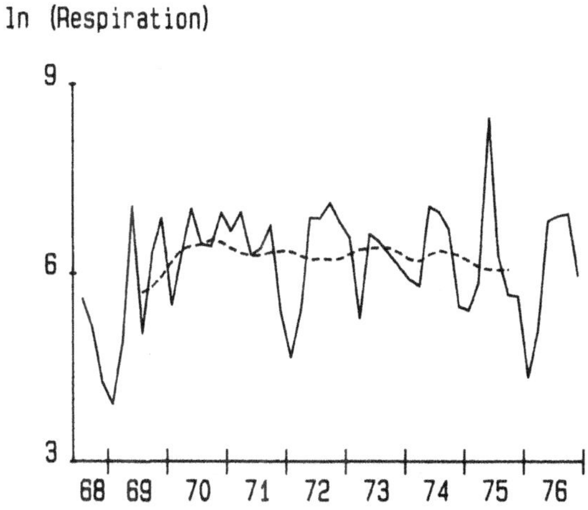

Fig. 9. Total respiration of the copepods in the community studied: original series (full line) and 2-yr. moving average (dashed line).

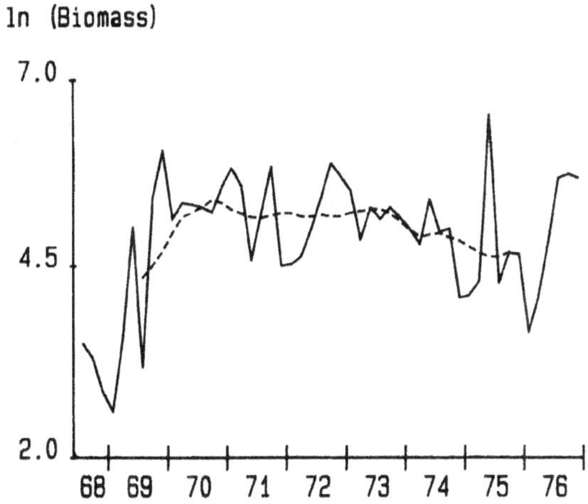

Fig. 8. Total biomass of the copepods in the community studied: original series (full line) and 2-yr. moving average (dashed line).

Table 2. Comparison of the coefficient of variation (CV = s/x* 100%) between the density series of three meiobenthic copepods, and three parameters of the meiobenthic copepod community.

Series	CV (%)
P. nanus density	32
T. discipes density	94
C. perplexa density	49
Total density	17
Biomass	17
Respiration	15

4. *An inefficient sampling scheme for* P. nanus

Due to storms, unavailability of research vessels, broken cables, damage to the winches, administrative or financial problems, lack of time and other miscellaneous problems, it is not always possible to follow a strict sampling scheme (e.g. monthly) for a station at sea. As an example we took the roughly seasonal sampling scheme that was realized for a few stations in the Belgian coastal zone between 1977 and 1983 (Herman R., unpublished). We transposed the dates 7 years back, and took from the fortnightly series of Dievengat samples of *P. nanus* the one that was next to each date. These samples were subsequently taken as representative for the season in which they fell (Winter, Spring, Summer, Fall). Values for unsampled seasons were linearly interpolated.

The resulting series is shown in Fig. 10 (Compare with Fig. 2). Surprisingly, the major features of the original series are well preserved. After extraction of the trend, the 2 yr. periodicity (Fig. 10B) is apparent. It is very similar in form to that extracted from the whole series (Fig. 2C). Even the 1.5 yr. cycle can be clearly seen in the residuals. The compound harmonic part of the series (Fig. 10C) is very comparable to the one in the complete series (compare Fig. 2E).

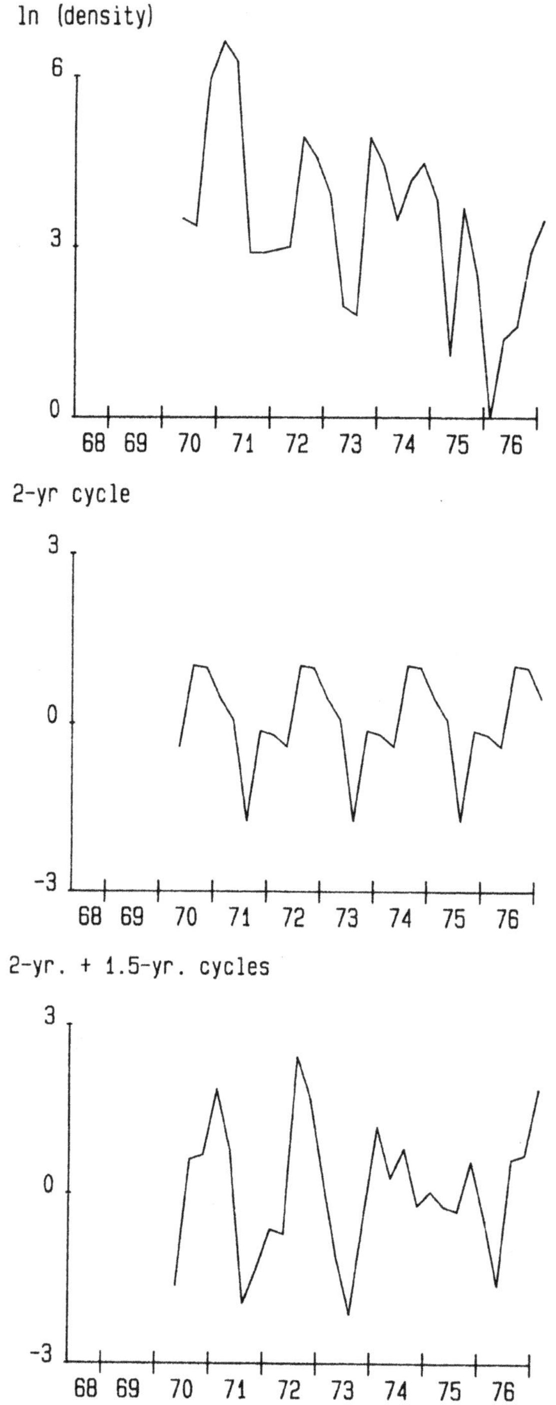

Fig. 10. Paronychocamptus nanus: samples taken from the original series according to a (realized) irregular sampling scheme (see text). A: density, B: extracted 2-yr. cycle, C: sum of the extracted 2-yr. and 1.5 yr. cycles.

Discussion

Of the three population density time-series discussed here, two have an important part of their variance explained by cyclic terms. This part may, in principle, be usable for prediction purposes. This is certainly the case for seasonal (1 yr.) factors. The mechanism underlying seasonal variation is obvious, and one can be sure about the exact period, and the truly cyclic nature of the phenomena. It is, however, less sure for 2 yr. periods. The succession of high peaks by lower ones the next year, higher again the next year etc. may be produced by chance. However, if the peak's height is purely random, only in one case out of 64 an alternating pattern will be produced in 8 successive years. Therefore we consider the 2 yr. period cycle in *P. nanus* to be most likely a truly cyclic phenomenon.

What to do, however, with a 1.5 yr. period cycle? A 1.5 yr. period swing is undoubtedly present in the data. The problem in interpreting it as a long-lasting, true cycle is that we cannot conceive of a mechanism producing it. Moreover, although the 2 yr. and 1.5 yr. period cycles are observed 3 to 4 times in the series, the 'interference pattern' between them is repeated only once every 6 yr. This is about equal to the period of observation.

In contrast to the situation for *P. nanus* and *T. discipes,* in the *C. perplexa* series almost no cycling was present, compared to the importance of the long-term trend.

In all the series studied, no particular structure was left after extraction of cyclic components and trend. Trend in particular may render prediction difficult. We can take the trend in *P. nanus* as an example. This can be decomposed in a linear trend with negative slope, and a long-term swing around this line. MESA – spectral analysis indicated that the dominant period in this swing is 4.6 yr. (Herman & Heip, 1983; Heip & Herman, 1985). However, the determination of the period is unsure, as is the answer to the question if a cycle is really present. Obviously, extrapolation to the future depends on this answer.

The same is true for the linear trend. Extrapolation of this trend yields highly unrealistic values for 1980. The mean density in this year was about 350 ind. per 10 cm^2 (Herman & Heip, 1985), whereas the predicted value of the trend corresponds to about 9.5 ind. per 10 cm^2. Is the 'linear' trend part of an oscillation with a very long period, or is there random or pseudocyclic fluctuation in the yearly mean densities? One would need at least 20 year of data in order to choose between the alternatives.

Based on the data between 1968 and 1974, one would hardly suppose a negative trend at all. The best fitting line on the moving average between mid-1969 and mid-1973 has a slope of −0.10 per year, compared to −0.20 per year for the whole series.

The difficulty of interpreting patterns in trend is also illustrated by the *C. perplexa* series. MESA revealed a 3.8 yr. periodicity in the data between '70 and '76. This can still be found back in the trend (Fig. 5A), but the '68–'69 data clearly distort this pattern. Similarly, the harmonic extrapolation of the 3.8-years period 'cycle' would not have predicted the disappearance of the population in 1980.

Gray & Christie (1982) postulated that oscillations with periods of 6–7, 10–12,... yr. may be a rule in marine populations. However, the evidence for this is still very scarce. The mechanisms they propose are unlikely to have any influence in an enclosed pond such as the Dievengat (except, maybe, for the (in)famous 'sunspot cycle'). On the other hand, the fact that cyclicity is the only pattern emerging on shorter terms, may provide some kind of justification for the search for cycles, even on the longer time scales.

Heip & Herman (1985) showed that the community parameters had less linear trend and less long-term fluctuations than the densities of the constituent species. This greater stability of community parameters is also shown in the coefficients of variation (Table 2). Stability is most pronounced in the 'functional' parameter respiration. Considering the whole system in which we sampled the meiobenthic copepods, the function of the 'copepod black box' is probably best represented by the total respiration, as it is highly correlated with production (Herman *et al.,* 1984). Most of the variation in respiration is explained by cyclic components. Trend is almost completely absent, and the residual variance is small. These features make it a very predictable series, and a likely candidate for monitoring purposes. There are two drawbacks, however: 1) it is not known how sensitive a measure this parameter is for changes in the environment: will it remain unchanged even after dramatic changes in the community? 2) it is not easily measured. Total density and biomass are much easier to measure. If the sampling scheme is intensive enough to estimate seasonal or other cyclic components, total density seems to be the preferable parameter.

As shown by our exercise on *P. nanus,* even a very inefficient sampling scheme was able to reveal most of the structure in this relatively complex series. Monitoring does not seem to depend too heavily on a close interval sampling scheme. As a rule of thumb we would suggest that three to four years of sampling seasonally will already produce the most obvious characteristics of the time-series. This is longer than suggested by Coull & Dudley (1985). It should be noted, however, that the populations studied by these authors exhibit less long-term changes (such as the two-year cycles shown here). A too short sampling time will miss these components, which are shown here to be important in some cases. Whether this holds for other groups besides copepods remains to be studied.

Acknowledgments

Both authors acknowledge a grant of the Belgian National Fund for Scientific Research (NFWO). Part of this research was sponsored through the grant FKFO-2.9007.82 'Systematics and ecology of marine organisms'. We thank Wies Gyselinck for technical assistance.

References

Coull, B. C., 1985. Long-term variability of estuarine meiobenthos: an 11-year study. Mar. Ecol. Prog. Ser. 24: 205–218.

Coull, B. C. & B. W. Dudley, 1985. Dynamics of meiobenthic copepod populations: a long-term study (1973–1983). Mar. Ecol. Prog. Ser. 24: 219–229.

Gray, J. S. & H. Christie, 1983. Predicting long-term changes in marine benthic communities. Mar. Ecol. Prog. Ser. 13: 87–94.

Heip, C., 1980a. The influence of competition and predation on production of meiobenthic copepods. In: K. R. Tenore & B. C. Coull (eds), Marine benthic dynamics. The Belle W. Baruch Library in Marine Science, no. 11. University of South Carolina Press, pp. 167–177.

Heip, C., 1980b. Meiobenthos as a tool in the assessment of the quality of the marine environment. Rapp. P.-v. Réun. Cons. Int. Explor. Mer 179: 182–187.

Heip, C. & P. Engels, 1977. Spatial segregation in copepod species from a brackish water habitat. J. exp. mar. Biol. Ecol. 26: 77–96.

Heip, C. & P. M. J. Herman, 1985. The stability of a benthic copepod community. In: P. E. Gibbs (ed.), Proceedings of the 19th European Marine Biology Symposium. Cambridge Univ. Press, pp. 255–263.

Herman, P. M. J. & C. Heip, 1983a. Long-term dynamics of meiobenthic populations. Oceanologica Acta, Actes 17 European Marine Biology Symposium, pp. 109–112.

Herman, P. M. J. & C. Heip, 1983b. The respiration of five brackish-water harpacticoid species. J. exp. mar. Biol. Ecol. 71: 249–256.

Herman, P. M. J., C. Heip & B. Guillemijn, 1984. Production of Tachidius discipes Giesbrecht 1881 (Copepoda: Harpacticoida). Mar. Ecol. Prog. Ser. 17: 271–278.

Herman, P. M. J., G. Vranken & C. Heip, 1984. Problems in meiofauna energy-flow studies. Hydrobiologia 118: 21–28.

Kendall, M., 1976. Time-Series. 2nd Ed. Griffin & Co.

Palmer, M., 1984. Invertebrate drift: behavioral experiments with intertidal meiobenthos. Mar. Behav. Physiol. 10: 235–253.

Platt, T. & K. Denman, 1975. Spectral analysis in ecology. Ann. Rev. Ecol. Syst. 6: 189–210.

Warwick, R. M., 1984. Species size distributions in marine benthic communities. Oecologia (Berl.) 61: 32–41.

Hydrobiologia 142: 291–308 (1986)

Interpreting long-term changes in benthic community structure: a new protocol

Donald C. Rhoads & Joseph D. Germano
Science Applications International Corporation, Admiral's Gate, 221 Third Street, Newport, R. I. 02840-6669, USA

Keywords: benthic monitoring, animal-sediment, remote sensing

Abstract

Documentation of long-term change in benthic ecosystems is important for assessing and managing the effects of such change on: 1) secondary production, particularly leading to commercially important food webs, 2) pollutant transfer within the food web, 3) the ability of the 'new' assemblage to metabolically burn-off labile detritus that might otherwise accumulate, contributing to long-term hypoxia, and 4) recyling of nutrients from the seafloor back to primary producers.

Organism-sediment relationships which accompany benthic disturbances have predictable features. Although participating species may vary regionally or seasonally, their life-history attributes and functional relationships to the associated sediment appear to be universal. Pioneering seres are near-surface dwelling, productive, and are readily available to demersal predators. However, these taxa may be potential pollutant vectors. Dense tube mats may promote the deposition and retention of high BOD organic matter. Late successional stage seres are represented by deeply bioturbating 'head-down' deposit feeders. The deep cryptic infaunal habitat of these species may make them less important as prey for epifaunal predators. Sediments populated by these equilibrium assemblages are characteristically low in labile organic matter, sedimentary sulphides, and oxygen demand. Nutrients (N, P, Si) are returned to primary producers by biogenic irrigation of sediment pore water.

Mapping of successional mosaics is important for documenting major long-term change in benthic community structure and associated biogenic processes. Our mapping tool consists of a vessel-deployed sediment-profile camera; organism-sediment relationships can be imaged *in situ* with this instrument. Such a mapping protocol is not intended to replace traditional sampling. Rather, the successional maps are used to efficiently detect change in a system, design a cost-efficient sampling grid for obtaining geochemical and biological ground-truth samples, and to construct hypotheses about how the change might answer the four outlined management questions.

Introduction

The majority of environmental monitoring or assessment work over the last few decades has consisted of short-term research (i.e., less than one or two years) to provide short-term answers to problems recognized as 'urgent' at the time of the study. This is particularly true with marine studies, where costs associated with field sampling operations are fairly significant. History has shown that long-term, irreversible changes affect man most seriously; even the structure seen in natural, undisturbed communities is usually controlled by infrequent severe events that disrupt the community and return the successional process to an earlier stage (Green, 1984). Long-term studies are needed to obtain reliable estimates of both natural variation in undisturbed systems as well as interpreting the long-term effects of impacts on complex systems.

Four management issues are raised regarding the

significance of long-term change in benthic community structure on: 1) secondary production and its importance for commercially important species, 2) pollutant transfer within the food web, 3) the ability of the 'new' benthic assemblage to metabolically burn-off labile detritus that might otherwise accumulate contributing to long-term hypoxia, and 4) recycling of nutrients from the seafloor back to primary producers. It is our opinion that existing monitoring methods do not provide much useful information regarding these management questions.

It is our purpose to draw attention to the shortcomings of existing approaches to benthic monitoring and to offer an alternative approach which may provide more useful information to those charged with the responsibility of making management decisions about changes in benthic community structure. It is becoming increasingly clear that unless managers are provided with more useful paradigms and efficient monitoring methods, traditional monitoring approaches will be curtailed. This trend is already apparent in the United States and has had a serious effect on both pure and applied benthic research. In fact, the inadequacy of the type of information gained from traditional benthic sampling methods for use by environmental managers coupled with the associated expense and slow data return have most likely been responsible for the lack of long-term benthic monitoring efforts to date.

Our perspective will be focused on fine-grained subtidal sediments for three reasons: 1) this is the dominant facies in many estuaries and embayments; 2) particle-bound pollutants are often associated with muds; and, 3) this facies may be associated with hypoxia and methanogenesis.

Critique of existing methods

Over the past 15 years, a great deal of applied benthic work has been done in the context of environmental baseline studies, impact assessment, and monitoring. Most of this work, including our own, involved traditional benthic sampling methods, i.e. grab or box-core sampling followed by removal of the organisms from the sediment by sieving. The resulting samples are used to generate species abundance lists, diversity and faunal similarity indices, and community dendrograms. The resulting baseline statistics are then mapped in space or compared over time in order to characterize the range of natural variation. With these baseline data in hand, investigators attempt to infer pollution or disturbance events at those stations where a change in diversity (or other measure of structure) has occurred at a selected level of significance.

Two major problems with the above traditional approach have prompted us to abandon this type of monitoring. The first problem stems from the kind of information that is provided. For the last three decades, benthic investigators have been searching for means of extracting ecological information from species lists, attempting to convert descriptive, natural history observations into a format that would convey the impression of a 'hard' science (what Gould, 1981, has appropriately described as 'physics envy'). The calculation of species diversity indices popularized in the 1960's has persisted within the community of benthic ecologists, even though most terrestrial ecologists have abandoned their use years ago. Peet (1974) has pointed out that one of the strongest objections to using a diversity index for comparing a variety of habitats is that there is no valid basis for comparing the richness of a series of communities using a single index unless one can assume that the underlying species-individuals relationships are similar. A variety of other investigators have pointed out the many shortcomings of diversity indices (e.g., Green, 1979; Routledge, 1980; Wolda, 1981).

With the increased use of computers and application of multivariate techniques, benthic ecologists are now attempting to interpret changing patterns of species distributions in time as well as space, trying to extract ecologically meaningful information from the large volume of data that a benthic survey typically produces by using techniques such as factor and cluster analyses. Gould (1981) has pointed out that the decision to use techniques like these as the primary method to analyze large data sets implies a deep ignorance of principles and causes, in reality an elegant method of grasping for straws; these methods should be applied with the hope that they might provide suggestions for more fruitful lines of inquiry. Unfortunately, large scale benthic surveys rarely, if ever, progress beyong this, viewing the dendrograms and resolution of principal components as the end result.

Information statistics, per se, commonly provide little ecological insight. Beyond documenting that a change has occurred in benthic population structure, these data do not provide information that can be used by a manager to make judgements as to whether the documented change is desirable or will have a negative impact on the system in terms of resource management objectives.

Other than saying that diversity has changed, what other information can be provided to an environmental manager? In a management context, the definition of a desirable or negative impact is usually judged in terms of how a long-term change might ultimately effect man. The problem can be reduced to the question of what is intrinsically positive or negative (in an ecosystem sense) about a relatively high or low diversity facies or a system that experiences periodic change in diversity? The only unequivocal answer would be related to an event which resulted in a high or low diversity system reverting to an azoic state (zero diversity). This dramatic change would clearly result in a non-productive system (and would not require the calculation of diversity indices to illustrate the point). If maintenance or enhancement of production is the management goal, this change is clearly a negative one. However, this extreme case is a rare example; most monitoring involves the detection of more subtle gradients in species abundances, number of species, and biomass. The manager is then faced with the inevitable 'So what?' question once a change in population structure is demonstrated. This problem has also been noted by Gray (1976), who cites the inadequacy of existing theoretical frameworks for interpreting the significance of change in benthic community structure.

The second problem concerns techniques for sampling the benthos in order to detect long-term change. Sampling tools for large scale surveys traditionally have been limited to grab or box-core samplers. This method of sampling the bottom and associated on-deck processing results in various degrees of destruction of important *in situ* ecological information which is contained in the spatial and structural relationships of benthic organisms to one another and to the associated sedimentary matrix. This destruction can occur in a variety of ways, either from the pressure wave which preceeds the instrument as it approaches the bottom, from water passing through the sampler and washing away the sediment surface during recovery, or from the final process of sieving the sample, which destroys any remaining organism-sediment information. The ensuing sorting and faunal identification is extremely labor-intensive and expensive; there is characteristically a long time-lag between field collection and data return or final interpretation. While grabs and box corers will always be important for quantitative sampling, sampling technology needs to be advanced so that benthic gradients can be more rapidly and efficiently mapped *in situ* by remote methods. Reconnaissance mapping techniques can then be used to detect major changes in benthic communities; once a change has been detected, reconnaissance maps can serve to locate station positions for efficient sampling with traditional quantitative methods.

An organism-sediment successional paradigm

Organism-sediment relationships which accompany benthic disturbances have predictable features. The term disturbance is used here to define natural processes, such as seafloor erosion, changes in seafloor chemistry, foraging disturbances which cause major reorganization of the resident benthos, or anthropogenic impacts, such as dredged material or sewage sludge dumping, thermal effluents from power plants, deposition of drilling muds and cuttings, pollution impacts from industrial discharge, etc. Because the perspective of this paper is to address long-term change, these disturbances must be extraordinary, i.e. beyond the ability of the resident benthos to regulate or accomodate the perturbation. Infaunal succession in shallow nearshore clastic sediments commonly takes between 1 and 10 years: '... it appears that over a wide range of communities many successions take about as long as the average longevity of the dominants of the infrequently disturbed or "climax" community' (McCall & Tevesz, 1983).

Figure 1 shows the progressive development of the infaunal community over time following deposition of dredged materials (1A) and the shallow to deep spatial tiering of infauna along a pollution gradient (1B). Although these examples show different taxa participating in various stages of infaunalization, the organism-sediment relationships appear to be similar (see Rhoads & Germano, 1982, for details). This primary succession is defined as 'the predictable appearance of macrobenthic invertebrates belonging to specific functional types following a benthic disturbance. These invertebrates interact with the sediment in specific ways. Because functional types are the biological units of interest, our definition does not demand a sequential appearance of particular invertebrate species or genera.' (Rhoads & Boyer, 1982).

An important aspect of using this successional approach to interpret benthic monitoring results is relating organism-sediment relationships to the dynamical aspects of end-member seres. This involves deducing dynamics from structure, a technique pioneered by R. G. Johnson (1972) for marine soft-bottom habitats. The application of an inverse methods approach to benthic monitoring involves making *in situ* measurements of salient structural features of the organism-sediment relationships shown in Fig. 1. This can be accomplished with sediment profile photography (Rhoads & Cande, 1971; Rhoads & Germano, 1982). The organism-sediment relationships illustrated in Fig. 1 can be

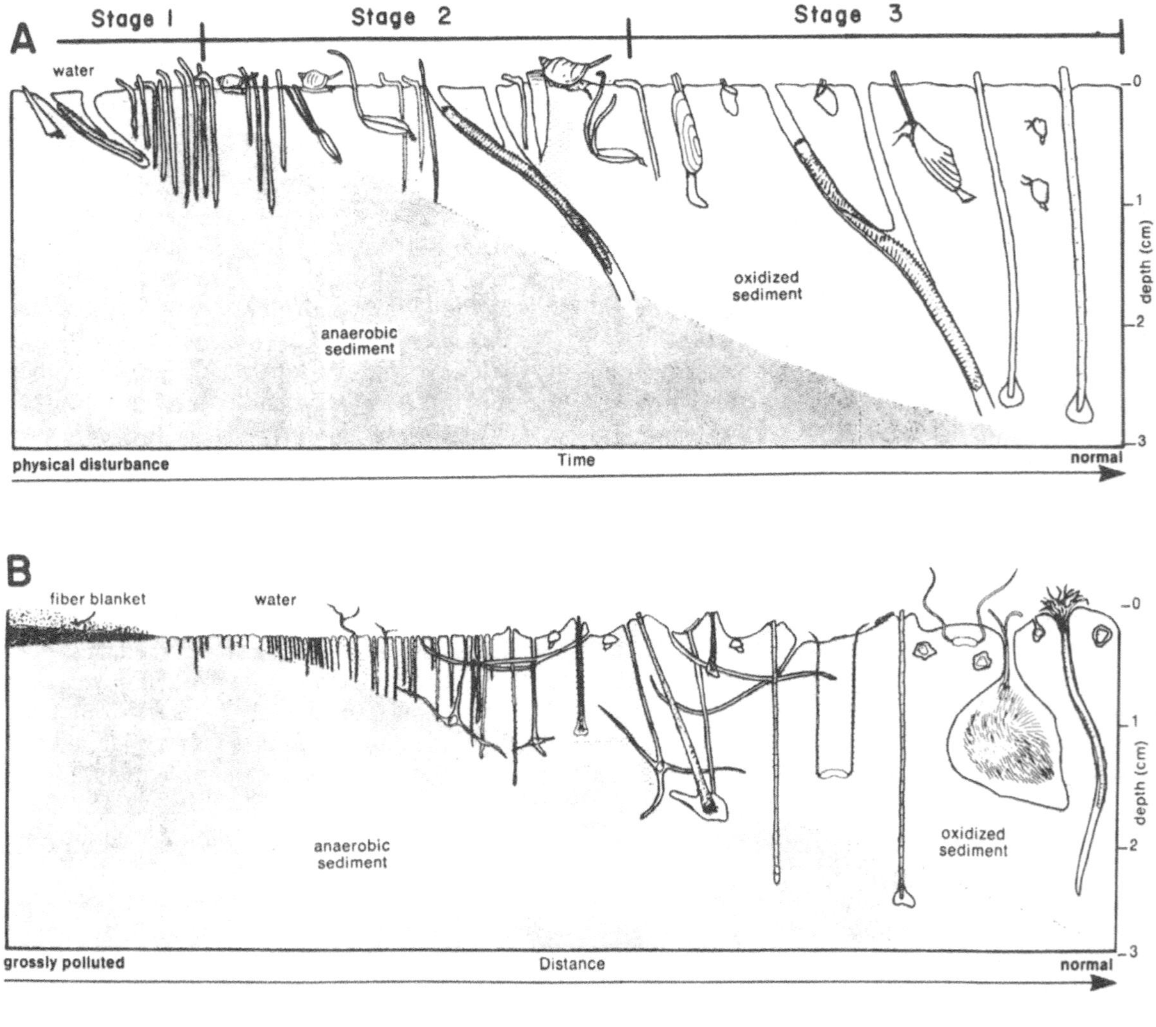

Fig. 1. (A) Development of organism-sediment relationships over time following a physical disturbance in Long Island Sound, USA. Subsequent mapping experience shows that the values on the depth axis also vary seasonally and with the dominant bioturbating taxa in the Stage III sere. Later work has also shown that in environments rich in labile detritus, Stage I polychaetes can co-exist with Stage II and Stage III seres. (B) Organism-sediment relationships associated with a pollution gradient associated with a pulp-mill effluent. Modified from Pearson and Rosenberg (1978). This figure is from Rhoads and Germano (1982). See text for additional discussion.

imaged *in situ* using this technique.

Long-term degradation of the benthic environment frequently involves the loss of equilibrium species (shown at the right-hand side of Fig. 1). These high-order seres are then replaced by pioneering seres (left-hand side of Fig. 1). Changes in organism-sediment relations and population dynamics accompanying this change are summarized below with examples (for a more complete discussion, the reader is referred to reviews by Pearson & Rosenberg, 1978; Rhoads & Germano, 1982; Rhoads & Boyer, 1982).

Many representatives of high-order seres (subsequently referred to as Stage III seres) are deeply burrowing errant or tube-dwelling infauna which feed head-down (conveyor-belt species *sensu* Rhoads, 1967). These infauna advect particles over vertical dimensions of several centimeters. The areas of subsurface feeding are recognized as water-filled pockets (feeding voids) which contain coarse sediment which apparently has been rejected in the feeding process. Dominants in these Stage III assemblages include, for example, maldanid, pectinid, and orbinid polychaetes, caudate holothurians, protobranch bivalves, some infaunal ophiuroids and irregular urchins (see above reviews for further discussion). These taxa typically have larger body sizes and longer mean life spans than early successional seres. The *in situ* mapping of Stage III seres requires the recognition of subsurface feeding pockets which identify the presence of these conveyor-belt species. Examples of Stage III seres from a wide geographic range of near-shore environments are shown in sediment-profile images in Fig. 2.

These stages are also associated with a thick near-surface zone of high-reflectance sediment related to deep bioturbation. This high reflectance layer is related both to the presence of ferric hydroxide coatings on particles and to low concentrations of pore-water sulphides. Although the presence of ferric hydroxide coatings on particles indicates that at some time these particles have been in a positive redox environment, one may not assume that pore waters are aerobic at the time of sampling from their presence alone unless microprobe redox measurements have been made (Revsbech *et al.*, 1979). Nevertheless, we have found measurements of the thickness of the apparent positive redox zone (as manifested in high sediment reflectance) to be correlated with the depth of infaunal bioturbation and useful in reconnaissance mapping of physical and biological disturbance gradients on the seafloor. The thickness of this zone can be expected to change seasonally in benthic environments where changing water temperatures affect rates of bioturbation. For this reason, maximum information from this reconnaissance mapping technique can be obtained during the summer and early autumn when bioturbation rates are high. Pearson & Stanley (1979) have also mapped redox depths as an efficient means of detecting pollution gradients.

The mapping of early or low-order successional stages, subsequently referred to as Stage I seres, requires the recognition of small tubiculous polychaetes or oligochaetes at the sediment surface. These may be very densely aggregated, particularly in eutrophic environments (Fig. 3). Stage I assemblages commonly are not quantitatively sampled in most benthic monitoring studies, particularly if the sieve mesh diameter that is used for processing samples is greater than 0.3 mm (Germano, 1983b). This successional sere is dominated by tubiculous worms that feed on labile detritus located at, or near, the sediment surface. Capitellid and spionid polychaetes are the commonly encountered members of this sere (Pearson & Rosenberg, 1978; Rhoads *et al.*, 1978); meiofauna may also dominate (Rudnick *et al.*, 1985). Because the exploited resources are located near the sediment surface, the feeding and bioturbation zone is thin. The thickness of the apparent redox zone therefore is also thin or may even be absent in habitats where dissolved oxygen is low in concentration (Fig. 3B).

The long-term recovery of a benthic habitat, starting at an azoic or Stage I sere, may involve a transitional stage (Stage II sere) before a Stage III (equilibrium) system is established. This may involve the appearance of a diverse assemblage of tubiculous amphipods, molluscs, and polychaetes. Some head-down feeders may appear at this stage, but a major fraction of the colonizing assemblage will tend to feed at, or near, the sediment-water interface. The thickness of the apparent positive redox layer is typically transitional between an early pioneering polychaete Stage I and a Stage III assemblage (Fig. 1). In our experience, Stage I taxa can persist and are commonly associated with Stage III seres. In the retrograde transition of

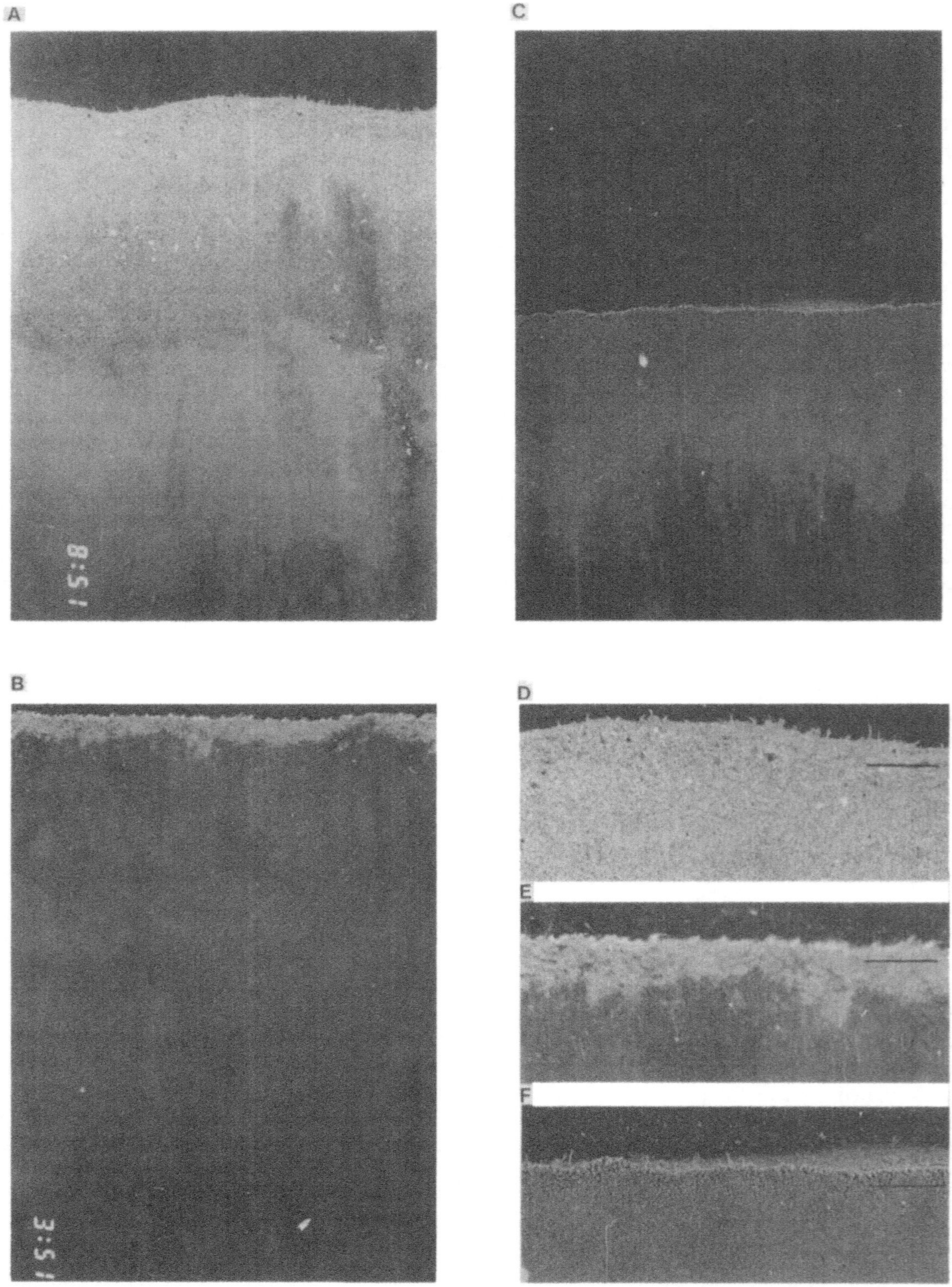

Fig. 3. Examples of Stage I seres from a wide range of subtidal sediments. Figures 3 A-C are 15 cm in width. Bar scales in D-F equal 1 cm. (A) A Stage I sere populating a sandy layer overlying mud on the inner continental shelf of the Gulf of Mexico. An enlarged view of these polychaetes is shown in (D). (B) A Stage I sere from western Long Island Sound showing a shallow apparent redox boundary, sulphidic sediment, and methane gas bubbles at the bottom of the photograph. An enlarged view of the surface is shown in (E). The profile camera was fitted with a dissolved oxygen probe located near the top of the optical window with a digital readout in the window. When this image was taken, the instantaneous dissolved oxygen reading 2 cm above the bottom was 2.9 mg/l. (C) A late Stage I sere from central Long Island Sound showing that the apparent redox boundary can be depressed to over 4 cm with this assemblage. An enlarged view is shown in (F).

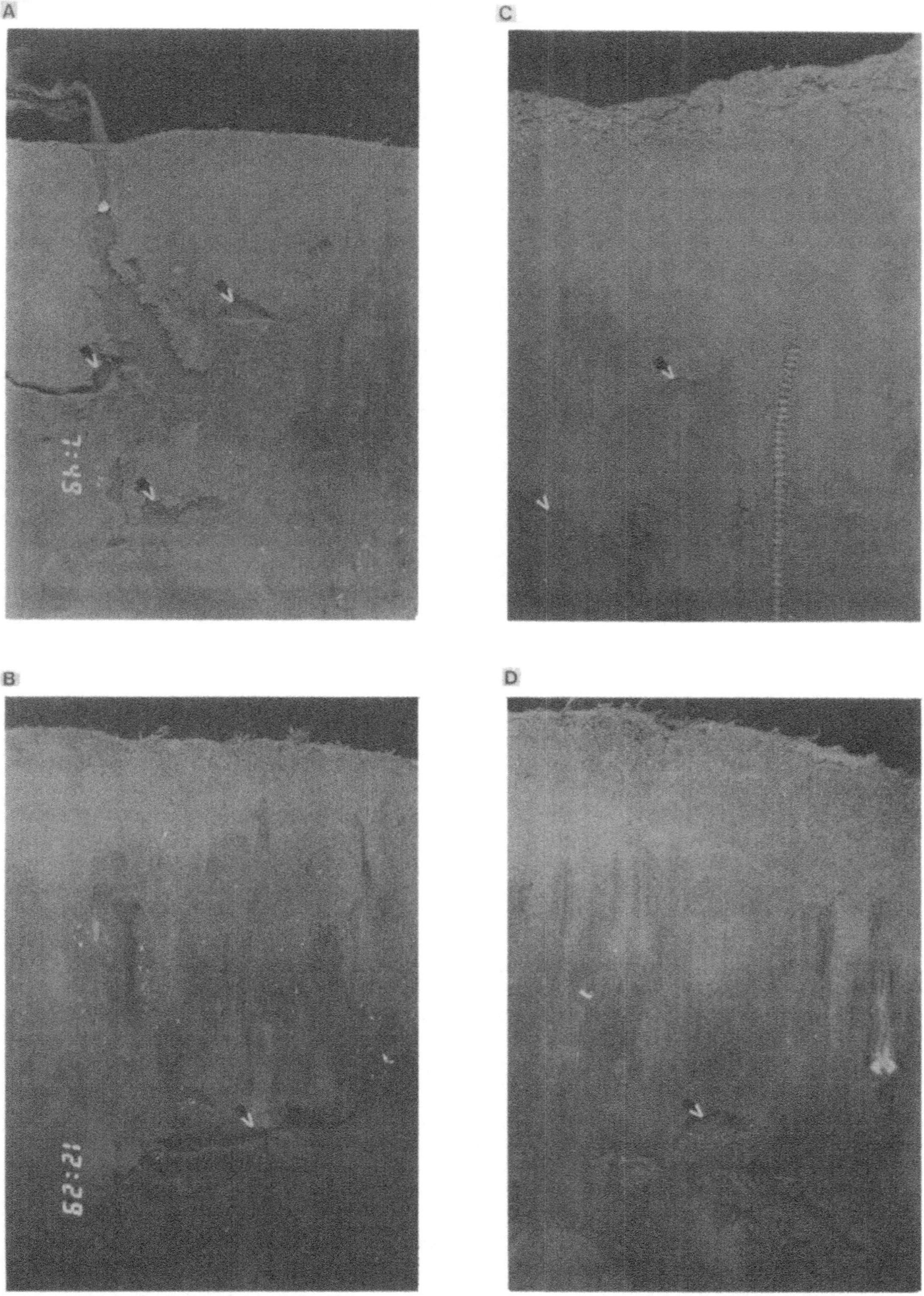

Fig. 2. Examples of Stage III seres from a wide range of subtidal muds. Arrows identify the presence of active feeding voids (i.e., open feeding pockets) produced by head-down conveyor-belt deposit feeders. The width of each photograph is 15 cm. (A) Subsurface feeding by maldanid polychaetes in the lower Chesapeake Bay. (B) A Stage III sere from the inner continental shelf of the Gulf of Mexico. Note Stage I taxa at the sediment surface. Species producing the subsurface feeding voids is not known. (C) Stage III feeding voids from the mouth of the Bay of Fundy. (D) A stage III community consisting of the bioturbating caudate holothurian *Molpadia oolitica* and maldanid polychaetes from Cape Cod Bay, Massachusetts. Note the presence of Stage I polychaetes at the surface (*Euchone incolor*). (D) has been reproduced from Rhoads & Young (1971).

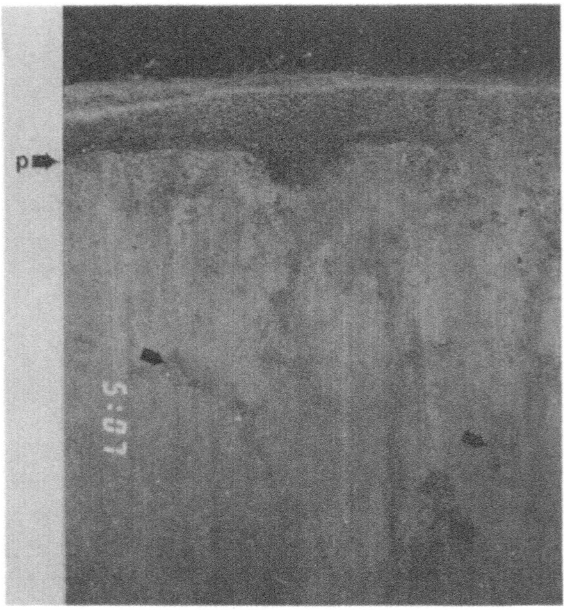

Fig. 4. A sediment profile image showing the record of a retrograde succession in the upper Chesapeake Bay. The symbol 'P' identifies the interface between a prehypoxic interval (below) and a hypoxic interval (above). The prehypoxic interval consists of high reflectance sediment (low pore-water sulphides) and the remnants of relict, i.e. collpased and inactive, feeding zones for Stage III taxa (arrows). The hypoxic interval consists of reduced fecal pellets of Stage I taxa.

Stage III to Stage I, it is sometimes possible to recognize the presence of relict (i.e. collapsed and inactive) feeding voids (Fig. 4).

Deducing dynamics from structure

The structural and spatial relationships of organisms to their sedimentary habitat, as described above, can be used to infer dynamical aspects about benthic secondary production, food web contamination, bottom water hypoxia, sediment reworking, and nutrient recycling. These structure-process relationships are summarized in Table 1.

Secondary Production: The relationship of the successional paradigm to secondary production is based on field work which has shown that populations of pioneering species have very high intrinsic rates of increase, and that annual or subannual disturbances may enhance secondary production by stimulating repopulation of newly opened space (Odum, 1969; Rhoads *et al.*, 1978; McCall & Tevesz, 1983). There is also some preliminary evidence that disturbances which promote the diffusion of hydrogen sulphide from bottom sediments into the water column may attract pioneering species to such disturbance sites (Cuomo, 1985). Because the biomass of pioneering seres is concentrated at, or near, the sediment surface, this productive stage

Table 1. Benthic ecosystem attributes associated with pioneering and late stage seres.

System attribute	Successional stage		Selected references
	Early (Stage I)	*Late (Stage III)*	
Secondary production	High potential for r-selected taxa	Lower potential for K-selected taxa	Odum (1969); Rhoads et al. (1978); McCall & Tevesz (1983).
Prey availability	High as prey are concentrated near the surface	Lower as infauna are deep burrowing*	Becker & Chew (1983); Smith (1985).
Potential for food-web contamination	Highest for suspended or recently sedimented particulates. Body burdens may be low related to short mean life-spans(?)	Highest for deeply buried contaminents. Longer mean life spans may lead to significant body burdens(?)	Bryan (1985); Lee & Swartz (1980); Swartz & Lee (1980); Phelps (1967).
Nutrient recycling	Limited to solutes in ≤ 3 cm	Solutes exchanged over distances to 20 cm or deeper	Aller (1978), (1980), (1982), (1985); Lee & Swartz (1980); Rice (1986).
Potential for bottom water hypoxia	High, storage systems for labile detritus	Low, a recycling or 'purging' system	Untested hypothesis

* Non-lethal predation of distal ends of siphons or caudal segments may be important for some predator species (DeVlas, (1979); Trevallion et al. (1970); Woodin (1982), (1984)).

may provide food for commercially important secondary consumers (Rhoads *et al.*, 1978; Becker & Chew, 1983).

Stage III seres, in contrast, are populated by species having relatively low ontogenetic and population growth rates. These infaunal species live cryptically within the sediment and commonly do not show significant seasonal changes in abundance or biomass (McCall & Tevesz, 1983). Stage III taxa may therefore prove to be less important as prey than Stage I species. However, Woodin (1985) has shown that non-lethal predation (presumably by demersal fish) of the caudal ends of maldanid polychaetes (a common Stage III taxon) can occur.

Once successional mosaics have been mapped, questions involving the importance of each sere to commercial fin-fisheries and potential pollutant transfer within the benthic food web can be investigated employing a Benthic Resources Assessment Technique (BRAT) (Lunz & Kendall, 1982). This involves quantitative sampling of demersal fish and potential invertebrate prey species from the different infaunal successional stages that are mapped. The size-class distribution of major taxa of ingested prey is determined from stomach contents and is compared with the size-class distribution of prey species as a function of depth in the sediment for each type of successional assemblage sampled. This is accomplished by separately sieving strata removed from vertically partitioned box cores (Fig. 5). Identification of the available zone or maximum feeding depth in the sediments used by a specific demersal predator can be accomplished by comparing relative prey size distribution patterns in a predator's diet with environmental size-depth patterns from the box core samples. These data can then be used to identify a subset of species for subsequent tissue analysis of both predators and their prey relative to the range of pollutants in local sediments that may be of concern. The success of such a BRAT analysis depends on the investigator's ability to establish the food spectrum of different year-classes of fish and seasonal or diel differences in foraging and prey availability.

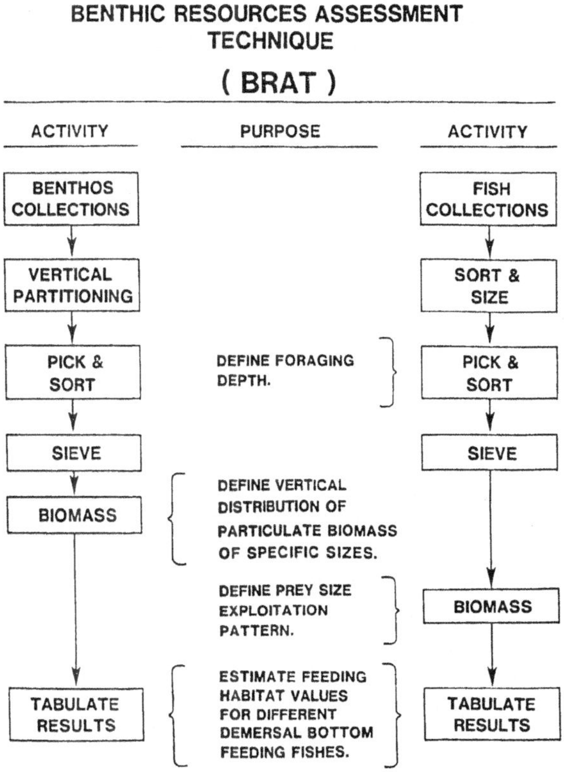

Fig. 5. A flow chart showing the operational procedures involved in relating the resource value of different benthic successional stages to commercially important predators through the BRAT technique; see Lunz & Kendall (1982) for additional details.

Food Web Contamination: Because the feeding and bioturbation depths of Stage I pioneering seres are not as great as Stage III seres, pioneering taxa are exposed to contaminants associated with the near sediment surface zone. Phelps (1967) showed that partitioning of stable elements (Fe, Zn, Sc, and Sm) within a benthic community could be related to the vertical stratification of feeding. Iron is preferentially partitioned into tissues of near surface feeders while Zn is concentrated in deep burrowers. Partitioning of toxic metals, radionuclides, and organic toxins may also be related to feeding depths, local redox conditions and quality of organic substrates.

The difference in mean life-spans between Stage I and Stage III taxa suggests that bioaccumulation of some pollutants may be less of a problem with early successional stages than with late stages. This hypothesis has not been adequately tested, because the small polychaetes or oligochaetes associated with disturbed habitats are commonly not sampled. In those studies where they have been sampled, tissue analysis has not been done because of the small size of individuals. Nevertheless, a general appreciation of these relationships is leading to the practice of capping polluted dredged materials with a clean layer of sediment that exceeds the mean bioturbation depth of local Stage III taxa (Sumeri, 1984). Additional field data concerning this issue are sorely needed, because existing laboratory data do not lend themselves to making long-term predictions (Bryan, 1985).

Nutrient Recycling: Recent work has shown that bioturbation is a quantitatively important process for moving water and dissolved constituents into and out of sedments; this process may be orders-of-magnitude greater than simple diffusion rates (Lee & Schartz, 1980; Aller, 1980; Aller, 1982). Sediment irrigation can be a very important issue in a management context when the change in benthic community structure ultimately affects primary production by controlling the flux rate of nutrients (N, particularly in the form of ammonium, P, and Si) from the bottom back into the water column. For example, sediment irrigation has been estimated to provide from 10 to 30% of the daily N and P requirements of phytoplankton in Long Island Sound (Aller, 1980). The control of pore-water chemistry by macrofaunal colonizers has been shown in detailed studies of early chemical diagenesis of disposed dredged material (Rhoads *et al.*, 1977). The early colonization phase is characterized by diffusional profiles of pore-water sulphate, ammonium, and phosphate. Near-surface regions of these profiles are progressively influenced by biogenic irrigation as colonization progresses. The flux rate of pore-water solutes is not only a function of the cumulative pumping rate of infauna, but the flux and reaction rates are also affected by three-dimensional diffusion geometry (Aller, 1982). Detailed descriptions of this phenomenon has been described for protobranch bivalves (Aller, 1978), and maldanid (Aller, 1978), capitellid (Aller & Yingst, 1985), orbiniid (Rice, 1986), and terebellid polychaetes (Aller & Yingst, 1978). Burrows and tubes effectively increase the three-dimensional surface area of diffusion boundaries. This geometry is best developed in Stage III seres and can be quantitatively important for recycling nutrients back to primary producers.

The relative importance of Stage I successional seres in nutrient cycling is not well documented. However, many pioneering polychaetes live in short, cylindrical tubes which may have compositional properties that serve to decrease diffusion coefficients across the tube wall (Tenore *et al.*, 1982: 747). The biogenically mixed zone is relatively shallow. Although sediments populated by pioneering seres may contain large inventories of nutrients, the return of these nutrients to the water column will be largely controlled by molecular diffusion rates. In this sense, sediments associated with pioneering seres have the potential to be storage systems, while those populated with Stage III seres may be considered as purging or recycling systems. From these relationships, we hypothesize that a long-term change in community structure involving the replacement of a Stage III sere with a pioneering stage may result in a decrease in bioturbational nutrient cycling from bottom sediments to primary producers in the water column.

Bottom-water Hypoxia: Eutrophication of coastal areas often results in the sedimentation of labile planktic detritus at a rate which exceeds the ability of the benthic system to metabolize this detritus by aerobic pathways. Excess labile organic matter therefore accumulates, producing a high demand for oxygen which (depending on oxygen supply rates) may result in bottom water hypoxia. This phenomenon is particularly acute in the summer period when the combined potential for high water temperatures, low oxygen solubility, little wind mixing, and water column stratification exists (Officer *et al.*, 1984). Although hydrographic factors are of first-order importance for the supply of oxygen, Stage III seres may play an important role in preventing the build-up of labile organic matter in sediments. Deep biogenic irrigation of the sediment column increases pore-water oxygen and maintains positive Eh conditions up to several centimeters below the surface (Lee & Swartz, 1980). Figure 6 shows that under subcritical input rates of labile detritus, Stage III macrofauna may be able to balance the potential deleterious effects of inputs of labile organic matter by stimulating microbial processes. According to Aller (1981), this is done by: 1) increasing particle surface areas, 2) grazing, which keeps microbial populations in log-phase growth, 3) flushing metabolites from pore waters, thereby keeping electron acceptors continually supplied by irrigation, particle reworking, and multidimensional diffusion, 4) secreting mucus to provide new reactive substrates, and 5) subducting labile surface organic matter to be metabolized at depth. A major research issue in benthic ecology that needs to be addressed is the determination of the critical organic loading rate for Stage III seres, i.e., the rate which causes their local extinction. Once this is known, the datum may prove valuable for managing anthropogenic inputs of nutrients and labile organics.

Once the critical organic loading rate is exceeded, Stage III taxa are locally eliminated, and organic enrichment species (*sensu* Pearson & Rosenberg, 1978) dominate (Fig. 7). This may be related to the accumulation of antibiotic decomposition products (Bader, 1954), low dissolved oxygen (Rhoads & Morse, 1971), sulphides, or a combination of these factors. As described earlier, the influence of Stage I seres on nutrient cycling and aeration of the sediment column may be very limited. Hence, sulphate reduction and methanogenesis predominate over oxidative metabolism. These metabolites, sulphides, ammonium, and methane contribute further to bottom water oxygen demand as they diffuse into the water column. The loss of a Stage III sere from an area and its replacement by a Stage I sere may be expected to be accompanied by a major change in both the depth and rate of biogenic processing of bottom sediments. Evidence of such a retrograde succession is an early warning sign of the potential for developing hypoxic or anoxic conditions in the near future. An important attribute of Stage III seres is that they apparently serve to prevent the accumulation of labile organic matter; Stage I seres, in contrast, tend to be associated with storage areas of high BOD organic matter. Dense tube mats associated with some pioneering seres may in themselves promote the sedimentation of fine organic particulates by changing flow conditions near the bottom (Lee & Swartz, 1980; Nowell *et al.*, 1981; Rhoads & Boyer, 1982). Spatial gradients between Stage I and Stage III seres can be very sharp, suggesting that once a critical organic loading rate is exceeded, Stage III taxa are completely eliminated (eg. Fig. 9C).

Methods of in situ *successional mapping*

In order to obtain *in situ* data about organism-sediment relationships, we have developed a new method of obtaining information about the seafloor. This has been accomplished with sediment-profile imaging. The instrument consists of an optical prism operated like an inverted periscope which cuts a vertical profile of up to 20 cm of the sediment column. This optical corer transects the bottom at a fall rate of about 6 cm sec^{-1} in order to minimize physical disturbance. The prism is filled with distilled or filtered water so that ambient bottom turbidity does not compromise image quality. Optical resolution is approximately 0.06 mm. The instrument is deployed from a vessel, and once lowered to the bottom, an image can be taken every 40 seconds. An example of a sediment-profile camera is shown in Fig. 8. A survey taken with such a sediment-profile camera can generate up to 200 images in one survey day. Efficient data analysis and recording is accomplished with computer digital image analysis, permitting systematic measurement of up to 22 variables per image in about 4–6 minutes. For mapping purposes, several of these parameters are considered together, expressed as an Organism-Sediment Index (Table 2). In some cases, tentative taxonomic identifications can be made from imaged epifaunal species or from diagnostic structures produced by infauna. More detailed descriptions of the instruments and the theory of image interpretation are given in Rhoads & Germano (1982) and Germano (1983a).

An example of how reconnaisance maps can be generated by a sediment-profile camera survey and

Table 2. Method of calculating the Organism-Sediment Index. This index reflects the apparent mean redox potential discontinuity depth in the sediment, the presence or absence of sedimentary methane, dissolved oxygen over the bottom, and successional stage.

Chemical parameters		Biological parameters	
Mean apparent RPD depth, (cm)	Index value	Successional stage (Primary succession)	Index value
0	0		
>0 –0.75	1	Azoic	−4
0.76–1.50	2	Stage 1	1
1.51–2.25	3	Stage 1–2	2
2.26–3.00	4	Stage 2	3
3.01–3.75	5	Stage 2–3	4
>3.75	6	Stage 3	5
Methane present	−2	(Secondary succession)	
No/low O_2	−4	Stage 1 on Stage 3	5
		Stage 2 on Stage 3	

Benthic index = Sum of chemical and biological index values.
Potential index range: −10 to +11.

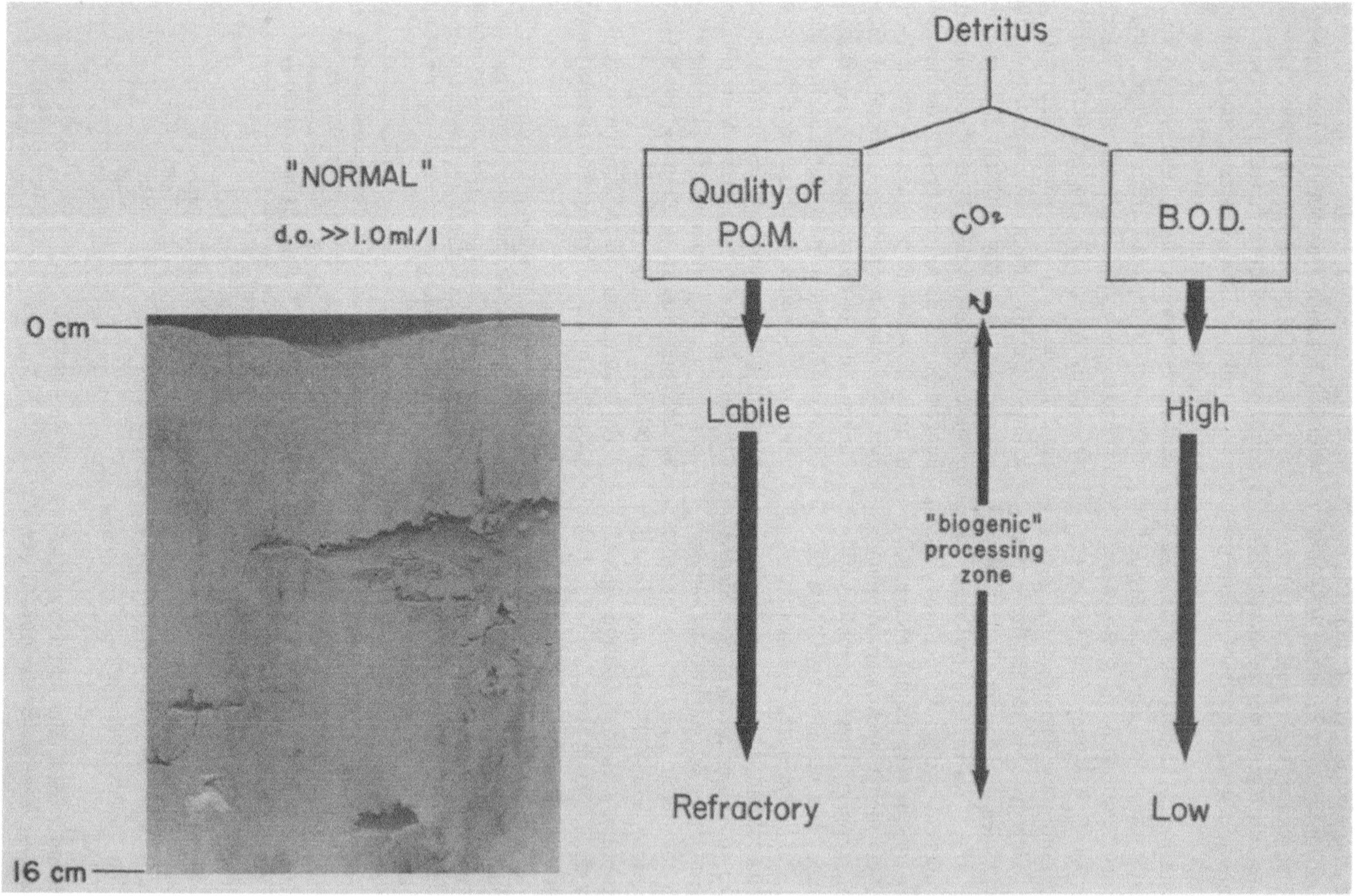

Fig. 6. Stage III seres appear to be able to prevent the build-up of pore-water nutrients and labile organic matter by their deep bioturbation and head-down feeding habit. This apparently can only take place below a critical organic loading rate, where the overlying water is maintained in an aerobic condition. This critical loading rate has not been documented for stage III systems. We infer these relationships from mapping enrichment gradients in the Chesapeake Bay and western Long Island Sound.

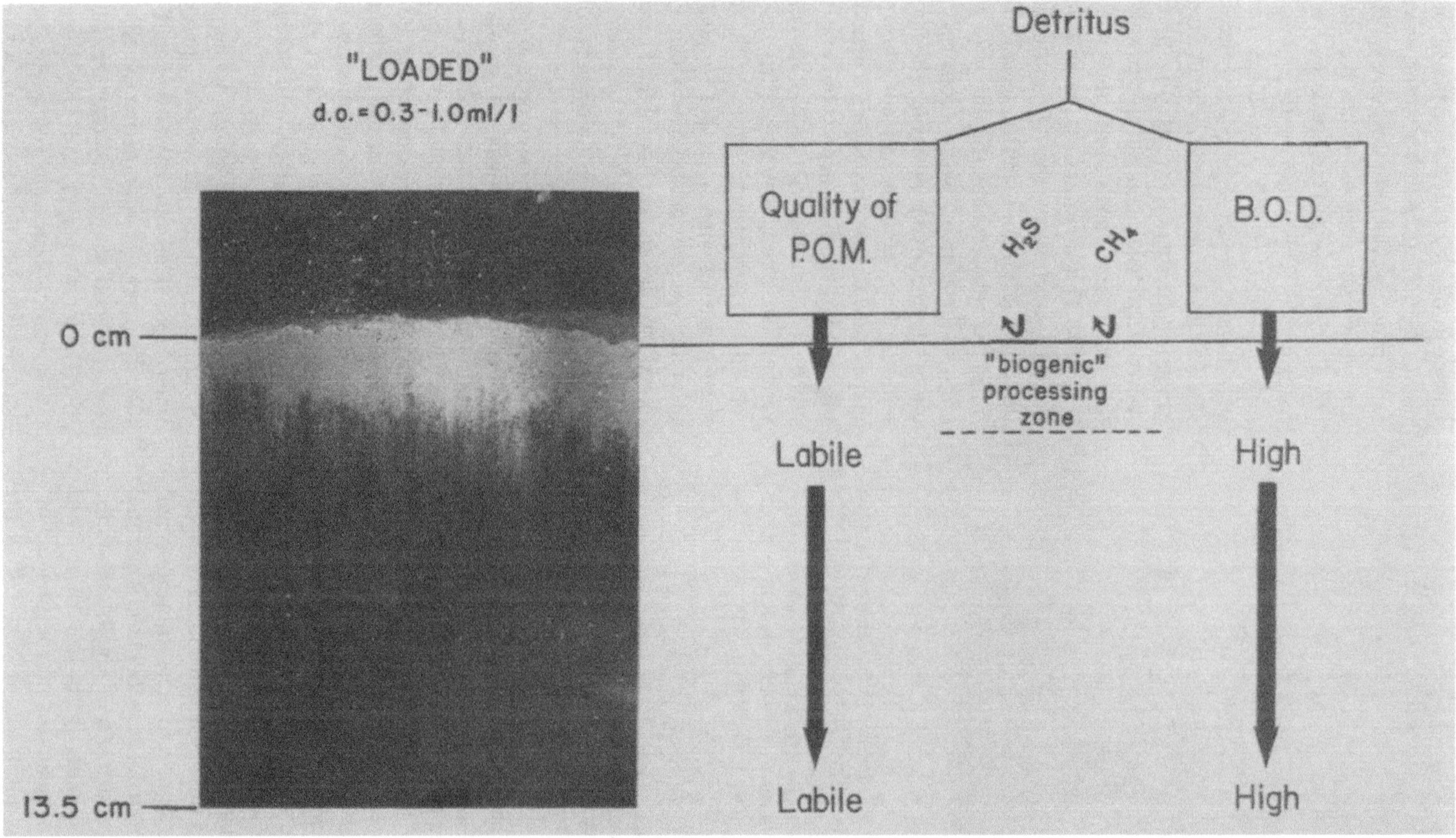

Fig. 7. Stage I seres are characterized by a shallow biogenic processing zone. This type of association is commonly well-developed in a partially aerated (dysaerobic) zone. Such bottom types may be periodically productive but the long-term storage of labile organic matter may result in anoxia and total loss of macrofaunal productivity.

Fig. 8. An example of a Sediment-Profile Camera. The arrow identifies a scale bar which equals 50 cm. The doors on the base are only used to deploy the instrument on very soft substrata. The optical prism is held above the bottom to avoid a pressure wave disturbance as the camera is lowered to the bottom. As tension is released on the winch wire, the inverted periscope cuts a vertical profile of the bottom at a slow rate of fall. Several station replicates are taken by pulling the camera a meter off the bottom and setting it back onto the bottom. Photographic sequencing is done automatically. This system is operable to 4000 meters. Photograph provided by Benthos, Inc., North Falmouth, Massachusetts.

used to make management decisions is shown in Fig. 9 for an 3.2 sq. km area in the lower Chesapeake Bay off the mouth of the Rappahannock River. This survey involved occupying 30 stations (3 replicates each) on a 6×5 orthogonal sampling grid. The survey was completed in one day and the maps were available to managers within 30 days. The purpose of the survey was to evaluate this area (as well as three others of similar size) as potential sites for dredged material disposal. The goals of the project were to find site(s) where disposed sediment would be contained (no sediment transport) and where the disposal operation would have a minimal impact on the ambient benthos. Fig. 9A shows the distribution of sediment grain-size determined through analysis of sediment-profile images; organic-rich silt-clays are in the western half of the area and fine to medium rippled sands are in the eastern half. The sands were observed in the profile images to overlay muds near the sand-mud facies boundary, indicating that a strong kinetic gradient exists from east (high) to west (low). Profile images also show that stations within the silt-clay facies contain methane and are associated with shallow redox depths. This is apparently an area of high organic loading (Fig. 9B). The silt-clay facies consists of a mosaic of azoic or Stage I seres, while the sandy areas are populated by either Stage I or both Stage I and Stage III seres (Fig. 9C). The contoured Organism-Sediment Index values indicate that the area consists of three major zones (Fig. 9D). This division is based on past survey experience which shows that Organism-Sediment Index values less than +6 reflect recently disturbed or otherwise stressed habitats.

From these maps, we can construct hypotheses about the underlying causes for the mapped distributions. The bathymetry, not shown on these maps, shows that the silt-clay area (zone A of Fig. 9D) lies within a depression, and sandy areas (zones B & C of Fig. 9D) occupy a topographic elevation. The strong kinetic gradient is apparently related to water depth. Azoic or Stage I seres in zone A are apparently related to organic enrichment and the seasonal hypoxia which occurs annually in the Chesapeake Bay. Stage I seres in zone C may be related to massive sediment movement which precludes the establishment of Stage III taxa.

From these maps we can conclude the following: 1) Containment of disposed dredged material would probably be assured in Zone A; it is a natural, low-kinetic, depositional area. 2) Impacts to the ambient benthic population would be minimized by disposal in Zone A. The existing benthic ecosystem is already stressed, as evidenced by the presence of Stage I seres or azoic conditions. 3) Filling-in of this low kinetic depression with dredged material may improve both the quality and productivity of zone A by bringing the bottom up above the depth of stratification for the seasonal hypoxic bottom water layer. This can be verified and impacts to fisheries assessed by comparing the potential of the site as a food source for demersal fish through a BRAT analysis before and after disposal operations. Standard benthic community analyses for documenting key prey species or for testing other hypotheses about bottom processes can also be per-

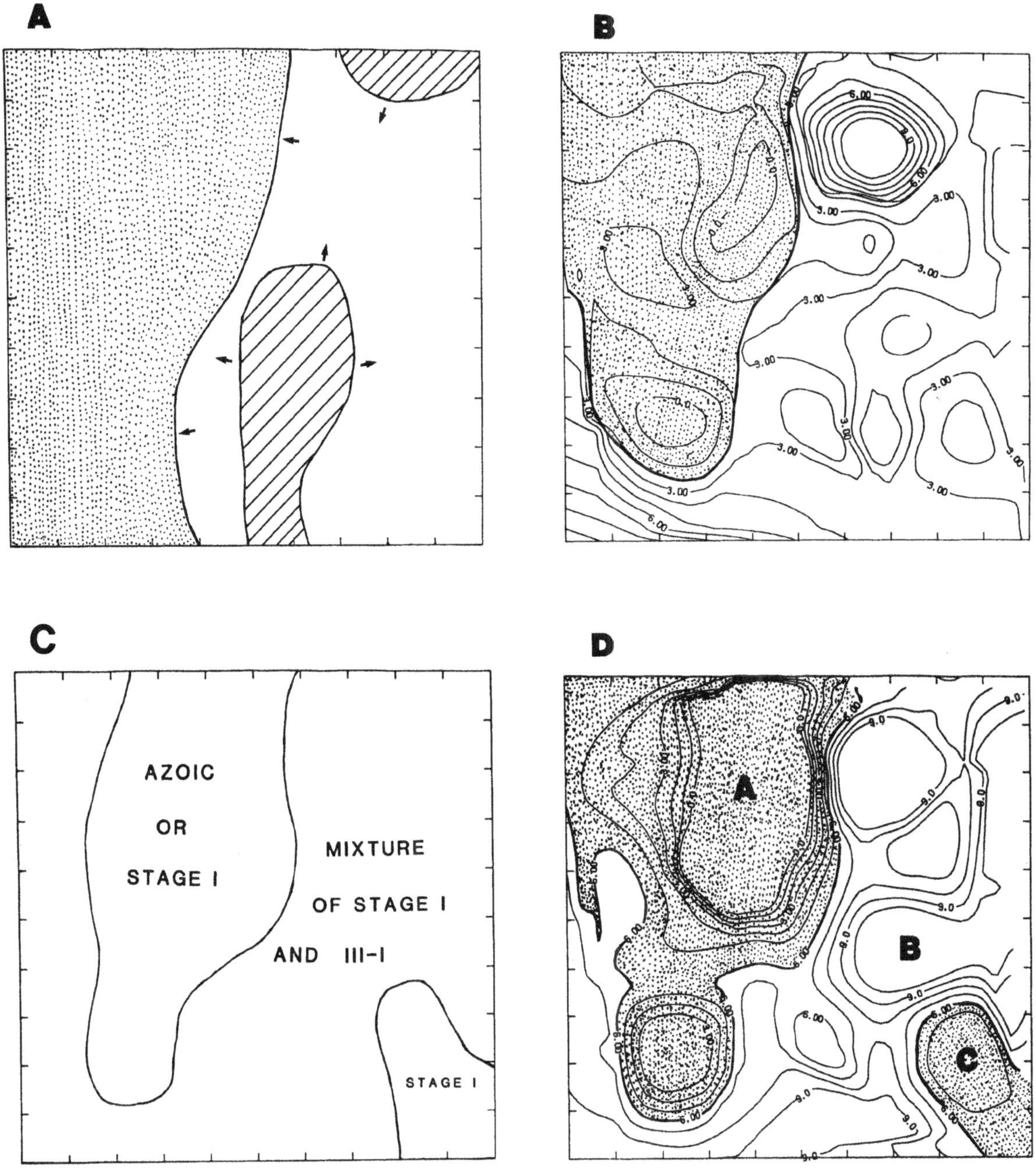

Fig. 9. An example of remotely acquired benthic data from a one-day reconnaisance mapping exercise using a sediment-profile camera. The four maps (A-D) are of the same 3.2 sq. km area in the lower part of the Chesapeake Bay, south of the Rappahannock River. Sampling was done in August, 1983 on a 6×5 station matrix. The mapped data are from three station replicates. (A) Modal grain-size shows the western half consists of silt-clay (>4 phi is stippled) and the eastern half consists of fine-sand (3−2 phi is white) or medium sand (2−1 phi is cross-hatched). Arrows show the direction of movement of the sand as deduced from profile images showing ripples and the superposition of sand-over-mud. (B) Contoured values of the mean apparent redox depth as measured from density-sliced images. Images showing the presence of thin (or no) apparent redox boundary or methanogenic sediment were obtained from the stippled area. (C) The distribution of successional seres as inferred from sediment-profile images. (D) The contoured values of Organism-Sediment Indices (see Table II). Values less than 6 are contained within stippled areas. For purposes of ground-truth sampling, this part of the seafloor consists or three major benthic facies (A, B, and C). See text for further explanation.

formed with the box core samples taken as part of the BRAT investigation.

Summary and conclusions

Four key major management questions have been identified regarding the significance of long-term changes in benthic community structure. These questions focus on how such change might potentially affect secondary production leading to a commercially important food web, the transfer of pollutants within the food web, the flux of N, P, and Si from the seafloor back into the water column, and the potential of the changed community to experience hypoxia. In order to address these questions efficiently, traditional monitoring methods and perspectives require revision.

An organism-sediment successional conceptual model has been used to predict the relative importance of pioneering (Stage I) seres relative to high-order (Stage III) seres on these management issues. Although these relationships are imperfectly known, existing data are sufficient to guide management decisions until future research provides more insight into these problems.

Organism-sediment dynamics may be inferred from structures observed in the upper 20 centimeters of fine-grained sediments. *In situ* imaging of these structures through remote sensing can be accomplished by utilizing a sediment-profile camera. Up to 200 images per day can be taken with such an instrument, allowing reconnaisance mapping of several square kilometers of seafloor in one survey data.

Reconnaisance maps of successional seres, depth of the apparent redox boundary, sediment modal grain-size, disturbance gradients inferred from Organism-Sediment Index values, information about erosional or depositional regimes, sedimentary methane, or bedforms can be prepared for large data sets within a few weeks after a survey. The effect of the documented changes in benthic community structure on higher food webs may be evaluated through a Benthic Resource Assessment Technique (BRAT) which relates benthic prey availability to commercially-important foraging predators. If extraordinary change is detected in organism-sediment relations, these reconnaisance maps may be used to locate stations for further documentation using traditional benthic sampling techniques. Continued reliance on traditional marine sampling techniques alone will merely perpetuate the 'after-the-fact' remedial management actions necessitated by the 3–12 month time lag on data return associated with many of these methods. Rapid return of data to both the scientist and environmental manager is important for detecting and evaluating changes in the benthic ecosystem so that monitoring data can be used for making real-time management decisions.

Acknowledgements

The authors wish to thank E. Revelas, J. Hill, R. Diaz, and D. Ward for their valuable assistance in the Chesapeake Bay survey. We are also indebted to J. Lunz and D. Kendall for their cooperation and support with data from the Benthic Resource Assessment Technique developed at the Waterways Experiment Station of the U.S. Army Corps of Engineers.

References

Aller, R. C., 1978. Experimental studies of changes produced by deposit feeders on pore water, sediment, and overlying water chemistry. Am. J. Sci. 278: 1185–1234.

Aller, R. C., 1980. Diagenetic processes near the sediment-water interface of Long Island Sound. I. Decomposition and nutrient element geochemistry (S, N, P). In B. Saltzman (ed.), Estuarine Physics and Chemistry: Studies in Long Island Sound. Advances in Geophysics, Academic Press, New York; London; Toronto; Sydney; San Francisco: 238–350.

Aller, R. C., 1982. The effects of macrobenthos on chemical properties of marine sediments and overlying waters. In P. L. McCall & M. J. S. Tevesz (eds), Animal-Sediment Relations: The Biogenic Alteration of Sediments. Plenum Press, New York: 53–102.

Aller, R. C. & J. Y. Yingst, 1978. Biogeochemistry of tube-dwellings: a study of the sedentary polychaete *Amphitrite ornata* (Leidy). J. mar. Res. 36: 201–254.

Aller, R. C. & J. Y. Yingst, 1985. Effects of the marine deposit-feeders *Heteromastus filiformis* (polychaeta), *Macoma balthica* (bivalvia), and *Tellina texana* (bivalvia) on averaged sedimentary solute transport, reaction rates, and microbial distributions. J. mar. Res. 43: 615–645.

Bader, R. G., 1954. The role of organic matter in determining the distribution of pelecypods in marine sediments. J. mar. Res. 13: 32–47.

Becker, D. S. & K. K. Chew, 1983. Fish-benthos coupling in sewage enriched marine environments. NOAA Final Report, Project NA80RAD00050, School of Fisheries, University of Washington, Seattle, Washington: 78 pp.

Bryan, G. W., 1985. Bioavailability and effects of heavy metals in marine deposits. In B. H. Ketchum, J. M. Capuzzo, W. V. Burt, I. W. Duedall, P. K. Park & D. R. Kester (eds), Wastes in the Ocean. John Wiley & Sons, New York: 41–79.

Cuomo, M. C., 1985. Sulphide as a larval settlement cue for *Capitella* sp I. Biogeochemistry 1: 169–181.

De Vlas, J., 1979. Annual food intake by plaice and flounder in a tidal flat area in the Dutch Wadden Sea, with special reference to consumption of regenerating parts of macrobenthic prey. Neth. J. Sea Res. 13: 117–153.

Germano, J. D., 1983a. High resolution sediment profiling with Remots® camera system. Sea Technology 24: 35–41.

Germano, J. D., 1983b. Infaunal succession in Long Island Sound: animal-sediment interactions and the effects of predation. Unpublished PhD. Thesis, Yale University, New Haven, Ct.: 142 pp.

Gould, S. J., 1981. The Mismeasure of Man. W. W. Norton & Company, N.Y.: 352 pp.

Gray, J. S., 1976. Are marine base-line surveys worth while? New Scientist 70: 219–221.

Green, R. H., 1979. Sampling design and statistical methods for environmental biologists. J. Wiley & Sons, N.Y.: 257 pp.

Green, R. H., 1984. Some guidelines for the design of biological monitoring programs in the marine environment. In H. H. White (ed.), Concepts in Marine Pollution Measurements. University of Maryland, College Park, (Maryland): 647–655.

Johnson, R. G., 1972. Conceptual models of benthic marine communities. In T. J. M. Schopf (ed.), Models in Paleobiology. Freeman, Cooper, and Co. San Francisco: 148–159.

Lee, H. & R. C. Swartz, 1980. Biological processes affecting the distribution of pollutants in marine sediments. Part II. Biodeposition and bioturbation. In R. A. Baker (ed.), Contaminants and Sediments. Ann Arbor Science Publishers, Ann Arbor, Mich. 2: 555–606.

Lunz, J. D. & D. R. Kendall, 1982. Benthic Resources Assessment Technique, a method for quantifying the effects of benthic community changes on fish resources. Conference Proceedings of the Marine Pollution Sessions, Oceans '82, NOAA, Office of Marine Pollution Assessment, Rockville, Maryland: 1021–1027.

McCall, P. L. & M. J. S. Tevesz, 1983. Soft-bottom succession and the fossil record. In M. J. S. Tevesz & P. L. McCall (eds), Biotic Interactions in Recent and Fossil Benthic Communities. Plenum Press, New York; London: 157–194.

Nowell, A. R. M., P. A. Jumars & J. E. Eckman, 1981. Effects of biological activity on the entrainment of marine sediments. Mar. Geol. 42: 133–153.

Officer, C. B., R. B. Biggs, J. L. Taft, L. E. Cronin, M. A. Tyler & W. R. Boynton, 1984. Chesapeake Bay anoxia: origin, development, and significance. Science 223: 22–27.

Odum, E. P., 1969. The strategy of ecosystem development. Science 16: 262–270.

Pearson, T. H. & R. Rosenberg, 1978. Macrobenthic succession in relation to organic enrichment and pollution of the marine environment. Oceanogr. mar. Biol. A. Rev. 16: 229–311.

Pearson, T. H. & S. D. Stanley, 1979. Comparative measurement of the redox potential of marine sediments as a rapid means of assessing the effect of organic pollution. Mar. Biol. 53: 371–379.

Peet, R. K., 1974. The measurement of species diversity. Ann. Rev. Ecol. Syst. 5: 285–307.

Phelps, D. K., 1967. Partitioning of the stable elements Fe, Zn, Sc, and Sm within a benthic community, Anasco Bay, Puerto Rico. In B. Aberg & F. P. Hungate (eds), Radioecological Concentration Processes. Pergamon Press, New York: 721–734.

Revsbech, N. P., B. B. Jørgensen & T. H. Blackburn, 1979. Oxygen in the sea bottom measured with a microelectrode. Science 207: 1355–1356.

Rhoads, D. C., 1967. Biogenic reworking of intertidal and subtidal sediments in Barnstable Harbor and Buzzards Bay, Massachusetts. J. Geol. 75: 461–476.

Rhoads, D. C. & L. F. Boyer, 1982. The effects of marine benthos on physical properties of sediments. In P. L. McCall & M. J. S. Tevesz (eds), Animal-Sediment Relations. Plenum Press, New York; London: 3–52.

Rhoads, D. C. & S. Cande, 1971. Sediment profile camera for *in situ* study of organism-sediment relations. Limnol. Oceanogr. 16: 110–114.

Rhoads, D. C. & J. D. Germano, 1982. Characterization of organism-sediment relations using sediment profile imaging: an efficient method of remote ecological monitoring of the seafloor (REMOTS® System). Mar. Ecol. prog. Ser. 8: 115–128.

Rhoads, D. C. & J. W. Morse, 1971. Evolutionary and ecological significance of oxygen-deficient marine basins. Lethaia 4: 413–428.

Rhoads, D. C. & D. K. Young, 1971. Animal-sediment relations in Cape Cod Bay, Massachusetts. II. Reworking by *Molpadia oolitica* (Holothuroidea). Mar. Biol. 11: 255–261.

Rhoads, D. C., R. C. Aller & M. Goldhaber, 1977. The influence of colonizing benthos on physical properties and chemical diagenesis of the estuarine seafloor. In B. C. Coull (ed.), Ecology of the Marine Benthos. Belle Baruch Library in Marine Sciences, Univ. S. Carolina Press, Columbia, S.C. 6: 113–138.

Rhoads, D. C., P. L. McCall & J. Y. Yingst, 1978. Disturbance and production on the estuarine seafloor. Am. Sci. 66: 577–586.

Rice, D. L., 1986. Early diagenesis in bioadvected sediments: relationships between the diagenesis of Beryllium-7, sediment reworking rates, and the abundance of conveyor-belt deposit-feeders. J. mar. Res. 44: 149–184.

Routledge, R. D., 1980. Bias in estimating the diversity of large, uncensused communities. Ecology 61: 276–281.

Rudnick, D. T., R. Elmgren & J. B. Frithsen, 1985. Meiofaunal prominence and benthic seasonality in a coastal marine ecosystem. Oecologia 67: 157–168.

Smith, L. D., 1985. Juvenile spot (*Leiostomus xanthurus*) predation on meiofauna in muddy and sandy substrates. Unpublished M. S. Thesis, Univ. S. Carolina, Columbia, S.C.: 46 pp.

Sumeri, A., 1984. Operational aspects of capped in-water disposal of contaminated dredged material. Environmental Effects of Dredging, Waterways, Experiment Station, Vicksburg, D-84-5.

Swartz, R. C. & H. E. Lee II, 1980. Biological processes affecting the distribution of pollutants in marine sediments. Part I. Accumulation, trophic transfer, biodegradation and migration. In R. A. Baker (ed.), Contaminants and Sediments. Ann Arbor Science Publishers, Ann Arbor, Mich. 2: 533–553.

Tenore, K. R., L. F. Boyer, J. Corral, C. Garcia-Fernandez, N. Gonzalez, E. G. Gurrian, R. B. J. Iglesias, M. Krom, E.

Lopez-Jamar, J. McClain, M. Pamatmat, A. Perez, D. C. Rhoads, R. M. Rodriguez, G. Santiago, J. Tietjen, J. Westrich & H. L. Windom, 1982. Coastal upwelling in the Rias Bajas, NW Spain: contrasting the benthic regimes of the Rias de Arosa and de Muros. J. mar. Res. 40: 701–772.

Trevaillon, A., R. R. C. Edwards & J. H. Steele, 1970. Dynamics of a benthic bivalve. Steele (ed.), Marine Food Chains. Univ. of California Press, Berkeley, Ca.: 285–295.

Wolda, H., 1981. Similarity indices, sample size, and diversity. Oecologia 50: 296–302.

Woodin, S. A., 1982. Browsing: Important in marine sedimentary environments? Spionid polychaete examples. J. exp. mar. Biol. Ecol. 60: 35–45.

Woodin, S. A., 1984. Effects of browsing predators: Activity changes in infauna following tissue loss. Biol. Bull. 166: 558–573.

Hydrobiologia 142: 309–325 (1986)

Environmental influences on long-term variability in marine plankton

J. M. Colebrook
Natural Environment Research Council, Institute for Marine Environmental Research, Prospect Place, The Hoe, Plymouth, United Kingdom

Keywords: plankton, north-east Atlantic, seasonal cycles, annual variations, climate

Abstract

The major patterns of geographical and seasonal variability of the plankton of the north-east Atlantic and the North Sea are described to provide the background to a presentation of the dominant patterns of year-to-year fluctuations in the abundance of the plankton of the area for the period 1948 to 1984. A feature of the variability is a marked similarity both between species and between areas. The main pattern of year-to-year change has the form of a quasi linear downward trend in abundance with, superimposed on this, an element of variability with a periodicity of about three years. There is a complex relationship between the plankton and an estimate of changes in the frequency of westerly weather which can be interpreted in terms of influences acting over limited periods of the seasonal cycle coupled with persistence in the stocks of zooplankton. Relationships between year-to-year variations in the abundance of phytoplankton and zooplankton can be interpreted in terms of a response by the zooplankton to variations in food supply coupled with feed-back to the phytoplankton involving in situ nutrient regeneration.

Introduction

The plankton is of obvious interest to this meeting as it provides an important source of supply of carbon energy to the benthos, and the information about the plankton derived from the Continuous Plankton Recorder survey will, it is hoped, be of particular interest in the context of the COST 647 project because these data provide information about long-term changes, over a period of several decades, in the abundance and distribution of the plankton. It will become apparent that it is not yet possible to give detailed explanations, in terms of mechanisms and processes, of the causes of these long-term changes. However some of the environmental influences have been identified in the realm of long-term changes in climate, and it is possible to say something about the nature of the response of the plankton to such changes.

The main patterns of geographical distribution of the plankton of the North Atlantic ocean, the European shelf and the North Sea will be reviewed briefly and seasonality will be considered in greater detail as this is a very important element in the dynamics of the plankton in temperate waters. Geography provides a more or less static pattern while seasonality provides a rapidly changing pattern and it is within both of these elements that the long-term changes in the abundance of the plankton have their origin.

Methods

A routine, monthly, synoptic survey of the plankton of the North Atlantic Ocean and the North Sea has been carried out since 1948 using Continuous Plankton Recorders (Fig. 1) towed by merchant ships and Ocean Weather Ships on regular routes (Fig. 2). Samples are collected at the standard depth of 10 m and, as far as possible, the organisms are identified to species, however, for

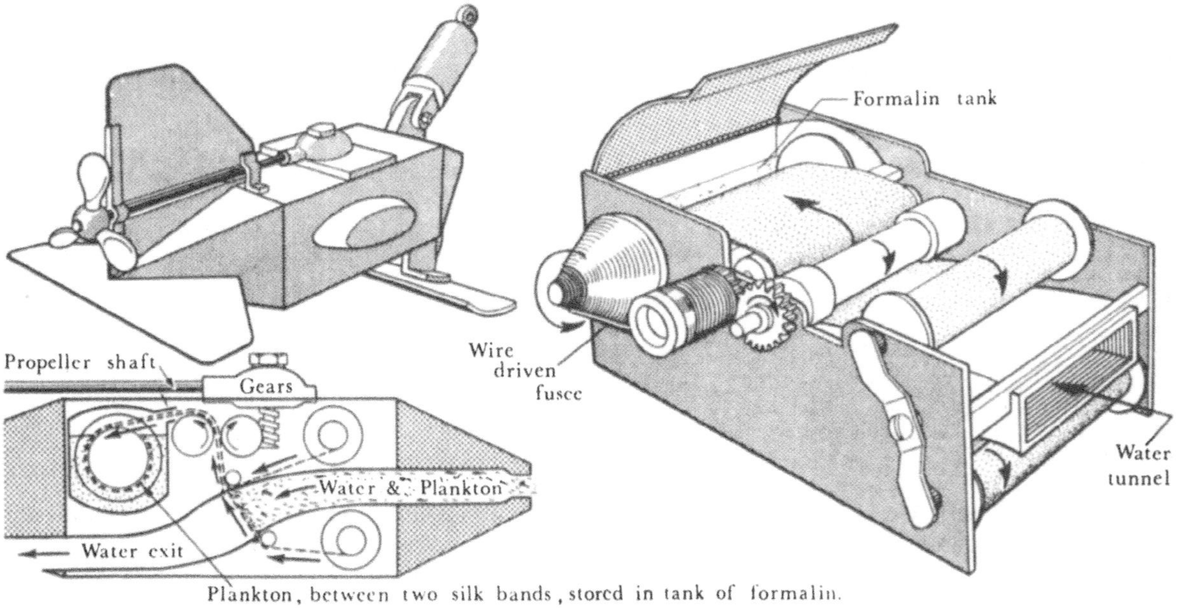

Fig. 1. The Continuous Plankton Recorder. The whole towed vehicle is shown top left. The filtering mechanism which fits inside the vehicle is shown on the right and a schematic diagram of how it works is shown bottom left.

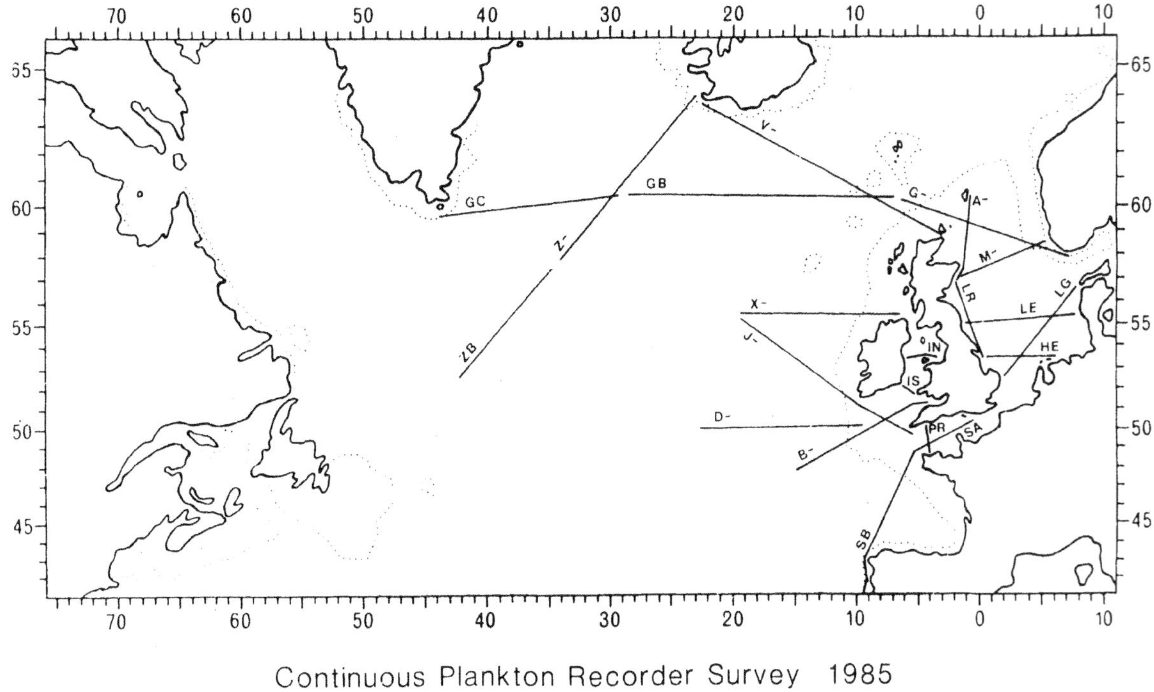

Fig. 2. A chart of the routes along which Continuous Plankton Recorders were towed in 1985. As far as possible, tows are made at monthly intervals along each route.

some taxonomic groups, identification is to genus, family or even higher categories (Rae, 1952, Colebrook, 1960). The data from the survey provides information about the geographical, seasonal and year-to-year changes in the abundance of the plankton (Glover, 1967).

Geographical distributions

There are six main patterns of geographical distribution which embrace most of the holoplanktonic species of the North Atlantic (Colebrook, 1972). These can be interpreted in terms of north-south and oceanic-neritic coordinates. Starting with a clearly neritic pattern (Fig. 3), these species tend to be confined to the shallow seas and shelf waters. Two groups of species show distributions intermediate between neritic and oceanic. One (Fig. 3) showing a spread from the North Sea into southern

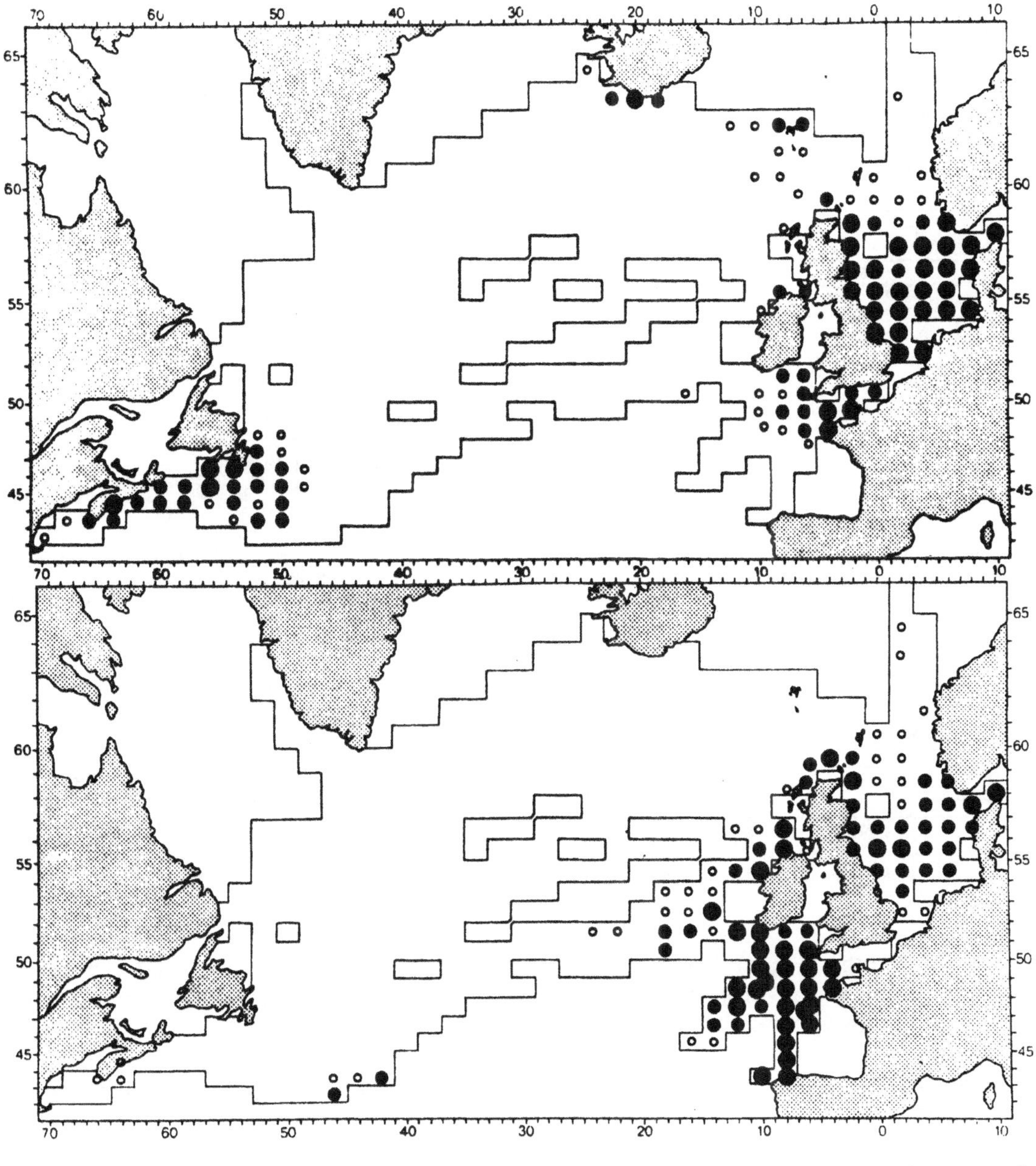

Fig. 3. Patterns of geographical distribution. Top: neritic, bottom: south-east intermediate. The method of derivation of these diagrams is described in Colebrook (1972).

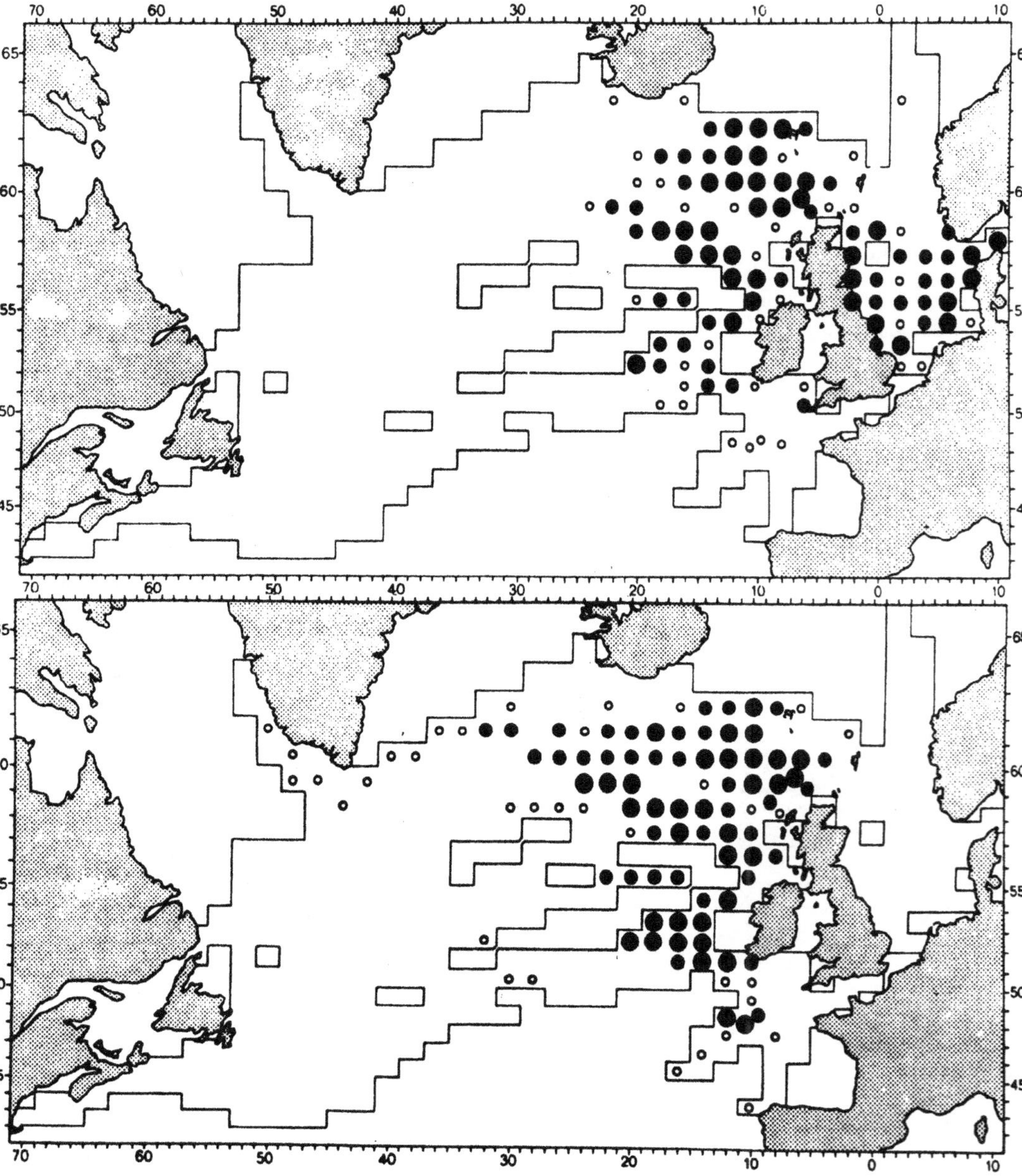

Fig. 4. Patterns of geographical distribution. Top: north-east intermediate, bottom: north-east oceanic.

oceanic waters, the other (Fig. 4) showing a spread into northern oceanic waters. There are three more strictly oceanic groups with distributions centered in north-eastern, north-western and southern waters respectively (Figs 4 & 5).

There are two anomalies in these groups of species. The north-eastern oceanic group consists, with one exception, of phytoplankton species only, while the southern oceanic group consists, almost exclusively, of zooplankton. The group contains a number of rare species of phytoplankton, mostly dinoflagellates, but no abundant ones.

It must be stressed that these distributions reflect differences in relative abundance. They are not based on clear presence or absence boundaries and the anomalies do not imply any significant im-

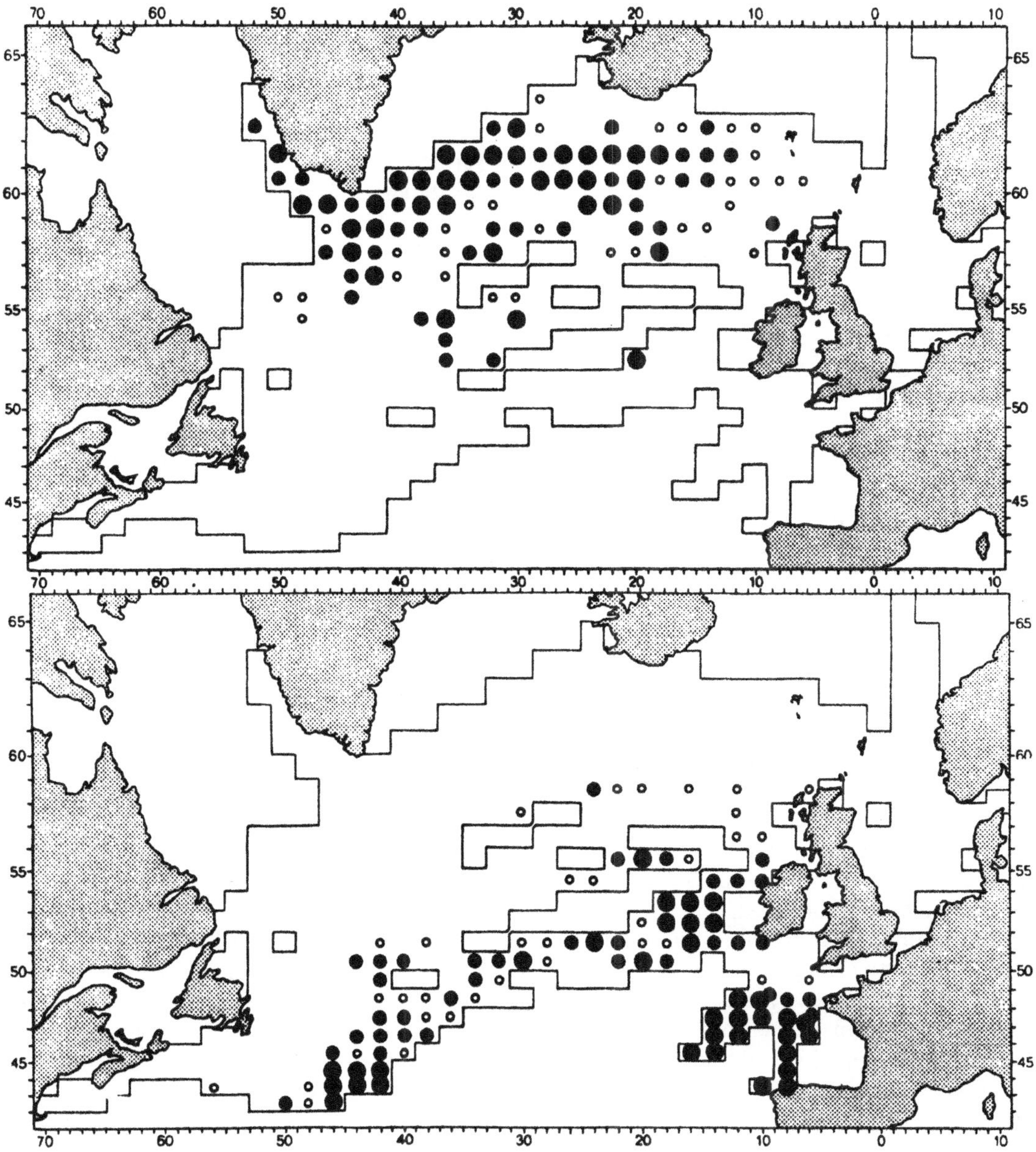

Fig. 5. Patterns of geographical distribution. Top: north-west oceanic, bottom: southern oceanic.

balance in the abundance of zooplankton in relation to phytoplankton in these areas. The north-east oceanic anomaly can be explained in the context of seasonal variations but the reason for the lack of abundant phytoplankton species showing southern oceanic distributions is, as yet, not clear.

Seasonal variations

As might be expected for temperate waters and organisms with relatively short life cycles, there are considerable seasonal variations in the abundance of nearly all the species (Colebrook, 1979, 1982b,

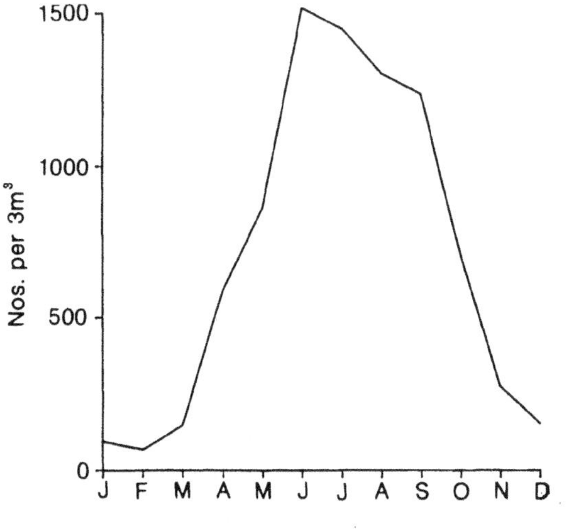

Fig. 6. A graph of the average (1948 to 1984) seasonal variations in the abundance of the total of all copepod species for the central North Sea (areas C1 and C2 in Fig. 10).

1984). As an example, Fig. 6 shows the seasonal variations in the total of all the species of copepods in the central North Sea.

The timing of the seasonal peaks in abundance vary from species to species. Fig. 7 contains a contoured plot showing the seasonal variations in the abundance of 24 species of zooplankton, the timing of the peak abundance ranges more or less continuously from May for the earliest species to October-November for the latest.

Fig. 7 also contains a similar diagram for the phytoplankton. In this diagram there is a clear discontinuity in the sequence, indicating a clear differentiation between those species associated with the spring bloom, with peaks in abundance in April and May, and those with peaks later in the year, from July to October. Graphs of the average seasonal cycles of each of these groups of species are also included in Fig. 7.

The phytoplankton also show marked geographical variations in the pattern of the seasonal cycle. Fig. 8 contains graphs, for each of the areas given in the chart in Fig. 10, of the average of the spring and summer groups of species. The most significant differences between the seasonal cycles can be seen clearly in the middle row of graphs. Comparing the oceanic area (C5) with the North Sea areas (C2 & C1) it is clear that the spring bloom is about a month earlier in the North Sea than in the ocean, while the summer species start to increase earlier in the open ocean than in the North Sea.

Both of these differences can be attributed to temperature. The onset of the spring bloom is determined to a large extent by the establishment of vertical thermal stratification associated with vertical stability of the water column and this happens earlier in the relatively shallow waters of the North Sea. At the same time, in the winter and early spring, the open ocean waters are warmer than those of the North Sea and once growth starts it is faster there than in the North Sea.

This can be seen very clearly in the diagram in Fig. 9 (left) which is a contour diagram based on a table of data with a column for each area and a row for each month. The data in the diagram refer to the spring group of species but the data for the summer group looks much the same. In the winter, stocks are higher in the North Sea than in the open ocean. As soon as growth starts there is a rapid shift in the centre of distribution from the North Sea in March, through the shelf in April and out into the open ocean by May, reflecting the faster growth in this area.

The data presented in this and the previous diagram suggests that the phytoplankton achieve something approaching a steady state in their response to the seasonal variations in temperature and stratification. In marked contrast to this most of the zooplankton show much less pronounced seasonal variations in geographical distribution. The diagram, in Fig. 9 (right), has the same format as that for the phytoplankton but the data are averages for all the abundant zooplankton species. It is clear that the range of seasonal variation in geographical distribution is much less than that of the phytoplankton. To quote from Colebrook (1984), in which this diagram first appeared. 'It is suggested, therefore, that the limited seasonal variation in geographical distribution shown by most zooplankton is due to inherent limitations in population growth rates, and that the pattern of distribution of zooplankton is strongly influenced by the distributions of overwintering stocks'.

These limitations provide an explanation for the virtual absence of zooplankton species showing north-eastern oceanic distributions. In this area the overwintering stocks are low and therefore the general levels of stocks are relatively low. There are two exceptions, the copepod *Calanus finmarchicus*,

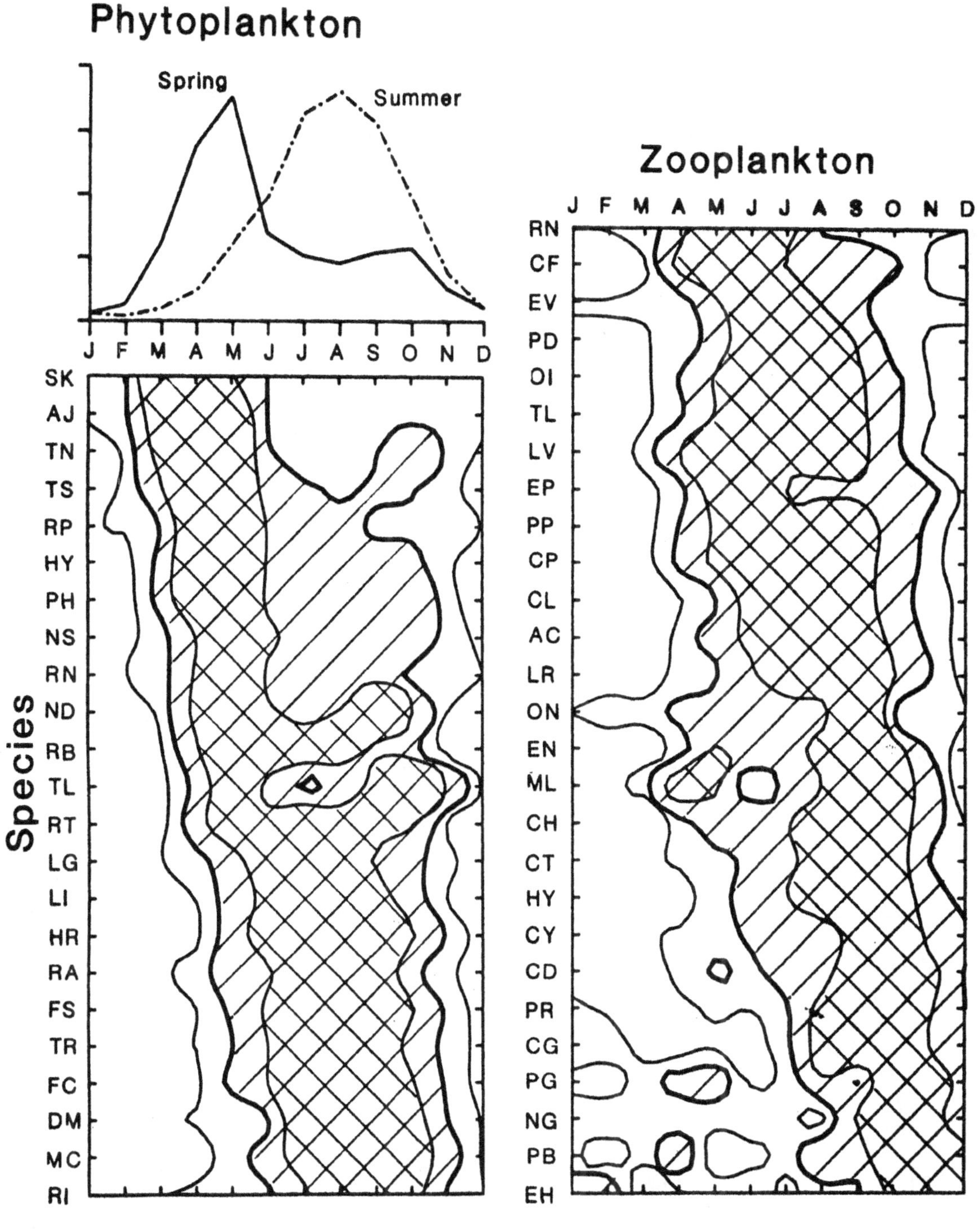

Fig. 7. Top left: graphs of the average (1959 to 1984) seasonal cycles of abundance of phytoplankton for all the areas shown in Fig. 10 with the species differentiated into spring and summer groups. The y-axis scale is arbitrary.

The contour diagrams illustrate the seasonal variations in the timing of the seasonal cycles of species of phytoplankton (left) and zooplankton (right). The species are ranked in the order of the timing of their seasonal cycles with early species at the top and late species at the bottom. Differences in abundance between species have been removed by transforming the data for each species to a uniform, arbitrary scale. A key to the species is given in Table 1.

Table 1. Alphabetical key to species for Fig. 7.

Asterionella japonica	AJ	*Acartia clausi*	AC
Ceratium furca	FC	*Calanus finmarchicus,* V-VI	CF
Ceratium fusus	FS	*Calanus helgolandicus,* V-VI	CH
Ceratium horridum	HR	*Candacia armata*	CD
Ceratium lineatum	LI	*Centropages typicus*	CT
Ceratium longipes	LG	Chaetognatha	CG
Ceratium macroceros	MC	*Clione limacina*	CL
Ceratium tripos	TR	*Corycaeus* spp.	CY
Chaetoceros (Hyalochaete) spp.	HY	*Euchaeta hebes*	EH
Chaetoceros (Phaeoceros) spp.	PH	*Euchaeta norvegica*	EN
Dactyliosolen mediterraneus	DM	Euphausiacea	EP
Nitzschia delicatissima	ND	*Evadne* spp.	EV
Nitzschia seriata	NS	Hyperiidea	HY
Rhizosolenia alata alata	RA	Larvacea	LV
Rhizosolenia alata indica	RI	*Limacina retroversa*	LR
Rhizosolenia alata inermis	RN	*Metridia lucens*	ML
Rhizosolenia hebatata semispina	RP	*Neocalanus gracilis*	NG
Rhizosolenia imbricata shrubsolei	RB	*Oithona* spp.	OI
Rhizosolenia styliformis	RT	*Oncaea* spp.	ON
Skeletonema costatum	SK	*Pleuromamma borealis*	PB
Thalassionema nitzschioides	TN	*Pleuromamma gracilis*	PG
Thalassiosira spp.	TS	*Pleuromamma robusta*	PR
Thalassiothrix longissima	TL	*Podon* spp.	PD
		Pseudocalanus elongatus	PP
		Rhincalanus nasutus	RN
		Temora longicornis	TL
		Total Copepoda	CP

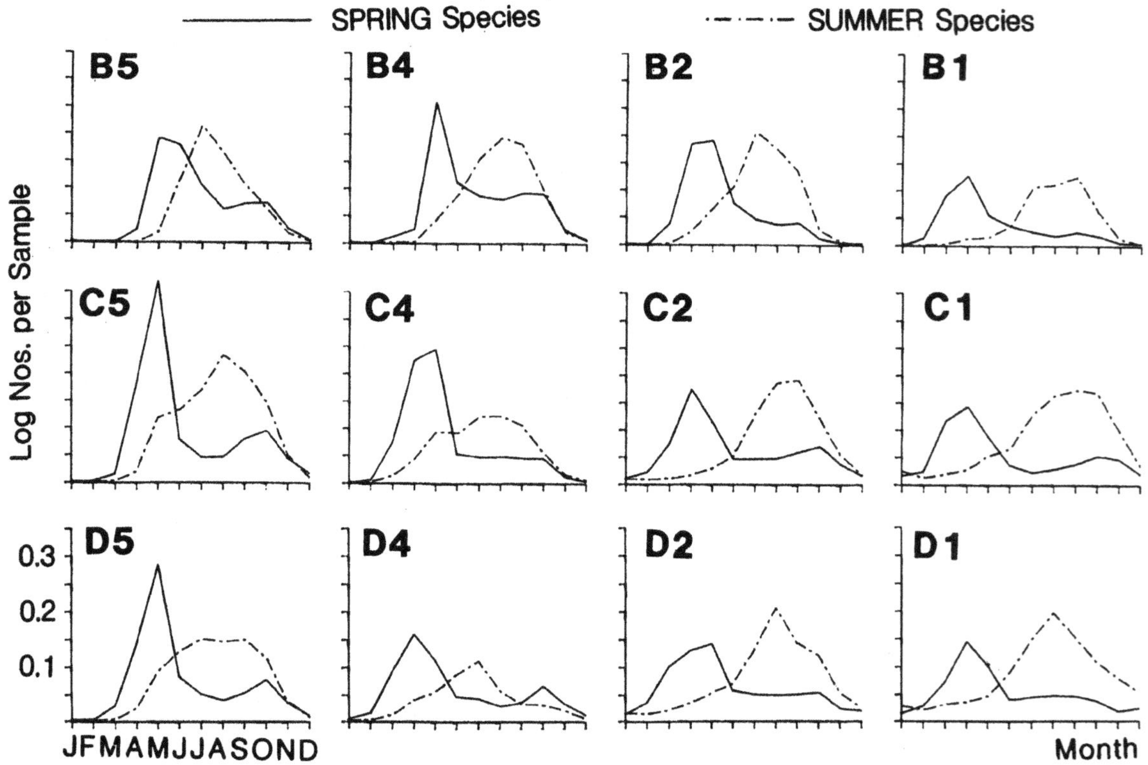

Fig. 8. Graphs of the average (1958 to 1984) seasonal cycles in the abundance of phytoplankton for each of the areas shown in Fig. 10, with the species differentiated into spring and summer groups.

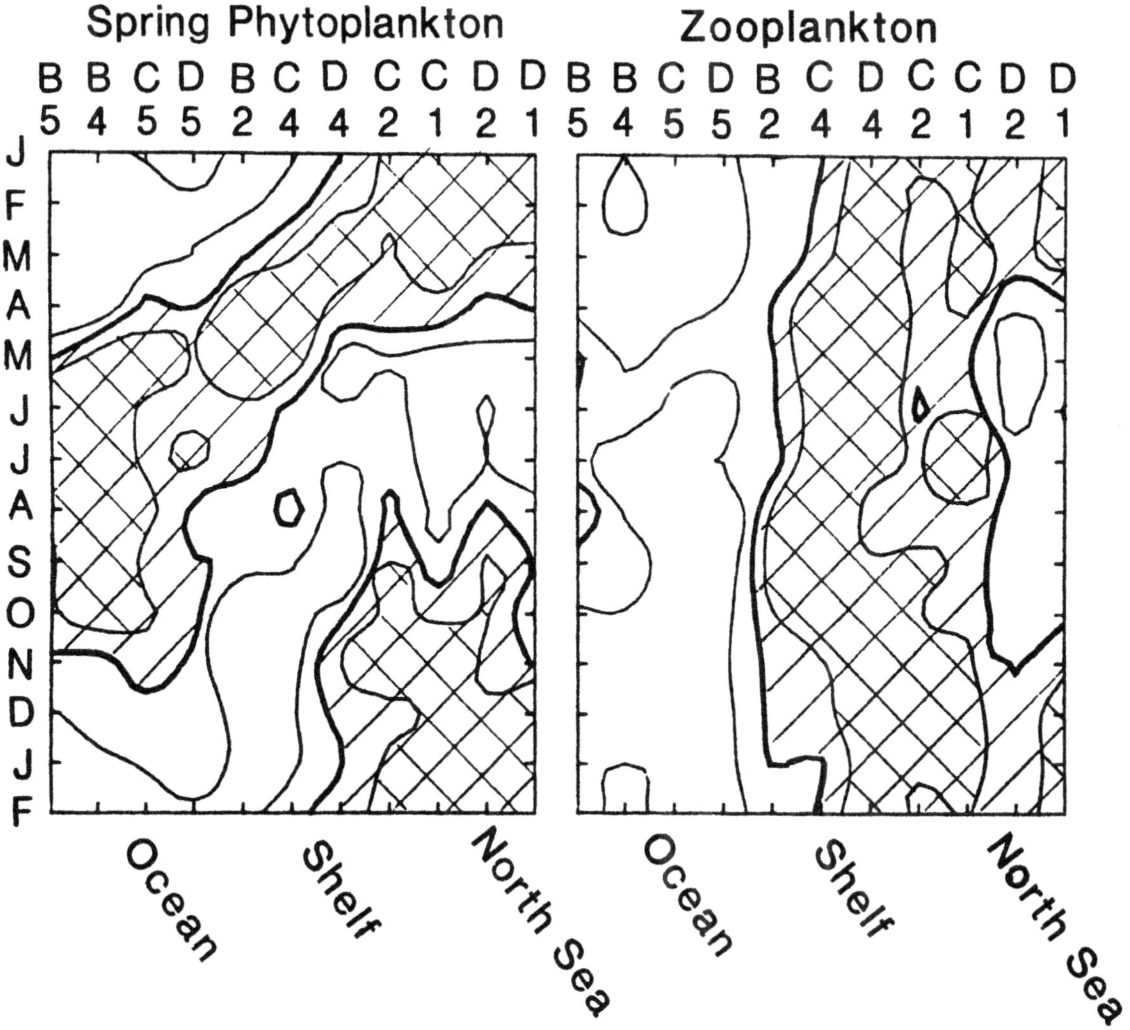

Fig. 9. Contour diagrams illustrating the extent of seasonal variations in geographical distribution. In these diagrams the columns refer to the areas shown in Fig. 10, ranked from oceanic (left) through shelf to North Sea (right) within each category the areas are ranked from north to south. The rows refer to months. Differences in abundance between areas and between months have been removed by standardization, leaving in the diagrams only the seasonal changes in relative geographical distribution. The contour levels are, therefor, arbitrary.

which overwinters in relatively large numbers by migrating down into deep cold water where they can survive on food reserves through the winter, and another copepod *Acartia clausi*, which appears to have a higher potential population growth rate than most other zooplankton species.

With these exceptions it would appear that, on the whole, the stocks of zooplankton species, in contrast to the phytoplankton, do not achieve anything like a steady state response with respect to the seasonal variations in temperature and food supply and, since these are the dominant patterns in the dynamics of the zooplankton, it would seem to imply that any variability due to interactions between zooplankton species resulting from factors such as competition or predation will also not exhibit anything approaching a steady state and will probably be difficult to identify and characterise.

These features of the response to seasonal forcing functions, the contrast between phytoplankton and zooplankton and the role of the distribution of overwintering stocks also have to be considered when looking at the response of the plankton to longer term environmental variability.

Data are available from the Continuous Plankton Recorder survey for some 23 species of phytoplankton and 18 of zooplankton for each of a set of areas (Fig. 10) covering the north-east Atlantic and the North Sea, for the period from 1948 to 1984 for the zooplankton and 1958 to 1984 for the phytoplankton.

Thus, the data base contains nearly 500 time-series and it is obviously quite impossible to describe each of these in detail. Fig. 11 contains one example of the monthly data, for the copepod *Pseudocalanus elongatus* in the central North Sea. It is obvious that the repeated seasonal cycle is very clear and its amplitude is much larger than that of the year-to-year changes. With data like these it is reasonable to remove the seasonal cycle by calculating annual means.

Fig. 12 shows graphs of the annual means in abundance of the same species in each of the north-east Atlantic and North Sea areas shown in Fig. 10.

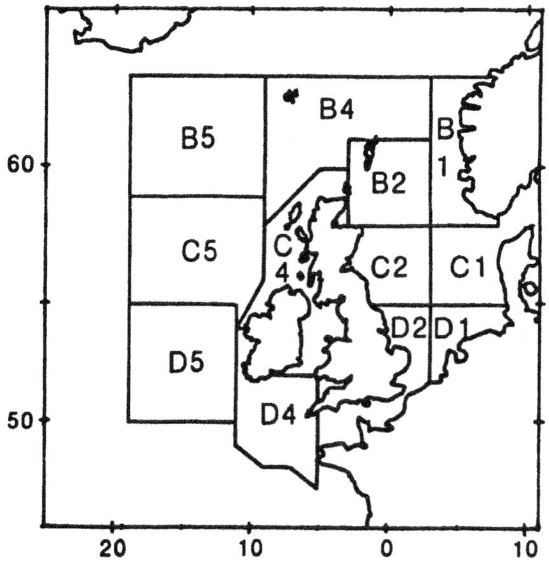

Fig. 10. A chart showing the division of the survey into sub-areas used in Figs 6, 8, 9, 11, 12, 13, 14.

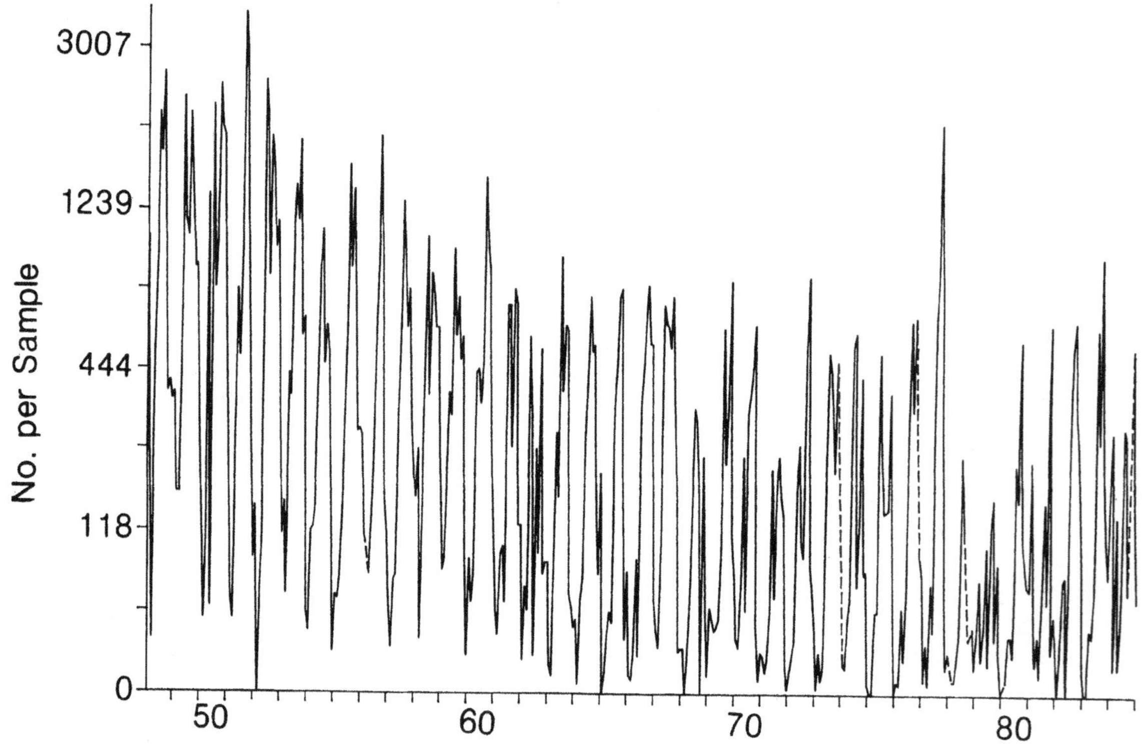

Fig. 11. A graph of the monthly variations in abundance of the copepod *Pseudocalanus elongatus* for area C1 (Fig. 10), for the years 1948 to 1984.

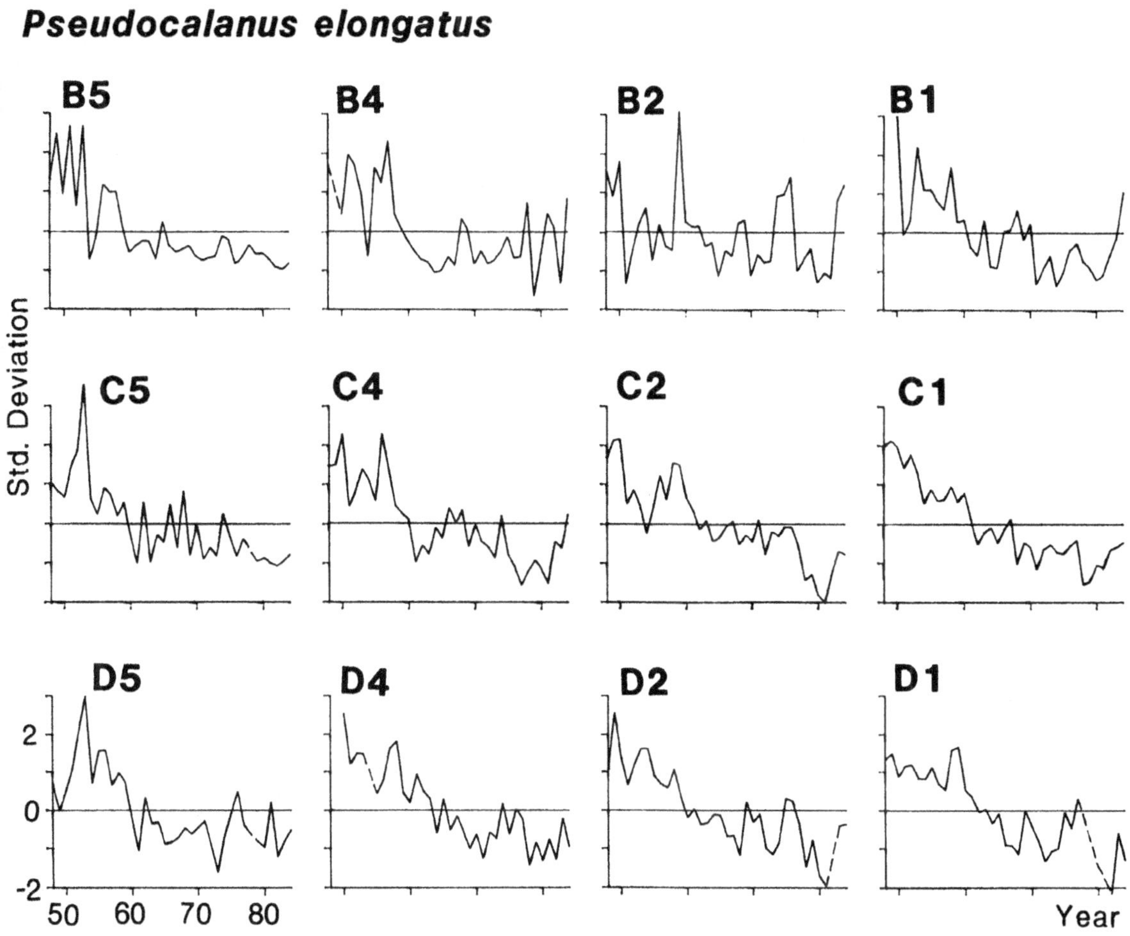

Fig. 12. Graphs of annual means of abundance of *Pseudocalanus elongatus* for the years 1948 to 1984 for each of the areas shown in Fig. 10. The data in each graph are standardized to zero mean and unit variance.

In these plots, differences in abundance between areas have been removed by standardizing each graph to zero mean and unit variance. It is clear that there are marked similarities between the graphs, and an expression of the common pattern can be obtained by using Principal Components Analysis. Fig. 13 shows a graph of the first component and a scatter plot of the first two vectors. The component can be regarded as being the best possible single representation of all the variables included in the previous figure and the scatter plot provides a spatial representation of the relationships between the areas.

By processes such as these the whole data base can be reduced to manageable proportions, while still retaining the dominant patterns of variability in representations of year-to-year changes in abundance.

Fig. 14 shows graphs of the first principal components, for each of the 12 north-east Atlantic and North Sea areas (Fig. 10), of the year-to-year fluctuations of all the zooplankton species. Each graph can be considered as the best possible representation of the year-to-year changes in the abundance of the zooplankton as a whole.

It is obvious that there are marked similarities between the patterns in the different areas and that this pattern involves a clear structure in the form of a more or less continuous downward trend. The extent of the similarity between the areas is remarkable considering the varying hydrographic conditions represented in the areas which range from

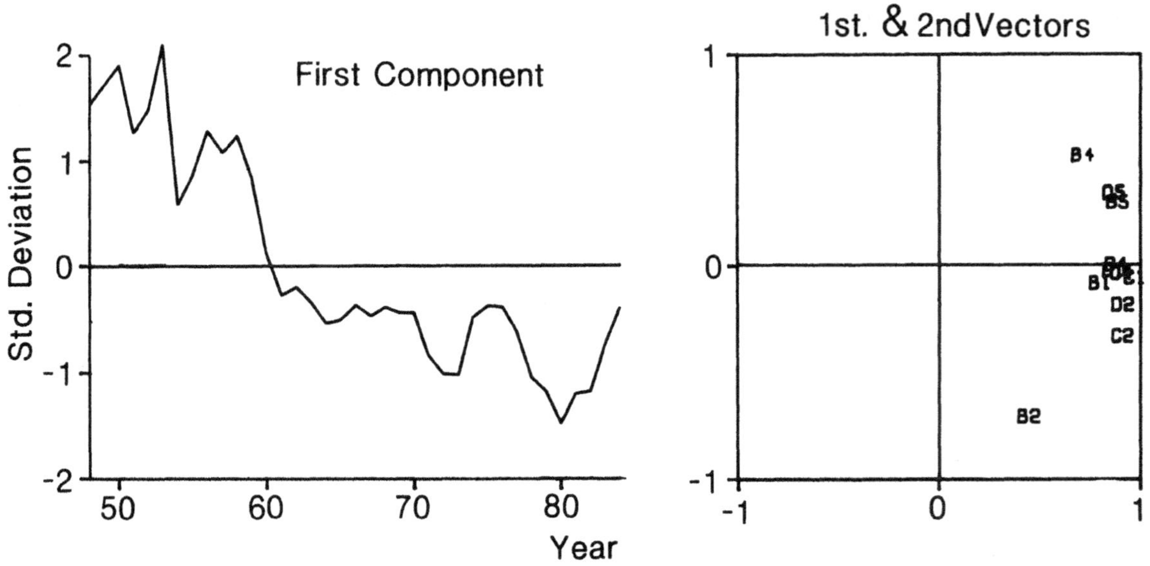

Fig. 13. The time-series graph (left) is the first principal component of the data presented in Fig. 12. Right is a scatter plot of the first two vectors of the same data.

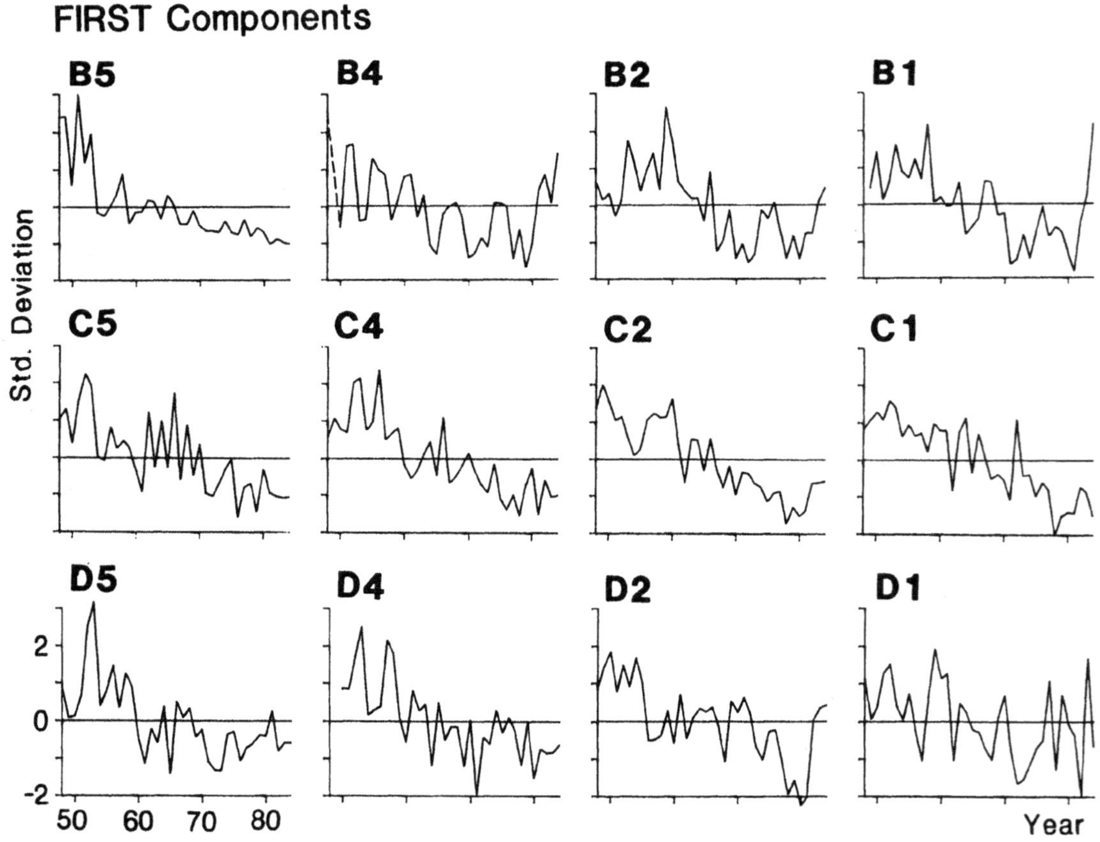

Fig. 14. Graphs of the first principal components for each of the areas shown in Fig. 10 of the year-to-year variations in abundance of all the zooplankton species.

shallow, unstratified waters in the southern North Sea to deep, seasonaly stratified waters in the open ocean.

The search for processes that might determine patterns of long-term variability in the zooplankton clearly has to be limited to systems which operate over large areas, and the most obvious of these are related to changes in climate.

It has already shown that with respect to seasonal forcing, the stocks of zooplankton species exhibit persistence and do not achieve a steady state in their response to this forcing. The same question has to be asked with respect to longer term variability and, in these data the presence of a trend and clear associations between successive years suggests that persistence from year to year is a possibility which may have to be considered in searching for relationships with environmental variables (Colebrook, 1981, 1982a).

So far, two variables have been identified which show convincing relationships with the year-to-year changes in the abundance of the zooplankton and these will be illustrated in relation to the fluctuations in the abundance of the total of all the copepod species averaged for the North Sea.

The two variables are, firstly, an index of year-to-year variations in the frequency of occurrence of westerly weather over the British Isles, and secondly, the sea surface temperature in the North Sea averaged for the months of February to June.

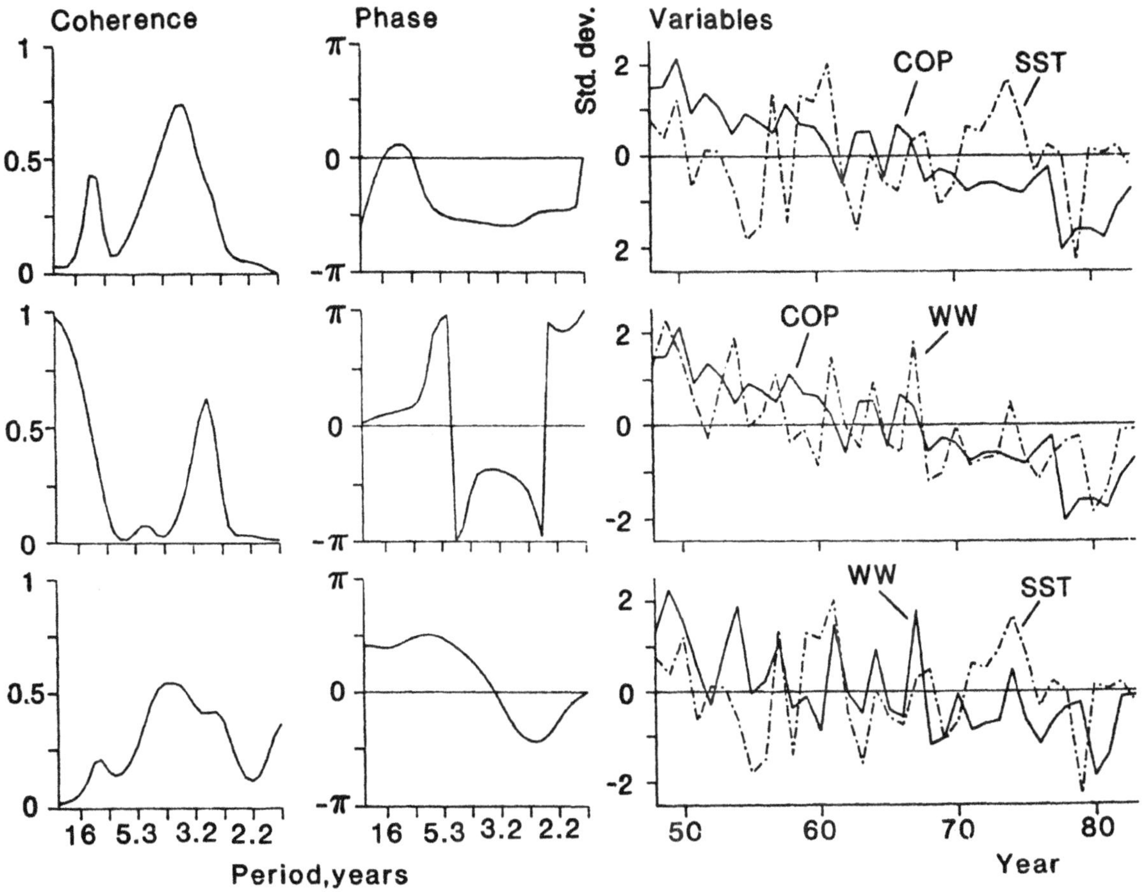

Fig. 15. Graphs (right) of the year-to-year variability, for 1948 to 1983, of pairs of variables. COP – total copepods, SST – sea surface temperature, WW – frequency of westerly weather. The plots are all standardized to zero mean and unit variance. The plankton and temperature data are for the North Sea. The graphs, left and centre, are plots of coherence and phase derived from maximum entropy cross spectral analyses of the corresponding pairs of variables. For details see text.

The relationships of these variables with each other and with the zooplankton (Colebrook, 1985b) have been examined using maximum entropy cross spectral analysis. The products of this method of analysis are plots of Coherence and Phase. The coherence is roughly equivalent to a set of correlations between the two variables across the spectrum of frequencies in the variables. The phase measures the extent of any time-lag between the variables across the same spectrum. The wavelengths included in the analysis range from infinity to a period of two years. The maximum entropy method involves the derivation of filters that convert the original time-series to white noise and the coherence and phase spectra are estimated in relation to the frequency responses of these filters.

Fig. 15 shows plots of the pairs of variables: copepods and westerly weather, copepods and temperature, westerly weather and temperature, together with corresponding plots of coherence and phase. Copepods and westerly weather show a high coherence at the longest wavelengths, reflecting the clear similarity in the downward trend in each case. There is also a peak in coherence with a wavelength of about 3 years. The relationship between copepods and temperature shows only the peak in coherence at about 3 years, as does the relationship between westerly weather and temperature.

Looking at the phase relationships with respect to the 3-year peak, westerly weather and temperature are in phase, while the only sensible interpretation of the relationship with the plankton implies a negative correlation in both cases with a time-lag of about a quarter wavelength, roughly nine months.

The relationship between the plankton and the westerly weather, therefore, contains two components, a positive relationship with respect to the long-term trend and a negative relationship with respect to the 3-year waveband. This obviously presents problems of interpretation.

There is evidence to suggest that the long-term trend has its origins in winter and its presence through the year is a function of persistence (Colebrook, 1985a). Fig. 16 shows graphs giving a seasonal breakdown of the year-to-year changes in abundance of four species of zooplankton in the North Sea. Differences in abundance between the seasons and the species have been removed by standardization. It can be seen that all the species show an element of downward trend in winter, and for most of the species this persists throughout the

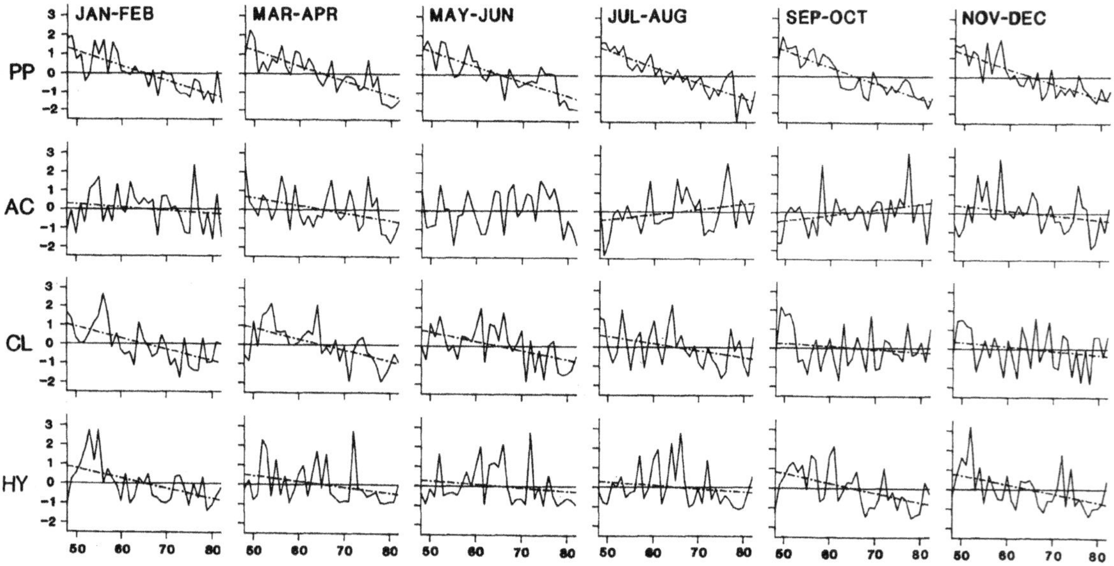

Fig. 16. Graphs illustrating seasonal variations in the pattern of year-to-year fluctuations in abundance of zooplankton for the period 1948 to 1982. The data are for the North Sea. PP – *Pseudocalanus elongatus,* AC – *Acartia clausi,* CL – *Calanus* spp., HY – Hyperiidea. All the graphs are standardized to zero mean and unit variance and the direction and extent of linear trends are indicated by broken lines.

year, but it tends to be most pronounced in winter, and for one species, *Acartia clausi*, the trend reverses in summer. *Acartia* has already been mentioned as a species with an apparently higher potential rate of population increase than most of the zooplankton. The impression gained from looking at these data is reinforced by a study of vector plots derived from principal components analyses of the data which indicate that the species are generally more coherent in winter than in summer.

Therefore it seems possible that the long-term trend in the abundance of the zooplankton is produced by forcing which is effective during a limited period of the seasonal cycle, in this case in winter.

The seasonal breakdown of the year-to-year changes in abundance inevitably results in an increase in the 'noise' levels in these data and it is not possible to detect any convincing seasonal differentiation with respect to the 3-year periodicity. However, this is possible for the temperatures, and in these data the 3-year periodicity is present only from February to June and possibly July. Again it seems possible that the relationship with the plankton with respect to the 3-year periodicity is produced by forcing effective during a limited period of the seasonal cycle, in this case in spring and early summer.

If we assume that the forcing due to the westerly weather is related to processes involving vertical mixing of the water column then it seems reasonable to expect these processes to have different effects in winter when the waters are unstratified compared with the period of temperature stratification.

This example provides a useful indication of some of the problems which can arise in the empirical approach to relationships between environment and populations. In particular, it highlights the value of relatively long time-series of observations in the identification and characterisation of such relationships.

Looking back over the studies of the year-to-year variability in the abundance of the plankton based on the data from the Continuous Plankton Recorder survey, the beginnings of patterns of coherence between species and between areas were indentified with 12 years of data (Colebrook & Robinson, 1964). The long-term, quasi linear trend was identified with about 20 years of data (Colebrook, 1972). The 3-year periodicity was first noticed with 25–26 years of data and, at about the same time, the relationship with westerly weather was identified (Colebrook, 1978). But, a resonably reliable characterisation of the relationship did not emerge until the study described here, based on 36 years of data.

With respect to year-to-year changes in the phytoplankton, consistent data are available only from 1958, due to a change in the methods of counting the samples in the survey. But, since zooplankton eat phytoplankton it would seem logical to look at the phytoplankton in the search for variables related to the zooplankton and, at first sight it looks as though a simple relationship is involved. Fig. 17 shows graphs of the year-to-year variations in the abundance of the zooplankton and the phytoplankton, combining the data for all areas and species.

The variables are obviously correlated, and a major element in the correlation is the long-term trend. But, in view of the interpretation proposed for the relationship between zooplankton and westerly weather, there are problems in viewing this relationship between the zooplankton and the phytoplankton in terms of a simple trophic interaction. It has been suggested, for the zooplankton, that the long-term trend has its origins in winter, through differential survival of over-wintering stocks, and these differences persist throughout the rest of the year. This argument is difficult to apply

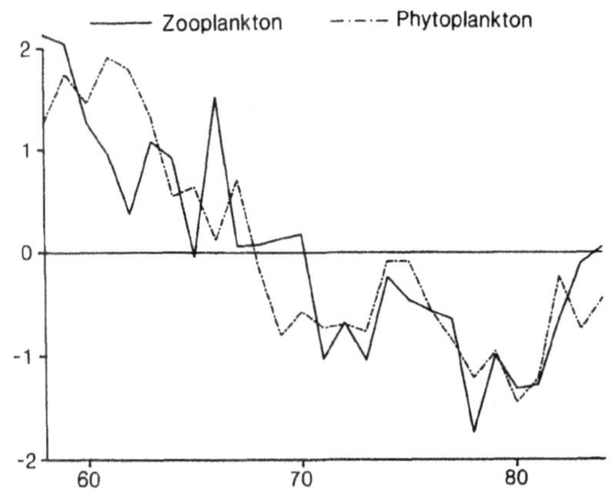

Fig. 17. Time-series of the abundance of zooplankton and phytoplankton for 1958 to 1984 derived from principal components analyses of the data for all species and all the areas shown in Fig. 10.

to the phytoplankton which have already been shown to achieve something like a steady state response to the seasonal cycles of light, temperature and vertical stability. In contrast to most of the zooplankton, for most of the phytoplankton there is little of no relationship between the geographical distributions of their overwintering stocks and their geographical distributions in summer. It is therefore most unlikely that year-to-year variations in the levels of overwintering stocks will persist through the year. And yet, the patterns of year-to-year fluctuations in the abundance of the phytoplankton looks persistent and in fact contains about the same proportion of long wavelength variability as for the zooplankton.

It seems possible that the relationship between the phytoplankton and the zooplankton has to be interpreted not just as a response by the zooplankton to variations in food supply but as also involving feed-back from the zooplankton to the phytoplankton to the extent of influencing the year-to-year variations in the abundance of the phytoplankton. A likely mechanism for the feed-back is *in situ* nutrient recycling. There has been a growing realisation, in recent years, of the role of recycled nutrients in the growth of phytoplankton (Eppley, 1981; Jawed, 1973). The form of the relationship between phytoplankton and zooplankton in the north-east Atlantic and the North Sea does appear to be consistent with these observations and extends their scope to a consideration of the influence of nutrient recycling on the long-term variability of the plankton.

It is difficult to see how a feed-back system, based on nutrient recycling, could have much impact on the dynamics of the spring bloom when the phytoplankton are utilising nutrients mixed into the surface waters during the winter. The information on the limited extent of seasonal variation in geographical distribution of most of the zooplankton, as compared with the phytoplankton suggests that there is a lag between phytoplankton growth and grazing pressure in the spring indicating that, at least in the open ocean and to some extent in the North Sea, the spring bloom of phytoplankton is nutrient limited. The implication is that in the spring a significant proportion of the phytoplankton stock is not eaten by zooplankton and presumably sinks out of the euphotic zone and much of it, at least in the shallower waters, must finish up on the bottom and provide a significant input to the benthos. There is growing evidence, based of direct observations of the bottom and from sediment traps, that this does in fact happen (Davies & Payne, 1984). It seems possible that much of the work which has been done on the spring bloom of the plankton may have more significance for the benthos than it does for the plankton during the post-bloom period.

Thus it is possible to derive a doubtless oversimplified, conceptual model of relationships between the plankton and the benthos in terms of a partitioning of the nutrient pool between two annual cycles, one confined to the euphotic zone and determined by the dynamics of the zooplankton, the other involving the benthos and the spring bloom of the phytoplankton.

It would be good to see the results of the COST 647 project contributing to our knowledge of the interaction between these two nutrient cycles and to a better understanding of the relationships between the plankton and the benthos.

Acknowledgements

This work forms part of the programme of the Institute for Marine Environmental Research, which is a component of the Natural Environment Research Council: it was commissioned in part by the Ministry of Agriculture, Fisheries and Food.

References

Colebrook, J. M., 1960. Continuous plankton records: methods of analysis 1950–1958. Bull. mar. Ecol. 5: 51–64.

Colebrook, J. M., 1972. Variability in the distribution and abundance of the plankton. Spec. Publ. Int. Comm. NW Atlant. Fish. 8: 167–186.

Colebrook, J. M., 1978. Continuous plankton records: zooplankton and environment, north-east Atlantic and North Sea, 1948–1975. Oceanol. Acta 1: 9–23.

Colebrook, J. M., 1979. Continuous plankton records: seasonal cycles of phytoplankton and copepods in the North Atlantic Ocean and the North Sea. Mar. Biol. 51: 23–32.

Colebrook, J. M., 1981. Continuous plankton records: persistence in time-series of annual means of abundance of zooplankton. Mar. Biol. 61: 143–149.

Colebrook J. M., 1982a. Continuous plankton records: persistence in time-series and the population dynamics of *Pseudocalanus elongatus* and *Acartia clausi*. Mar. Biol. 66: 289–294.

Colebrook, J. M., 1982b. Continuous plankton records: seasonal variations in the distribution and abundance of plankton in the North Atlantic Ocean and the North Sea. J. Plankton Res. 4: 435–462.

Colebrook, J. M., 1984. Continuous plankton records: relationships between species of phytoplankton and zooplankton in the seasonal cycle. Mar. Biol. 84: 261–265.

Colebrook, J. M., 1985a. Continuous plankton records: overwintering and annual fluctuations in the abundance of zooplankton. Mar. Biol. 84: 261–265.

Colebrook, J. M., 1985b. Sea surface temperature and zooplankton, North Sea, 1948 to 1983. J. Cons. perm. int. Explor. Mer 42: 179–185.

Colebrook, J. M. & G. A. Robinson, 1964. Continuous plankton records: annual variations of abundance of plankton in 1948–1960. Bull. mar. Ecol. 6: 52–59.

Davies, J. M. & R. Payne, 1984. Supply of organic matter to the sediment in the northern North Sea during a spring phytoplankton bloom. Mar. Biol. 78: 313–324.

Eppley, R. W., 1981. Autotrophic production of particulate matter. In: Analyses of marine ecosystems. Ed. A.R. Longhurst. Academic Press: 343–361.

Glover, R. S., 1967. The continuous plankton recorder survey of the North Atlantic. Symp. zool. Soc. Lond. 19: 189–210.

Jawed, M., 1973. Ammonia excretion by zooplankton and its significance to primary production during summer. Mar. Biol. 23: 115–120.

Rae, K. M., 1952. Continuous plankton records: explanation and methods 1946–1949. Hull Bull. mar. Ecol. 3: 135–155.

List of posters

Ibanez, M., J. Pena & J. Feliu. Reproduction of *Patella* spp. on the Basque coast of Spain.

White, M. The reproductive biology of *Dendrodoa grossularia* (Van Bededen) (Tunicata: Ascidiacea) at intertidal and subtidal locations on the West Coast of Ireland.

Larsson, C. S., T. Lundälv & L. Axelsson. Ecophysiological studies of *Halidrys siliquosa* in relation to observed changes in natural populations.

Bouchet, J. M. *Abra alba* populations in and off Arcachon Bay (France). Mean seasonal evolution of benthic temperatures and salinities from 1971 to 1984.

Dinneen, P. & S. Neiland. Spatial and temporal distribution patterns within an *Amphiura filiformis – Abra alba* community.

Costelloe, M. Population dynamics of *Thyasira flexuosa* (Bivalvia: Thyasiridae) in inner Galway Bay, West Coast of Ireland.

Grehan, A. Population dynamics of the ampharetid polychaete *Melinna palmata* Grube in Inner Galway Bay, West Coast of Ireland.

Groupe d'etude des milieux estuariens et littoraux. Etude comparative de trois estuaires de Manche: Baie des Veys, estuaire de Seine, Baie de Somme.

Seip, P., R. Brand, J. van der Meer, A. C. Smaal, F. Creutzberg, A. Engelberts & K. Hoek. A Benthos Inventory in the Zeeland coastal zone (The Netherlands).

Coosen, J. Biomass and density fluctuations of the macrozoobenthos of the intertidal flats in the Oosterschelde, The Netherlands.

Madsen, P. B. Dynamics of the dominating macrozoobenthos in the Danish Wadden Sea 1980–1985.

Bullimore, B. Skomer Marine Reserve Subtidal Monitoring Project.

Hydrobiologia 142: 327 (1986)

Reproduction of *Patella* spp. on the Basque coast of Spain

Miguel Ibanez, Junkal Pena & Jaime Feliu
INSUB., Museo de Oquendo, Apartado de Correos 3.031, San Sebastian, Spain

Four species occur on the Basque coast: *P. lusitanica* (*P. rustica*) in the upper levels of exposed areas, *P. depressa* and *P. vulgata* throughout mid-levels, with the former the most abundant and showing much polymorphism, and *P. aspera* which is common only on the lower shore. Gonad indices were compiled from monthly samples of at least 50 individuals of each species taken from February 1983 until August 1985 at Fuenterrabia, San Sebastian and Zumaya.

P. lusitanica, at virtually its northern limit of distribution, has a short breeding period beginning in July, with first spawnings in Aug/Sept and gonads declining rapidly in Oct/Nov or Nov/Dec.

In *P. vulgata*, which reaches its southern limit in Portugal, the period of gonad activity has varied between three and five months over late spring/early summer. Gonad indices reached maxima in October and November, but also showed some evidence of re-ripening in Jan/Feb before final declines in March/May.

Bij contrast, *P. depressa* and *P. aspera* which are in the centre of their range show irregular, high levels of activity for most of the year. There is a tendency for least activity in June/July, this being the most marked in *P. aspera*. The irregularities in the indices suggest continuous development and frequent small spawnings.

These data for *P. vulgata* and *P. depressa* are similar to those of Miyares (1980) for the Asturian coast of Spain, and also to the Portuguese data (including *P. aspera*) of Guerra & Gaudencio (see this volume). They contrast strongly with data from Britain and Ireland (Orton & Southward, 1961; Thompson, 1979; Bowman, 1985; Bowman & Lewis, this volume) where gonad activity is of much shorter duration and usually confined to midsummer and/or early autumn.

The decreased duration northwards may reflect a minimum temperature requirement by gonads, and it is possibly significant that *P. lusitanica* has as short a breeding period here, at its northern limit, as *P. aspera* has at its northern limit in Britain. The time changes support the expectation of a latitudinal progression in breeding/recruitment from midsummer/early autumn in the north, to the colder half of the year in the south. Problems remain, however; does the year-round activity in *P. depressa* in Spain contract towards its southern limit? When, during the apparently extensive breeding period in Spain, is recruitment most successful?

References

Bowman, R. S., 1985. The biology of the limpet *Patella vulgata* L. in the British Isles. in P. G. Moore & R. Seed (eds) The Ecology of Rocky Coasts. Hodder & Stroughton, London: 178–193.

Bowman, R. S. & J. R. Lewis, 1986. Geographical variation in the breeding cycles and recruitment of *Patella* spp. Hydrobiologia.

Guerra, M. T. & M. J. Gaudencio, 1986. Aspects of the ecology of *Patella* spp. on the Portuguese coast. Hydrobiologia.

Miyares, M. P., 1980. Biologia de *Patella intermedia* y *P. vulgata* (Mollusca, Gasteropoda) en el litoral asturiano (N. de Espana) durante un ciclo anual (Diciembre de 1978 a Noviembre 1979). Bol. Cienc. Nat. I.D.E.A. 26: 173–192.

Orton, J. H. & A. J. Southward, 1961. Studies on the biology of limpets. IV. The breeding of *Patella depressa* Pennant on the north Cornish coast. J. mar. biol. Ass. U.K. 41: 653–662.

Thompson, G. B., 1979. Distribution and population dynamics of the limpet *Patella aspera* (Lamarck) in Bantry Bay. J. exp. mar. Biol. Ecol. 40: 115–135.

Hydrobiologia 142: 328 (1986)

The reproductive biology of *Dendrodoa grossularia* (Van Beneden) (Tunicata:Ascidiacea) at intertidal and subtidal locations on the West Coast of Ireland

Mark White
Zoology Department, University College, Galway, Ireland

The reproductive biology of *Dendrodoa grossularia* (Van Beneden) was examined in the broader context of a study of succession and development in ascidian dominated communities, under the umbrella of the COST 647 Programme. Samples were collected from an intertidal location in the inner part of Galway Bay, and from a subtidal site at Ard Bay, on an approximately monthly basis over the period August, 1983 to June, 1985.

The intertidal site is prone to periodic depression in salinity (down to $5^0/_{00}$), under the influence of a major river in the area. Salinity at the subtidal site never fell below $32^0/_{00}$ and temperature at both sites fluctuated seasonally between 3 °C and 18.5 °C. Fecundity was significantly higher at the subtidal site where there were two distinct peaks in the reproductive cycle annually, in Spring and Autumn, both of roughly equal intensity. At the intertidal site, on the other hand, although there were two peaks annually, the second in Autumn was greatly reduced. The differences between the two populations were attributed to a combination of factors, among the most important of which were suppressed salinity and tidal emmersion at the intertidal site.

Hydrobiologia 142: 329 (1986)

Ecophysiological studies of *Halidrys siliquosa* in relation to observed changes in natural populations

C. S. Larsson*, T. Lundälv & L. Axelsson
Kristineberg Marine Biological Station, 5-450 34 Fiskebëckskil, Sweden
**also University of Göteborg, Dept. of Marine Botany*

An ecophysiological study of the brown algae *H. siliquosa* has been initiated in order to evaluate if changes in abiotic conditions could be responsible for a massive decrease in the populations of this algal species. Such a decrease occurred in several localities during the years 1979–81 along the Swedish west coast. The first purpose was to study factors determining the northern limit of the alga's distribution. *H. siliquosa*, with its main distribution to the south, has been compared to the closely related and morphologically similar *Ascophyllum nodosum*, f *scorpioides*, which has a more northern distribution. This comparison revealed the following differences between the two algal species:

Irradiation in stagnant water resulted in much higher photooxidative damage on *H. siliquosa* (both reversible and irreversible). This alga is thus more sensitive to a combination of high irradiation, low temperature and low water motion.

H. siliquosa showed lower rates of bicarbonate uptake, especially at lower temperatures, and in stagnant water. In addition, *H. siliquosa* lacked a light buffering system which occurs in other members of Fucales growing on the Swedish west coast. Also, in the northern regions, the alga occurs only close to the surface. These last facts suggest that the conservation of light energy is the critical factor.

Hydrobiologia 142: 330 (1986)

Abra alba populations in and off Arcachon bay (France): mean seasonal evolution of benthic temperatures and salinities from 1971 to 1984

Jean-Marie Bouchet
Institut Universitaire de Biologie Marine d'Arcachon (Université de Bordeaux I), 2, rue du Pr. Jolyet, 33120 Arcachon (France)

Abra alba populations occur on fine sands outside the Arcachon bay and in several biotopes of the bay, according to the distance from the ocean and the variability of temperatures and salinities. Numerous sectional elevations at the time of high spring tides, show the temperature stratification and the thermal differentiation between the open sea and the inner part of the bay. Only superficial waters penetrate into the lagoon. For example, during summer (July 1983) on the offshore *Abra alba* community, at depths greater than 30 m, temperatures are 13 °C. A thermal amplitude of 8 degrees separates the two extreme biotopes (distance 27 kilometers), the inner attaining 22 °C. There exists from March to May, and from October to November, a relative thermal homogenization between the bay and the ocean. Salinities are always stratified, and may be very heterogeneous, especially during spring (April and May), when precipitation is particularly frequent. The Eire river, and its numerous tributaries located on the south east of the bay, are the principal sources responsible for lowering salinities.

Five diagrams corresponding to the main biotopes, show the mean seasonal evolution of bottom temperatures and salinities. Months are characterized by fan-shaped curves showing strong tendencies to overlap from year to year. Curves also represent a relative prediction of T and S evolution. A serious cut exists between the ocean and the bay. In the bay itself, another cut marks the Eire estuary. The corresponding diagram may overlap considerably for low salinities during spring and early summer.

Abra alba populations never extend beyond the geographic boundary shown on this last diagram. Their disappearance from the inner part of the Bay occurs during October. Stocks may persist or disappear (Arguin bank, at the mouth of the bay), during winter. Recruitment takes place during spring (March) from the ocean (not yet proven), or from a surviving stock in the bay.

Hydrobiologia 142: 331–332 (1986)

Spatial and temporal distribution patterns within an *Amphiura filiformis* – *Abra alba* community

Patricia Dinneen & Sinead Neiland
Zoology Department, University College, Galway, Ireland

Keywords: macrobenthos, sublittoral infauna, community composition, spatial distribution, temporal distribution, Ireland coast

The benthic macrofauna of Kinsale Harbour on the south coast of Ireland was investigated by members of the Zoology Department. This involved:

(a) a qualitative survey at 68 sites in 1978;
(b) a quantitative characterisation of the infauna at 50 soft-bottom sites in 1979–81;
(c) an evaluation of the stability of faunal assemblages by means of annual recharacterising surveys at a reduced, i.e. relative to the initial, number of selected sites;
(d) monthly monitoring, over a two year period, of assemblage structure and development together with the population dynamics and reproduction of some numerically important species. This presentation gives an outline of the infaunal assemblages found to characterise the area over the period 1979–81 to 1984. More specifically, an intercomparison of the species dominance patterns within four faunal groupings is made with respect to the years 1979–81 and 1984.

The study area comprises the euhaline zone of the Bandon River estuary with a mean river discharge of 15.25 $m^3 s^{-1}$ which equals or exceeds 50 $m^3 s^{-1}$ for less than 5% of the time. Sediments in the inner harbour area (Group 1) are heterogenous muddy sands with some coarse material. Clean fine sands characterise the middle ground (Group 2) while homogenous muddy sands prevail in the outer harbour (Groups 3).

Wind and tidal induced currents turbate the bottom sediments in all but the innermost part of the harbour.

In terms of faunal composition four faunal assemblages were identified for the area using classificatory analyses. Affirming that the faunal groups are no more than convenient abstractions from continua, Group 1 approximates an admixture of *Abra alba* and *Venus striatula* communities with elements of an *Amphiura* assemblage too; Group 2a resembles the *Tellina fabula* facies of the *Venus striatula* community; Group 2b approximates the classical *Venus striatula* community; Group 3 is equivalent to the *Amphiura filiformis* community.

The broadscale spatial pattern of the assemblages (Groups) essentially remained constant over the six year period. However, some intra-group fluctuations in species abundances took place. Of the numerically dominant species within the various faunal assemblages, between 1979–81 and 1984:

(i) *Abra alba* (Wood), *Mysella bidentata* (Montagu), *Owenia fusiformis* delle Chiaje, *Euclymene oerstedii* (Claparède), *Scoloplos armiger* (O. F. Müller), *Anaitides maculata/mucosa, Pariambus typicus* (Kröyer), *Magelona mirabilis* (Johnston), *Chaetozone spp., Spisula spp., Amphiura filiformis* (O. F. Müller), *Lumbrineris gracilis* Ehlers, *Nucula turgida* Leckenby and Marshall, and *Cylichna cylindracea* (Pennant) maintained their status;

(ii) *Mediomastus fragilis* Rasmussen, *Myriochele* cf *oculata* Zachs, *Spiophanes bombyx* (Claparède), *Pholoe minuta* (Fabricius), *Magelona filiformis* Wilson, *Exogone hebes* (Webster & Benedict), *Urothoe elegans* (Bate), *Corophium crassicorne* Bruzelius, *Mya arenaria* L., *Harpinia antennaria* Meinert, *Tellina fabula* Gmelin, *Spio filicornis* (O. F. Müller), *Magelona minuta* Eliason and *Spiophanes kroyeri* Grube decreased in abundance;

(iii) *Nephtys hombergi* Audouin & Milne-Edwards, *Melinna palmata* Grube, *Ampelisca tenui-*

cornis Lilljeborg, *Venerupis pullastra* (Montagu), *Sthenelais boa* (Johnston), *Amphipholis squamata* (delle Chiaje), *Harmothoe impar* (Johnston), *Eteone flava/longa, Venus striatula* (da Costa), *Hiatella arctica* (L.), *Iphinoe trispinosa* (Goodsir), *Ensis* sp., *Gari fervensis* (Gmelin), *Dosinia lupinus* (L.), *Scalibregma inflatum* Rathke, *Glycera tridactyla* Schmarda and *Thyasira flexuosa* (Montagu) increased in importance. The reasons for these changes in abundance patterns are not known.

Hydrobiologia 142: 333 (1986)

Population dynamics of *Thyasira flexuosa* (Bivalvia: Thyasiridae) in inner Galway Bay, West Coast of Ireland

Mark Costelloe
Zoology Dept., University College Galway, Ireland

Thyasira flexuosa (Montagu) occurred in high densities over a restricted area of Inner Galway Bay. As part of a broader study being carried out in the bay, the population dynamics of this bivalve were investigated, at a single site, over the two year period November 1982 to October 1984. Monthly sampling (five 0.1 m^2 van Veen grabs washed on a 0.5 mm sieve) revealed a mean population density of $72.6 \pm 42.4/0.1\ m^2$.

Recruitment took place between October and July of each year sampled and the growth rate of the Y_O year class (1983/'84) was followed from length frequency histograms.

During the course of the study there was a major change in the *Thyasira* population at the sampling station. From November 1982 to September 1983, the size-frequency structure did not alter greatly, as was the case from November 1979 to April 1981 (Conneely, 1983). In October 1983, the numbers fell dramatically, but a similar drop had been noted in February 1983 and it was assumed that this reflected a patchiness in the bivalves microdistribution. In January 1984, however, numbers were down again on the previous months and this intensified to a point in October 1984 where the majority of individuals recovered were those that had settled in the previous months.

The reasons for this population crash are, as yet, unknown. Temperature and salinity data over the sampling period show the same seasonal trend since 1979. Sediment data, on the other hand, did indicate some change.

This work is continuing and the crash of *T. flexuosa* will be viewed in context of the returns for the total macrofauna over the same study period.

References

Conneely, M. E., 1983. Benthic ecological studies in Inner Galway Bay (West Coast of Ireland) with particular reference to Bivalvia. (Unpubl. Ph.D. Thesis). University College Galway. National University of Ireland, 144 pp.

Hydrobiologia 142: 334 (1986)

Population dynamics of the ampharetid polychaete, *Melinna palmata* Grube, in Inner Galway Bay, west coast of Ireland*

Anthony Grehan
Department of Zoology, University College, Galway, Ireland

The ampharetid, *Melinna palmata* is a common member of the polychaete infauna occurring along the north shore of Galway Bay, (O'Connor, 1981). A high density population at one location was the focus for a two year study (April, 1983 to March, 1985). Five samples were taken each month with a modified 0.12 m^2 van Veen grab (hinged 0.5 mm mesh doors fitted to the top to reduce bow wave). Samples were sieved on a 0.5 mm mesh and fixed in 10% buffered formalin. They were subsequently divided into 2 mm, 1 mm and 0.5 mm fractions. *M. palmata* were sorted from the 2 mm and 1 mm fractions and measured (10th setiger width) for construction of size frequency histograms. Monthly examination of coelomic contents from anaesthetised worms revealed the principal events of the reproductive cycle.

Melinna palmata in Galway Bay was shown to be polytelic with a breeding season extending from May to July. This is earlier than reported for the Breton coast (Retiere, 1979; Guillou & Hily, 1983). Adult worms were gravid in May with spawning taking place until August when water temperatures were at their maxima, (approx. 17 °C). Simultaneously, proliferation of the following season's population of gametes commenced and increased proportionally during the breeding period as spawning was completed. At the end of the spawning period, residual oocytes were reabsorbed and by October only new previtellogenic oocytes were present. Little change in oocyte diameter was apparent during the winter when water temperatures were at a minimum, (approx. 5 °C), but a rapid increase in oocyte diameter was observed in February coinciding with increasing water temperature.

Recruitment took place each year of the study as confirmed by additional samples taken in the post-spawning period with a Muus sampler (quantitative meiofaunal samples from the top 2 cm of sediment with an area of *ca.* 160 cm^2). At least three year classes could be discerned from the size frequency histograms.

*This work was carried out within the framework of the COST 47, core programme for the study of the 'target species', *Melinna palmata* Grube (Guillou & Pearson, 1981).

References

Guillou, M. & C. Hilly, 1983. Dynamics and biological cycle of a *Melinna palmata* (Ampharetidae) population during the recolonisation of a dredged area in the vicinity of the harbour of Brest (France). Marine Biology 74: 43–50.

Guillou, M. & T. Pearson, 1982. Core programme: Annexe ID. *Melinna palmata*. COST 47 Newsletter. No. 2. National Board for Science and Technology, Dublin, 13 pp.

O'Connor, B., 1981. Benthic macrofaunal studies in the Galway Bay Area. Volume II. (Unpublished Ph.D. Thesis). University College, Galway. National University of Ireland, 292 pp.

Retiére, C., 1979. Contribution à la connaissance des peuplements benthiques du Golfe normanno-bréton. Thèse d'Etat, Université de Rennes, 421 pp.

Hydrobiologia 142: 335–336 (1986)

Etude comparative de trois estuaires de manche: Baie des Veys, estuaire de Seine, Baie de Somme

GEMEL: Groupe d'Etude des Milieux Estuariens et Littoraux
Faculté des Sciences, B.P. 67, 76130 Mont Saint Aignan

Les chercheurs du GEMEL contribuent depuis plusieurs années à la connaissance des caractéristiques physiques et biologiques de trois estuaires de Manche centrale (Baie des Veys, Seine) et orientale (Baie de Somme). Leurs études concernent plus particulièrement les relations qui unissent macrofaune benthique endogée et sédiment.

La confrontation des résultats obtenus sur ces trois estuaires a permis de dégager certaines conclusions concernant:
– l'etat coenotique de ces trois estuaires,
– l'évolution pluriannuelle de leurs biocoenoses,
– les conséquences des aménagements (endiguements, poldérisation),
– l'utilisation potentielle de ces milieux (aquaculture).

1. Etat coenotique

Les trois estuaires présentent des états coenotiques comparables avec une dissymétrie des biofaciès qui a son origine dans la dissymétrie de leurs conditions hydrodynamiques locales respectives.

L'application aux résultats de l'analyse factorielle des correspondances met clairement en évidence ce rôle primordial de l'hydrodynamisme local qui se traduit par l'influence prépondérante qu'exercent sur la répartition des peuplements benthiques des facteurs comme la teneur en particules fines et en matière organique du sédiment, le degré d'exposition ou les conditions de salinité.

Ces facteurs expliquent l'existence de faciès diversifiés et appauvris, tant dans le secteur marin qu'estuarien des trois sites.

Le facteur 'qualité de l'eau' (salinité mais aussi teneur en polluants) ne joue un rôle déterminant sur le benthos que dans l'estuaire de la Seine; les faibles débits de la Vire et de la Somme minimisent l'influence de ce facteur et justifient ainsi l'appellation de Baie réservée à ces deux estuaires (ainsi en Baie des Veys, la salinité est toujours supérieure à $25^{0}/_{00}$).

2. Evolution des biocoenoses

L'état coenotique de ces estuaires n'est pas figé mais en constante évolution. – En Baie des Veys, l'évolution des biocoenoses est suivie depuis dix ans; elle se traduit par la migration vers l'aval de certaines espèces (*Pygospio elegans, Abra tenuis*), la régression et le changement de rive d'autres espèces (*Macoma balthica*), enfin par l'installation d'espèces marines (*Lanice conchilega, Urothoe grimaldii*). Ces modifications des biocoenoses sont la conséquence directe de l'évolution sédimentologique liée à la poldérisation (§3).

– Dans l'estuaire de la Seine, l'évolution des biocoenoses est la conséquence des nombreux aménagements industriels et portuaires (§3 et communic. DESPREZ-DUPONT).

– En Baie de Somme, l'évolution 'naturelle' des biocoenoses est liée à l'exhaussement général de la Baie; comme dans l'estuaire de la Seine, on assiste à une migration vers l'aval des faciès, avec notamment une forte progression du schorre au détriment des biofaciès estuairiens typiques (communauté à *Macoma*, faciès à *Nereis* et *Scrobicularia*).

Ces biocoenoses évoluent également de façon artificielle sous l'influence d'importants apports en nitrates et phosphates provenant du ressuyage des terrains agricoles. La principale conséquence est la prolifération des Polychètes, essentiellement aux dépens des Bivalves. Un exemple intéressant de cette explosion des Polychètes est fournie par le Spionidé *Pygospio elegans* dont les colonies erratiques constituent des banquettes qui accélèrent l'exhaussement de la Baie, permettent l'implantation d'une plante pionnière (la Spartine) et contribuent à la disparition de biofaciès originaux tels les sables envasés à *Cerastoderma*.

3. Enseignements des aménagements

Une évolution des biofaciès liée à des aménagements a pu être étudiée tant en Baie des Veys (poldérisation) qu'en Baie de Somme et dans l'estuaire de la Seine (endiguements).

– En Baie des Veys, la poldérisation qui se poursuit actuellement provoque la suppression d'importantes surfaces de piégeage des particules fines en fond d'estuaire. L'envasement se trouve par conséquent reporté vers l'aval, plus particulièrement au niveau de la corne du chenal d'Isigny, secteur qui a connu en 10 ans (1970–1980) un exhaussement supérieur à 4 mètres! C'est dans ces faciès vaseux que s'est installée la communauté à *Macoma balthica* qui avait disparu du flanc ouest de la Baie à cause de l'induration du sédiment par le *Pygospio*.

– En estuaire de Seine et Baie de Somme, des digues submersibles ont été construites pour stabiliser les chenaux de navigation. Ces aménagements ont eu pour conséquences annexes de servir de support à la sédimentation et permettre l'extension des schorres, et le provoquer le comblement des anciens chenaux.

Dans l'estuaire de la Seine, la création récente de déversoirs a pour but de minimiser ces conséquences secondaires négatives des digues submersibles.

En modifiant les conditions hydrodynamiques locales, les aménagements réalisés ont une profonde répercussion sur les biofaciès voisins: modification des conditions de salinité, modification de la texture des sédiments, modification de l'équilibre des peuplements benthiques qui intègrent les variations de tous les paramètres du milieu, modification de la physiologie des espèces...

L'ampleur des conséquences biosédimentologiques de ce type d'aménagements confirme le rôle primordial des conditions hydrodynamiques locales dans la répartition des biofaciès estuariens (§1).

4. Recommandations aux utilisateurs potentiels du milieu

Ces études de base permettent aujourd'hui de mettre à la disposition des aménageurs et autres utilisateurs potentiels du milieu (notamment dans le domaine de l'aqua-culture), une quantité de renseignements utiles en matière de:

– stabilité du milieu: en Baie des Veys, l'application de l'analyse en composantes principales aux résultats du suivi saisonnier, permet de mettre en évidence, dans certains secteurs de la Baie, l'existence de profondes perturbations des peuplements benthiques, liées à des conditions météorologiques exceptionnelles (tempêtes de nord-est).

– qualité du milieu: en Baie de Somme et estuaire de Seine, un indice de qualité biologique et un indice de charge de pollution ont permis de 'noter' la qualité de ces deux estuaires. Il ressort de ces deux indices que la charge polluante de l'estuaire de la Seine est environ le double de celle de la Somme et que les teneurs en hydrocarbures et en phosphates sont problématiques dans les deux sites.

– productivité du milieu: les études de biomasses réalisées en Baie des Veys et en Baie de Somme ont permis de quantifier la productivité naturelle de ces estuaires et de la comparer à celle d'autres estuaires européens. Si les biomasses enregistrées en Baie de Somme sont comparables à celles de la Mer des Wadden ou de l'estuaire de la Gironde (25–30 $g/m^2/an$), celles de la Baie des Veys sont inférieures (10–28 $g/m^2/an$); ces dernières valeurs restent cependant nettement supérieures à celles des peuplements subtidaux (< 2 $g/m^2/an$ en Mer du Nord).

Etant donné l'intérêt croissant que suscitent les projets aquacoles en milieu estuarien (cf. programmes IFREMER en Baie de Somme) l'importance de cette recherche de base se fait plus évidente encore.

Hydrobiologia 142: 337 (1986)

A benthos inventory in the Zeeland Coastal zone (the Netherlands)

P. Seip[1], R. Brand[1], J. v.d. Meer[2], A. C. Smaal[2], F. Creutzberg[1], A. Engelberts[1] & K. Hoek[2]
[1]*NIOZ: The Netherlands Institute for Sea Research, Texel*
[2]*RWS: Rijkswaterstaat, Tidal Water Division, Middelburg*

In 1984 The Delta Department of Rijkswaterstaat (Ministry of Transport and Public Works) initiated a project in cooperation with the Netherlands Institute for Sea Research (NIOZ) and The State University of Gent-Belgium to estimate species abundance and distribution of macro- and meiobenthos in the Coastal Waters of Zeeland (< 15 m depth).

In this area considerable hydrological and morphological changes occur due to the coastal engineering projects performed in the Rhine-Scheldt delta (the Delta project).

The benthos program is part of an integrated research project to describe and monitor the ecological development of this coastal zone. Surveys of meio- and macrofauna have been carried out in sept-nov. 1984, april-may 1985 and sept-oct. 1985, by sampling with a Reineck box corer (0.068 m^2) or a Van Veen grab (0.2 m^2).

At the start selection of 65 stations was based on available knowledge of abiotic conditions. At each station 5 sample-units have been taken. Using the macrofauna density data of this first survey the macrofauna group was able to discern TWINSPAN-cluster members (stations) in spatial groups (strata).

As a consequence one decided to use a stratified random sampling strategy for the second and third survey: ± 200 sample-units were distributed over respectively 6 and 10 strata. So in all cases information from the preceeding survey was used to refine the sampling strategy of the following.

This strategy yields statistically reliable estimates for the densities of the more common species not concerning stations but concerning the strata and the whole area.

At every sample point sediments have been sampled for among others grain-size analysis.

On the poster, besides a few pictures of sampling and processing macro-benthos, the ordered sample-species matrices and the corresponding spatial distribution of the cluster members (sample-units) together with the deduced strata have been shown.

Hydrobiologia 142: 338 (1986)

Biomass and density fluctuations of the macrozoobenthos of the intertidal flats in the Oosterschelde, the Netherlands

J. Coosen

As part of the BALANS-project the benthic fauna of the Oosterschelde was studied at 8 intertidal stations situated at 3 of the major tidal flats of the area: 3 at the Roggenplaat (in the western part of the Oosterschelde), 1 at the Hooge Kraayer and 4 at the mudflats of the Verdronken Land van Zuid-Beveland (eastern part of the Oosterschelde). At each station 15 core samples of 0.0083 m^2 were taken to a depth of 30 cm and washed in the field through a 1 mm sieve. To establish the density and biomass of the larger sparsely distributed animals the top layer (10 cm) of a two square meters sampling area was collected and washed through a 3 mm sieve. The sampling area was further dug out to a depth of 50 cm and the organisms were sorted by hand. The formalin preserved residues were sorted in the laboratory under a stereo-microscope and all animals were identified to species level, except for the oligochaetes. Ash-Free Dry Weights were determined by standard methods.

Samples were taken from spring 1983 until winter 1984 at three month intervals. Biomass and density-figures for the dominant species are analysed. These species are, in sequence of decreasing biomass: *Cerastoderma edule, Arenicola marina, Lanice conchilega, Hydrobia ulvae, Littorina littorea, Macoma balthica, Nereis diversicolor, Scrobicularia plana, Mya arenaria, Nephtys hombergii, Heteromastus filiformis* and *Scoloplos armiger.* They contribute more than 96% of the total biomass, *Mytilus edulis* excluded. Mussels are grown on culture plots with high biomass per m^2, in the shallow parts of the Oosterschelde, on the slopes of some of the gullies and on some intertidal flats. These were not sampled.

In general biomass values are lowest in February for all stations (10 – 40 g AFDW/m^2). Highest biomass values vary considerably from station to station but are generally reached in August-September (25 – 88 g AFDW/m^2). In several stations biomass tends to be higher in 1983 than in 1984. As far as data are available a comparison will be made with stations sampled in 1979 till 1981 in the Oosterschelde. Suspension feeders (*Cerastoderma edule, Mya arenaria*) contribute most to the biomass in about half of the samples, especially from the stations near the mouth of the Oosterschelde. Deposit feeders (*Arenicola marina, Macoma balthica* and *Scrobicularia plana*) contribute most in 3 stations, 2 located in the innermost part of the Oosterschelde. Some stations low in the intertidal zone have a high share of filter feeders and some stations high in the intertidal zone have a high share of deposit feeders, but this is no general rule.

Densities vary likewise from station to station and from season to season. Species like *Cerastoderma edule* have low densities in February/May (50 – 500 ind./m^2) and reach their maximum in August/September (500 – 2000 ind./m^2). *Arenicola marina* shows little variance over the year except for juveniles in spring on stations high in the intertidal zone, where *Heteromastus filiformis* and *Macoma balthica* juveniles also have their highest densities. Other species that will be discussed are: *Scoloplos armiger, Nephtys hombergii* and *Nereis diversicolor.*

Finally, distribution, biomass and density of these species is correlated with some abiotic factors like position in the intertidal zone (emersion time), sediment type and organic fraction in the sediment.

Hydrobiologia 142: 339 (1986)

Dynamics of the dominating macrozoobenthos in the Danish Wadden Sea 1980–1985

Poul Brinch Madsen
Marin ID, Skodsborg Strandvej 85, DK-2942 Skodsborg, Denmark

Sampling of macrozoobenthos was carried out by use of a core sampler and a 1 mm sieve at 6 localities in the Danish Wadden Sea 2–4 times per year (1980–1985). Each locality was sampled as a transect from MHWL to MLWL.

The mean abundance (N/m^2) and biomass (g ADW/m^2) of the 10 dominating species (excl. *Mytilus edulis* and *Hydrobia* sp.) were calculated for each transect, as was a mean for all 6 transects.

From the beginning of 1980 both abundance and biomass were more than halved during the first 2/3 of the year. In 1981 and 1982 both increased slowly to the early 1980 level (late in 1982), and later increased to more than 2.5 times this value from 1983 to 1985. In all these years abundance and biomass was higher and more unstable in the eutrophicated Ho Bay than around the island of Rømø where the water was more clean.

In 1985 sampling was only carried out once – in the autumn. This sampling showed a rather different species composition compared to the previous years. 5 of the normally dominating species as well as 6 of the normally rare species increased greatly in abundance.

Most of these species are often declared as indicator species of eutrophication.

The winter 1984–1985 was extremely severe with thick ice cover for more than 2 months.

The most probable explanation of these changings in 1985 seems to be a recolonisation after the severe winter, but an increased eutrophication cannot be ignored either.

Conclusion

All these results indicate that a better knowledge of the natural fluctuations of both the dominating and the non-dominating species is necessary for a better understanding of the results of short term pollution monitoring. The long term project COST-647 can provide this knowledge.

Hydrobiologia 142: 340 (1986)

Skomer marine reserve subtidal monitoring project

B. Bullimore
6 Mill Street, Penclawdd, Gower, West Glamorgan, SA4 3XJ, Swansea, United Kingdom

The causes and extent of natural fluctuations in subtidal rocky epibenthic communities in British waters is largely unknown and a project to monitor some representative communities in the voluntary Skomer Island Marine Reserve, S.W. Wales, was begun in 1982 with the encouragement of the reserve's management committee.

The non-destructive sampling technique of stereophotography chosen for data gathering was based on proven techniques in use in Scandinavia. The stereocamera system used between 1982 and early 1984 was, owing to severe financial limitations, designed for use with a single Nikonos camera and standard lens. Four sites, of three 0.2 m^2 quadrats each, at 10, 15, 20 and 25 m below chart datum, were established during July 1982 at a location on the north coast of Skomer Island exposed to strong tidal streams and occasional heavy wave action. The 'typical' vertical bedrock communities at this 'North Wall' station were sampled irregularly until early 1984 when funding by the Nature Conservancy Council enabled expansion of the project and facilitated a more regular sampling programme. Further sites were established in order to monitor several long-lived and/or lusitanian species of scientific interest. Adjacent to the existing North Wall sites, new sites included colonies of *Alcyonium glomeratum, Pentapora foliacea, Eunicella verrucosa* and *Gymnangium montagui.* A second station was established on the wave exposed south coast of the island where the communities sampled include *Balanophyllia regia, Isozoanthus sulcatus* and several species of sponges and ascidians. A Welsh Water Authority grant also enabled a more efficient stereocamera system to be constructed utilising two cameras with wide angle lenses.

The 20 and 25 m North Wall sites are dominated by *Corynactis viridis* and a short bryozoan turf partially overlying barnacles and *Salmacina dysteri.* Seasonal variation in the bryozoan turf was observed between 1982 and 1984. During 1985 the 25 m site was heavily grazed by *Echinus esculentus* leaving areas of bare rock suitable for colonisation. The communities at the 10 and 15 m sites are dominated by *Alcyonium digitatum* and seasonal growths of hydroids and bryozoans. An understory of *Corynactis, Caryophyllia smithii,* sponges and barnacles is only apparent during the winter. The density of colonies of the bryozoan *Cellaria* spp., which generally appear to survive two seasons, have remained fairly constant, although the actual distribution has varied considerably. Considerable growth has been observed in the sampled *Pentapora* and a seasonal growth of epifauna on the colonies noted. No growth in *Eunicella* has been recorded and no apparent change in either the *Balanophyllia* or *A. glomeratum* communities noted during the short time since observations began. The inherent difficulties in estimating the size of highly contractile organisms, such as *A. glomeratum*, are recognised. The *Gymnangium* colony was sampled once in 1982, before inclusion in the regular programme, and since that time has moved across the rock face as new hydrocauli have grown on the 'leading edge' of the colony.

In addition to the above work, a second investigator is monitoring subtidal algal communities at the North Wall station and both the S.M.R.S.M.P. workers are studying the impact of scallop dredges on, and monitoring the recovery of, the benthos of a mixed sediment seabed elsewhere in the reserve.

GPSR Compliance

The European Union's (EU) General Product Safety Regulation (GPSR) is a set of rules that requires consumer products to be safe and our obligations to ensure this.

If you have any concerns about our products, you can contact us on ProductSafety@springernature.com

In case Publisher is established outside the EU, the EU authorized representative is:

Springer Nature Customer Service Center GmbH
Europaplatz 3
69115 Heidelberg, Germany

Batch number: 10762191

Printed by Printforce, the Netherlands